Lecture Notes in Artificial Intelligence 7413

Subseries of Lecture Notes in Computer Science

LNAI Series Editors

Randy Goebel
University of Alberta, Edmonton, Canada
Yuzuru Tanaka
Hokkaido University, Sapporo, Japan
Wolfgang Wahlster
DFKI and Saarland University, Saarbrücken, Germany

LNAI Founding Series Editor

Joerg Siekmann
DFKI and Saarland University, Saarbrücken, Germany

JingTao Yao Yan Yang Roman Słowiński
Salvatore Greco Huaxiong Li
Sushmita Mitra Lech Polkowski (Eds.)

Rough Sets and Current Trends in Computing

8th International Conference, RSCTC 2012
Chengdu, China, August 17-20, 2012
Proceedings

Springer

Series Editors

Randy Goebel, University of Alberta, Edmonton, Canada
Jörg Siekmann, University of Saarland, Saarbrücken, Germany
Wolfgang Wahlster, DFKI and University of Saarland, Saarbrücken, Germany

Volume Editors

JingTao Yao
University of Regina, Canada
E-mail: jtyao@cs.uregina.ca

Yan Yang
Southwest Jiaotong University, Chengdu, China
E-mail: yyang@swjtu.edu.cn

Roman Słowiński
Poznan University of Technology, Poland
E-mail: roman.slowinski@cs.put.poznan.pl

Salvatore Greco
University of Catania, Italy
E-mail: salgreco@unict.it

Huaxiong Li
Nanjing University, China
E-mail: huaxiongli@nju.edu.cn

Sushmita Mitra
Indian Statistical Institute, Kolkata, India
E-mail: sushmita@isical.ac.in

Lech Polkowski
Polish-Japanese Institute of Information Technology, Warsaw, Poland
E-mail: polkow@pjwstk.edu.pl

ISSN 0302-9743 e-ISSN 1611-3349
ISBN 978-3-642-32114-6 e-ISBN 978-3-642-32115-3
DOI 10.1007/978-3-642-32115-3
Springer Heidelberg Dordrecht London New York

Library of Congress Control Number: 2012942867

CR Subject Classification (1998): F.4.1, I.2.4, I.2.6, H.2.8, I.2, G.1.2, G.1.6, F.4, H.4, H.3

LNCS Sublibrary: SL 7 – Artificial Intelligence

Typesetting: Camera-ready by author, data conversion by Scientific Publishing Services, Chennai, India

Printed on acid-free paper

Springer is part of Springer Science+Business Media (www.springer.com)

Preface

This volume contains the papers selected for presentation at RSCTC 2012: The 8th International Conference on Rough Sets and Current Trends in Computing (RSCTC) held during August 17–20, 2012, in Chengdu, China, one of the co-located conferences of the 2012 Joint Rough Set Symposium (JRS 2012). JRS 2012 consists of RSCTC 2012 and the 7th International Conference on Rough Sets and Knowledge Technology (RSKT 2012).

RSCTC has been held biannually since 1998. It aims to present the state of the art in rough set theory, current computing methods and their applications. It intends to bring together researchers and practitioners from universities, laboratories, and industry, to facilitate dialogue and cooperation. The first RSCTC was held in 1998 in Warsaw, Poland, followed by RSCTC 2000 in Banff, Canada, RSCTC 2002 in Malvern, USA, RSCTC 2004 in Uppsala, Sweden, RSCTC 2006 in Kobe, Japan, RSCTC 2008 in Akron, USA, and RSCTC 2010 in Warsaw, Poland.

JRS 2012 received 292 papers and competition submissions from 56 countries and regions, including Afghanistan, Antarctica, Antigua and Barbuda, Argentina, Australia, Austria, Bangladesh, Belgium, Brazil, Canada, China, Colombia, Croatia, Cuba, Denmark, Egypt, Finland, France, Germany, Ghana, Greece, Hong Kong, Hungary, India, Indonesia, Iran, Israel, Italy, Japan, Jordan, Korea, Lebanon, Mexico, The Netherlands, New Zealand, Pakistan, Poland, Portugal, Reunion, Romania, Russian Federation, Rwanda, Saudi Arabia, Serbia and Montenegro, Singapore, Slovenia, Spain, Sweden, Switzerland, Taiwan, Tunisia, Ukraine, UK, USA, Venezuela, and Vietnam.

Following the tradition of the previous RSCTC and RSKT conferences, all submissions underwent a very rigorous reviewing process. Every submission was reviewed by at least two Program Committee(PC) members and at least one external domain expert. On average, each submission received 3.6 reviews. About ten papers received more than six reviews each. Finally, the PC selected 55 papers (including 34 regular papers and 21 short papers), based on their originality, significance, correctness, relevance, and clarity of presentation, to be included in this volume of the proceedings. Revised camera-ready submissions were further reviewed by PC Chairs. Some authors were requested to make additional revisions. We would like to thank all the authors for submitting their papers for consideration for presentation at the conference. We also wish to congratulate those authors whose papers were selected for presentation and publication in the proceedings. Their contribution was crucial for the quality of this conference.

The JRS 2012 program was further enriched by four keynote speeches. We are grateful to RSKT keynote speakers, Andrzej Skowron and Zhi-Hua Zhou, as well as RSCTC keynote speakers, Yiyu Yao and Bo Zhang, for their inspiring talks on rough sets, knowledge technology, and current trends in computing.

The JRS 2012 program included one workshop, Advances in Granular Computing 2012, and five Special Sessions, Decision-Theoretic Rough Set Model and Applications, Intelligent Decision-Making and Granular Computing, Mining Complex Data with Granular Computing, Formal Concept Analysis and Granular Computing, and Rough Set Foundations. In addition, we selected papers written by the winners of the JRS 2012 Data Mining Competition: Topical Classification of Biomedical Research Papers.

This data mining competition was a special event associated with the JRS 2012 conference. It was organized by a research team from the University of Warsaw and co-funded by organizers of JRS 2012, Southwest Jiaotong University, and the SYNAT project. A task in this challenge was related to the problem of predicting topical classification of scientific publications in the field of biomedicine. It was an interactive on-line competition, hosted on the TunedIT platform (http://tunedit.org). The JRS 2012 Data Mining Competition attracted participants from 50 different countries across six continents. There were 126 active teams who submitted at least one solution to the leaderboard. Apart from submitting solutions, participants were asked to deliver short descriptions of their approaches. The most interesting of these reports were extended to conference papers and included in the RSCTC proceedings.

JRS 2012 would not have been successful without the support of many people and organizations. We wish to thank the members of the Steering Committee for their invaluable suggestions and support throughout the organization process. We are indebted to the PC members and external reviewers for their effort and engagement in providing a rich and rigorous scientific program. We express our gratitude to the Special Session Chairs (Mihir Kr. Chakraborty, Degang Chen, Davide Ciucci, Qinghua Hu, Andrzej Janusz, Xiuyi Jia, Adam Krasuski, Huaxiong Li, Jiye Liang, Tsau Young Lin, Dun Liu, Xiaodong Liu, Fan Min, Hung Son Nguyen, Jianjun Qi, Dominik Slezak, Sebastian Stawicki, Lidong Wang, Xizhao Wang, Ling Wei, JingTao Yao, Yiyu Yao, and Hong Yu) for selecting and coordinating the sessions on very interesting topics. Thanks also go to the Tutorial Chairs (Chris Cornelis and Qinghua Hu), Special Session/Workshop Chairs (Davide Ciucci and Wei-Zhi Wu), Publicity Chairs (Jianchao Han, Pawan Lingras, Dun Liu, Duoqian Miao, Mikhail Moshkov, Shusaku Tsumoto), and Organizing Chairs (Hongmei Chen, Yan Yang and Qinghua Zhang).

We are also grateful to Anping Zeng, Chuan Luo, Shaoyong Li, Jie Hu, Shengjiu Liu, and Junbo Zhang from Southwest Jiaotong University, whose great effort ensured the success of the conference. We greatly appreciate the co-operation, support, and sponsorship of various institutions, companies, and organizations, including Southwest Jiaotong University, the University of Regina, the University of Warsaw, the International Rough Set Society, the Rough Sets and Soft Computation Society, the Chinese Association for Artificial Intelligence, Infobright, the Chongqing Institute of Green and Intelligent Technology, the Chinese Academy of Sciences, Section of Intelligent Decision Support Systems and

Granular Computing of the Computer Science Committee of the Polish Academy of Sciences. In addition, we would like to give special thanks for the support of the National Science Foundation of China (Funding Numbers: 61175047, 61170111, 61100117 and 61073146).

We acknowledge the use of the EasyChair conference system for paper submission, review, and editing of the proceedings. Its new feature of editing LNCS volumes is especially useful. We are thankful to Alfred Hofmann and the excellent LNCS team at Springer for their support and cooperation in publishing the proceedings as a volume of the *Lecture Notes in Computer Science.*

May 2012

JingTao Yao
Yan Yang
Roman Slowinski
Salvatore Greco
Huaxiong Li
Sushmita Mitra
Lech Polkowski

Organization

JRS 2012 Conference Committee

Honorary Chairs	Lotfi A. Zadeh, Bo Zhang
Conference Chairs	Roman Slowinski, Guoyin Wang
Program Chairs	Tianrui Li, Hung Son Nguyen, JingTao Yao
RSKT PC Co-chairs	Jerzy Grzymała-Busse, Ryszard Janicki, Aboul Ella Hassanien, Hong Yu
RSCTC PC Co-chairs	Salvatore Greco, Huaxiong Li, Sushmita Mitra, Lech Polkowski
Tutorial Chairs	Qinghua Hu, Chris Cornelis
Special Session/ Workshop Chairs	Davide Ciucci, Wei-Zhi Wu
Publicity Chairs	Jianchao Han, Pawan Lingras, Dun Liu, Duoqian Miao, Mikhail Moshkov, Shusaku Tsumoto
Organizing Chairs	Hongmei Chen, Yan Yang, Qinghua Zhang
Steering Committee Chairs	Jiye Liang, Andrzej Skowron, Dominik Slezak, Yiyu Yao
Secretary-General	Anping Zeng
Secretaries	Jie Hu, Shaoyong Li, Shengjiu Liu, Chuan Luo, Junbo Zhang

Steering Committee

James F. Peters
Malcolm Beynon
Hans-Dieter Burkhard
Gianpiero Cattaneo
Nicholas Cercone
Mihir K. Chakraborty
Juan-Carlos Cubero
Didier Dubois
Ivo Duentsch
Salvatore Greco
Jerzy Grzymala-Busse
Aboul E. Hassanien
Masahiro Inuiguchi
Etienne Kerre
Yuefeng Li
Tsau Young Lin
Jiming Liu
Qing Liu
Jie Lu
Stan Matwin
Ernestina Menasalvas-Ruiz
Duoqian Miao
Sadaaki Miyamoto
Masoud Nikravesh
Ewa Orlowska
Sankar K. Pal
Witold Pedrycz
Lech Polkowski
Wladyslaw Skarbek
Roman Slowinski
Zbigniew Suraj
Shusaku Tsumoto
Julio V. Valdes
Guoyin Wang
Hui Wang
S.K. Michael Wong
Bo Zhang
Wen-Xiu Zhang
Ning Zhong
Wojciech Ziarko

Program Committee

Aijun An
Qiusheng An
Mohua Banerjee
Jan Bazan
Theresa Beaubouef
Jerzy Blaszczynski
Zbigniew Bonikowski
Maciej Borkowski
Cory Butz
Gianpiero Cattaneo
Nick Cercone
Mihir K. Chakraborty
Chien-Chung Chan
Hongmei Chen
Jiaxing Cheng
Davide Ciucci
Chris Cornelis
Krzysztof Cyran
Andrzej Czyzewski
Jianhua Dai
Martine De Cock
Dayong Deng
Ivo Düntsch
Lin Feng
Yang Gao
Anna Gomolinska
Xun Gong
Salvatore Greco
Jerzy Grzymała-Busse
Jianchao Han
Aboul Ella Hassanien
Jun He
Christopher Henry
Daryl Hepting
Joseph Herbert
Shoji Hirano
Jie Hu
Qinghua Hu
Xiaohua Hu
Masahiro Inuiguchi
Lakhmi Jain
Ryszard Janicki
Andrzej Janusz
Jouni Jarvinen
Richard Jensen
Xiuyi Jia
Chaozhe Jiang
Janusz Kacprzyk
John A. Keane
C. Maria Keet
Jan Komorowski
Jacek Koronacki
Bozena Kostek
Abd El-Monem Kozae
Krzysztof Krawiec
Marzena Kryszkiewicz
Yasuo Kudo
Henry Leung
Daoguo Li
Deyu Li
Fanchang Li
Huaxiong Li
Jinjin Li
Longshu Li
Tianrui Li
Yuefeng Li
Jiye Liang
Tsau Young Lin
Pawan Lingras
Dun Liu
Qing Liu
Qun Liu
Xiaodong Liu
Amjad Mahmood
Pradipta Maji
Benedetto Matarazzo
Lawrence Mazlack
Ernestina Menasalvas
Jusheng Mi
Duoqian Miao
Pabitra Mitra
Sushmita Mitra
Sadaaki Miyamoto
Mikhail Moshkov
Murice Mulvenna
Som Naimpally
Michinori Nakata
Sinh Hoa Nguyen
Ewa Orlowska
Hala Own
Sankar K. Pal
Krzysztof Pancerz
Neil Mac Parthalain
Puntip Pattaraintakorn
Witold Pedrycz
Bo Peng
Alberto Guillen Perales
Georg Peters
James Peters
Lech Polkowski
Yuhua Qian
Keyun Qin
Guofang Qiu
Anna Maria Radzikowska
Vijay Raghavan
Sheela Ramanna
C. Raghavendra Rao
Zbigniew Ras
Henryk Rybinski
Hiroshi Sakai
Lin Shang
Qiang Shen
Kaiquan Shi
Arul Siromoney
Władysław Skarbek
Andrzej Skowron
Dominik Slezak
Roman Slowinski
Hung Son Nguyen
Urszula Stanczyk
John Stell
Jaroslaw Stepaniuk
Zbigniew Suraj
Marcin Szczuka
Marcin Szellag
Fei Teng
Li-Shiang Tsay
Shusaku Tsumoto

Muhammad Zia Ur Rehman
Julio Valdes
Aida Vitoria
Alicja Wakulicz-Deja
Krzysztof Walczak
Guoyin Wang
Hongjun Wang
Hui Wang
Xin Wang
Anita Wasilewska
Piotr Wasilewski
Richard Weber
Ling Wei
Paul Wen
Szymon Wilk
Marcin Wolski
Tao Wu
Wei-Zhi Wu
Xiaohong Wu
Jiucheng Xu
Ronald Yager
Yan Yang
Yingjie Yang
Yong Yang
JingTao Yao
Yiyu Yao
Dongyi Ye
Hong Yu
Jian Yu
Slawomir Zadrozny
Xianhua Zeng
Bo Zhang
Ling Zhang
Qinghua Zhang
Yanping Zhang
Shu Zhao
Ning Zhong
Shuigeng Zhou
William Zhu
Wojciech Ziarko

Additional Reviewers

Stefano Aguzzoli
Piotr Artiemjew
Nouman Azam
Paweł Betliński
Bingzhen Sun
Long Chen
Xiaofei Deng
Martin Dimkovski
Yasunori Endo
Joanna Golinska-Pilarek
Dianxuan Gong
Przemysław Górecki
Feng He
Qiang He
Feng Hu
Md. Aquil Khan
Beata Konikowska
Adam Krasuski
Qingguo Li
Wen Li
Lihe Guan
Fan Min
Amit Mitra
Piero Pagliani
Wangren Qiu
Ying Sai
Mingwen Shao
Sebastian Stawicki
Lin Sun
Nele Verbiest
Jin Wang
Junhong Wang
Lidong Wang
Nan Zhang
Yan Zhang
Yan Zhao
Bing Zhou
Wojciech Świeboda

Table of Contents

Keynote

Part I: Rough Sets and Its Applications

Part II: Current Trends in Computing

Part III: Special Session: Decision-Theoretic Rough Set Model and Applications

Part IV: Special Session: Formal Concept Analysis and Granular Computing

Part V: Special Session: Mining Complex Data with Granular Computing

Part VI: Special Session: Data Mining Competition

An Outline of a Theory of Three-Way Decisions

Yiyu Yao

Department of Computer Science, University of Regina,
Regina, Saskatchewan, Canada S4S 0A2
yyao@cs.uregina.ca

Abstract. A theory of three-way decisions is constructed based on the notions of acceptance, rejection and noncommitment. It is an extension of the commonly used binary-decision model with an added third option. Three-way decisions play a key role in everyday decision-making and have been widely used in many fields and disciplines. An outline of a theory of three-way decisions is presented by examining its basic ingredients, interpretations, and relationships to other theories.

1 Introduction

The concept of three-way decisions was recently proposed and used to interpret rough set three regions [52, 54, 55]. More specifically, the positive, negative and boundary regions are viewed, respectively, as the regions of acceptance, rejection, and noncommitment in a ternary classification. The positive and negative regions can be used to induce rules of acceptance and rejection; whenever it is impossible to make an acceptance or a rejection decision, the third noncommitement decision is made [54]. It can be shown that, under certain conditions, probabilistic three-way decisions are superior to both Palwak three-way decisions and two-way (i.e., binary) decisions [55]. Many recent studies further investigated extensions and applications of three-way decisions [1, 7–10, 12, 13, 17–21, 23–29, 31, 45, 46, 56, 60–62, 64–66].

The essential ideas of three-way decisions are commonly used in everyday life [32] and widely applied in many fields and disciplines, including, for example, medical decision-making [30, 37, 38], social judgement theory [39], hypothesis testing in statistics [42], management sciences [5, 44], and peering review process [43]. However, a close examination surprisingly reveals that there still does not exist a unified formal description. To extend the concept of three-way decisions of rough sets to a much wider context, this paper outlines a theory of three-way decisions.

2 A Description of Three-Way Decisions

The essential ideas of three-way decisions are described in terms of a ternary classification according to evaluations of a set of criteria.

Suppose U is a finite nonempty set of objects or decision alternatives and C is a finite set of conditions. Each condition in C may be a criterion, an objective, or

J.T. Yao et al. (Eds.): RSCTC 2012, LNAI 7413, pp. 1–17, 2012.

a constraint. For simplicity, in this paper we refer to conditions in C as criteria. Our decision task is to classify objects of U according to whether they satisfy the set of criteria. In widely used two-way decision models, it is assumed that an object either satisfies the criteria or does not satisfy the criteria. The set U is divided into two disjoint regions, namely, the positive region POS for objects satisfying the criteria and the negative region NEG for objects not satisfying the criteria. There are usually some classification errors associated with such a binary classification. Two main difficulties with two-way approaches are their stringent binary assumption of the satisfiability of objects and the requirement of a dichotomous classification.

In many situations, it may happen that an object only satisfies the set of criteria to some degree. Even if an object may actually either satisfy or not satisfy the criteria, we may not be able to identify without uncertainty the subset of objects that satisfy the criteria due to uncertain or incomplete information. Consequently, we are only able to search for an approximate solution. Instead of making a binary decision, we use thresholds on the degrees of satisfiability to make one of three decisions: (a) accept an object as satisfying the set of criteria if its degree of satisfiability is at or above a certain level; (b) reject the object by treating it as not satisfying the criteria if its degree of satisfiability is at or below another level; and (c) neither accept nor reject the object but opt for a noncommitment. The third option may also be referred to as a deferment decision that requires further information or investigation. From the informal description, we give a formal definition.

The Problem of Three-Way Decisions. Suppose U is a finite nonempty set and C is a finite set of criteria. The problem of three-way decisions is to divide, based on the set of criteria C, U into three pair-wise disjoint regions, POS, NEG, and BND, called the positive, negative, and boundary regions, respectively.

Corresponding to the three regions, one may construct rules for three-way decisions. In our previous studies [52, 54], we used three types of rules, namely, rules for acceptance, rejection, and noncommitment, respectively. It now appears to us that only rules for acceptance and rules for rejection are meaningful and sufficient. That is, the noncommitment set is formed by those objects to which neither a rule for acceptance nor a rule for rejection applies. It is not necessary to have, and in many cases may be impossible to construct, rules for noncommitement.

To formally describe the satisfiability of objects, rules for acceptance and rules for rejection, we need to introduce the notion of evaluations of objects and designated values for acceptance and designated values for rejection. Evaluations provide the degrees of satisfiability, designated values for acceptance are acceptable degrees of satisfiability, and designated valued for rejection are acceptable degrees of non-satisfiability. They provide a basis for a theory of three-way decisions.

A theory of three-way decisions must consider at least the following three issues regarding evaluations and designated values:

1. Construction and interpretation of a set of values for measuring satisfiability and a set of values for measuring non-satisfiability. The former is used by an evaluation for acceptance and the latter is used by an evaluation for rejection. In many cases, a single set may be used by both. It is assumed that the set of evaluation values is equipped with an ordering relation so that we can compare at least some objects according to their degrees of satisfiability or non-satisfiability. Examples of a set evaluation values are a poset, a lattice, a set of a finite numbers of grades, the set of integers, the unit interval, and the set of reals. Social judgement theory uses latitudes of acceptance, rejection, and noncommitment [6, 39], which is closely related to our formulation of three-way decisions.
2. Construction and interpretation of evaluations. An evaluation depends on the set of criteria and characterizes either satisfiability or non-satisfiability of objects in U. Evaluations for the purposes of acceptance and rejection may be either independent or the same. Depending on particular applications, evaluations may be constructed and interpreted in terms of more intuitive and practically operable notions, including costs, risks, errors, profits, benefits, user satisfaction, committee voting, and so on. Based on the values of an evaluation, one can at least compare some objects.
3. Determination and interpretation of designated values for acceptance and designated values for rejection. The sets of designated values must meaningfully reflect an intuitive understanding of acceptance and rejection. For example, we can not accept and reject an object simultaneously. This requires that the set of designated values for acceptance and the set of designated value for rejection are disjoint. The designated values for acceptance should lead to monotonic decisions; if we accept an object x then we should accept all those objects that have the same or larger degrees of satisfiability than x. It is also desirable if we can systematically determine the sets of designated values on a semantically sound basis.

By focusing on these issues, we examine three classes of evaluations. Evaluations are treated as a primitive notion for characterizing the satisfiability or desirability of objects. Their concrete physical interpretations are left to particular applications.

3 Evaluation-Based Three-Way Decisions

We assume that evaluations for acceptance and rejection can be constructed based on the set of criteria. This enables us to focus mainly on how to obtain three-way decisions according to evaluations. The problem of constructing and interpreting evaluations is left to further studies and specific applications. A framework of evaluation-based three-way decisions is proposed and three models are introduced and studied.

3.1 Three-Way Decisions with a Pair of Poset-Based Evaluations

For the most general case, we consider a pair of (may be independent) evaluations, one for the purpose of acceptance and the other for rejection.

Definition 1. *Suppose U is a finite nonempty set and $(L_a, \preceq_a)$ $(L_r, \preceq_r)$ are two posets. A pair of functions $v_a : U \longrightarrow L_a$ and $v_r : U \longrightarrow L_r$ is called an acceptance evaluation and a rejection evaluation, respectively. For $x \in U$, $v_a(x)$ and $v_r(x)$ are called the acceptance and rejection values of x, respectively.*

In real applications, the set of possible values of acceptance may be interpreted based on more operational notions such as our confidence of an object satisfying the given set of criteria, or cost, benefit, and value induced by the object. For two objects $x, y \in U$, if $v_a(x) \preceq_a v_a(y)$, we say that x is less acceptable than y. By adopting a poset $(L_a, \preceq_a)$, we assume that some objects in U are incomparable. Similar interpretation can be said about the possible values of an evaluation for rejection. In general, acceptance and rejection evaluations may be independent.

To accept an object, its value $v_a(x)$ must be in a certain subset of L_a representing the acceptance region of L_a. Similarly, we need to define the rejection region of L_r. By adopting a similar terminology of designated values in many-valued logics [4], these values are called designated values for acceptance and designated values for rejection, respectively. Based on the two sets of designated values, one can easily obtain three regions for three-way decisions.

Definition 2. *Let $\emptyset \neq L_a^+ \subseteq L_a$ be a subset of L_a called the designated values for acceptance, and $\emptyset \neq L_r^- \subseteq L_r$ be a subset of L_r called the designated values for rejection. The positive, negative, and boundary regions of three-way decisions induced by (v_a, v_r) are defined by:*

$$\begin{aligned}
\mathrm{POS}_{(L_a^+, L_r^-)}(v_a, v_r) &= \{x \in U \mid v_a(x) \in L_a^+ \wedge v_r(x) \notin L_r^-\},\\
\mathrm{NEG}_{(L_a^+, L_r^-)}(v_a, v_r) &= \{x \in U \mid v_a(x) \notin L_a^+ \wedge v_r(x) \in L_r^-\},\\
\mathrm{BND}_{(L_a^+, L_r^-)}(v_a, v_r) &= (\mathrm{POS}_{(L_a^+, L_r^-)}(v_a, v_r) \cup \mathrm{NEG}_{(L_a^+, L_r^-)}(v_a, v_r))^c\\
&= \{x \in U \mid (v_a(x) \notin L_a^+ \wedge v_r(x) \notin L_r^-) \vee\\
&\qquad (v_a(x) \in L_a^+ \wedge v_r(x) \in L_r^-)\}. \qquad (1)
\end{aligned}$$

The boundary region is defined as the complement of the union of positive and negative regions. The conditions in the definition of the positive and negative regions make sure that they are disjoint. Therefore, the three regions are pairwise disjoint. The three regions do not necessarily form a partition of U, as some of them may be empty. In fact, two-way decisions may be viewed as a special case of three-way decisions in which the boundary region is always empty.

By the interpretation of the orderings $\preceq_a$ and $\preceq_r$, the designated values L_a^+ for acceptance and the designated values L_r^- for rejection must satisfy certain properties. If L_a has the largest element $\mathbf{1}$, then $\mathbf{1} \in L_a^+$. If $w \preceq_a u$ and $w \in L_a^+$, then $u \in L_a^+$. That is, if $v_a(x) \preceq_a v_a(y)$ and we accept x, then we must accept y. Similarly, if L_r has the largest element $\mathbf{1}$, then $\mathbf{1} \in L_r^-$. If $w \preceq_r u$ and $w \in L_r^-$, then $u \in L_r^-$.

3.2 Three-Way Decisions with One Poset-Based Evaluation

In some situations, it may be more convenient to combine the two evaluation into a single acceptance-rejection evaluation. In this case, one poset $(L, \preceq)$ is used and two subsets of the poset are used as the designated values for acceptance and rejection, respectively.

Definition 3. *Suppose $(L, \preceq)$ is a poset. A function $v : U \longrightarrow L$ is called an acceptance-rejection evaluation. Let $L^+, L^- \subseteq L$ be two subsets of L with $L^+ \cap L^- = \emptyset$, called the designated values for acceptance and the designated values for rejection, rspectively. The positive, negative, and boundary regions of three-way decisions induced by v is defined by:*

$$\begin{aligned} \mathrm{POS}_{(L^+,L^-)}(v) &= \{x \in U \mid v(x) \in L^+\}, \\ \mathrm{NEG}_{(L^+,L^-)}(v) &= \{x \in U \mid v(x) \in L^-\}, \\ \mathrm{BND}_{(L^+,L^-)}(v) &= \{x \in U \mid v(x) \notin L^+ \wedge v(x) \notin L^-\}. \end{aligned} \tag{2}$$

The condition $L^+ \cap L^- = \emptyset$ ensures that the three regions are pair-wise disjoint. A single evaluation v may be viewed as a special case of two evaluations in which $\preceq_a = \preceq$ and $\preceq_r = \succeq$. In this way, acceptance is related to rejection in the sense that the reverse ordering of acceptance is the ordering for rejection. To ensure the meaningfulness of L^+ and L^-, it is required that $\neg(w \preceq u)$ for all $w \in L^+$ and $u \in L^-$. In other words, L^+ contains larger elements of L and L^- contains smaller elements of L.

3.3 Three-Way Decisions with an Evaluation Using a Totally Ordered Set

Consider now an evaluation based on a totally ordered set $(L, \preceq)$ where $\preceq$ is a total order. That is, $\preceq$ is a partial order and any two elements of L are comparable. This is in fact a widely used approach. For example, L is either the set of real numbers or the unit interval $[0, 1]$ and $\preceq$ is the less-than-or-equal relation $\leq$. For a total order, it is possible to define the sets of designated values for acceptance and rejection by a pair of thresholds.

Definition 4. *Suppose $(L, \preceq)$ is a totally ordered set, that is, $\preceq$ is a total order. For two elements α, β with $\beta \prec \alpha$ (i.e., $\beta \preceq \alpha \wedge \neg(\alpha \preceq \beta)$), suppose that the set of designated values for acceptance is given by $L^+ = \{t \in L \mid t \succeq \alpha\}$ and the set of designated values for rejection is given by $L^- = \{b \in L \mid b \preceq \beta\}$. For an evaluation function $v : U \longrightarrow L$, its three regions are defined by:*

$$\begin{aligned} \mathrm{POS}_{(\alpha,\beta)}(v) &= \{x \in U \mid v(x) \succeq \alpha\}, \\ \mathrm{NEG}_{(\alpha,\beta)}(v) &= \{x \in U \mid v(x) \preceq \beta\}, \\ \mathrm{BND}_{(\alpha,\beta)}(v) &= \{x \in U \mid \beta \prec v(x) \prec \alpha\}. \end{aligned} \tag{3}$$

Although evaluations based on a total order are restrictive, they have a computational advantage. One can obtain the three regions by simply comparing the evaluation value with a pair of thresholds. It is therefore not surprising to find that many studies in fact use a total order.

3.4 Comments on Evaluations and Designated Values

Construction and interpretation of evaluations and designated values are vital for practical applications of three-way decisions. At a theoretical level, it may be only possible to discuss required properties of evaluations. It is assumed that an evaluation is determined by a set of criteria, representing costs, benefits, degrees of desirability, objectives, constraints, and so on. Further studies on evaluations may be a fruitful research direction.

As an illustration, consider a simple linear model for constructing an evaluation. Suppose $C = \{c_1, c_2, \ldots, c_m\}$ are a set of m criteria. Suppose $v_{c_i} : U \longrightarrow \Re$ denotes an evaluation based on criterion v_i, $1 \leq i \leq m$. An overall evaluation function $v : U \longrightarrow \Re$ may be simply defined by a linear combination of individual evaluations:

$$v(x) = v_{c_1}(x) + v_{c_2}(x) + \ldots + v_{c_m}(x). \tag{4}$$

Details of this linear utility model and other models can be found in literature of multi-crieria and multi-objective decision making [14].

Construction and interpretation of designated values may be explained in terms of benefits or risks of the resulting three regions of three-way decisions. For example, consider the model that uses a total order. Let $R_P(\alpha, \beta)$, $R_N(\alpha, \beta)$ and $R_B(\alpha, \beta)$ denote the risks of the positive, negative, and boundary regions, respectively. It is reasonable to require that the sets of designated values are chosen to minimize the following overall risks:

$$R(\alpha, \beta) = R_P(\alpha, \beta) + R_N(\alpha, \beta) + R_B(\alpha, \beta). \tag{5}$$

That is, finding a pair of thresholds can be formulated as the following optimization problem:

$$\arg \min_{(\alpha,\beta)} R(\alpha, \beta). \tag{6}$$

As a concerte example, R may be understood as uncertainty associated with three regions, by minimizing the overall uncertainty one can obtain the set of designed values in a probabilistic rough set model [2]. Two additional examples will be given in the next section when reviewing decision-theoretic rough sets [50, 57, 58] and shadowed sets [34, 35].

4 Models of Three-Way Decisions

We show that many studies on three-way decisions can be formulated within the framework proposed in the last section. For simplicity and as examples, we focus on the concept of concepts in a set-theoretical setting. In the classical view of concepts [40, 41], every concept is understood as a unit of thought consisting of two parts, the intension and the extension of the concept. Due to uncertain or insufficient information, it is not always possible to precisely have a set of objects as the extension of a concept. Consequently, many generalizations of sets have been proposed and studied.

4.1 Interval Sets and Three-Valued Logic

Interval sets provide a means to describe partially known concepts [47, 53]. On the one hand, it is assumed that an object may actually be either an instance or not an instance of a concept. On the other hand, due to a lack of information and knowledge, one can only express the state of instance and non-instance for some objects, instead of all objects. That is, one has a partially known concept defined by a lower bound and upper bound of its extension.

Formally, a closed interval set is a subset of 2^U of the form,

$$[A_l, A_u] = \{A \in 2^U \mid A_l \subseteq A \subseteq A_u\}, \tag{7}$$

where it is assumed that $A_l \subseteq A_u$, and A_l and A_u are called the lower and upper bound, respectively. Any set $X \in [A_l, A_u]$ may be the actual extension of the partially known concept. Constructive methods for defining interval sets can be formulated within an incomplete information table [16, 22].

An interval set is an interval of the power set lattice 2^U; it is also a lattice, with the minimum element A_l, the maximum element A_u, and the standard set-theoretic operations.

Interval-set algebra is related to Kleene's three-valued logic [15, 36], in which a third truth value is added to the standard two-valued logic. The third value may be interpreted as unknown or undeterminable. Let $L = \{F, I, T\}$ denote the set of truth values with a total order $F \preceq I \preceq T$. An interval set $[A_l, A_u]$ can be equivalently defined by an acceptance-rejection evaluation as,

$$v_{[A_l,A_u]}(x) = \begin{cases} F, & x \in (A_u)^c, \\ I, & x \in A_u - A_l, \\ T, & x \in A_l. \end{cases} \tag{8}$$

Suppose the sets of designated values for acceptance and rejection are defined by a pair of thresholds (T, F), namely, $L^+ = \{a \in L \mid T \preceq a\} = \{T\}$ and $L^- = \{b \in L \mid b \preceq F\} = \{F\}$. According to Definition 4, an interval set provides the following three-way decisions:

$$\begin{aligned} \mathrm{POS}_{(T,F)}([A_l, A_u]) &= \{x \in U \mid v_{[A_l,A_u]}(x) \succeq T\} = A_l, \\ \mathrm{NEG}_{(T,F)}([A_l, A_u]) &= \{x \in U \mid v_{[A_l,A_u]}(x) \preceq F\} = (A_u)^c, \\ \mathrm{BND}_{(T,F)}([A_l, A_u]) &= \{x \in U \mid F \prec v_{[A_l,A_u]}(x) \prec T\} = A_u - A_l. \end{aligned} \tag{9}$$

Although the re-expression of an interval set in terms of three-way decisions is somewhat trivial, it does provide a new view to look at interval sets.

4.2 Pawlak Rough Sets

Pawlak rough set theory deals with approximations of a concept based on a family of definable concepts [33].

Let $E \subseteq U \times U$ denote an equivalence relation on U, that is, E is reflexive, symmetric, and transitive. The equivalence class containing x is defined by

$[x]_E = [x] = \{y \in U \mid xEy\}$, which is a set of objects equivalent to x. The family of all equivalence classes of E is called the quotient set induced by E, denoted as U/E. In an information table, an equivalence class is a definable set that can be defined by the conjunction of a family of attribute-value pairs [49].

For a subset $A \subseteq U$, the Pawlak rough set lower and upper approximations of A are defined by:

$$\begin{aligned} \underline{apr}(A) &= \{x \in U \mid [x] \subseteq A\}, \\ \overline{apr}(A) &= \{x \in U \mid [x] \cap A \neq \emptyset\} \\ &= \{x \in U \mid \neg([x] \subseteq A^c)\}. \end{aligned} \tag{10}$$

In the definition, we use an equivalent condition $\neg([x] \subseteq A^c)$ so that both lower and upper approximations are defined uniformly by using set inclusion $\subseteq$. According to the pair of approximations, the Pawlak positive, negative and boundary regions are defined by:

$$\begin{aligned} \mathrm{POS}(A) &= \underline{apr}(A), \\ &= \{x \in U \mid [x] \subseteq A\}; \\ \mathrm{NEG}(A) &= U - \overline{apr}(A), \\ &= \{x \in U \mid [x] \subseteq A^c\}; \\ \mathrm{BND}(A) &= \overline{apr}(A) - \underline{apr}(A), \\ &= \{x \in U \mid \neg([x] \subseteq A^c) \wedge \neg([x] \subseteq A)\} \\ &= (\mathrm{POS}(A) \cup \mathrm{NEG}(A))^c. \end{aligned} \tag{11}$$

Again, these regions are defined uniformly by using set inclusion. The three regions are pair-wise disjoint. Conversely, from the three regions, we can compute the pair of approximations by:

$$\begin{aligned} \underline{apr}(A) &= \mathrm{POS}(A) \\ \overline{apr}(A) &= \mathrm{POS}(A) \cup \mathrm{BND}(A). \end{aligned}$$

Therefore, rough set theory can be formulated by either a pair of approximations or three regions.

Three-way decisions with rough sets can be formulated as follows. Let $L_a = L_r = \{F, T\}$ with $F \preceq T$, and let $L_a^+ = L_r^- = \{T\}$. All objects in the same equivalence class have the same description. Based on descriptions of objects, we have a pair of an acceptance evaluation and a rejection evaluation:

$$v_{(a,A)}(x) = \begin{cases} T, & [x] \subseteq A, \\ \\ F, & \neg([x] \subseteq A); \end{cases} \qquad v_{(r,A)}(x) = \begin{cases} T, & [x] \subseteq A^c, \\ \\ F, & \neg([x] \subseteq A^c). \end{cases} \tag{12}$$

According to Definition 2, for a set $A \subseteq U$, we can make the following three-way decisions:

$$\begin{aligned} \mathrm{POS}_{(\{T\},\{T\})}(A) &= \{x \in U \mid v_{(a,A)}(x) \in \{T\} \wedge v_{(r,A)}(x) \notin \{T\}\} \\ &= \{x \in U \mid v_{(a,A)}(x) = T\} \\ &= \{x \in U \mid [x] \subseteq A\}, \end{aligned}$$

$$
\begin{aligned}
\mathrm{NEG}_{(\{T\},\{T\})}(A) &= \{x \in U \mid v_{(a,A)}(x) \notin \{T\} \wedge v_{(r,A)}(x) \in \{T\}\}, \\
&= \{x \in U \mid v_{(r,A)}(x) = T\} \\
&= \{x \in U \mid [x] \subseteq A^c\}, \\
\mathrm{BND}_{(\{T\},\{T\})}(A) &= (\mathrm{POS}(v_a, v_r) \cup \mathrm{NEG}(v_a, v_r))^c \\
&= \{x \in U \mid \neg([x] \subseteq A) \wedge \neg([x] \subseteq A^c)\}. \qquad (13)
\end{aligned}
$$

The reformulation of rough set three regions based uniformly on set inclusion provides additional insights into rough set approximations. It explicitly shows that acceptance is based on an evaluation of the condition $[x] \subseteq A$ and rejection is based on an evaluation of the condition $[x] \subseteq A^c$. By those two conditions, both decisions of acceptance and rejection are made without any error. Whenever there is any doubt, namely, $\neg([x] \subseteq A) \wedge \neg([x] \subseteq A^c)$, a decision of noncommitment is made.

4.3 Decision-Theoretic Rough Sets

Decision-theoretic rough sets (DTRS) [48, 50, 51, 57, 58] are a quantitative generalization of Pawlak rough sets by considering the degree of inclusion of an equivalence class in a set.

The acceptance-rejection evaluation used by a DTRS model is the conditional probability $v_A(x) = Pr(A|[x])$, with values from the totally ordered set $([0,1], \leq)$. Given a pair of thresholds (α, β) with $0 \leq \beta < \alpha \leq 1$, the sets of designated values for acceptance and rejections are $L^+ = \{a \in [0,1] \mid \alpha \leq a\}$ and $L^- = \{b \in [0,1] \mid b \leq \beta\}$. According to Definition 4, a DTRS model makes the following three-way decisions: for $A \subseteq U$,

$$
\begin{aligned}
\mathrm{POS}_{(\alpha,\beta)}(A) &= \{x \in U \mid v_A(x) \succeq \alpha\} \\
&= \{x \in U \mid Pr(A|[x]) \geq \alpha\}, \\
\mathrm{NEG}_{(\alpha,\beta)}(A) &= \{x \in U \mid v_A(x) \preceq \beta\} \\
&= \{x \in U \mid Pr(A|[x]) \leq \beta\}, \\
\mathrm{BND}_{(\alpha,\beta)}(A) &= \{x \in U \mid \beta \prec v_A(x) \prec \alpha\} \\
&= \{x \in U \mid \beta < Pr(A|[x]) < \alpha\}. \qquad (14)
\end{aligned}
$$

Three-way decision-making in DTRS can be easily related to incorrect acceptance error and incorrect rejection error [55]. Specifically, incorrect acceptance error is given by $Pr(A^c|[x]) = 1 - Pr(A|[x]) \leq 1 - \alpha$, which is bounded by $1 - \alpha$. Likewise, incorrect rejection error is given by $Pr(A|[x]) \leq \beta$, which is bounded by β. Therefore, the pair of thresholds can be interpreted as defining tolerance levels of errors.

A main advantage of a DTRS model is its solid foundation based on Bayesian decision theory. In addition, the pair of thresholds can be systematically computed by minimizing overall ternary classification cost [55].

Bayesian decision theory [3] can be applied to the derivation of DTRS as follows. We have a set of 2 states and a set of 3 actions for each state. The set of states is given by $\Omega = \{A, A^c\}$ indicating that an object is in A and not in A, respectively. For simplicity, we use the same symbol to denote both a subset A and the corresponding state. With respect to the three regions, the set of actions with respect to a state is given by $\mathcal{A} = \{a_P, a_N, a_B\}$, where a_P, a_N, and a_B represent the three actions in classifying an object x, namely, deciding $x \in \mathrm{POS}(A)$, deciding $x \in \mathrm{NEG}(A)$, and deciding $x \in \mathrm{BND}(A)$, respectively. The losses regarding the risk or cost of those classification actions with respect to different states are given by the 3×2 matrix:

	A (P)	A^c (N)
a_P	λ_{PP}	λ_{PN}
a_N	λ_{NP}	λ_{NN}
a_B	λ_{BP}	λ_{BN}

In the matrix, λ_{PP}, λ_{NP} and λ_{BP} denote the losses incurred for taking actions a_P, a_N and a_B, respectively, when an object belongs to A, and λ_{PN}, λ_{NN} and λ_{BN} denote the losses incurred for taking the same actions when the object does not belong to A

To determine a pair of thresholds for three-way decisions, one can minimize the following overall risk [12, 55]:

$$R(\alpha, \beta) = R_P(\alpha, \beta) + R_N(\alpha, \beta) + R_B(\alpha, \beta), \tag{15}$$

where

$$\begin{aligned} R_P(\alpha, \beta) &= \sum_{Pr(A|[x]) \geq \alpha} [\lambda_{PP} Pr(A|[x]) + \lambda_{PN} Pr(A^c|[x])] Pr([x]), \\ R_N(\alpha, \beta) &= \sum_{Pr(A|[x]) \leq \beta} [\lambda_{NP} Pr(A|[x]) + \lambda_{NN} Pr(A^c|[x])] Pr([x]), \\ R_B(\alpha, \beta) &= \sum_{\beta < Pr(A|[x]) < \alpha} [\lambda_{BP} Pr(A|[x]) + \lambda_{BN} Pr(A^c|[x])] Pr([x]), \end{aligned} \tag{16}$$

represent, risks incurred by acceptance, rejection, and noncommitment, and the summation is over all equivalence classes. It can be shown [50, 55] that under the following conditions,

$$\begin{aligned} &(c_1) \quad \lambda_{PP} < \lambda_{BP} < \lambda_{NP}, \qquad \lambda_{NN} < \lambda_{BN} < \lambda_{PN}, \\ &(c_2) \quad (\lambda_{PN} - \lambda_{BN})(\lambda_{NP} - \lambda_{BP}) > (\lambda_{BN} - \lambda_{NN})(\lambda_{BP} - \lambda_{PP}), \end{aligned} \tag{17}$$

a pair of threshold (α, β) with $0 \leq \beta < \alpha \leq 1$ that minimizes R is given by:

$$\begin{aligned} \alpha &= \frac{(\lambda_{PN} - \lambda_{BN})}{(\lambda_{PN} - \lambda_{BN}) + (\lambda_{BP} - \lambda_{PP})}, \\ \beta &= \frac{(\lambda_{BN} - \lambda_{NN})}{(\lambda_{BN} - \lambda_{NN}) + (\lambda_{NP} - \lambda_{BP})}. \end{aligned} \tag{18}$$

That is, the pair of thresholds can be computed from the loss function.

Other models for determining the pair of thresholds include a game-theoretic framework [1, 9, 11], a multi-view decision model [17, 66], and the minimization of uncertainty of the three regions [2]. The conditional probability required by DTRS can be estimated based on a naive Bayesian rough set model [59] or a regression model [25].

4.4 Three-Valued Approximations in Many-Valued Logic and Fuzzy Sets

Three-valued approximations in many-valued logics are formulated based on the discussion given by Gottwald [4] on positively designated truth degrees and negatively designated truth degrees.

In many-valued logic, the set of truth degrees or values is normally an ordered set $(L, \preceq)$ and contains the classical truth values F and T (often coded by 0 and 1) as its minimum and maximum elements, namely, $\{F, T\} \subseteq L$ and for any $u \in L$, $F \preceq u \preceq T$. It is also a common practice to use a subset L^+ of positively designated truth degrees to code the intuitive notion of truth and to use another subset L^- of negatively designated truth degrees to code the opposite. For the two sets to be meaningful, the following conditions are normally assumed [4]:

$$
\begin{aligned}
&\text{(i)} && T \in L^+, \\
& && F \in L^-, \\
&\text{(ii)} && L^+ \cup L^- \subseteq L, \\
& && L^+ \cap L^- = \emptyset, \\
&\text{(iii)} && w \preceq u \wedge w \in L^+ \Longrightarrow u \in L^+, \\
& && w \preceq u \wedge u \in L^- \Longrightarrow w \in L^-.
\end{aligned}
$$

Three-valued approximations of a many-valued logic derive from three-way decisions based on the two designated sets. We accept a truth degree as being true if it is in the positively designated set, reject it as being true if it is the negatively designated set, and neither accept nor reject if it is not in any of the two sets. By so doing, we can have a new three-valued logic with the set of truth values $L_3 = \{L^-, L - (L^- \cup L^+), L^+\}$ under the ordering $L^- \preceq_3 L - (L^- \cup L^+) \preceq_3 L^+$, which is an approximation of the many-valued logic.

To a large extent, our formulation of three-way decisions, as given in the last section, draws mainly from such a consideration. Specifically, we borrowed the notions of designated truth degrees from studies of many-valued logic to introduce the notions of designated values for acceptance and rejection in the theory of three-way decisions.

A fuzzy set $\mathcal{A}$ is characterized by a mapping from U to the unit interval, namely, $\mu_{\mathcal{A}} : U \longrightarrow [0, 1]$. The value $\mu_{\mathcal{A}}(x)$ is called the degree of membership of the object $x \in U$. Fuzzy sets may be interpreted in terms of a many-valued logic with the unit interval as its set of truth degrees. According to the three-valued approximations of a many-valued logic, one can similarly formulate three-valued approximations of a fuzzy set. This formulation was in fact given by Zadeh [63] in his seminal paper on fuzzy sets and was shown to be related to Kleene's three-valued logic.

Given a pair of thresholds (α, β) with $0 \leq \beta < \alpha \leq 1$, one can define the designated sets of values for acceptance and rejection as $L^+ = \{a \in [0,1] \mid \alpha \leq a\}$ and $L^- = \{b \in [0,1] \mid b \leq \beta\}$. According to Definition 4, if a fuzzy membership function $\mu_\mathcal{A}$ is used as an acceptance-rejection evaluation, namely, $v_{\mu_\mathcal{A}} = \mu_\mathcal{A}$, we have the following three-way decisions,

$$\begin{aligned}
\mathrm{POS}_{(\alpha,\beta)}(\mu_\mathcal{A}) &= \{x \in U \mid v_{\mu_\mathcal{A}}(x) \succeq \alpha\} \\
&= \{x \in U \mid \mu_\mathcal{A}(x) \geq \alpha\}, \\
\mathrm{NEG}_{(\alpha,\beta)}(\mu_\mathcal{A}) &= \{x \in U \mid v_{\mu_\mathcal{A}}(x) \preceq \beta\} \\
&= \{x \in U \mid \mu_\mathcal{A}(x) \leq \beta\}, \\
\mathrm{BND}_{(\alpha,\beta)}(\mu_\mathcal{A}) &= \{x \in U \mid \beta \prec v_{\mu_\mathcal{A}}(x) \prec \alpha\} \\
&= \{x \in U \mid \beta < \mu_\mathcal{A}(x) < \alpha\}. \qquad (19)
\end{aligned}$$

Zadeh [63] provided an interpretation of this three-valued approximations of a fuzzy set: one may say that (1) x belongs to $\mathcal{A}$ if $\mu_\mathcal{A}(x) \geq \alpha$; (2) x does not belong to $\mathcal{A}$ if $\mu_\mathcal{A}(x) \leq \beta$; and (3) x has an indeterminate status relative to $\mathcal{A}$ if $\beta < \mu_\mathcal{A}(x) < \alpha$. This interpretation explicitly uses the notions of acceptance and rejection and is consistent with our three-way decisions.

4.5 Shadowed Sets

In contrast to decision-theoretic rough sets in which the pair of thresholds can be interpreted by classification errors, there is a difficulty in interpreting thresholds in three-valued approximations of a fuzzy sets. The introduction of a shadowed set induced by a fuzzy set attempts to address this problem [34, 35].

A shadowed set $\mathbb{A}$ is defined as a mapping, $S_\mathbb{A} : U \longrightarrow \{0, [0,1], 1\}$, from U to a set of three truth values. It is assumed that the three values are ordered by $0 \preceq [0,1] \preceq 1$. The value $[0,1]$ represents the membership of objects in the shadows of a shadowed set. Like the interval-set algebra, shadowed-set algebra is also related to Kleene's three-valed logic. Shadowed sets provide another model of three-way decisions.

Unlike an interval set, a shadowed set is constructed from a fuzzy set $\mu_\mathcal{A} : U \longrightarrow [0,1]$ as follows:

$$S_\mathbb{A}(x) = \begin{cases} 0, & \mu_\mathcal{A}(x) \leq \tau, \\ [0,1], & \tau < \mu_\mathcal{A}(x) < 1-\tau, \\ 1, & \mu_\mathcal{A}(x) \geq 1-\tau, \end{cases} \qquad (20)$$

where $0 \leq \tau < 0.5$ is a threshold. Given a pair of thresholds $(1, 0)$ for the set of truth values $\{0, [0,1], 1\}$, by Definition 4 and equations (19) and (20), we have the following three-way decision for a shadowed set:

$$\begin{aligned}
\mathrm{POS}_{(1,0)}(S_\mathbb{A}) &= \{x \in U \mid v_{S_\mathbb{A}}(x) \succeq 1\} \\
&= \{x \in U \mid \mu_\mathcal{A}(x) \geq 1-\tau\} \\
&= \mathrm{POS}_{(1-\tau,\tau)}(\mu_\mathcal{A}),
\end{aligned}$$

$$\begin{aligned}
\mathrm{NEG}_{(1,0)}(S_{\mathbb{A}}) &= \{x \in U \mid v_{S_{\mathbb{A}}}(x) \preceq 0\} \\
&= \{x \in U \mid \mu_{\mathcal{A}}(x) \leq \tau\}, \\
&= \mathrm{NEG}_{(1-\tau,\tau)}(\mu_{\mathcal{A}}), \\
\mathrm{BND}_{(1,0)}(S_{\mathbb{A}}) &= \{x \in U \mid 0 \prec v_{S_{\mathbb{A}}}(x) \prec 1\} \\
&= \{x \in U \mid \tau < \mu_{\mathcal{A}}(x) < 1-\tau\} \\
&= \mathrm{BND}_{(1-\tau,\tau)}(\mu_{\mathcal{A}}).
\end{aligned} \tag{21}$$

That is, a shadowed set is a three-valued approximation of a fuzzy set with $(\alpha, \beta) = (1 - \tau, \tau)$. In general, one can also consider shadowed set by a pair of thresholds (α, β) with $0 \leq \beta < \alpha \leq 1$ on a fuzzy set $\mu_{\mathcal{A}}$.

As shown in [34, 35], the threshold τ for constructing a shadowed set can be determined by minimizing the following function,

$$\Omega(\tau) = \mathrm{abs}(\Omega_r(\tau) + \Omega_e(\tau) - \Omega_s(\tau)), \tag{22}$$

where abs(·) stands for the absolute value and

$$\begin{aligned}
\Omega_r(\tau) &= \sum_{\{x \in U \mid \mu_{\mathcal{A}}(x) \leq \tau\}} \mu_{\mathcal{A}}(x), \\
\Omega_e(\tau) &= \sum_{\{y \in U \mid \mu_{\mathcal{A}}(y) \geq 1-\tau\}} (1 - \mu_{\mathcal{A}}(y)), \\
\Omega_s(\tau) &= \mathrm{card}(\{z \in U \mid \tau < \mu_{\mathcal{A}}(z) < 1 - \tau\}),
\end{aligned} \tag{23}$$

are, respectively, the total of reduced membership values from $\mu_{\mathcal{A}}(x)$ in the fuzzy set to 0 in the shadowed set (i.e., $\mu_{\mathcal{A}}(x) - 0 = \mu_{\mathcal{A}}(x)$), the total of elevated membership values from $\mu_{\mathcal{A}}(y)$ in the fuzzy set to 1 in the shadowed set (i.e., $1 - \mu_{\mathcal{A}}(y)$), and the cardinality of the shadows of the shadowed set. The minimization of $\Omega(\tau)$ may be equivalently formulated as finding a solution to the equation,

$$\Omega_r(\tau) + \Omega_e(\tau) = \Omega_s(\tau), \tag{24}$$

if it has a solution. Although the problem of finding the threshold τ is formulated precisely, the meaning of the objective function $\Omega(\tau)$ still needs further investigation. It is interesting to note that the objective function $\Omega(\tau)$ shares some similarity to the objective function $R(\alpha, \beta)$ of a DTRS model, which may shed some light on the problem of determining the threshold in shadowed sets.

5 Conclusions

The concept of three-way decisions provides an appealing interpretation of three regions in probabilistic rough sets. The positive and negative regions are sets of accepted objects and rejected objects, respectively. The boundary region is the set of objects for which neither acceptance nor rejection is possible, due to uncertain or incomplete information. A close examination of studies and applications of three-way decisions shows that (a) essential ideas of three-way decisions are general applicable to a wide range of decision-making problems; (b) we routinely make three-way decisions in everyday life; (c) three-way decisions appear across many fields and disciplines; and (d) there is a lack of formal theory for three-way decisions. These findings motivate a study of a theory of three-way decisions in its own right.

We outline a theory of three-way decisions based on the notions of evaluations for acceptance and evaluations for rejection. One accepts or rejects an object when its values from evaluations fall into some designated areas; otherwise, one makes a decision of noncommitment. We propose and study three classes of evaluations. We demonstrate that the proposed theory can describe and explain three-way decisions in many-valued logics and generalizations of set theory, including interval sets, rough sets, decision-theoretic rough sets, fuzzy sets, and shadowed sets. As future research, we plan to investigate three-way decisions in other settings.

Acknowledgements. This work is partially supported by a Discovery Grant from NSERC Canada.

References

1. Azam, N., Yao, J.T.: Multiple Criteria Decision Analysis with Game-theoretic Rough Sets. In: Li, T., Nguyen, H.S., Wang, G., Grzymala-Busse, J.W., Janicki, R., Hassanien, A.E., Yu, H. (eds.) RSKT 2012. LNCS (LNAI), vol. 7414, pp. 399–408. Springer, Heidelberg (2012)
2. Deng, X.F., Yao, Y.Y.: An Information-Theoretic Interpretation of Thresholds in Probabilistic Rough Sets. In: Li, T., Nguyen, H.S., Wang, G., Grzymala-Busse, J.W., Janicki, R., Hassanien, A.E., Yu, H. (eds.) RSKT 2012. LNCS (LNAI), vol. 7414, pp. 369–378. Springer, Heidelberg (2012)
3. Duda, R.O., Hart, P.E.: Pattern Classification and Scene Analysis. Wiley, New York (1973)
4. Gottwald, S.: A Treatise on Many-Valued Logics. Research Studies Press, Baldock (2001)
5. Goudey, R.: Do statistical inferences allowing three alternative decision give better feedback for environmentally precautionary decision-making. Journal of Environmental Management 85, 338–344 (2007)
6. Granberg, D., Steele, L.: Procedural considerations in measuring latitudes of acceptance, rejection, and noncommitment. Social Forces 52, 538–542 (1974)
7. Grzymała-Busse, J.W.: Generalized Parameterized Approximations. In: Yao, J., Ramanna, S., Wang, G., Suraj, Z. (eds.) RSKT 2011. LNCS (LNAI), vol. 6954, pp. 136–145. Springer, Heidelberg (2011)
8. Grzymala-Busse, J.W., Yao, Y.Y.: Probabilistic rule induction with the LERS data mining system. International Journal of Intelligent Systems 26, 518–539 (2011)
9. Herbert, J.P., Yao, J.T.: Learning Optimal Parameters in Decision-Theoretic Rough Sets. In: Wen, P., Li, Y., Polkowski, L., Yao, Y., Tsumoto, S., Wang, G. (eds.) RSKT 2009. LNCS (LNAI), vol. 5589, pp. 610–617. Springer, Heidelberg (2009)

10. Herbert, J.P., Yao, J.T.: Game-theoretic rough sets. Fundamenta Informaticae 108, 267–286 (2011)
11. Herbert, J.P., Yao, J.T.: Analysis of Data-Driven Parameters in Game-Theoretic Rough Sets. In: Yao, J., Ramanna, S., Wang, G., Suraj, Z. (eds.) RSKT 2011. LNCS (LNAI), vol. 6954, pp. 447–456. Springer, Heidelberg (2011)
12. Jia, X.Y., Li, W.W., Shang, L., Chen, J.J.: An Optimization Viewpoint of Decision-Theoretic Rough Set Model. In: Yao, J., Ramanna, S., Wang, G., Suraj, Z. (eds.) RSKT 2011. LNCS (LNAI), vol. 6954, pp. 457–465. Springer, Heidelberg (2011)
13. Jia, X.Y., Zhang, K., Shang, L.: Three-way Decisions Solution to Filter Spam Email: An Empirical Study. In: Yao, J., Yang, Y., Slowinski, R., Greco, S., Li, H., Mitra, S., Polkowski, L. (eds.) RSCTC 2012. LNCS (LNAI), vol. 7413, pp. 287–296. Springer, Heidelberg (2012)
14. Keeney, R.L., Raiffa, H.: Decisions With Multiple Objectives: Preferences and Value Tradeoffs. John Wiley and Sons, New York (1976)
15. Kleene, S.C.: Introduction to Mathematics. Groningen, New York (1952)
16. Li, H.X., Wang, M.H., Zhou, X.Z., Zhao, J.B.: An interval set model for learning rules from incomplete information table. International Journal of Approximate Reasoning 53, 24–37 (2012)
17. Li, H.X., Zhou, X.Z.: Risk decision making based on decision-theoretic rough set: A three-way view decision model. International Journal of Computational Intelligence Systems 4, 1–11 (2011)
18. Li, H.X., Zhou, X.Z., Li, T.R., Wang, G.Y., Miao, D.Q., Yao, Y.Y. (eds.): Decision-Theoretic Rough Sets Theory and Recent Research. Science Press, Beijing (2011) (in Chinese)
19. Li, H., Zhou, X., Zhao, J., Huang, B.: Cost-Sensitive Classification Based on Decision-Theoretic Rough Set Model. In: Li, T., Nguyen, H.S., Wang, G., Grzymala-Busse, J.W., Janicki, R., Hassanien, A.E., Yu, H. (eds.) RSKT 2012. LNCS (LNAI), vol. 7414, pp. 379–388. Springer, Heidelberg (2012)
20. Li, H.X., Zhou, X.Z., Zhao, J.B., Liu, D.: Attribute Reduction in Decision-Theoretic Rough Set Model: A Further Investigation. In: Yao, J., Ramanna, S., Wang, G., Suraj, Z. (eds.) RSKT 2011. LNCS (LNAI), vol. 6954, pp. 466–475. Springer, Heidelberg (2011)
21. Li, W., Miao, D.Q., Wang, W.L., Zhang, N.: Hierarchical rough decision theoretic framework for text classification. In: Proceedings of the 9th IEEE International Conference on Cognitive Informatics, pp. 484–489 (2009)
22. Lipski Jr., W.: On semantic issues connected with incomplete information databases. ACM Transactions on Database Systems 4, 269–296 (1979)
23. Liu, D., Li, H.X., Zhou, X.Z.: Two decades' research on decision-theoretic rough sets. In: Proceedings of the 9th IEEE International Conference on Cognitive Informatics, pp. 968–973 (2010)
24. Liu, D., Li, T.R., Li, H.X.: A multiple-category classification approach with decision-theoretic rough sets. Fundamenta Informaticae 115, 173–188 (2012)
25. Liu, D., Li, T.R., Liang, D.C.: A New Discriminant Analysis Approach under Decision-Theoretic Rough Sets. In: Yao, J., Ramanna, S., Wang, G., Suraj, Z. (eds.) RSKT 2011. LNCS (LNAI), vol. 6954, pp. 476–485. Springer, Heidelberg (2011)
26. Liu, D., Li, T., Liang, D.: Decision-theoretic Rough Sets with Probabilistic Distribution. In: Li, T., Nguyen, H.S., Wang, G., Grzymala-Busse, J.W., Janicki, R., Hassanien, A.E., Yu, H. (eds.) RSKT 2012. LNCS (LNAI), vol. 7414, pp. 389–398. Springer, Heidelberg (2012)

27. Liu, D., Li, T.R., Ruan, D.: Probabilistic model criteria with decision-theoretic rough sets. Information Sciences 181, 3709–3722 (2011)
28. Liu, D., Yao, Y.Y., Li, T.R.: Three-way investment decisions with decision-theoretic rough sets. International Journal of Computational Intelligence Systems 4, 66–74 (2011)
29. Liu, J., Min, F., Liao, S., Zhu, W.: Minimal Test Cost Feature Selection with Positive Region Constraint. In: Yao, J., Yang, Y., Slowinski, R., Greco, S., Li, H., Mitra, S., Polkowski, L. (eds.) RSCTC 2012, vol. 7413, pp. 259–266. Springer, Heidelberg (2012)
30. Lurie, J.D., Sox, H.C.: Principles of medical decision making. Spine 24, 493–498 (1999)
31. Ma, X., Wang, G., Yu, H.: Multiple-category Attribute Reduct Using Decision-Theoretic Rough Set Model. In: Yao, J., Yang, Y., Slowinski, R., Greco, S., Li, H., Mitra, S., Polkowski, L. (eds.) RSCTC 2012. LNCS (LNAI), vol. 7413, pp. 267–276. Springer, Heidelberg (2012)
32. Marinoff, L.: The Middle Way, Finding Happiness in a World of Extremes. Sterling, New York (2007)
33. Pawlak, Z.: Rough sets. International Journal of Computer and Information Sciences 11, 341–356 (1982)
34. Pedrycz, W.: Shadowed sets: Representing and processing fuzzy sets. IEEE Transactions on Systems, Man, and Cybernetics, Part B: Cybernetics 28, 103–109 (1998)
35. Pedrycz, W.: From fuzzy sets to shadowed sets: Interpretation and computing. International Journal of Intelligent Systems 24, 48–61 (2009)
36. Rescher, N.: Many-valued Logic. McGraw-Hill, New York (1969)
37. Pauker, S.G., Kassirer, J.P.: The threshold approach to clinical decision making. The New England Journal of Medicine 302, 1109–1117 (1980)
38. Schechter, C.B.: Sequential analysis in a Bayesian model of diastolic blood pressure measurement. Medical Decision Making 8, 191–196 (1988)
39. Sherif, M., Hovland, C.I.: Social Judgment: Assimilation and Contrast Effects in Communication and Attitude Change. Yale University Press, New Haven (1961)
40. Smith, E.E.: Concepts and induction. In: Posner, M.I. (ed.) Foundations of Cognitive Science, pp. 501–526. The MIT Press, Cambridge (1989)
41. van Mechelen, I., Hampton, J., Michalski, R.S., Theuns, P. (eds.): Categories and Concepts, Theoretical Views and Inductive Data Analysis. Academic Press, New York (1993)
42. Wald, A.: Sequential tests of statistical hypotheses. The Annals of Mathematical Statistics 16, 117–186 (1945)
43. Weller, A.C.: Editorial Peer Review: Its Strengths and Weaknesses. Information Today, Inc., Medford (2001)
44. Woodward, P.W., Naylor, J.C.: An application of Bayesian methods in SPC. The Statistician 42, 461–469 (1993)
45. Yang, X.P., Song, H.G., Li, T.J.: Decision Making in Incomplete Information System Based on Decision-Theoretic Rough Sets. In: Yao, J., Ramanna, S., Wang, G., Suraj, Z. (eds.) RSKT 2011. LNCS (LNAI), vol. 6954, pp. 495–503. Springer, Heidelberg (2011)
46. Yang, X.P., Yao, J.T.: Modelling multi-agent three-way decisions with decision-theoretic rough sets. Fundamenta Informaticae 115, 157–171 (2012)
47. Yao, Y.Y.: Interval-set algebra for qualitative knowledge representation. In: Proceedings of the 5th International Conference on Computing and Information, pp. 370–374 (1993)

48. Yao, Y.Y.: Probabilistic approaches to rough sets. Expert Systems 20, 287–297 (2003)
49. Yao, Y.Y.: A Note on Definability and Approximations. In: Peters, J.F., Skowron, A., Marek, V.W., Orłowska, E., Słowiński, R., Ziarko, W.P. (eds.) Transactions on Rough Sets VII. LNCS, vol. 4400, pp. 274–282. Springer, Heidelberg (2007)
50. Yao, Y.Y.: Decision-Theoretic Rough Set Models. In: Yao, J., Lingras, P., Wu, W.-Z., Szczuka, M.S., Cercone, N.J., Ślęzak, D. (eds.) RSKT 2007. LNCS (LNAI), vol. 4481, pp. 1–12. Springer, Heidelberg (2007)
51. Yao, Y.Y.: Probabilistic rough set approximations. International Journal of Approximation Reasoning 49, 255–271 (2008)
52. Yao, Y.Y.: Three-Way Decision: An Interpretation of Rules in Rough Set Theory. In: Wen, P., Li, Y., Polkowski, L., Yao, Y., Tsumoto, S., Wang, G. (eds.) RSKT 2009. LNCS (LNAI), vol. 5589, pp. 642–649. Springer, Heidelberg (2009)
53. Yao, Y.Y.: Interval sets and interval-set algebras. In: Proceedings of the 8th IEEE International Conference on Cognitive Informatics, pp. 307–314 (2009)
54. Yao, Y.Y.: Three-way decisions with probabilistic rough sets. Information Sciences 180, 341–353 (2010)
55. Yao, Y.Y.: The superiority of three-way decisions in probabilistic rough set models. Information Sciences 181, 1080–1096 (2011)
56. Yao, Y.Y., Deng, X.F.: Sequential three-way decisions with probabilistic rough sets. In: Proceedings of the 10th IEEE International Conference on Cognitive Informatics & Cognitive Computing, pp. 120–125 (2011)
57. Yao, Y.Y., Wong, S.K.M.: A decision theoretic framework for approximating concepts. International Journal of Man-machine Studies 37, 793–809 (1992)
58. Yao, Y.Y., Wong, S.K.M., Lingras, P.: A decision-theoretic rough set model. In: Ras, Z.W., Zemankova, M., Emrich, M.L. (eds.) Methodologies for Intelligent Systems, vol. 5, pp. 17–24. North-Holland, New York (1990)
59. Yao, Y.Y., Zhou, B.: Naive Bayesian Rough Sets. In: Yu, J., Greco, S., Lingras, P., Wang, G., Skowron, A. (eds.) RSKT 2010. LNCS (LNAI), vol. 6401, pp. 719–726. Springer, Heidelberg (2010)
60. Yu, H., Chu, S.S., Yang, D.C.: Autonomous knowledge-oriented clustering using decision-theoretic rough set theory. Fundamenta Informaticae 115, 141–156 (2012)
61. Yu, H., Liu, Z.G., Wang, G.Y.: Automatically Determining the Number of Clusters Using Decision-Theoretic Rough Set. In: Yao, J., Ramanna, S., Wang, G., Suraj, Z. (eds.) RSKT 2011. LNCS (LNAI), vol. 6954, pp. 504–513. Springer, Heidelberg (2011)
62. Yu, H., Wang, Y.: Three-way Decisions Method for Overlapping Clustering. In: Yao, J., Yang, Y., Slowinski, R., Greco, S., Li, H., Mitra, S., Polkowski, L. (eds.) RSCTC 2012. LNCS (LNAI), vol. 7413, pp. 277–286. Springer, Heidelberg (2012)
63. Zadeh, L.A.: Fuzzy sets. Information and Control 8, 338–353 (1965)
64. Zhou, B.: A New Formulation of Multi-category Decision-Theoretic Rough Sets. In: Yao, J., Ramanna, S., Wang, G., Suraj, Z. (eds.) RSKT 2011. LNCS (LNAI), vol. 6954, pp. 514–522. Springer, Heidelberg (2011)
65. Zhou, B., Yao, Y.Y., Luo, J.G.: A Three-Way Decision Approach to Email Spam Filtering. In: Farzindar, A., Kešelj, V. (eds.) Canadian AI 2010. LNCS (LNAI), vol. 6085, pp. 28–39. Springer, Heidelberg (2010)
66. Zhou, X.Z., Li, H.X.: A Multi-View Decision Model Based on Decision-Theoretic Rough Set. In: Wen, P., Li, Y., Polkowski, L., Yao, Y., Tsumoto, S., Wang, G. (eds.) RSKT 2009. LNCS (LNAI), vol. 5589, pp. 650–657. Springer, Heidelberg (2009)

A Multi-granulation Model under Dominance-Based Rough Set Approach

Shaoyong Li, Tianrui Li, and Chuan Luo

School of Information Science and Technology, Southwest Jiaotong University, Chengdu, 610031, China
meterer@163.com, trli@swjtu.edu.cn, luochuan9@gmail.com

Abstract. Multi-granulation rough set is a generalization of Rough Set Theory (RST) in order to adapt to cases that there are multiple relations in the universe. To allow Dominance-based Rough Set Approach (DRSA) being applied in multiple relations cases, we propose a multi-granulation model based on DRSA. A numerical example is employed to validate the rationality and feasibility of our model.

Keywords: Rough sets, Multi-granulation, Dominance relation.

1 Introduction

In many cases, the experts deal with problems based on imperfect information systems. These information are often incomplete, inconsistence and uncertain. To obtain right decision making from these information, people always concentrate their attention on exploring new techniques to come their aims true. Of course, people have obtained some expertise from their untiring study. Rough Set Theory (RST) proposed by Pawlak is one of these brilliant achievements [1]. It is an excellent mathematical tool for processing imperfect information systems and has been applied widely in artificial intelligence, data mining [2, 4] and so on.

Qian et al. noticed that RST and it's generalization make information granulated based on a binary relation [5–8]. They believed that there may exist many different viewpoints to a decision problem. Thus, in the universe, information may be granulated by many kinds of relations into some different families of granules. Hence, they built a multi-granulation model based on RST to adapt to multiple relations environments.

Dominance-based Rough Set Approach (DRSA) proposed by Greco et al. is one of RST's generalization [9–11]. The main feature of DRSA different to RST is the substitution of the indiscernibility relation by a dominance relation. It is a powerful tool to process information with preference-ordered attribute domains.

In this paper, we aim to build a multi-granulation model based on DRSA. Following Qian et al.'s idea, we also will take into account several families of granules in the process of calculating upper and lower approximations of unions of decision classes in DRSA. Furthermore, we will use a support ratio α to generalize Qian et al.'s two standpoints in their multi-granulation model: Optimistic

J.T. Yao et al. (Eds.): RSCTC 2012, LNAI 7413, pp. 18–25, 2012.

standpoint and pessimistic standpoint [5, 6]. α belongs to a range from $\frac{1}{m}$ to 1, where m is the number of decision-makers. m also means the number of dominance relations adopted by decision-makers in decision making. The case $\alpha = \frac{1}{m}$ is similar to Qian et al.'s optimistic standpoint and the case $\alpha = 1$ is similar to their pessimistic standpoint. When $\frac{1}{m} < \alpha < 1$, this case presents the standpoint that are lower than pessimistic standpoint but higher than optimistic standpoint.

The remainder of this paper is organized as follows. We present basic notions of DRSA and multi-granulation rough sets in Section 2. We introduce a multi-granulation model based on DRSA in Section 3. In Section 4, we give an illustrative example to validate our approach. This paper ends with conclusions and further research topics in Section 5.

2 Preliminaries

In this section, we briefly review some concepts, notations and results of DRSA [9–11] and multi-granulation rough sets [5–8].

A decision information system is a 4-tuple $S = (U, C \cup \{d\}, V, f)$, where U is a non-empty finite set of objects, called as the universe; C is a non-empty finite set of condition attributes, d is the decision attribute; V is regarded as the domain of all attributes; $f : U \times C \cup \{d\} \to V$ is an information function such that $f(x, a) \in V_a$, $\forall a \in C \cup \{d\}$ and $x \in U$, where V_a is a domain of attribute a.

$\forall a \in C$, there is a preference relation on the set of objects with respect to attribute a, denoted by $\succeq_a$. $\forall x, y \in U$, $x \succeq_a y$ means " x is at least as good as y with respect to attribute a ". For a nonempty finite attribute set P, $P \subseteq C$, if $x \succeq_a y$ for all $a \in P$, we say that x dominates y with respect to P, denoted by xD_Py. Therefore, there are the following two sets:

- A set of objects dominating x, called P-dominating set, $D_P^+(x) = \{y \in U : yD_Px\}$;
- A set of objects dominated by x, called P-dominated set, $D_P^-(x) = \{y \in U : xD_Py\}$.

Decision attribute d makes a partition of U into a finite number of classes. Let $\boldsymbol{Cl} = \{Cl_t, t \in T\}$, $T = \{1, \cdots, n\}$, be a set of these classes that are ordered. $\forall r, s \in T$ such that $r > s$, the objects from Cl_r are preferred to the objects from Cl_s. The sets to be approximated are an upward union and a downward union of classes such that

$$Cl_t^{\geq} = \bigcup_{t' \geq t} Cl_{t'}, \quad Cl_t^{\leq} = \bigcup_{t' \leq t} Cl_{t'}, \quad \forall t, t' \in T.$$

Here, $x \in Cl_t^{\geq}$ means "x belongs to at least class Cl_t", and $x \in Cl_t^{\leq}$ means "x belongs to at most class Cl_t".

Let S be a decision information system, $P \subseteq C$, $t \in T$. The lower and upper approximations of $Cl_t^{\geq}$ are defined respectively as:

$$\underline{P}(Cl_t^{\geq}) = \{x \in U : D_P^+(x) \subseteq Cl_t^{\geq}\},$$
$$\overline{P}(Cl_t^{\geq}) = \{x \in U : D_P^-(x) \cap Cl_t^{\geq} \neq \emptyset\}.$$

The lower and upper approximations of $Cl_t^{\leq}$ are defined respectively as:

$$\underline{P}(Cl_t^{\leq}) = \{x \in U : D_P^-(x) \subseteq Cl_t^{\leq}\},$$
$$\overline{P}(Cl_t^{\leq}) = \{x \in U : D_P^+(x) \cap Cl_t^{\leq} \neq \emptyset\}.$$

The upper and lower approximations of upward and downward unions of decision classes have a complementarity property,

$$\underline{P}(Cl_t^{\geq}) = U - \overline{P}(Cl_{n-1}^{\leq}) \ \ and \ \ \overline{P}(Cl_t^{\geq}) = U - \underline{P}(Cl_{n-1}^{\leq}), \ \ t = 2, \cdots, n$$

$$\underline{P}(Cl_t^{\leq}) = U - \overline{P}(Cl_{n+1}^{\geq}) \ \ and \ \ \overline{P}(Cl_t^{\leq}) = U - \underline{P}(Cl_{n+1}^{\geq}), \ \ t = 1, \cdots, n-1$$

All the objects belonging to $Cl_t^{\geq}$ and $Cl_t^{\leq}$ with some ambiguity constitute the P-boundary of $Cl_t^{\geq}$ and $Cl_t^{\leq}$, denoted by $Bn_P(Cl_t^{\geq})$ and $Bn_P(Cl_t^{\leq})$, respectively. They can be represented in terms of upper and lower approximations as follows:

$$Bn_P(Cl_t^{\geq}) = \overline{P}(Cl_t^{\geq}) - \underline{P}(Cl_t^{\geq})$$
$$Bn_P(Cl_t^{\leq}) = \overline{P}(Cl_t^{\leq}) - \underline{P}(Cl_t^{\leq})$$

Qian et al. proposed two different multi-granulation rough set models from two different standpoints: Optimistic multi-granulation rough set model and pessimistic multi-granulation rough set model. They introduced definitions of these two models as follows: $S = (U, AT, f)$ is an information system, $A_1, \cdots, A_m \subseteq AT$, $X \subseteq U$. With respect to attribute sets $A_1, \cdots, A_m$, optimistic lower and upper approximations of X are denoted by $\underline{\sum_{i=1}^m A_i}^O(X)$ and $\overline{\sum_{i=1}^m A_i}^O(X)$, respectively, where

$$\underline{\sum_{i=1}^m A_i}^O(X) = \bigcup_{i=1}^m \{x \in U | [x]_{A_i} \subseteq X\}$$
$$\overline{\sum_{i=1}^m A_i}^O(X) = \sim \underline{\sum_{i=1}^m A_i}^O(\sim X)$$

and optimistic boundary region is

$$Bn^O_{\sum_{i=1}^m A_i}(X) = \overline{\sum_{i=1}^m A_i}^O(X) \setminus \underline{\sum_{i=1}^m A_i}^O(X)$$

With respect to attribute sets $A_1, \cdots, A_m$, pessimistic lower and upper approximations of X are denoted by $\underline{\sum_{i=1}^m A_i}^P(X)$ and $\overline{\sum_{i=1}^m A_i}^P(X)$, respectively, where

$$\underline{\sum_{i=1}^m A_i}^P(X) = \{x \in U | [x]_{A_1} \subseteq X \wedge [x]_{A_2} \subseteq X \wedge \cdots \wedge [x]_{A_m} \subseteq X\}$$
$$\overline{\sum_{i=1}^m A_i}^P(X) = \sim \underline{\sum_{i=1}^m A_i}^P(\sim X)$$

and pessimistic boundary region is

$$Bn^{P}_{\sum_{i=1}^{m} A_i}(X) = \overline{\sum_{i=1}^{m} A_i}^{P}(X) \setminus \underline{\sum_{i=1}^{m} A_i}^{P}(X)$$

3 A Multi-granulation Model Based on DRSA

We assume that there are several decision-makers who have different viewpoints on selecting attributes (criteria) to make decisions according to a decision table. Thus it may lead to the existence of several dominance relations in the universe. Based on these different dominance relations, the information system will be granulated into different families of granules. Since decision making is based on different families of granules, results from decision-makers may be different. For an overall estimation, people used to summarize all results according to a kind of special requirements and make a conclusion. In the following, we present a definition of the multi-granulation model based on DRSA according to the above process of decision making.

Definition 1. *$S = (U, C \cup \{d\}, V, f)$ is a decision system. Let $\boldsymbol{P} = \{P_1, \cdots, P_m\}$ be a family of attribute sets, where $P_i \subseteq C$, $i \in \{1, \cdots, m\}$. For a support ratio α, $\alpha \in [\frac{1}{m}, 1]$, we have $\alpha-$lower and $\alpha-$upper approximations of $Cl_t^{\geq}$. They are denoted by $\underline{\sum_{i=1}^{m} P_i}^{\alpha}(Cl_t^{\geq})$ and $\overline{\sum_{i=1}^{m} P_i}^{\alpha}(Cl_t^{\geq})$, respectively, where*

$$\underline{\sum_{i=1}^{m} P_i}^{\alpha}(Cl_t^{\geq}) = \{x \in U : \frac{|P(x, Cl_t^{\geq})|}{m} \geq \alpha\} \tag{1}$$

$$\overline{\sum_{i=1}^{m} P_i}^{\alpha}(Cl_t^{\geq}) = \sim \underline{\sum_{i=1}^{m} P_i}^{\alpha}(\sim Cl_t^{\geq}) \tag{2}$$

$\alpha-$boundary region of $Cl_t^{\geq}$ is

$$Bn^{\alpha}_{\sum_{i=1}^{m} P_i}(Cl_t^{\geq}) = \overline{\sum_{i=1}^{m} P_i}^{\alpha}(Cl_t^{\geq}) \setminus \underline{\sum_{i=1}^{m} P_i}^{\alpha}(Cl_t^{\geq}) \tag{3}$$

$\alpha-$lower and $\alpha-$upper approximations of $Cl_t^{\leq}$ are denoted by $\underline{\sum_{i=1}^{m} P_i}^{\alpha}(Cl_t^{\leq})$ and $\overline{\sum_{i=1}^{m} P_i}^{\alpha}(Cl_t^{\leq})$, respectively, where

$$\underline{\sum_{i=1}^{m} P_i}^{\alpha}(Cl_t^{\leq}) = \{x \in U : \frac{|P(x, Cl_t^{\leq})|}{m} \geq \alpha\} \tag{4}$$

$$\overline{\sum_{i=1}^{m} P_i}^{\alpha}(Cl_t^{\leq}) = \sim \underline{\sum_{i=1}^{m} P_i}^{\alpha}(\sim Cl_t^{\leq}) \tag{5}$$

$\alpha-$boundary region of $Cl_t^{\leq}$ is

$$Bn^{\alpha}_{\sum_{i=1}^{m} P_i}(Cl_t^{\leq}) = \overline{\sum_{i=1}^{m} P_i}^{\alpha}(Cl_t^{\leq}) \setminus \underline{\sum_{i=1}^{m} P_i}^{\alpha}(Cl_t^{\leq}) \tag{6}$$

where $P(x, Cl_t^{\geq}) = \{P_i \in \boldsymbol{P} | D_{P_i}^{+}(x) \subseteq Cl_t^{\geq}\}$, $P(x, Cl_t^{\leq}) = \{P_i \in \boldsymbol{P} | D_{P_i}^{-}(x) \subseteq Cl_t^{\leq}\}$ and $|\cdot|$ indicates the cardinality.

When $\alpha = \frac{1}{m}$, the model will be an optimistic standpoint model according to Qian et al.'s viewpoint. Of course, it will be an pessimistic standpoint model when $\alpha = 1$.

$\alpha-$upper approximations of decision classes union may also be obtained by another way as the following Proposition 1.

Proposition 1

$$\overline{\sum_{i=1}^{m} P_i}^{\alpha}(Cl_t^{\geq}) = \{x \in U : \frac{|Q(x, Cl_t^{\geq})|}{m} > \alpha\} \tag{7}$$

$$\overline{\sum_{i=1}^{m} P_i}^{\alpha}(Cl_t^{\leq}) = \{x \in U : \frac{|Q(x, Cl_t^{\leq})|}{m} > \alpha\} \tag{8}$$

where $Q(x, Cl_t^{\geq}) = \{P_i \in \boldsymbol{P} | D_{P_i}^{-}(x) \cap Cl_t^{\geq} \neq \emptyset\}$, $Q(x, Cl_t^{\leq}) = \{P_i \in \boldsymbol{P} | D_{P_i}^{+}(x) \cap Cl_t^{\leq} \neq \emptyset\}$.

Proof: According to equation (2), we have
$\overline{\sum_{i=1}^{m} P_i}^{\alpha}(Cl_t^{\geq}) =\sim \underline{\sum_{i=1}^{m} P_i}^{\alpha}(\sim Cl_t^{\geq}) =\sim \underline{\sum_{i=1}^{m} P_i}^{\alpha}(Cl_{t-1}^{\leq}) =\sim \{x \in U : \frac{|P(x, Cl_{t-1}^{\leq})|}{m} \geq \alpha\} = \{x \in U : \frac{|P(x, Cl_{t-1}^{\leq})|}{m} < \alpha\}$.

As we known $\overline{\sum_{i=1}^{m} P_i}^{\alpha}(Cl_t^{\geq}) = U$ for $t = 1$. Hence we do not care about the case $t = 1$. If $D_{P_i}^{-}(x) \nsubseteq Cl_{t-1}^{\leq}$, then $D_{P_i}^{-}(x) \cap Cl_t^{\geq} \neq \emptyset$. Therefore, $\{x \in U : \frac{|P(x, Cl_{t-1}^{\leq})|}{m} < \alpha\} = \{x \in U : \frac{|Q(x, Cl_t^{\geq})|}{m} > \alpha\}$. Hence, $\overline{\sum_{i=1}^{m} P_i}^{\alpha}(Cl_t^{\geq}) = \{x \in U : \frac{|Q(x, Cl_t^{\geq})|}{m} > \alpha\}$.

Similarly, equation (8) holds. □

4 A Numerical Example

Given a decision system $S = (U, C \cup \{d\}, V, f)$ in Table 1, where $U = \{x_1, x_2, \cdots, x_{10}\}$, $C = \{a_1, a_2, a_3, a_4\}$, $V_{a_1} = V_{a_2} = V_{a_3} = V_{a_4} = V_d = \{1, 2, 3\}$, $P_1 = \{a_1, a_2\}, P_2 = \{a_2, a_3\}, P_3 = \{a_1, a_4\}$. Let $\alpha = \frac{2}{3}$. Decision classes are $Cl_1 = \{x_1, x_4, x_5\}$, $Cl_2 = \{x_2, x_3, x_6, x_8\}$ and $Cl_3 = \{x_7, x_9, x_{10}\}$. Upward unions of decision classes are $Cl_1^{\geq} = U$, $Cl_2^{\geq} = \{x_2, x_3, x_6, x_7, x_8, x_9, x_{10}\}$ and $Cl_3^{\geq} = \{x_7, x_9, x_{10}\}$. Downward unions of decision classes are $Cl_3^{\leq} = U$, $Cl_2^{\leq} = \{x_1, x_2, x_3, x_4, x_5, x_6, x_8\}$ and $Cl_1^{\leq} = \{x_1, x_4, x_5\}$.

Table 1. A decision table

object	a_1	a_2	a_3	a_4	d
x_1	2	2	1	3	1
x_2	3	2	1	2	2
x_3	2	3	1	1	2
x_4	1	2	3	1	1
x_5	1	1	2	3	1
x_6	1	2	2	1	2
x_7	3	3	1	2	3
x_8	3	2	2	2	2
x_9	2	2	3	1	3
x_{10}	3	2	3	3	3

We calculate P_i-dominating and P_i-dominated sets according to the attribute set P_i, where $i = 1, 2, 3$, respectively.

(1) P_1-dominating and P_1-dominated sets

$D^+_{P_1}(x_1) = \{x_1, x_2, x_3, x_7, x_8, x_9, x_{10}\}$, $D^-_{P_1}(x_1) = \{x_1, x_4, x_5, x_6, x_9\}$;
$D^+_{P_1}(x_2) = \{x_2, x_7, x_8, x_{10}\}$, $D^-_{P_1}(x_2) = \{x_1, x_2, x_4, x_5, x_6, x_8, x_9, x_{10}\}$;
$D^+_{P_1}(x_3) = \{x_3, x_7\}$, $D^-_{P_1}(x_3) = \{x_1, x_3, x_4, x_5, x_6, x_9\}$;
$D^+_{P_1}(x_4) = \{x_1, x_2, x_3, x_4, x_6, x_7, x_8, x_9, x_{10}\}$, $D^-_{P_1}(x_4) = \{x_4, x_5, x_6\}$;
$D^+_{P_1}(x_5) = \{x_1, x_2, x_3, x_4, x_5, x_6, x_7, x_8, x_9, x_{10}\}$, $D^-_{P_1}(x_5) = \{x_5\}$;
$D^+_{P_1}(x_6) = \{x_1, x_2, x_3, x_4, x_6, x_7, x_8, x_9, x_{10}\}$, $D^-_{P_1}(x_6) = \{x_4, x_5, x_6\}$;
$D^+_{P_1}(x_7) = \{x_7\}$, $D^-_{P_1}(x_7) = \{x_1, x_2, x_3, x_4, x_5, x_6, x_7, x_8, x_9, x_{10}\}$;
$D^+_{P_1}(x_8) = \{x_2, x_7, x_8, x_{10}\}$, $D^-_{P_1}(x_8) = \{x_1, x_2, x_4, x_5, x_6, x_8, x_9, x_{10}\}$;
$D^+_{P_1}(x_9) = \{x_1, x_2, x_3, x_7, x_8, x_9, x_{10}\}$, $D^-_{P_1}(x_9) = \{x_1, x_4, x_5, x_6, x_9\}$;
$D^+_{P_1}(x_{10}) = \{x_2, x_7, x_8, x_{10}\}$, $D^-_{P_1}(x_{10}) = \{x_1, x_2, x_4, x_5, x_6, x_8, x_9, x_{10}\}$.

(2) P_2-dominating and P_2-dominated sets

$D^+_{P_2}(x_1) = \{x_1, x_2, x_3, x_4, x_6, x_7, x_8, x_9, x_{10}\}$, $D^-_{P_2}(x_1) = \{x_1, x_2\}$;
$D^+_{P_2}(x_2) = \{x_1, x_2, x_3, x_4, x_6, x_7, x_8, x_9, x_{10}\}$, $D^-_{P_2}(x_2) = \{x_1, x_2\}$;
$D^+_{P_2}(x_3) = \{x_3, x_7\}$, $D^-_{P_2}(x_3) = \{x_1, x_2, x_3, x_7\}$;
$D^+_{P_2}(x_4) = \{x_4, x_9, x_{10}\}$, $D^-_{P_2}(x_4) = \{x_1, x_2, x_4, x_5, x_6, x_8, x_9, x_{10}\}$;
$D^+_{P_2}(x_5) = \{x_4, x_5, x_6, x_8, x_9, x_{10}\}$, $D^-_{P_2}(x_5) = \{x_5\}$;
$D^+_{P_2}(x_6) = \{x_4, x_6, x_8, x_9, x_{10}\}$, $D^-_{P_2}(x_6) = \{x_1, x_2, x_5, x_6, x_8\}$;
$D^+_{P_2}(x_7) = \{x_2, x_7\}$, $D^-_{P_2}(x_7) = \{x_1, x_2, x_3, x_7\}$;
$D^+_{P_2}(x_8) = \{x_4, x_6, x_8, x_9, x_{10}\}$, $D^-_{P_2}(x_8) = \{x_1, x_2, x_5, x_6, x_8\}$;
$D^+_{P_2}(x_9) = \{x_4, x_9, x_{10}\}$, $D^-_{P_2}(x_9) = \{x_1, x_2, x_4, x_5, x_6, x_8, x_9, x_{10}\}$;
$D^+_{P_2}(x_{10}) = \{x_4, x_9, x_{10}\}$, $D^-_{P_2}(x_{10}) = \{x_1, x_2, x_4, x_5, x_6, x_8, x_9, x_{10}\}$.

(3) P_3-dominating and P_3-dominated sets

$D^+_{P_3}(x_1) = \{x_1, x_{10}\}$, $D^-_{P_3}(x_1) = \{x_1, x_3, x_4, x_5, x_6, x_9\}$;
$D^+_{P_3}(x_2) = \{x_2, x_7, x_8, x_{10}\}$, $D^-_{P_3}(x_2) = \{x_2, x_3, x_4, x_6, x_7, x_8, x_9\}$;

$D^{+}_{P_3}(x_3) = \{x_1, x_2, x_3, x_7, x_8, x_9, x_{10}\}$, $D^{-}_{P_3}(x_3) = \{x_3, x_4, x_6, x_9\}$;
$D^{+}_{P_3}(x_4) = U$, $D^{-}_{P_3}(x_4) = \{x_4, x_6\}$;
$D^{+}_{P_3}(x_5) = \{x_1, x_5, x_{10}\}$, $D^{-}_{P_3}(x_5) = \{x_4, x_5, x_6\}$;
$D^{+}_{P_3}(x_6) = U$, $D^{-}_{P_3}(x_6) = \{x_4, x_6\}$;
$D^{+}_{P_3}(x_7) = D^{+}_{P_3}(x_8) = \{x_2, x_7, x_8, x_{10}\}$,
$D^{-}_{P_3}(x_7) = D^{-}_{P_3}(x_8) = \{x_2, x_3, x_4, x_6, x_7, x_8, x_9\}$;
$D^{+}_{P_3}(x_9) = \{x_1, x_2, x_3, x_7, x_8, x_9, x_{10}\}$, $D^{-}_{P_3}(x_9) = \{x_3, x_4, x_6, x_9\}$;
$D^{+}_{P_3}(x_{10}) = \{x_{10}\}$, $D^{-}_{P_3}(x_{10}) = U$.

(4) $\alpha-$lower approximations, $\alpha-$upper approximations and $\alpha-$boundary regions of decision classes unions.

$\underline{\sum_{i=1}^{4} P_i}^{\alpha}(Cl_1^{\geq}) = U$, $\overline{\sum_{i=1}^{4} P_i}^{\alpha}(Cl_1^{\geq}) = U$, $Bn^{\alpha}_{\sum_{i=1}^{4} P_i}(Cl_1^{\geq}) = \emptyset$;
$\underline{\sum_{i=1}^{4} P_i}^{\alpha}(Cl_2^{\geq}) = \{x_2, x_3, x_7, x_8, x_{10}\}$,
$\overline{\sum_{i=1}^{4} P_i}^{\alpha}(Cl_2^{\geq}) = \{x_1, x_2, x_3, x_4, x_6, x_7, x_8, x_9, x_{10}\}$,
$Bn^{\alpha}_{\sum_{i=1}^{4} P_i}(Cl_2^{\geq}) = \{x_1, x_4, x_6, x_9\}$;
$\underline{\sum_{i=1}^{4} P_i}^{\alpha}(Cl_3^{\geq}) = \emptyset$, $\overline{\sum_{i=1}^{4} P_i}^{\alpha}(Cl_3^{\geq}) = \{x_1, x_2, x_3, x_7, x_8, x_9, x_{10}\}$,
$Bn^{\alpha}_{\sum_{i=1}^{4} P_i}(Cl_3^{\geq}) = \{x_1, x_2, x_3, x_7, x_8, x_9, x_{10}\}$;
$\underline{\sum_{i=1}^{4} P_i}^{\alpha}(Cl_1^{\leq}) = \{x_5\}$, $\overline{\sum_{i=1}^{4} P_i}^{\alpha}(Cl_1^{\leq}) = \{x_1, x_4, x_5, x_6, x_9\}$,
$Bn^{\alpha}_{\sum_{i=1}^{4} P_i}(Cl_1^{\leq}) = \{x_1, x_4, x_6, x_9\}$;
$\underline{\sum_{i=1}^{4} P_i}^{\alpha}(Cl_2^{\leq}) = \{x_4, x_5\}$, $\overline{\sum_{i=1}^{4} P_i}^{\alpha}(Cl_2^{\leq}) = U$,
$Bn^{\alpha}_{\sum_{i=1}^{4} P_i}(Cl_2^{\leq}) = \{x_1, x_2, x_3, x_6, x_7, x_8, x_9, x_{10}\}$;
$\underline{\sum_{i=1}^{4} P_i}^{\alpha}(Cl_3^{\leq}) = U$, $\overline{\sum_{i=1}^{4} P_i}^{\alpha}(Cl_3^{\leq}) = U$.
$Bn^{\alpha}_{\sum_{i=1}^{4} P_i}(Cl_3^{\leq}) = \emptyset$;

5 Conclusions

In this paper, we built a multi-granulation model based on DRSA according to the principles adopted by people in decision making. It allows DRSA to be applied to situations that many kinds of relations exist in the universe. A numerical example validated the rationality and feasibility of our model. In the future, we will continue to discuss properties of the proposed model and study approaches for updating approximations of this model in the dynamic environment.

Acknowledgements. This work is supported by the National Science Foundation of China (Nos. 60873108, 61175047, 61100117), the Youth Social Science Foundation of the Chinese Education Commission (No. 11YJC630127), the Fundamental Research Funds for the Central Universities (Nos. SWJTU11ZT08, SWJTU12CX091) and the Scientific Research Foundation of Sichuan Provincial Education Department (No. 10ZB049).

References

1. Pawlak, Z.: Rough sets. International Journal of Computer and Information Sciences 11(5), 341–356 (1982)
2. Peters, J., Suraj, Z., Shan, S.: Classification of meteorological volumetric radar data using rough set methods. Pattern Recognition Letters 24(6), 911–920 (2003)
3. Polkowski, L., Lin, T., Tsumoto, S.: Rough set methods and applications: New developments in knowledge discovery in information systems. Physica-Verlag, Heidelberg (2000)
4. Shen, L., Loh, H.: Applying rough sets to market timing decisions. Decision Support Systems 37(4), 583–597 (2004)
5. Qian, Y.H., Liang, J.Y.: Rough set method based on multi-granulations. In: Proceedings of 5th IEEE Conference on Cognitive Informatics, vol. I, pp. 297–304 (2006)
6. Qian, Y.H., Liang, J.Y., Yao, Y.Y., Dang, C.Y.: Incomplete mutigranulation rough set. IEEE Transactions on Systems, Man and Cybernetics, Part A 20, 420–430 (2010)
7. Qian, Y.H., Liang, J.Y., Wei, W.: Pessimistic rough decision. In: 2nd International Workshop on Rough Sets Theory, pp. 19–21 (2010)
8. Qian, Y.H., Liang, J.Y., Yao, Y.Y., Dang, C.Y.: MGRS: A multi-granulation rough set. Information Sciences 180, 949–970 (2010)
9. Greco, S., Matarazzo, B., Slowinski, R.: Rough sets theory for multicriteria decision analysis. European Journal of Operational Research 129, 1–47 (2001)
10. Greco, S., Matarazzo, B., Slowinski, R.: Rough approximation by dominance relations. International Journal of Intelligent Systems 17(2), 153–171 (2002)
11. Slowinski, R., Greco, S., Matarazzo, B.: Axiomatization of utility, outranking and decision rule preference models for multiple-criteria classification problems under partial inconsistency with the dominance principle. Control and Cybernetics 31(4), 1005–1035 (2002)

A Refined Rough k-Means Clustering with Hybrid Threshold

Hailiang Wang and Mingtian Zhou

School of Computer Science and Engineering,
University of Electronic Science and Technology of China,
Chendu, 610051, P.R. China
wanghailiang2009@gmail.com, mtzhaou@uestc.du.cn

Abstract. In this paper, we propose a new type of adaptive weight based on the definiteness of rough clusters and a hybrid threshold by combining the difference and distance threshold. And then, we refine the algorithm for assigning objects based on the hybrid thresholds to ensure that the outliers in inline positions and rectangle positions to be represented reasonably. At last, some experiments are provided to compare this refined RCM with the original RCM.

Keywords: Rough sets, Clustering, Approximation Accuracy, Hybrid Threshold.

1 Introduction

Clustering is an unsupervised pattern classification algorithm which can be considered as a task of dividing a finite data set into several subgroups so that the objects in the same subgroup are as similar as possible, while the objects in different subgroups are as dissimilar as possible.

K-means [1] clustering is an important partitive clustering algorithms in which each object must be assigned to exactly one subgroup. However, in a real-life data set, there often exist some objects which cannot be assigned into any subgroup certainly. For dealing with the clusters without crisp boundaries, some fuzzy clustering algorithms based on the fuzzy sets theory [2], including probabilistic fuzzy c-means clustering (FCM) [3], [4], [5] and possibilistic fuzzy c-means clustering (PCM), were proposed [6], [7].

The rough sets theory proposed by Palawk [8], [9] is another approach to handle vagueness, uncertainty and inconsistency in data. Lingras, et al. integrated the rough sets theory and classical k-means clustering algorithm and proposed the rough c-means clustering (RCM) [12], in which the boundary regions of clusters are introduced to describe the overlapping areas. Later, Peters discussed the handling of outliers in RCM [18], and evaluated the Lingras RCM with respect to the objective function, numerical stability, the stability of the clusters and others, and then suggested some refinements to Lingras RCM algorithm [19].

Recently, some extensions to the original RCM are proposed. In [11], Zhou, et al. substitute the distances of objects to centers with kernel functions. In [10],

J.T. Yao et al. (Eds.): RSCTC 2012, LNAI 7413, pp. 26–35, 2012.

Mitraa, et al. proposed the shadowed c-means clustering. In [13], [14], the rough sets and fuzzy sets have been integrated with the classical k-means clustering algorithm. In [15], Maji and Pal proposed a generalized hybrid clustering algorithm termed as rough-fuzzy PCM (RFPCM) by combining the k-means clustering, rough sets, and probabilistic and possibilistic memberships of fuzzy sets. Furthermore, Yao, et al. pointed out that, strictly speaking, the Lingras RCM is not part of classical rough sets theory but a two-layer interval clustering approach [16].

In RCM, some parameters are required, including the threshold for partitioning objects and the weights for calculating the centroid of a cluster. Several resolutions to select these parameters have been proposed. In [17], Zhou, et al. proposed a RCM algorithm with adaptive parameters. Małyszko and Stepaniuk introduced the rough entropy, which is proposed in [21] to quantify the roughness of rough sets, to estimate these parameters during the process of clustering [22]. Mitra proposed the evolutionary RCM algorithm in which a genetic algorithm is employed to optimize these parameters [20].

However, there are still some problems about the RCM algorithm that have not been resolved very well. First, the weights of lower and upper approximations are same for all rough clusters. The properties of different clusters have not been taken into account. Second, the outliers in data set have not been appropriate described. In this paper, we propose a refined RCM algorithm with the relative weights and hybrid threshold.

The remainder of this paper is structured as follows: section 2 summarizes the basic conceptions of RCM; section 3 proposes the relative weights and hybrid thresholds and presents the rule for partitioning objects by using the hybrid threshold. Experiments are given in section 4, and this paper is concluded with section 5.

2 Rough c-Means Clustering Algorithm

2.1 The Lingras Rough c-Means Clustering

RCM proposed by Lingras et al. is an integration of rough sets theory and classical k-means algorithm [12]. In RCM, a rough cluster C_i is described as a lower approximation $\underline{C}_i$, which contains the objects definitively belonging to C_i, and an upper approximation $\overline{C}_i$, which contains the objects probably belonging to C_i. Objects in RCM have the properties as follows:

- An object x can be part of exactly one lower approximation;
- If $x \in C_i$, then $x \in C_i$;
- If object x is not part of any lower approximation, then x belongs to at least two upper approximations.

A pair of weights w_l and w_u are introduced to describe the different effects to cluster centers of the objects in lower and upper approximations respectively.

The mean function of RCM is as that in Eq.(1):

$$m_j = \begin{cases} w_l \frac{\sum_{x \in \underline{C}_j} x}{|\underline{C}_j|} + w_u \frac{\sum_{x \in (\overline{C}_j - \underline{C}_j)} x}{|\overline{C}_j - \underline{C}_j|} & \overline{C}_j - \underline{C}_j \neq \emptyset \\ w_l \frac{\sum_{x \in \underline{C}_j} x}{|\underline{C}_j|} & otherwise \end{cases} \quad (1)$$

where the w_l and w_u have $w_l + w_u = 1$.

Given threshold ε and object x_i, if there exist tow different clusters C_l and C_m, which have $|d(x_i, C_l) - d(x_i, C_m)| < \varepsilon$, where the $d(x_i, C_m)$ is the distance from x_i to the centroid of cluster C_m, then x_i should be assigned to the upper approximations of both clusters C_l and C_m. Otherwise, x_i should be assigned to the lower approximation of the closest cluster.

As the threshold ε is compared with the difference values between the distances to different centers, it is also refereed as a difference threshold.

2.2 The Peters Rough C-Means Clustering

In [19], Peters evaluated the Lingras RCM and refined it in several aspects:

- Modify the mean function

Peters suggested to calculate the center of a cluster as follow:

$$m_j = w_l \frac{\sum_{x \in \underline{C}_j} x}{|\underline{C}_j|} + w_u \frac{\sum_{x \in \overline{C}_j} x}{|\overline{C}_j|} \quad (2)$$

where $w_l + w_u = 1$.

When the lower and upper approximations of a cluster are identical, Eq.(2) boils down to the classical k-means.

- Refine the rule of assigning objects

The algorithm is refined to ensure that each cluster has at least one representative member.

- Substitute the absolute threshold with a relative threshold

Peters suggested to substitute the absolute threshold with a relative threshold defined as follow:

$$T' = \left\{ t : \frac{d(\boldsymbol{X}_n, \boldsymbol{m}_k)}{d(\boldsymbol{X}_n, \boldsymbol{m}_h)} \leq \zeta \wedge h \neq k \right\} \quad (3)$$

2.3 The Selection of Parameters

Because RCM heavily depends on the threshold and weights, it is important to select appropriate values for them. An intuitive approach is to repeat running RCM on the same data set with different parameters and then compare the results. In [12], the authors ran RCM with various weights (w_l, w_u) ranging from $(0.95, 0.05)$ to $(0.55, 0.45)$, and found out that when (w_l, w_u) was set to $(0.75, 0.25)$, the result is better than that of other weights. Małyszko and Stepaniuk introduced rough entropy to select weights [22]. In [17], Zhou, et al. proposed adaptive weights and threshold calculated as follows:

$$w_l = cos^2\left(\frac{w_l}{times^{\frac{1}{n}}}\right), w_u = 1 - w_l \tag{4}$$

where *times* is the iterative numbers and n is the attenuation coefficient. Generally, $n = 2$.

$$\varepsilon = \varepsilon + \frac{1}{times^n} \tag{5}$$

where n is the attenuation coefficient. Generally, $n = 2$.

3 The Proposed Rough c-Means Clustering

3.1 The Adaptive Weights Based on Approximation Accuracy

Obviously, a good rough clustering should assign as many as possible objects into lower approximations. Consider a rough cluster C_j, if most of the objects in C_j are assigned to the lower approximation $\underline{C}_j$, the position of centroid of C_j should be hold on in the next iteration. At this time, the objects in C_j should take more important. The more objects are set to C_j, the more important the C_j should take.

Specially, if all the objects belonging to C_j are assigned to the $\underline{C}_j$, the rough cluster C_j becomes a crisp cluster and the centroid of C_j should be calculated absolutely depending on the objects in $\underline{C}_j$. In this case, the weight for the lower approximation $\underline{C}_j$ should be set to 1, and the weight for the upper approximation $\overline{C}_j$ should be set to 0.

In other words, for a rough cluster, the more definitive it is, the more important the lower approximation should take. Accuracy of approximation is an extensively used measure to describe the certainty of a rough set [9]. So, we propose a new type of adaptive weights based on the accuracy of rough clusters.

Furthermore, we extend the weights $\{w_l, w_u\}$ to vectors $\overrightarrow{w_l} = \{w_l^1, w_l^2, \cdots, w_l^k\}$ and $\overrightarrow{w_u} = \{w_u^1, w_u^2, \cdots, w_u^k\}$ relating to the properties of different clusters, where $w_l^j + w_u^j = 1, i = \{1, 2, \cdots, k\}$, and k is the number of clusters.

For a rough cluster $C_j = [\underline{C}_j, \overline{C}_j]$, the accuracy of approximation is calculated as:

$$\alpha_j = \frac{|\underline{C}(j)|}{|\overline{C}(j)|} \tag{6}$$

The lower weight of rough cluster C_j is calculated as:

$$w_l^j = \alpha_j^{\frac{1}{m}} = \left(\frac{|\underline{C}_j|}{|\overline{C}_j|}\right)^{\frac{1}{m}} \tag{7}$$

where the m is a constant greater than 1. Generally, $m = 2$.

The weight of upper approximation C_j is calculated as: $w_u^j = 1 - w_l^j$. Obviously, when $\underline{C}_j = \overline{C}_j \neq \emptyset$, $w_l^j = 1$ and $w_u^j = 0$; when $\overline{C}_j \neq \emptyset$ and $\underline{C}_j = \emptyset$, $w_l^j = 0$ and $w_u^j = 1$.

The centroid of cluster C_j is calculated as:

$$m_j = \begin{cases} w_l^j \dfrac{\sum\limits_{x \in \underline{C}_j} x}{|\underline{C}_j|} + w_u^j \dfrac{\sum\limits_{x \in (\overline{C}_j - \underline{C}_j)} x}{|\overline{C}_j - \underline{C}_j|} & \overline{C}_j - \underline{C}_j \neq \emptyset \\ w_l^j \dfrac{\sum\limits_{x \in \underline{C}_j} x}{|\underline{C}_j|} & otherwise \end{cases} \tag{8}$$

The relative importance of the objects in lower approximations increases with the increasing of m. When $m \to +\infty$, the $w_l^j \to 1$. At this time, only the objects in lower approximations are considered when calculating the centers of clusters.

On the other hand, if m is set to 1, the centroid of cluster C_j is calculated as:

$$m_j = \frac{|\underline{C}_j|}{|\overline{C}_j|} \cdot \frac{\sum\limits_{x \in \underline{C}_j} x}{|\underline{C}_j|} + \frac{|\overline{C}_j| - |\underline{C}_j|}{|\overline{C}_j|} \cdot \frac{\sum\limits_{x \in (\overline{C}_j - \underline{C}_j)} x}{|\overline{C}_j| - |\underline{C}_j|} = \frac{\sum\limits_{x \in \underline{C}_j} x}{|\overline{C}_j|} + \frac{\sum\limits_{x \in (\overline{C}_j - \underline{C}_j)} x}{|\overline{C}_j|} = \frac{\sum\limits_{x \in \overline{C}_j} x}{|\overline{C}_j|}$$

In this case, the objects in lower approximations take the same importance as the objects in upper approximations. The Eq. (8) boils down to the mean function of classical k-means.

3.2 The Hybrid Threshold

In original RCM, an object is assigned to exactly one lower approximation or at least two upper approximations. However, in some situations, it is more reasonable to assign an object to only one upper approximation.

Consider the following two-dimensional data set X which has an outlier in the inline position:

$$X = \begin{pmatrix} -1.1 & 1.0 & 0.8 & 0.9 & 2.2 & 3.0 & 2.9 & 3.2 \\ -1.2 & 1.2 & 1.1 & 1.0 & 2.3 & 3.1 & 3.0 & 3.1 \end{pmatrix}$$

Amuse X is required to be divided into two parts, and the initial centers are set as follows:

$$M = \begin{pmatrix} 1.0 & 3.0 \\ 1.0 & 3.0 \end{pmatrix}$$

For the absolute difference threshold, we set $\varepsilon = 1$ to ensure that x_2, x_3, x_4 are assigned into $\underline{C}_1$, x_6, x_7, x_8 are assigned into $\underline{C}_2$ and x_5 is assigned into $\overline{C}_1$ and $\overline{C}_2$. At this time, x_1 is assigned into $\underline{C}_1$. The result is shown in Fig. 1.

For the relative difference threshold, when $\zeta = 1.8$, we will get the same result as that shown in Fig. 1. But, if ζ is set to 2.0, x_1 will be assigned into $\overline{C}_1$ and $\overline{C}_2$. The result with $\zeta = 2.00$ is shown in Fig. 2.

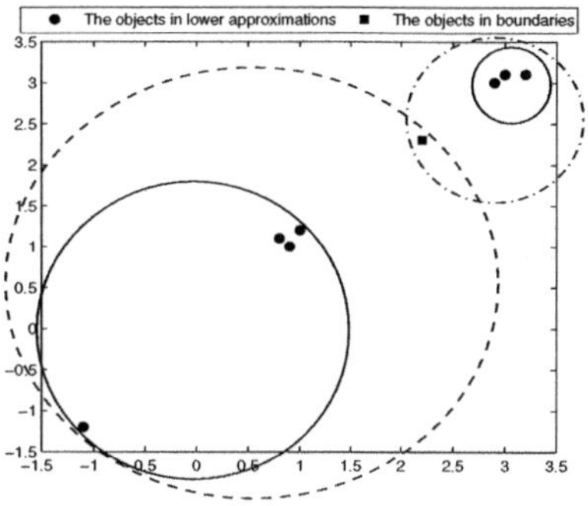

Fig. 1. The result for the absolute difference threshold on X with $\varepsilon = 1$

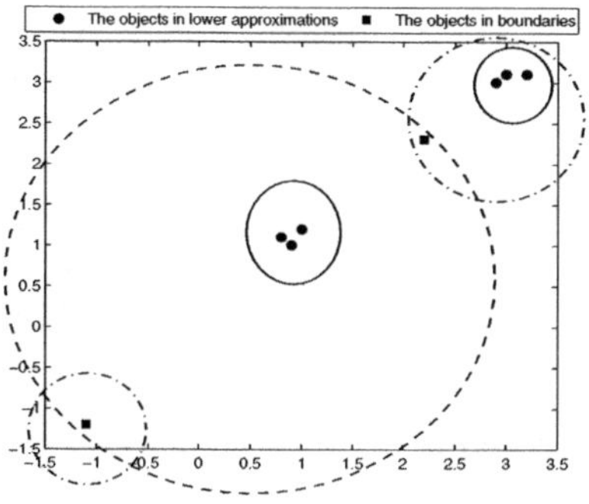

Fig. 2. The result for the relative difference threshold on X with $\zeta = 2.00$

However, as x_1 is significantly further from the centers of C_1 and C_2 than other objects in $\underline{C}_1$ and $\overline{C}_2$, it is more reasonable to assign x_1 only to $\overline{C}_1$.

In this study, we propose a new type of hybrid threshold by integrating the relative difference and distance thresholds. The steps of assigning objects are as follows:

Given an object x_i, set it to the upper approximation of the closest cluster C_l. Then, compare the different distance $|d(x_i, C_l) - |d(x_i, C_m)|(l \neq m)$ with the relative difference threshold ζ_{diff} to determine whether x_i should be set to the upper approximations of any other clusters. If x_i only belongs to $\overline{C}_l$, compare the distance $d(x_i, C_l)$ with the relative distance threshold ζ_{dis} to determine if x_i should be assigned into $\underline{C}_l$. The detailed process for the relative hybrid threshold is shown in algorithm 1.

We find that the following relative difference and distance threshold work well.

- The relative difference threshold:

$$\zeta_{diff} = \frac{\alpha}{n} \sum_{i=1}^{n} |d(x_i, C_l) - d(x_i, C_m)|, l \neq m \tag{9}$$

- The relative distance threshold:

$$\zeta_{dis} = \frac{1-\alpha}{k} \sum_{m=1}^{k} d(x_i, C_m) \tag{10}$$

Input: data set $X = \{x_1, x_2, \ldots, x_n\}$, prototype of cluster
$C = \{C_1, C_2, \cdots, C_k\}$, relative threshold α
Output: interval sets $[\underline{C}_j, \overline{C}_j]$ of each cluster

```
foreach object x_i in the object set X do
    Determine the closest cluster C_l to x_i
    Set x_i to the upper approximation \overline{C}_l
    Calculate the relative difference threshold ζ_diff as that in Eq.(9)
    foreach cluster C_m ≠ C_l do
        if |d(x_i, C_m) − d(x_i, C_l)| <= ζ_diff then
            Set x_i to the upper approximation \overline{C}_m
        end
    end
    if x_i is not assigned to any other upper approximation except \overline{C}_l then
        Calculate the relative distance threshold ζ_dis as that in Eq.(10)
        if |d(x_i, C_l)| <= ζ_dis then
            Set x_i to lower approximation \underline{C}_l
        end
    end
end
```

Algorithm 1. The algorithm of assigning objects according to the relative hybrid threshold

The parameter α lying in $[0, 1]$ denotes the bandwidth of boundary regions between different clusters. The greater the *alpha* is, the more objects are assigned to the boundary regions.

Finally, an example is provided to illustrate how to use this new threshold.

Consider the data set X, x_1 is assigned to $\overline{C}_1$, because C_1 is the closest cluster to x_1. Then the relative difference threshold for x_1 to C_1 is calculated as: $\zeta_{diff} = \frac{0.5}{8}\sum_{i=1}^{8} |d(x_i, C_1) - d(x_i, C_2)| = 1.24$, and the relative distance threshold for x_1 to C_1 is calculated as: $\zeta_{dis} = \frac{0.5}{2}\sum_{l=1}^{2} d(x_1, C_l) = 2.23$. So, x_1 should be only assigned to $\overline{C}_1$ because $|d(x_1, C_1) - d(x_1, C_2)| = 2.83 > \zeta_{diff} = 1.24$ and $d(x_1, C_1) = 3.04 > \zeta_{dis} = 2.23$. This result is shown in Fig. 3. Obviously, it is more reasonable than those showed in Fig. 1 and Fig. 2.

Furthermore, we can get similar results for the outliers in rectangle positions. Consider the following two-dimensional data set X':

$$X' = \begin{pmatrix} -1.1 & 1.0 & 0.8 & 0.9 & 2.2 & 3.0 & 2.9 & 3.2 \\ 3.2 & 1.2 & 1.1 & 1.0 & 2.3 & 3.1 & 3.0 & 3.1 \end{pmatrix}$$

We partition the data set X' into two rough clusters, and the initial centers of these clusters are set as:

$$M' = \begin{pmatrix} 1.0 & 3.0 \\ 1.0 & 3.0 \end{pmatrix}$$

If the relative threshold α is set to 0.5, the result for the relative hybrid thresholds is shown in Fig. 4.

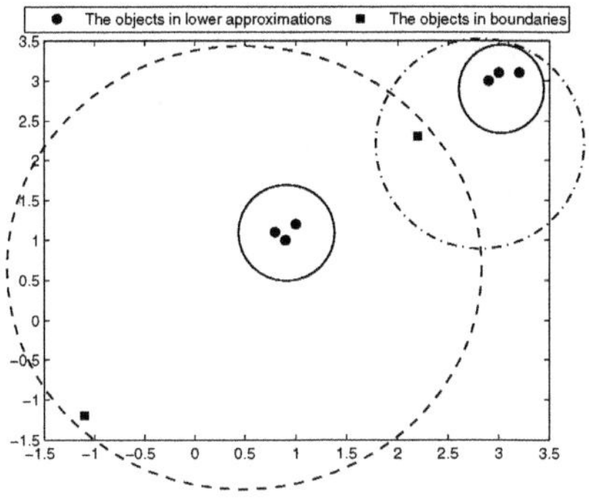

Fig. 3. The result for the hybrid threshold on X with $\alpha = 0.5$

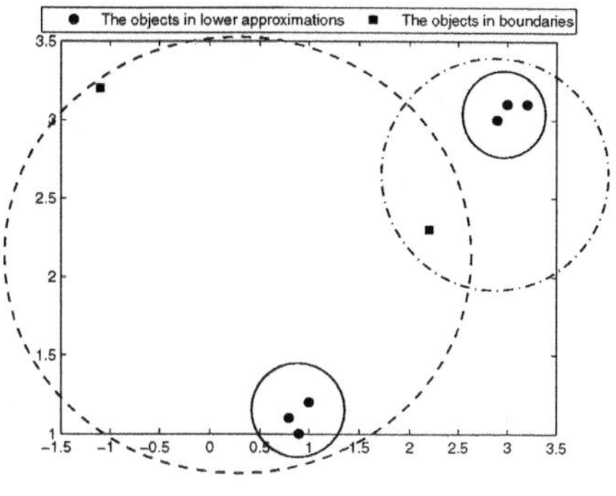

Fig. 4. Fig. 4. The result for the hybrid threshold on X' with $\alpha = 0.5$

4 Experiments

In this section, we compare the proposed RCM with the Lingras and Peters RCM. The following experiments are carried out on the same benchmark data sets like Wine recognition, Iris plants, and Ionosphere. All of them are downloaded from the UCI machine learning repository `http://archive.ics.uci.edu/ml/`.

The experimental results are compared in the terms of correctness and accuracy. The correctness of rough clustering is defined as the correct rate of lower approximations, because the objects in the lower approximation of a rough cluster are the representative members of this cluster.

- The correctness of rough clustering:

$$r = \frac{\sum_{j=1}^{k} |\underline{C}_j \cap C_j^*|}{\sum_{j=1}^{k} |\underline{C}_j|} \tag{11}$$

Here, $\underline{C}_j$ is the lower approximation of cluster j, and C_j^* is the set of objects labeled as class j.

On the other hand, the accuracy describe the classification ability of rough clustering. A rough clustering with higher correctness and accuracy is better than that with lower values.

- The accuracy of rough clustering:

$$a = \frac{\sum_{j=1}^{k} |\underline{C}_j|}{\sum_{j=1}^{k} |\overline{C}_j|} \tag{12}$$

As the RCM algorithm run into local optimums, the experiments are repeated 20 times with random initial centers and then calculate the average of correctness and accuracy to exclude any influence of different selections of the initial centers.

For the Lingras RCM, the parameters are selected as: $w_l = 0.7, w_u = 0.3$, and $\varepsilon = 0.3$. For the Peters RCM, the parameters are selected as: $w_l = 0.7, w_u = 0.3$, and ȩÎ= 1.3. For the proposed RCM, the parameters are selected as: $\zeta = 0.3$. The correctness and accuracy are shown in Table 1.

Table 1. The results for various RCM

	Wine		Iris		Ionosphere	
	r	a	r	a	r	a
Lingras RCM	0.7816	0.1453	0.8601	0.2423	0.7314	0.1197
Peters RCM	0.6288	0.7676	0.8771	0.7876	0.6803	0.6373
Proposed RCM	0.7011	0.7701	0.9494	0.7971	0.8340	0.6098

The result shows that the accuracy of Lingras RCM is low. This means that only a few objects are assigned to lower approximations. As mentioned above, a good rough clustering should assign as many as possible objects into lower approximations. So the result for Lingras RCM is worse than other two kinds of RCM.

Moreover, the comparison between the Peters RCM and the proposed RCM shows that the RCM with hybrid threshold is better than Peters RCM in all the three data set only except the accuracy on Ionosphere. This experiment proves that the proposed RCM is better than Lingras and Peters RCM in most situations

5 Conclusions

In this paper, we propose a new type of adaptive weights based on the rough accuracy relating to different rough clusters and a new hybrid threshold by combining the difference and distance thresholds, refine the algorithm for assigning objects into lower and upper approximations. In this new algorithm, it is possible to assign an outlier to only one upper approximation which is more reasonable in some special conditions.

Moreover, some experiments are performed on the same benchmark data sets to compare the proposed RCM with the Lingras RCM and RCM. The results prove that it is better than Lingras and Peters RCM in most situations.

References

1. Hartigan, J.A., Wong, M.A.: A K-Means Clustering Algorithm. Journal of Applied Statistics 28, 100–108 (1979)
2. Zadeh, L.A.: Fuzzy sets. Inform. Control 8, 338–353 (1965)

3. Ruspini, E.: A new approach to clustering. Inform. Control 15, 22–32 (1969)
4. Dunn, J.C.: A fuzzy relative of the ISODATA process and its use in detecting compact well-separated clusters. J. Cybernet. 3, 32–57 (1974)
5. Bezdek, J.C.: Pattern Recognition With Fuzzy Objective Function Algorithm. Kluwer, New York (1981)
6. Krishnapuram, R., Keller, J.M.: A possibilistic approach to clustering. IEEE Trans. Fuzzy Syst. 1(2), 98–110 (1993)
7. Pal, N.R., Pal, K., Keller, J.M., Bezdek, J.C.: A possibilistic fuzzy c-means clustering algorithm. IEEE Trans. Fuzzy Syst. 13(4), 517–530 (2005)
8. Pawlak, Z.: Rough Sets. International Journal of Information and Computer Sciences 11, 145–172 (1982)
9. Pawlak, Z.: Some Issues on Rough Sets. In: Peters, J.F., Skowron, A., Grzymała-Busse, J.W., Kostek, B.z., Świniarski, R.W., Szczuka, M.S. (eds.) Transactions on Rough Sets I. LNCS, vol. 3100, pp. 1–58. Springer, Heidelberg (2004)
10. Mitraa, S., Pedryczb, W., Barmanc, B.: Shadowed c-means: Integrating fuzzy and rough clustering. Pattern Recognition 43(4), 1282–1291 (2010)
11. Zhou, T., Zhang, Y., Lu, H., Deng, F., Wang, F.: Rough Cluster Algorithm Based on Kernel Function. In: Wang, G., Li, T., Grzymala-Busse, J.W., Miao, D., Skowron, A., Yao, Y. (eds.) RSKT 2008. LNCS (LNAI), vol. 5009, pp. 172–179. Springer, Heidelberg (2008)
12. Lingras, P., West, J.: Interval set clustering of web users with rough k-means. Journal of Intelligent Information Systems 23(1), 5–16 (2004)
13. Mitra, S., Banka, H., Pedrycz, W.: Rough-fuzzy collaborative clustering. IEEE Trans. Syst., Man, Cybern. B, Cybern. 36(4), 795–805 (2006)
14. Maji, P., Pal, S.K.: RFCM: A Hybrid Clustering Algorithm Using Rough and Fuzzy Sets. Fundamenta Informaticae 80(4), 477–498 (2007)
15. Maji, P., Pal, S.K.: Rough Set Based Generalized Fuzzy C-Means Algorithm and Quantitative Indices. IEEE Trans. Systems, Man, and Cybernetics 37(6), 1529–1540 (2007)
16. Yao, Y., Lingras, P., Wang, R., Miao, D.: Interval Set Cluster Analysis: A Reformulation. In: Sakai, H., Chakraborty, M.K., Hassanien, A.E., Ślęzak, D., Zhu, W. (eds.) RSFDGrC 2009. LNCS, vol. 5908, pp. 398–405. Springer, Heidelberg (2009)
17. Zhou, T., Zhang, Y.N., Lu, H.L.: Rough k-means Cluster with Adaptive Parameters. In: 6th Int. Conf. Machine Learning and Cybernetics, Hong Kong, China, pp. 3063–3068 (2007)
18. Peters, G.: Outliers in Rough k-Means Clustering. In: Pal, S.K., Bandyopadhyay, S., Biswas, S. (eds.) PReMI 2005. LNCS, vol. 3776, pp. 702–707. Springer, Heidelberg (2005)
19. Peters, G.: Some refinement of rough k-means clustering. Pattern Recognition 39, 1481–1491 (2006)
20. Mitra, S.: An evolutionary rough partitive clustering. Pattern Recognition Letters 25, 1439–1449 (2004)
21. Pal, S.K., Shankar, B.U., Mitra, P.: Granular computing, rough entropy and object extraction. Pattern Recognition Letters 26, 2509–2517 (2005)
22. Małyszko, D., Stepaniuk, J.: Rough Entropy Based *k*-Means Clustering. In: Sakai, H., Chakraborty, M.K., Hassanien, A.E., Ślęzak, D., Zhu, W. (eds.) RSFDGrC 2009. LNCS, vol. 5908, pp. 406–413. Springer, Heidelberg (2009)

An Incremental Approach for Updating Approximations of Rough Fuzzy Sets under the Variation of the Object Set

Anping Zeng[1,2], Tianrui Li[1,⋆], Junbo Zhang[1], and Dun Liu[3]

[1] School of Information Science and Technology,
Southwest Jiaotong University, Chengdu 610031, China
zengap@126.com, trli@swjtu.edu.cn, JunboZhang86@163.com
[2] School of Computer and Information Engineering,
Yibin University, Yibin 644007, China
[3] School of Economics and Management,
Southwest Jiaotong University, Chengdu 610031, China
newton83@163.com

Abstract. The lower and upper approximations are basic concepts in rough set theory, and the approximations will change dynamically over time. Incremental methods for updating approximations in rough set theory and its extension has been received much attention recently. This paper presents an approach for incrementally updating approximations of fuzzy rough sets in dynamic fuzzy decision systems when a single object immigrating and emigrating. Examples are employed to illustrate the proposed approach.

Keywords: Rough Fuzzy Sets, Approximations, Incremental Learning.

1 Introduction

The theory of rough sets was introduced by Pawlak [1] as an extension of set theory for the study of intelligent systems characterized by insufficient and incomplete information. In Pawlak's Rough Set Theory (RST), the values of decision attributes are assumed to be determinate. However, this model fails to deal with the case that the values of the decision attributes are fuzzy in practical databases. Therefore, an extended model of RST, rough fuzzy sets, was proposed to deal with fuzzy values in the information systems [2, 3].

In real-life applications, information systems often vary with time in most cases. Some researchers have paid attention to the problem of how to handle updating approximations of RST and its extension incrementally in dynamic information systems [4–18]. For example, Li et al. proposed approaches for incremental updating approximations and extracting rules when attributes vary in the information system [4–6]. In rough fuzzy sets, Cheng proposed approaches for incremental updating approximations when the attribute set evolves over

⋆ Corresponding author.

J.T. Yao et al. (Eds.): RSCTC 2012, LNAI 7413, pp. 36–45, 2012.

time [8]. As for the variation of the object set, some incremental approaches have been proposed in RST and its extension [9–12]. In addition, Wang et al. proposed an incremental rule acquisition algorithm based on variable precision rough set model while inserting new objects into the information system [13]. Zhang et al. proposed an incremental rule acquisition algorithm based on neighborhood rough sets when the object set evolves over time [14]. However, the incremental approach for updating approximations based on rough fuzzy set under the variation of objects has not been taken into account until now.

The paper is organized as follows. In Section 2, some basic concepts of RST and rough fuzzy sets. In Section 3, the updating principles for upper and lower approximations are also analyzed when object immigrating or emigrating. In Section 4, some examples are used to illustrate the proposed approach. In Section 5, the paper conclusion and the future research directions are presented.

2 Preliminaries

In this section, we briefly introduce basic concepts of rough sets and rough fuzzy sets [1,3].

Definition 1. *[1] Let (U, R) be a Pawlak approximation space. The universe $U \neq \emptyset$. $R \subseteq U \times U$ is an equivalence relation on U. U/R denotes the family of all equivalence classes R, and $[x]_R$ denotes an equivalence class of R containing an element $x \in U$. For any $X \subseteq U$, the lower approximation and the upper approximation of X are defined respectively as follows:*

$$\begin{aligned} \underline{R}X &= \{x \in U | [x]_R \subseteq X\}; \\ \overline{R}X &= \{x \in U | [x]_R \cap X \neq \emptyset\}. \end{aligned} \tag{1}$$

In order to describe a fuzzy concept in a crisp approximation space, Dubois and Prade introduced an extended model of RST, called as rough fuzzy sets [2,3].

Definition 2. *[3] Let (U, R) be a Pawlak approximation space, and R be an equivalence relation on U. If X is a fuzzy set on U, $X(x)$ denotes the degree of membership of x in U. The lower and upper approximations of X are a pair fuzzy sets in U, and their membership functions are defined as follows:*

$$\begin{aligned} (\underline{R}X)(x) &= \inf\{X(y) | y \in [x]_R\}; \\ (\overline{R}X)(x) &= \sup\{X(y) | y \in [x]_R\}. \end{aligned} \tag{2}$$

Example 1. Table 1 is a decision table with condition attributes a, b, c and a fuzzy decision attribute d. The partitions generated by the attribute a are:

$$U/a = \{E_1, E_2, E_3\} = \{\{x_1, x_2\}, \{x_3, x_4, x_5\}, \{x_6, x_7, x_8\}\}.$$

Let X be a fuzzy set on U. $X = \{x_1/0.7, x_3/0.7, x_4/0.8, x_6/0.5, x_7/0.2, x_8/0.6\}$.

According to Definiton 1, without considering the degrees of membership, the lower and upper approximations of X on $R = \{a\}$ can be obtained.

Table 1. Decision table with a fuzzy decision attribute

U	a	b	c	fuzzy attribute d
x_1	2	2	1	0.7
x_2	2	2	1	1
x_3	3	3	1	0.7
x_4	3	3	2	0.8
x_5	3	2	2	0
x_6	1	1	4	0.5
x_7	1	2	3	0.2
x_8	1	1	3	0.6

$$\underline{R}X = \{x_6, x_7, x_8\};$$
$$\overline{R}X = \{x_1, x_2, x_3, x_4, x_5, x_6, x_7, x_8\}.$$

According to Definiton 2, the degrees of membership are calculated as follows.

$$(\underline{R}X)(x_6) = \inf\{X(y)|y \in [x_6]_R\} = 0.5 \wedge 0.2 \wedge 0.6 = 0.2;$$
$$(\underline{R}X)(x_7) = (\underline{R}X)(x_8) = (\underline{R}X)(x_6);$$
$$(\overline{R}X)(x_1) = (\overline{R}X)(x_2) = \sup\{X(y)|y \in [x_1]_R\} = 0.7 \vee 1 = 1;$$
$$(\overline{R}X)(x_3) = (\overline{R}X)(x_5) = (\overline{R}X)(x_4) = 0.7 \vee 0.8 \vee 0 = 0.8;$$
$$(\overline{R}X)(x_6) = (\overline{R}X)(x_7) = (\overline{R}X)(x_8) = 0.5 \vee 0.2 \vee 0.6 = 0.6.$$

Therefore, the lower and upper approximations of X on $R = \{a\}$ are as follows.

$$\underline{R}X = \{x_6/0.2, x_7/0.2, x_8/0.2\};$$
$$\overline{R}X = \{x_1/1, x_2/1, x_3/0.8, x_4/0.8, x_5/0.8, x_6/0.6, x_7/0.6, x_8/0.6\}.$$

3 Incremental Methods for Updating Approximations of Rough Fuzzy Sets under the Variation of Objects

We discuss the variation of approximations in Fuzzy Decision Systems (FDS) when the object set evolves over time. Given a FDS $= (U, C \cup D, V, f)$ at time t, $U \neq \emptyset$ and $C \cap D = \emptyset$. For each fuzzy set $X \subseteq U$, the lower and upper approximations are denoted by $\underline{R_C}(X)$ and $\overline{R_C}(X)$, respectively. Let $\overline{x}$ denote the object immigrating into U at time $t+1$, $\overline{E}$ be the partition which $\overline{x}$ immigrates to and $\overline{E}' = \overline{E} \cup \{\overline{x}\}$. Let $\widetilde{x}$ be the object emigrating out U, $\widetilde{E}$ be the partition which $\widetilde{x}$ emigrate from and $\widetilde{E}' = \widetilde{E} - \{\widetilde{x}\}$. After immigrating and emigrating objects, the FDS will be changed into FDS$'$=$(U', C' \cup D', V', f')$. For each fuzzy set $X \subseteq U'$, the lower and upper approximations are denoted by $\underline{R'_C}X$ and $\overline{R'_C}X$, respectively.

3.1 The Immigration of a New Object

Suppose one object $\overline{x}$ enters into the FDS at time $t+1$. So $U' = U \cup \{\overline{x}\}$.

Proposition 1. *When a new object $\overline{x}$ enters into FDS $= (U, C \cup D, V, f)$, $c \in C$. One of the following results holds:*

(1) If $\exists x \in U, f(\overline{x}, c) = f(x, c)$, $\overline{x}$ *cannot form a new equivalence class;*
(2) If $\forall x \in U, f(\overline{x}, c) \neq f(x, c)$, $\overline{x}$ *can form a new equivalence class.*

Proposition 2. *For* $\underline{R'_C}X$, *if* $\overline{x}$ *cannot form a new class,* $\overline{x} \in \overline{E}$, *then*

$$\underline{R'_C}X = \begin{cases} \underline{R_C}X \cup \overline{x} & : \overline{E}' \subseteq X \\ \underline{R_C}X - \overline{E} & : \overline{E} \subseteq X \text{ and } \overline{E}' \nsubseteq X \\ \underline{R_C}X & : otherwise \end{cases}$$

$$where\ (\underline{R'_C}X)(x_i) = \begin{cases} (\underline{R_C}X)(x_i) \wedge X(\overline{x}) & : x_i \in \overline{E}' \\ (\underline{R_C}X)(x_i) & : otherwise \end{cases} \tag{3}$$

Proposition 3. *For* $\overline{R'_C}X$, *if* $\overline{x}$ *cannot form a new class, then*

$$\overline{R'_C}X = \begin{cases} \overline{R_C}X \cup \overline{E}' & : \overline{E}' \cap X \neq \emptyset \\ \overline{R_C}X & : otherwise \end{cases}$$

$$where\ (\overline{R'_C}X)(x_i) = \begin{cases} (\overline{R_C}X)(\overline{E}) \vee X(\overline{x}) & : x_i \in \overline{E}' \\ (\overline{R_C}X)(x_i) & : otherwise \end{cases} \tag{4}$$

Proposition 4. *For* $\underline{R'_C}X$ *and* $\overline{R'_C}X$, *if* $\overline{x}$ *form a new class, then*

$$\underline{R'_C}X = \begin{cases} \underline{R_C}X \cup \{\overline{x}\} & : \overline{x} \in X \\ \underline{R_C}X & : otherwise \end{cases}, \overline{R'_C}X = \begin{cases} \overline{R_C}X \cup \{\overline{x}\} & : \overline{x} \in X \\ \overline{R_C}X & : otherwise \end{cases}$$

$$where\ (\underline{R_C}X)(\overline{x}) = (\overline{R_C}X)(\overline{x}) = X(\overline{x}), if \overline{x} \in X. \tag{5}$$

The detailed process for updating approximations of a fuzzy concept when a single object immigrating is outlined in Algorithm 1.

3.2 The Emigration of One Object

Suppose there is one object $\widetilde{x}$ that gets out the FDS at time $t+1$. So $U' = U - \{\widetilde{x}\}$.

Proposition 5. *When one object* $\widetilde{x}$ *gets out of* $FDS = (U, C \cup D, V, f)$, $c \in C$. *One of the following results holds:*

(1) If $v = f(\widetilde{x}, c), |\widetilde{E}| = 1$, $\widetilde{x}$ *can eliminate an equivalence class with* v*;*
(2) If $v = f(\widetilde{x}, c), |\widetilde{E}| > 1$, $\widetilde{x}$ *can not eliminate an equivalence class.*

Proposition 6. *For* $\underline{R'_C}X$, *if* $\widetilde{x}$ *cannot eliminate an equivalence class, then*

$$\underline{R'_C}X = \begin{cases} \underline{R_C}X \cup \widetilde{E}' & : \widetilde{E}' \subseteq X \text{ and } \widetilde{E} \nsubseteq X \\ \underline{R_C}X & : otherwise \end{cases} \quad where,$$

$$(\underline{R'_C}X)(x_i) = \begin{cases} \inf\{X(y) | y \in \widetilde{E}'\} : & x_i \in \widetilde{E}' \text{ and } X(\widetilde{x}) \leq (\underline{R_C}X)(x_i) \\ (\underline{R_C}X)(x_i) & : otherwise \end{cases} \tag{6}$$

Algorithm 1. The Algorithm for Updating Approximations of a Fuzzy Concept when a Single Object Immigrating

---

Input:
    (1) FDS=$(U, C \cup D, V, f)$ at time t.
    (2) $\overline{x}$ is the immigrating object.
    (3) $\underline{R_C}X, \overline{R_C}X$.

Output:
    $\underline{R'_C}X, \overline{R'_C}X$ at time $t+1$.

Method:

01   Find the class $\overline{E}$ which $\overline{x}$ will immigrate to. If not find, $\overline{E} = null$;

02    $\overline{E}' = \overline{E} \cup \{\overline{x}\}$;

03   if ($\overline{E} = null$ and $\overline{x} \in X$)

04        $\underline{R'_C}X = \underline{R_C}(X) \cup \{\overline{x}\}, \overline{R'_C}X = \overline{R_C}(X) \cup \{\overline{x}\}$;

05   else if ($\overline{E} = null$)

06        $\underline{R'_C}X = \underline{R_C}(X), \ \overline{R'_C}X = \overline{R_C}(X)$;

07   else

08   if ($\overline{E}' \subseteq X$)

09            $\underline{R'_C}X = \underline{R_C}(X) \cup \{\overline{x}\}$;

10   else if ($\overline{E} \subseteq X$)

11            $\underline{R'_C}X = \underline{R_C}(X) - \overline{E}$;

12   else $\underline{R'_C}X = \underline{R_C}(X)$;

13   if ($\overline{E}' \cap X \neq \emptyset$)

14            $\overline{R'_C}X = \overline{R_C}(X) \cup \overline{E}'$;

15   else $\overline{R'_C}X = \overline{R_C}X$;

16   end if

17   for each $x_i \in \overline{E}'$

18        $(\underline{R'_C}X)(x_i) = (\underline{R_C}X)(\overline{E}) \wedge X(\overline{x})$;

19        $(\overline{R'_C}X)(x_i) = (\overline{R_C}X)(\overline{E}) \vee X(\overline{x})$;

20   end for

21   for each $x_i \notin \overline{E}'$

22        $(\underline{R'_C}X)(x_i) = (\underline{R_C}X)(x_i)$;

23        $(\overline{R'_C}X)(x_i) = (\overline{R_C}X)(x_i)$;

24   end for

25   output $\underline{R'_C}X, \overline{R'_C}X$;

Proposition 7. For $\overline{R'_C}X$, if $\widetilde{x}$ cannot eliminate an equivalence class, then

$$\overline{R'_C}X = \begin{cases} \overline{R_C}X - \widetilde{E} & : \widetilde{x} \in X \text{ and } \widetilde{E}' \cap X = \emptyset \\ \overline{R_C}X - \{\widetilde{x}\} & : \widetilde{E}' \cap X \neq \emptyset \\ \overline{R_C}X & : otherwise \end{cases}, where$$

$$(\overline{R'_C}X)(x_i) = \begin{cases} \sup\{X(y)|y \in \widetilde{E}'\} & : x_i \in \widetilde{E}' \text{ and } X(\widetilde{x}) \geq (\overline{R_C}X)(x_i) \\ (\overline{R_C}X)(x_i) & : otherwise \end{cases} \quad (7)$$

Proposition 8. *For $\underline{R'_C}X$ and $\overline{R'_C}X$, if $\widetilde{x}$ can eliminate an equivalence class, then*

$$\underline{R'_C}X = \begin{cases} \underline{R_C}X - \{\widetilde{x}\} : \widetilde{x} \in X \\ \underline{R_C}X \quad : otherwise \end{cases}, \overline{R'_C}X = \begin{cases} \overline{R_C}X - \{\widetilde{x}\} : \widetilde{x} \in X \\ \overline{R_C}X \quad : otherwise \end{cases} \quad (8)$$

The detailed process for updating approximations of a fuzzy concept when a single object emigrating is outlined in Algorithm 2.

4 Illustrations

(1) First, we consider the case when a single object immigrates into FDS.

Example 2. In Table 1, a new object $\overline{x} = (x_9, 1, 1, 4, 0.7)$ immigrates into the FDS, shown as Table 2. $U'/a = \{E'_1, E'_2, E'_3\} = \{E_1, E_2, E_3 \cup \{x_9\}\} = \{\{x_1, x_2\}, \{x_3, x_4, x_5\}, \{x_6, x_7, x_8, x_9\}\}$. Hence, $\overline{x}$ cannot form a new class.

Table 2. The immigration of a single object

U	a	b	c	fuzzy attribute d
x_1	2	2	1	0.7
x_2	2	2	1	1
x_3	3	3	1	0.7
x_4	3	3	2	0.8
x_5	3	2	2	0
x_6	1	1	4	0.5
x_7	1	2	3	0.2
x_8	1	1	3	0.6
$-- > x_9$	1	1	4	0.7

According to Proposition 2,

$$(\underline{R'_C}X)(x_9) = (\underline{R_C}(X))(E_3) \wedge X(x_9) = 0.2 \wedge 0.7 = 0.2;$$
$$(\underline{R'_C}X)(x_6) = (\underline{R'_C}X)(x_7) = (\underline{R'_C}X)(x_8) = (\underline{R'_C}X)(x_9).$$

$$(\overline{R'_C}X)(x_9) = (\overline{R_C}(X))(E_3) \vee X(x_9) = 0.6 \vee 0.7 = 0.7;$$
$$(\overline{R'_C}X)(x_6) = (\overline{R'_C}X)(x_7) = (\overline{R'_C}X)(x_8) = (\overline{R'_C}X)(x_9).$$

Therefore,

$$(\underline{R'_C}X) = \{x_6/0.2, x_7/0.2, x_8/0.2, x_9/0.2\};$$
$$(\overline{R'_C}X) = \{x_1/1, x_2/1, x_3/0.8, x_4/0.8, x_5/0.8, x_6/0.7, x_7/0.7, x_8/0.7, x_9/0.7\}.$$

Example 3. Shown as Table 3, another new object $\overline{x} = (x_{10}, 4, 1, 4, 0.7)$ immigrates into the FDS based on Example 2.

$U'/a = \{E_1, E_2, E_3, E'_4\} = \{\{x_1, x_2\}, \{x_3, x_4, x_5\}, \{x_6, x_7, x_8, x_9\}, \{x_{10}\}\}$. $\overline{x}$ forms a new class E'_4. Let $x_{10} \in X$.

Algorithm 2. the Algorithm for Updating Approximations of a Fuzzy Concept when a Single Object Emigrating

---

Input:
    (1) $FDS = (U, C \cup D, V, f)$ at time t.
    (2) $\widetilde{x}$ is the object that gets out the FDS.
    (3) $\underline{R_C}(X), \overline{R_C}(X)$.

Output:
    $\underline{R'_C}X, \overline{R'_C}X$ at time $t + 1$

Method:

01   Obtain the class $\widetilde{E}$ which $\widetilde{x}$ emigrates from, $\widetilde{E'} = \widetilde{E} - \{\widetilde{x}\}$;

02   if ($\widetilde{E'} = null$ and $\widetilde{x} \in X$)

03        $\underline{R'_C}X = \underline{R_C}(X) - \{\widetilde{x}\}$, $\overline{R'_C}X = \overline{R_C}(X) - \{\widetilde{x}\}$;

04   else if ($\widetilde{E'} = null$)

05        $\underline{R'_C}X = \underline{R_C}(X)$, $\overline{R'_C}X = \overline{R_C}(X)$;

06   else

07   if ($\widetilde{E'} \subseteq X$ and $\widetilde{E} \nsubseteq X$)

08            $\underline{R'_C}X = \underline{R_C}(X) \cup \widetilde{E'}$;

09   if ($X(\widetilde{x}) \leq \underline{R_C}X(\widetilde{E})$)

10                $\underline{R'_C}X(\widetilde{E'}) = \inf\{X(y)|y \in \widetilde{E'}\}$;

11   else

12                $\underline{R'_C}X(\widetilde{E'}) = \underline{R_C}X(\widetilde{E})$;

13   else $\underline{R'_C}X = \underline{R_C}(X)$;

14   end if

15   else

16   if ($\widetilde{x} \in X$ and $\widetilde{E'} \cap X = \emptyset$)

17            $\overline{R'_C}X = \overline{R_C}(X) - \widetilde{E}$;

18   else if ($\widetilde{E'} \cap X \neq \emptyset$)

19            $\overline{R'_C}X = \overline{R_C}(X) - \{\widetilde{x}\}$;

20   if ($X(\widetilde{x}) \geq \underline{R_C}X(\widetilde{E})$)

21                $\overline{R'_C}X(\widetilde{E'}) = \sup\{X(y)|y \in \widetilde{E'}\}$;

22   else

23                $\overline{R'_C}X(\widetilde{E'}) = \overline{R_C}X(\widetilde{E})$;

24   else $\overline{R'_C}X = \overline{R_C}X$

25   end if

26   end if

27   output $\underline{R'_C}X, \overline{R'_C}X$;

According to Proposition 4,

$$(\underline{R'_C}X)(x_{10}) = X(x_{10}) = 0.7, (\overline{R'_C}X)(x_{10}) = X(x_{10}) = 0.7.$$

Table 3. The immigration of a single object

U	a	b	c	fuzzy attribute d
x_1	2	2	1	0.7
x_2	2	2	1	1
x_3	3	3	1	0.7
x_4	3	3	2	0.8
x_5	3	2	2	0
x_6	1	1	4	0.5
x_7	1	2	3	0.2
x_8	1	1	3	0.6
x_9	1	1	4	0.7
$-->x_{10}$	4	1	4	0.7

Therefore,

$$(\underline{R'_C}X) = \{x_6/0.2, x_7/0.2, x_8/0.2, x_9/0.2, x_{10}/0.7\};$$

$$(\overline{R'_C}X) = \{x_1/1, x_2/1, x_3/0.8, x_4/0.8, x_5/0.8, x_6/0.7, x_7/0.7, x_8/0.7, x_9/0.7, x_{10}/0.7\}.$$

(2) We illustrate the phase of a single object emigrating from the FDS.

Example 4. In Table 1, an object $\widetilde{x} = x_1$ emigrates from the FDS, shown as Table 4. $U'/a = \{E'_1, E'_2, E'_3\} = \{E_1 - \{x_1\}, E_2, E_3\} = \{\{x_2\}, \{x_3, x_4, x_5\}, \{x_6, x_7, x_8\}\}$. $\widetilde{x}$ can not eliminate a class.

Table 4. The emigration can not eliminate a class

U	a	b	c	fuzzy attribute d
~~x_1~~	~~2~~	~~2~~	~~1~~	~~0.7~~
x_2	2	2	1	1
x_3	3	3	1	0.7
x_4	3	3	2	0.8
x_5	3	2	2	0
x_6	1	1	4	0.5
x_7	1	2	3	0.2
x_8	1	1	3	0.6

According to Proposition 6, Because $x_1 \in X, E_1 \not\subseteq X, \underline{R'_C}X = \underline{R_C}X$, and $\{E_1 - x_1\} \cap X = \emptyset, \overline{R'_C}X = \overline{R_C}X - E_1$. We have,

$$\underline{R'_C}X = \{x_6/0.2, x_7/0.2, x_8/0.2\};$$

$$\overline{R'_C}X = \{x_3/0.8, x_4/0.8, x_5/0.8, x_6/0.6, x_7/0.6, x_8/0.6\}.$$

Example 5. Shown as Table 5, another object $\widetilde{x} = x_2$ emigrates from the FDS based on Example 4. We have, $U'/a = \{E'_2, E'_3\} = \{E_2, E_3\} = \{\{x_3, x_4, x_5\}, \{x_6, x_7, x_8\}\}$. $\widetilde{x}$ eliminates class E_1.

According to Proposition 8, we have

$$(\underline{R'_C}X) = \{x_6/0.2, x_7/0.2, x_8/0.2\};$$

$$(\overline{R'_C}X) = \{x_3/0.8, x_4/0.8, x_5/0.8, x_6/0.6, x_7/0.6, x_8/0.6\}.$$

Table 5. The emigration can not eliminate a class

U	a	b	c	fuzzy attribute d
~~x_2~~	~~2~~	~~2~~	~~1~~	~~1~~
x_3	3	3	1	0.7
x_4	3	3	2	0.8
x_5	3	2	2	0
x_6	1	1	4	0.5
x_7	1	2	3	0.2
x_8	1	1	3	0.6

5 Conclusions

In the FDS, the objects generally need to be immigrated or emigrated. Updating principles of upper and lower approximations of fuzzy rough sets in the dynamic FDS were discussed in this paper. The corresponding algorithms for updating approximations incrementally were then presented. Several examples are used to illustrates the propose methods. Our future research work will focus on the case of the variation of multiple objects and validation of the proposed algorithms in real data sets.

Acknowledgement. This work is supported by the National Science Foundation of China (Nos. 60873108, 61175047, 61100117), the Youth Social Science Foundation of the Chinese Education Commission (No. 11YJC630127), the Fundamental Research Funds for the Central Universities (Nos. SWJTU11ZT08, SWJTU12CX117, SWJTU12CX091), the Scientific Research Foundation of Sichuan Provincial Education Department (No. 10ZB049) and the Scientific Research Fund of Yibin University (No. 2011Z15).

References

1. Pawlak, Z.: Rough sets. International Journal of Computer and Information Sciences 11(5), 341–356 (1982)
2. Dubois, D., Prade, H.: Rough fuzzy sets and fuzzy rough sets. International Journal of General Systems 17(2-3), 191–209 (1990)
3. Yao, Y.Y.: Combination of rough and fuzzy sets based on α-level sets. In: Lin, T.Y., Cercone, N. (eds.) Rough Sets and Data Mining: Analysis for Imprecise Data, pp. 301–321. Kluwer Academic Publishers, Boston (1997)
4. Li, T.R., Ruan, D., Wets, G., Song, J., Xu, Y.: A rough sets based characteristic relation approach for dynamic attribute generalization in data mining. Knowledge-Based Systems 20(5), 485–494 (2007)
5. Li, T.R., Xu, Y.: A generalized rough set approach to attribute generalization in data mining. Journal of Southwest Jiaotong University(English Edition) 8(1), 69–75 (2000)
6. Chan, C.C.: A rough set approach to attribute generalization in data mining. Information Sciences 107, 177–194 (1998)
7. Li, T.R., Yang, N., Xu, Y.: An incremental algorithm for mining classification rules in incomplete information system. In: Proc. 2004 Annual Meeting of the North American Fuzzy Information Processing Society, pp. 446–449. IEEE Press (2004)

8. Cheng, Y.: The incremental method for fast computing the rough fuzzy approximations. Data and Knowledge Engineering 70, 84–100 (2011)
9. Liu, D., Li, T.R., Ruan, D., Zhang, J.B.: Incremental learning optimization on knowledge discovery in dynamic business intelligent systems. Journal of Global Optimization, Journal of Global Optimization 51, 325–344 (2011)
10. Chen, H.M., Li, T.R., Hu, C.X., Ji, X.L.: An incremental updating principle for computing approximations in information systems while the object set varies with time. In: Proc. IEEE International Conference on Granular Computing, pp. 49–52. IEEE Press, Chengdu (2009)
11. Shusaku, T., Hiroshi, T.: Incremental learning of probabilistic rules from clinical database based on rough set theory. Journal of the American Medical Informations Association 4, 198–202 (1997)
12. Guan, Y., Wang, H.: Set-valued information systems. Information Sciences 176(17), 2507–2525 (2006)
13. Wang, L., Wu, Y., Wang, G.Y.: An incremental rule acquisition algorithm based on variable precision rough set model. Journal of Chongqing University of Posts and Telecommunications (Natural Science) 17(6), 709–713 (2005)
14. Zhang, J.B., Li, T.R., Ruan, D., Liu, D.: Neighborhood rough sets for dynamic data mining. International Journal of Intelligent Systems 27, 317–342 (2012)
15. Tong, L., An, L.: Incremental learning of decision rules based on rough set theory. In: The 4th World Congress on Intelligent Control and Automation(WCICA 2002), pp. 420–425. IEEE Press, Tianjin (2002)
16. Yong, L., Congfu, X., Yunhe, P.: A Parallel Approximate Rule Extracting Algorithm Based on the Improved Discernibility Matrix. In: Tsumoto, S., Słowiński, R., Komorowski, J., Grzymała-Busse, J.W. (eds.) RSCTC 2004. LNCS (LNAI), vol. 3066, pp. 498–503. Springer, Heidelberg (2004)
17. Guo, S., Wang, Z.Y., Wu, Z.C., Yan, H.P.: A novel dynamic incremental rules extraction algorithm based on rough set theory. In: Proc. Fourth International Conference on Machine Learning and Cybernetics, pp. 1902–1907. IEEE Press, GuangDong (2005)
18. Zheng, Z., Wang, G.Y.: A rough set and rule tree based incremental knowledge acquisition algorithm. Fundamenta Informaticae 59(2-3), 299–313 (2004)

How Good Are Probabilistic Approximations for Rule Induction from Data with Missing Attribute Values?

Patrick G. Clark[1], Jerzy W. Grzymala-Busse[1,2], and Zdzislaw S. Hippe[3]

[1] Department of Electrical Engineering and Computer Science, University of Kansas, Lawrence, KS 66045, USA
[2] Institute of Computer Science, Polish Academy of Sciences, 01–237 Warsaw, Poland
pclark@ku.edu, jerzy@ku.edu
[3] Department of Expert Systems and Artificial Intelligence, University of Information Technology and Management, 35-225 Rzeszow, Poland
zhippe@wsiz.rzeszow.pl

Abstract. The main objective of our research was to test whether the probabilistic approximations should be used in rule induction from incomplete data. Probabilistic approximations, well known for many years, are used in variable precision rough set models and similar approaches to uncertainty.

For our experiments we used five standard data sets. Three data sets were incomplete to begin with and two data sets had missing attribute values that were randomly inserted. We used two interpretations of missing attribute values: lost values and "do not care" conditions. Among these ten combinations of a data set and a type of missing attribute values, in one combination the error rate (the result of ten-fold cross validation) was smaller than for ordinary approximations; for other two combinations, the error rate was larger than for ordinary approximations.

1 Introduction

One of the fundamental concepts of rough set theory is an idea of lower and upper approximations. A generalization of such approximations, a probabilistic approximation, introduced in [1], was applied in variable precision rough set models, Bayesian rough sets and decision-theoretic rough set models [2–10]. The probabilistic approximation is associated with some parameter α (interpreted as a probability). If α is very small, say 0.001 (this number depends on the size of the data set), the probabilistic approximation is reduced to the upper approximation; if α is equal to 1.0, the probabilistic approximation becomes the lower approximation. The problem is how useful are *proper* probabilistic approximations (with α larger than 0.001 but smaller than 1.0). We studied usefulness of proper probabilistic approximations for inconsistent data sets [11],

J.T. Yao et al. (Eds.): RSCTC 2012, LNAI 7413, pp. 46–55, 2012.

where we concluded that proper probabilistic approximations are not frequently better than ordinary lower and upper approximations.

In this paper we study usefulness of the proper probabilistic approximations applied for rule induction from incomplete data. We will use two interpretations of missing attribute values, as *lost values* (the original attribute values are not longer accessible, for details see [12, 13]) and as *"do not care" conditions* (the original values were irrelevant, see [14, 15]).

For data sets with missing attribute values there exist many definitions of approximations [16], we use one of the most successful options (from the view point of rule induction) called *concept* approximations [16]. Concept approximations were generalized to concept probabilistic approximations in [17].

Our experiments on five data sets with two types of missing attribute values (altogether ten combinations) show that the proper concept probabilistic approximations are not very useful for rule induction from incomplete data sets: for one combination the error rate (result of ten-fold cross validation) was smaller than for ordinary concept approximations, for two combinations such error rate was larger than for ordinary concept approximations, for remaining seven combinations the error rate was neither smaller nor larger.

2 Incomplete Data Sets

The data sets are presented in the form of a *decision table*. Rows of the decision table represent *cases*, while columns are labeled by *variables*. The set of all cases will be denoted by U. In Table 1, $U = \{1, 2, 3, 4, 5, 6, 7, 8\}$. Independent variables are called *attributes* and a dependent variable is called a *decision* and is denoted by d. The set of all attributes will be denoted by A. In Table 1, $A = \{Wind, Humidity, Temperature\}$. The value for a case x and an attribute a will be denoted by $a(x)$.

In this paper we distinguish between two interpretations of missing attribute values: *lost values*, denoted by "?", and *"do not care" conditions*, denoted by "*". Table 1 present an incomplete data set affected by both lost values and "do not care" conditions.

One of the most important ideas of rough set theory [18, 19] is an indiscernibility relation, defined for complete data sets. Let B be a nonempty subset of A. The indiscernibility relation $R(B)$ is a relation on U defined for $x, y \in U$ as follows:

$$(x, y) \in R(B) \text{ if and only if } \forall a \in B \ (a(x) = a(y)).$$

The indiscernibility relation $R(B)$ is an equivalence relation. Equivalence classes of $R(B)$ are called *elementary sets* of B and are denoted by $[x]_B$. A subset of U is called *A-definable* if it is a union of elementary sets.

The set X of all cases defined by the same value of the decision d is called a *concept*. For example, a concept associated with the value *no* of the decision *Trip* is the set $\{1, 3, 5, 7\}$. The largest B-definable set contained in X is called the *B-lower approximation* of X, denoted by $\underline{appr}_B(X)$, and defined as follows

$$\cup\{[x]_B \mid [x]_B \subseteq X\}$$

Table 1. A decision table

	Attributes			Decision
Case	Wind	Humidity	Temperature	Trip
1	?	high	high	no
2	low	low	high	yes
3	low	*	low	no
4	*	low	low	yes
5	high	high	?	no
6	low	?	*	yes
7	high	high	low	no
8	high	low	low	yes

while the smallest B-definable set containing X, denoted by $\overline{appr}_B(X)$ is called the *B-upper approximation* of X, and is defined as follows

$$\cup\{[x]_B \mid [x]_B \cap X \neq \emptyset\}.$$

For a variable a and its value v, (a, v) is called a variable-value pair. A *block* of (a, v), denoted by $[(a, v)]$, is the set $\{x \in U \mid a(x) = v\}$ [20].

For incomplete decision tables the definition of a block of an attribute-value pair is modified in the following way.

- If for an attribute a there exists a case x such that $a(x) = ?$, i.e., the corresponding value is lost, then the case x should not be included in any blocks $[(a, v)]$ for all values v of attribute a,
- If for an attribute a there exists a case x such that the corresponding value is a "do not care" condition, i.e., $a(x) = *$, then the case x should be included in blocks $[(a, v)]$ for all specified values v of attribute a.

For the data set from Table 1 the blocks of attribute-value pairs are:

[(Wind, low)] = {2, 3, 4, 6},
[(Wind, high)] = {4, 5, 7, 8},
[(Humidity, high)] = {1, 3, 5, 7},
[(Humidity, low)] = {2, 3, 4, 8},
[(Temperature, high)] = {1, 2, 6},
[(Temperature, low)] = {3, 4, 6, 7, 8}.

For a case $x \in U$ and $B \subseteq A$, the *characteristic set* $K_B(x)$ is defined as the intersection of the sets $K(x, a)$, for all $a \in B$, where the set $K(x, a)$ is defined in the following way:

- If $a(x)$ is specified, then $K(x, a)$ is the block $[(a, a(x))]$ of attribute a and its value $a(x)$,

- If $a(x) = ?$ or $a(x) = *$ then the set $K(x, a) = U$, where U is the set of all cases.

For Table 1 and $B = A$,

$K_A(1) = \{1\}$,
$K_A(2) = \{2\}$,
$K_A(3) = \{3, 4, 6\}$,
$K_A(4) = \{3, 4, 8\}$,
$K_A(5) = \{5, 7\}$,
$K_A(6) = \{2, 3, 4, 6\}$,
$K_A(7) = \{7\}$,
$K_A(8) = \{4, 8\}$.

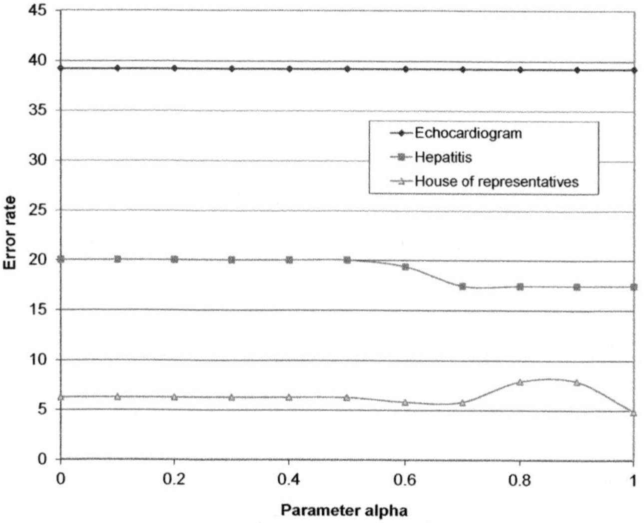

Fig. 1. Error rates for data sets *Echocardiogram*, *Hepatitis*, and *House of representatives* with lost values

Note that for incomplete data there is a few possible ways to define approximations [16], we use *concept* approximations [17]. A B-*concept lower approximation* of the concept X is defined as follows:

$$\underline{B}X = \cup\{K_B(x) \mid x \in X, K_B(x) \subseteq X\}.$$

A B-*concept upper approximation* of the concept X is defined as follows:

$$\overline{B}X = \cup\{K_B(x) \mid x \in X, K_B(x) \cap X \neq \emptyset\} = \\ = \cup\{K_B(x) \mid x \in X\}.$$

For Table 1, A-concept lower and A-concept upper approximations of the two concepts: {1, 3, 5, 7} and {2, 4, 6, 8} are:

$\underline{A}\{1, 3, 5, 7\} = \{1, 5, 7\}$,
$\underline{A}\{2, 4, 6, 8\} = \{2, 4, 8\}$,
$\overline{A}\{1, 3, 5, 7\} = \{1, 3, 4, 5, 6, 7\}$,
$\overline{A}\{2, 4, 6, 8\} = \{2, 3, 4, 6, 8\}$.

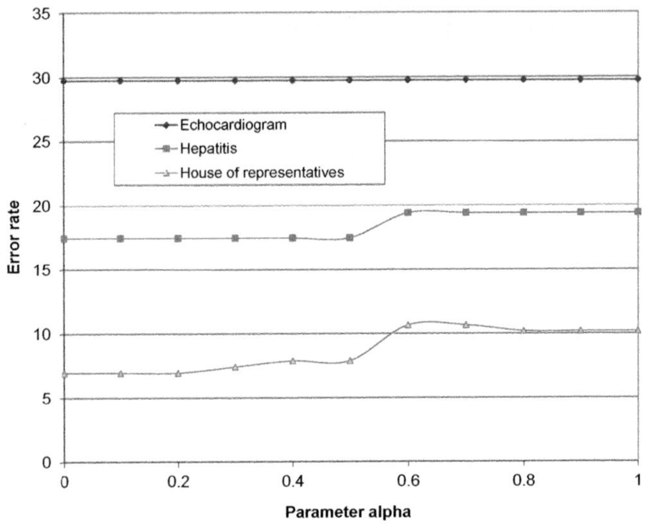

Fig. 2. Error rates for data sets *Echocardiogram*, *Hepatitis*, and *House of representatives* with "do not care" conditions

3 Probabilistic Approximations

In this paper we explore all probabilistic approximations that can be defined for a given concept X. For completely specified data sets a *probabilistic approximation* is defined as follows

$$appr_{\alpha}(X) = \cup\{[x] \mid x \in U, P(X \mid [x]) \geq \alpha\},$$

where $[x]$ is $[x]_A$ and α is a parameter, $0 < \alpha \leq 1$, see [17]. For discussion on how this definition is related to the value precision asymmetric rough sets see [11, 17].

Note that if $\alpha = 1$, the probabilistic approximation becomes the standard lower approximation and if α is small, close to 0, in our experiments it was 0.001, the same definition describes the standard upper approximation.

For incomplete data sets, a *B-concept probabilistic approximation* is defined by the following formula [17]

$$\cup\{K_B(x) \mid x \in X,\ Pr(X|K_B(x)) \geq \alpha\}.$$

For simplicity, we will denote $K_A(x)$ by $K(x)$ and the *A-concept probabilistic approximation* will be called a *probabilistic approximation*.

For Table 1 and the concept $X = [(Trip, no)] = \{1, 3, 5, 7\}$, for any characteristic set $K(x)$, $x \in U$, conditional probabilities $P(X|K_(x))$ are presented in Table 2.

Thus, for the concept $\{1, 3, 5, 7\}$ we may define only two distinct probabilistic approximations:

$$appr_{1.0}(\{1,3,5,7\}) = \{1,5,7\} \text{ and } appr_{0.333}(\{1,3,5,7\}) = \{1,3,4,5,6,7\}.$$

Table 2. Conditional probabilities

$K(x)$	{1}	{5, 7}	{7}	{3, 4, 6}	{3, 4, 8}	{2, 3, 4, 6}	{2}	{4, 8}
$P(\{1, 3, 5, 7\} \mid K(x))$	1.0	1.0	1.0	0.333	0.333	0.25	0	0

Table 3. Data sets used for experiments

Data set	Number of			Percentage of
	cases	attributes	concepts	missing attribute values
Echocardiogram	74	7	2	4.05
Hepatitis	155	19	2	5.67
House of Representatives	434	16	2	5.40
Image segmentation	210	19	7	70
Lymphography	148	18	4	70

4 Experiments

For our experiments we used five real-life data sets that are available on the University of California at Irvine *Machine Learning Repository*. Two of these data sets (*Image segmentation* and *Lymphography* were originally completely specified, i.e., they did not contain any missing attribute values. However, we replaced, randomly, 70% of existing attribute values by signs of missing attribute values, first by *lost* values and then we converted *lost values* to *"do not care" conditions*, see Table 3.

For rule induction we used the MLEM2 (Modified Learning from Examples Module version 2) rule induction algorithm, a component of the LERS (Learning from Examples based on Rough Sets) data mining system [20, 21].

The main objective of our research was to test whether proper probabilistic approximations are better than concept lower and upper approximations. We conducted experiments of a single ten-fold cross validation starting with 0.001 and then increasing the parameter α by 0.1 until reaching 1.0. For a given data set, in all of these eleven experiments we used identical ten pairs of larger (90%) and smaller (10%) data sets. Results of our experiments are shown in Figures 1–4. If during such a sequence of eleven experiments, the error rate was smaller than the minimum of the error rates for lower and upper approximations or larger than maximum of the error rated for lower and upper approximations, we selected more precise values of the parameter α and we conducted additional 30 experiments of ten-fold cross validation.

For example, for the *Echocardiogram* data set, affected by lost values, denoted by "?", the error rate was constant, so there is no need for additional 30 experiments, see Figure 1. Similarly, for the *Hepatitis* data set, also affected by *lost values*. But for the *House of representative* data set, affected by *lost values*, it is

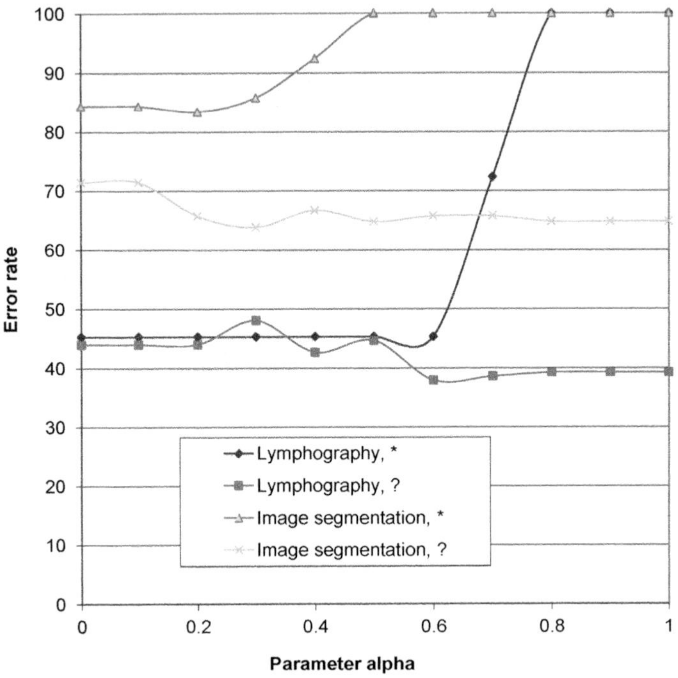

Fig. 3. Error rates for data sets *Image segmentation* and *Lymphography*

clear that we should look more closely at the parameter α around the values 0.65 and 0.85. Results are presented in Table 4. Using the standard statistical test for the difference between two averages (two tails and the significance level of 5%) we may conclude that there is no statistically significant difference between the probabilistic approximation associated with $\alpha = 0.65$ and the upper approximation ($\alpha = 0.001$). The same test indicates that the probabilistic approximation, associated with $\alpha = 0.85$ is worse than the upper approximation ($\alpha = 0.001$), as well as the lower approximation ($\alpha = 1.0$). Results of all remaining 30 experiments of ten-fold cross validation are presented in Tables 5–8.

In particular, for the *House of representatives* data set with "do not car" conditions as missing attribute values, for $\alpha = 0.65$, the corresponding probabilistic approximation is worse than both lower ($\alpha = 1.0$) and upper ($\alpha = 0.001$) approximations. On the other hand, for the *Image segmentation* data set with "do not care" conditions, for $\alpha = 0.2$, the error rate is significantly better than for both lower and upper approximations. In experiments reported in this paper this is the only situation of this type. For remaining data sets, no matter with lost values or "do not care" conditions, probabilistic approximations for α between 0.1 and 0.9 are neither worse than the worst for the two: lower and upper approximations nor better than the best of the two.

Table 4. Results of 30 experiments of ten-fold cross validation for *House of representatives*, lost values

α	Error rate	Standard deviation
0.001	6.59	0.6159
0.65	6.42	0.6396
0.85	7.31	0.7055
1.0	5.44	0.5885

Table 5. Results of 30 experiments of ten-fold cross validation for *House of representatives*, "do not care" conditions

α	Error rate	Standard deviation
0.001	5.97	0.5147
0.65	10.14	0.6819
1.0	9.72	0.7584

Table 6. Results of 30 experiments of ten-fold cross validation for *Image segmentation*, lost values

α	Error rate	Standard deviation
0.3	65.56	2.6567
1.0	63.44	2.5982

Table 7. Results of 30 experiments of ten-fold cross validation for *Image segmentation*, "do not care" conditions

α	Error rate	Standard deviation
0.001	85.20	1.1525
0.2	84.20	1.1191

Table 8. Results of 30 experiments of ten-fold cross validation for *Lymphography*, lost values

α	Error rate	Standard deviation
0.001	44.84	2.1767
0.3	44.64	2.4647
0.4	41.24	2.1031
1.0	37.61	2.2227

5 Conclusions

As follows from our experiments, the *proper* probabilistic approximations (ones with α between 0.1 and 0.9) were neither better nor worse than ordinary lower ($\alpha = 1.0$) and upper ($\alpha = 0.001$) approximations, except for three situations. In one of them (the *Image segmentation* data set with "do not care" conditions was better than ordinary approximations, in other two situations (both for the *House of representatives* data set, with lost values and "do not care" conditions) the proper probabilistic approximations were worse than ordinary approximations.

References

1. Wong, S.K.M., Ziarko, W.: INFER—an adaptive decision support system based on the probabilistic approximate classification. In: Proceedings of the 6th International Workshop on Expert Systems and their Applications, pp. 713–726 (1986)
2. Grzymala-Busse, J.W., Ziarko, W.: Data mining based on rough sets. In: Wang, J. (ed.) Data Mining: Opportunities and Challenges, pp. 142–173. Idea Group Publ., Hershey (2003)
3. Pawlak, Z., Skowron, A.: Rough sets: Some extensions. Information Sciences 177, 28–40 (2007)
4. Pawlak, Z., Wong, S.K.M., Ziarko, W.: Rough sets: probabilistic versus deterministic approach. International Journal of Man-Machine Studies 29, 81–95 (1988)
5. Ślęzak, D., Ziarko, W.: The investigation of the bayesian rough set model. International Journal of Approximate Reasoning 40, 81–91 (2005)
6. Yao, Y.Y.: Probabilistic rough set approximations. International Journal of Approximate Reasoning 49, 255–271 (2008)
7. Yao, Y.Y., Wong, S.K.M.: A decision theoretic framework for approximate concepts. International Journal of Man-Machine Studies 37, 793–809 (1992)
8. Yao, Y.Y., Wong, S.K.M., Lingras, P.: A decision-theoretic rough set model. In: Proceedings of the 5th International Symposium on Methodologies for Intelligent Systems, pp. 388–395 (1990)
9. Ziarko, W.: Variable precision rough set model. Journal of Computer and System Sciences 46(1), 39–59 (1993)
10. Ziarko, W.: Probabilistic approach to rough sets. International Journal of Approximate Reasoning 49, 272–284 (2008)
11. Clark, P.G., Grzymala-Busse, J.W.: Experiments on probabilistic approximations. In: Proceedings of the 2011 IEEE International Conference on Granular Computing, pp. 144–149 (2011)
12. Grzymala-Busse, J.W., Wang, A.Y.: Modified algorithms LEM1 and LEM2 for rule induction from data with missing attribute values. In: Proceedings of the Fifth International Workshop on Rough Sets and Soft Computing (RSSC 1997) at the Third Joint Conference on Information Sciences (JCIS 1997), pp. 69–72 (1997)
13. Stefanowski, J., Tsoukias, A.: Incomplete information tables and rough classification. Computational Intelligence 17(3), 545–566 (2001)
14. Grzymala-Busse, J.W.: On the unknown attribute values in learning from examples. In: Proceedings of the ISMIS 1991, 6th International Symposium on Methodologies for Intelligent Systems, pp. 368–377 (1991)

15. Kryszkiewicz, M.: Rough set approach to incomplete information systems. In: Proceedings of the Second Annual Joint Conference on Information Sciences, pp. 194–197 (1995)
16. Grzymala-Busse, J.W.: Rough set strategies to data with missing attribute values. In: Workshop Notes, Foundations and New Directions of Data Mining, in conjunction with the 3rd International Conference on Data Mining, pp. 56–63 (2003)
17. Grzymała-Busse, J.W.: Generalized Parameterized Approximations. In: Yao, J., Ramanna, S., Wang, G., Suraj, Z. (eds.) RSKT 2011. LNCS, vol. 6954, pp. 136–145. Springer, Heidelberg (2011)
18. Pawlak, Z.: Rough sets. International Journal of Computer and Information Sciences 11, 341–356 (1982)
19. Pawlak, Z.: Rough Sets. Theoretical Aspects of Reasoning about Data. Kluwer Academic Publishers, Dordrecht (1991)
20. Grzymala-Busse, J.W.: LERS—a system for learning from examples based on rough sets. In: Slowinski, R. (ed.) Intelligent Decision Support. Handbook of Applications and Advances of the Rough Set Theory, pp. 3–18. Handbook of Applications and Advances of the Rough Set Theory. Kluwer Academic Publishers, Dordrecht (1992)
21. Grzymala-Busse, J.W.: MLEM2: A new algorithm for rule induction from imperfect data. In: Proceedings of the 9th International Conference on Information Processing and Management of Uncertainty in Knowledge-Based Systems, pp. 243–250 (2002)

Induction of Ordinal Classification Rules from Incomplete Data

Jerzy Błaszczyński[1], Roman Słowiński[1,2], and Marcin Szeląg[1]

[1] Institute of Computing Science, Poznań University of Technology, 60-965 Poznań, Poland
[2] Systems Research Institute, Polish Academy of Sciences, 01-447 Warsaw, Poland
{jblaszczynski,rslowinski,mszelag}@cs.put.poznan.pl

Abstract. In this paper, we consider different ways of handling missing values in ordinal classification problems with monotonicity constraints within Dominance-based Rough Set Approach (DRSA). We show how to induce classification rules in a way that has desirable properties. Our considerations are extended to an experimental comparison of the postulated rule classifier with other ordinal and non-ordinal classifiers.

Keywords: Dominance-based Rough Set Approach, Ordinal classification with monotonicity constraints, Missing values, Decision rules.

1 Introduction

In data mining concerning classification problems, it is quite common to have missing values for attributes describing objects [15]. Thus, different ways of handling missing values, or more generally, incomplete data, have been proposed. The usual approach is to assume that some value(s) can represent correctly the missing one. Then, the missing values are replaced in some way by so-called representative values. In this case, the question is how to avoid data distortion [15].

Rough set approach to handling missing values avoids making changes in the data. The problem is addressed by a proper definition of the relation employed to form granules of knowledge. Extensions of the rough set model [18], that introduce relations forming granules of indiscernible or similar objects, include [14,13,17,20].

In this work, we focus on extensions of the Dominance-based Rough Set Approach (DRSA) [11,19] to handling missing values in *ordinal classification problems*. In these problems, decision classes are ordered. Then, it is often meaningful to consider *monotonicity constraints* (*monotonic relationships*) between ordered class labels and values of attributes expressed on ordinal or cardinal (numerical) scales [11,19]. The constraints result from background knowledge, e.g., "the higher the service quality and the lower the price, the higher the customer satisfaction" [12]. Objects violating such constraints are called *inconsistent*.

Some propositions of handling missing values in DRSA were given in [7,9,10]. We review these approaches and consider some new ones. Then, we present an

J.T. Yao et al. (Eds.): RSCTC 2012, LNAI 7413, pp. 56–65, 2012.

experimental validation of a classifier that employs the approach considered to be the best with respect to (w.r.t.) some desirable properties.

The rest of this paper is structured as follows. In Section 2, we present ways of handling missing values in DRSA. Section 3 explains how decision rules can be induced and applied in the presence of missing values. In Section 4, we present results of the experimental validation, and discussion of these results.

2 Different Ways of Handling Missing Values in DRSA

In this section, we briefly remind basics of DRSA [11,19], and then, we discuss some alternative ways of extending this approach to handle missing values.

2.1 Basics of DRSA

Data analyzed by DRSA concern a finite universe U of objects described by ordinal attributes from a finite set A. Moreover, A is divided into disjoint sets of condition attributes C and decision attributes Dec. The value set of $q \in C \cup Dec$ is denoted by V_q, and $V_P = \prod_{q=1}^{|P|} V_q$ is called P-evaluation space, where $P \subseteq C$. For simplicity, we assume that $Dec = \{d\}$. Values of d are ordered class labels.

We consider a given set $P \subseteq C$ of attributes. To simplify notation, where possible, we will skip P in all expressions valid for any $P \subseteq C$. Moreover, for any $q_i \in P$, we denote by $q_i(y)$ the evaluation of object $y \in U$ on attribute q_i, and we denote by $\succeq_{q_i}$ the weak preference relation over U confined to q_i.

Decision attribute d makes a partition of set U into n disjoint sets of objects, called *decision classes*. We denote this partition by $\mathcal{X} = \{X_1, \ldots, X_n\}$.

When there exists a monotonic relationship between evaluation of objects on condition attributes and their class labels, then, in order to make a meaningful representation of classification decisions, one has to consider the *dominance relation* D in the evaluation space. Given $y, z \in U$, object y dominates object z (and z is dominated by y), denoted by yDz, if and only if (iff) $y \succeq_{q_i} z$, for each $q_i \in P$. For any object $y \in U$, two dominance cones can be calculated in the P-evaluation space: positive dominance cone $D^+(y) = \{z \in U : zDy\}$, and negative dominance cone $D^-(y) = \{z \in U : yDz\}$.

The class labels are ordered, such that if $i < j$, then class X_i is considered to be worse than X_j. Moreover, rough approximations concern unions of decision classes: upward unions $X_i^{\geq} = \bigcup_{t \geq i} X_t$, and downward unions $X_i^{\leq} = \bigcup_{t \leq i} X_t$, where $i = 1, \ldots, n$ (technically, $X_1^{\geq}$, $X_n^{\leq}$ are not considered as $X_1^{\geq} = X_n^{\leq} = U$).

To simplify notation, where possible, we use a symbol X to denote a set of objects from $X_i^{\geq}$ or $X_i^{\leq}$ (when both unions of classes are considered jointly). We denote by $\neg X$ the set $U \setminus X$. Moreover, we denote by $E(y)$ the dominance cone "concordant" with X, and by $E^{-1}(y)$ the dominance cone "discordant" with X. Precisely, if in a given equation $X_i^{\geq}$ is substituted for X, then $D^+(y)$ should be substituted for $E(y)$, and $D^-(y)$ should be substituted for $E^{-1}(y)$; if in the same equation $X_i^{\leq}$ is substituted for X, then $D^-(y)$ should be substituted for $E(y)$, and $D^+(y)$ should be substituted for $E^{-1}(y)$. Finally, a missing attribute value is denoted by $*$.

In the classical DRSA, the *lower approximation* of set X is defined using strict inclusion relation between dominance cone $E(y)$ and approximated set X:

$$\underline{X} = \{y \in U : E(y) \subseteq X\}. \tag{1}$$

Moreover, the *upper approximation* of set X is defined as $\overline{X} = \{y \in U : E^{-1}(y) \cap X \neq \emptyset\}$. Definition (1) appears to be too restrictive in practical applications. It often leads to empty lower approximations of $X_i^{\geq}$ and $X_i^{\leq}$, preventing generalization of data in terms of decision rules. Therefore, we employ Variable Consistency DRSA (VC-DRSA) [4] which is a probabilistic extension of the classical DRSA. We use *object consistency measure* $\epsilon_X : U \rightarrow [0,1]$, introduced in [4], defined as:

$$\epsilon_X(y) = \frac{|E(y) \cap \neg X|}{|\neg X|}. \tag{2}$$

Value $\epsilon_X(y)$ reflects the consistency of object y w.r.t. X (or, the evidence for the membership of y to X). ϵ_X is a cost-type measure, which means that value zero denotes full consistency and the greater the value, the less consistent is a given object. Then, the *probabilistic lower approximation* of set X is defined as:

$$\underline{X} = \{y \in X : \epsilon_X(y) \leq \theta_X\}, \tag{3}$$

where threshold $\theta_X \in [0,1)$. When $\theta_X = 0$, approximation (3) boils down to approximation (1). *Probabilistic upper approximation* of X is defined using complementarity: $\overline{X} = U \setminus \underline{\neg X}$. In the following, we drop the word "probabilistic".

In [4], we introduced four *monotonicity properties* required from an object consistency measure: $(m1)$ – monotonicity w.r.t. growing set of attributes, $(m2)$ – monotonicity w.r.t. growing union of classes, $(m3)$ – monotonicity w.r.t. superunion of classes, and $(m4)$ – monotonicity w.r.t. dominance relation. We also proved that ϵ_X has properties $(m1)$, $(m2)$, and $(m4)$, sufficient in practical applications.

2.2 Extensions of DRSA to Handle Missing Values

The presence of missing values requires a proper adaptation of DRSA by redefinition of the dominance relation D. Once we fix this definition, we can proceed in a "usual" way by calculating approximations of unions of classes, and by induction of decision rules from these approximations [5,11,19]. We review some of the redefinitions of the dominance relation which are known from literature and we discuss a few other possibilities.

In the literature concerning rough set approaches to handling missing attribute values in classification data (see, e.g., [14,13]), one can find a proposal of a semantic distinction of missing values into "lost" and "do not care" values. The corresponding semantics is then used to define indiscernibility or similarity relation that is used to compare objects. Although it would be possible to adapt this distinction to the case of ordinal classification problems with monotonicity constraints, this would require additional knowledge about the nature of missing values in each particular problem. In this paper, instead of distinguishing a priori the semantics of the missing values, we propose to consider some desirable properties that dominance-based rough set approaches should have when handling missing values of any origins. Some of these properties involve, however, one of the two semantics, and this will be specified when characterizing particular approaches. These approaches, resulting from different definitions of the dominance relation, are denoted by DRSA-mv_i (where i stands for the version id). The considered properties are:

1. Property S (reflecting a specific kind of symmetry): DRSA-mv_i has property S iff $z \in D^{+}(y) \Leftrightarrow y \in D^{-}(z)$, for any $y, z \in U$.
2. Property R (reflecting reflexivity of dominance relation): DRSA-mv_i has property R iff yDy, for any $y \in U$.
3. Property T (reflecting transitivity of dominance relation): DRSA-mv_i has property T iff $wDy \wedge yDz \Rightarrow wDz$, for any $w, y, z \in U$.
4. Property P (reflecting precisiation of data): DRSA-mv_i has property P iff the lower approximations of unions $X_i^{\geq}$, $X_i^{\leq}$, $i = 1, \ldots n$, do not shrink when any missing attribute value is substituted by some not missing value.
5. Property RI (rough inclusion): DRSA-mv_i has property R iff $\underline{X} \subseteq X \subseteq \overline{X}$, for any $X \subseteq U$.
6. Property C (complementarity): DRSA-mv_i has property C iff $\underline{X} = U \setminus \overline{\neg X}$, for any $X \subseteq U$.
7. Property M_1 (monotonicity w.r.t. growing set of attributes): DRSA-mv_i has property M_1 iff the lower approximations of unions $X_i^{\geq}$, $X_i^{\leq}$, $i = 1, \ldots, n$, do not shrink when set P is extended by new attributes.
8. Property M_2 (monotonicity w.r.t. growing union of classes): DRSA-mv_i has property M_2 iff for any set $X \subseteq U$, the lower approximation of X does not shrink when this set is augmented by new objects.
9. Property M_3 (monotonicity w.r.t. super-union of classes): DRSA-mv_i has property M_3 iff given any two upward union of classes $X_i^{\geq}, X_j^{\geq}$, with $1 \leq i < j \leq n$, there is $\underline{X_i^{\geq}} \supseteq \underline{X_j^{\geq}}$, and, moreover, given any two downward union of classes $X_i^{\leq}, X_j^{\leq}$, with $1 \leq i < j \leq n$, there is $\underline{X_i^{\leq}} \subseteq \underline{X_j^{\leq}}$.
10. Property M_4 (monotonicity w.r.t. dominance relation): DRSA-mv_i has property M_4 iff for any upward and downward unions of classes $X_i^{\geq}$, $X_j^{\leq}$, with $i, j \in \{1, \ldots, n\}$, and for any $y, z \in U$ such that yDz, it is true that $\left((z \in \underline{X_i^{\geq}} \wedge y \in X_i^{\geq} \Rightarrow y \in \underline{X_i^{\geq}}) \text{ and } (y \in \underline{X_j^{\leq}} \wedge z \in X_j^{\leq} \Rightarrow z \in \underline{X_j^{\leq}})\right)$.

Note that there is a relationship between properties M_1, M_2, M_3, and M_4, concerning lower approximations of unions of classes, and properties (m1), (m2), (m3), and (m4), concerning object consistency measures used in VC-DRSA. However, when the dominance relation of VC-DRSA is redefined, this relationship is no longer one to one – for some $i \in \{1, \ldots, 4\}$, (mi) may be satisfied while M_i is not satisfied.

DRSA-mv_1 [9,10] considers dominance relation to be a directional statement where a subject is compared to a referent which cannot have missing values. Object y dominates referent z iff for each $q_i \in P$, $y \succeq_{q_i} z$ or $q_i(y) = *$; y is dominated by referent z iff for each $q_i \in P$, $z \succeq_{q_i} y$ or $q_i(y) = *$. Note that DRSA-mv_1 fails when all (or most of the) objects have a missing value. Moreover, dominance cones are defined only for objects without missing values. Thus, approximations of unions of classes do not contain objects with missing values.

DRSA-mv_2 was first proposed in [9,10], and then extended in [7] to handle imprecise evaluations on attributes and imprecise assignments to decision classes, both modeled by intervals. When considering missing values only, each $y \in U$ is assigned to a single decision class, and each missing attribute value corresponds to the interval spanning entire value set of this attribute. This results in the following definition of so-called *possible dominance relation* D: yDz iff for each $q_i \in P$, $y \succeq_{q_i} z$, or $q_i(y) = *$, or $q_i(z) = *$.

In DRSA-mv_3, object y dominates object z iff for each $q_i \in P$ such that $q_i(z) \neq *$, we have $q_i(y) \neq *$ and $y \succeq_{q_i} z$. Object y is dominated by object z iff for each $q_i \in P$ such that $q_i(z) \neq *$, we have $q_i(y) \neq *$ and $z \succeq_{q_i} y$.

DRSA-mv_4 (DRSA-mv_5) uses the *lower(upper)-end dominance relation* introduced in [7]. It boils down to treating each missing attribute value as the worst (best) value in the value set of this attribute. Then, the definition of dominance relation D is the same as in the case without missing values.

The properties of DRSA-mv_i, $i = 1, \ldots, 5$, are summarized in Table 1, where **T** and F denote presence and absence of a given property, respectively. Moreover, in case of two symbols $\cdot/\cdot$, the first (resp. second) one reflects only regular DRSA (resp. only VC-DRSA).

According to Table 1, DRSA-mv_1 is the less attractive due to lack of many important properties (like, e.g., *RI* and M_1). DRSA-mv_3 is dominated by DRSA-mv_4, and by DRSA-mv_5. However, the choice from among DRSA-mv_2, DRSA-mv_4, and DRSA-mv_5, depends on a particular application. Due to limited scope of this paper, we decided to apply in the experiment (Section 4) only DRSA-mv_2 since it has property P. This property guarantees that induced rules remain true when some of missing values become known. Thus, taking into account the semantics of missing values considered in [14,13], it can be said that DRSA-mv_2 treats missing values as "do not care" values. Moreover, the lack of property M_4, caused by the lack of property T, can be handled during induction of decision rules.

Table 1. Properties of DRSA-mv_i, $i = 1, \ldots, 5$

Property / Approach	DRSA-mv_1	DRSA-mv_2	DRSA-mv_3	DRSA-mv_4	DRSA-mv_5
S	F	**T**	F	**T**	**T**
R	F	**T**	**T**	**T**	**T**
T	**T**	F	**T**	**T**	**T**
P	**T**	**T**	F	F	F
RI	F	**T**	**T**	**T**	**T**
C	F	**T**	F/**T**	**T**	**T**
M_1	F	**T**	**T**	**T**	**T**
M_2	**T**	**T**	**T**	**T**	**T**
M_3	**T**/F	**T**/F	**T**/F	**T**/F	**T**/F
M_4	**T**	F	**T**	**T**	**T**

3 Induction and Application of Decision Rules

In VC-DRSA, decision rules are induced from lower approximations (3). In the experiment (Section 4), we employ to this end a two-fold adaptation of VC-DomLEM algorithm [5]. First, to handle the lack of property M_4 of DRSA-mv_2, we introduce cost-type *rule consistency measure* [5] $\widehat{\epsilon}_X : R_X \rightarrow [0,1]$, where R_X is the set of rules suggesting assignment to set X. Let us denote by $\Phi(r_X)$ and $\|\Phi(r_X)\|$, condition part of rule $r_X \in R_X$, and the set of objects covered by the rule, respectively. Then, measure $\widehat{\epsilon}_X$ is defined as:

$$\widehat{\epsilon}_X(r_X) = \frac{\left|\|\Phi(r_X)\| \cap \neg X\right|}{|\neg X|}. \tag{4}$$

Each rule r_X has to satisfy the same constraints on consistency as objects from $\underline{X}$ which serves as a basis for rule induction. Precisely, it is required that $\widehat{\epsilon}_X(r_X) \leq \theta_X$. Moreover, $\widehat{\epsilon}_X$ derives monotonicity properties from ϵ_X (2).

The second adaptation of VC-DomLEM concerns treatment of missing values in the context of elementary conditions. For each $q_i \in P$, only non-missing values can be used to create elementary conditions. Moreover, any elementary condition on q_i covers each $y \in U$ such that $q_i(y) = *$.

It is worth noting that the proposed adaptation of VC-DomLEM does not decrease its efficiency as measure $\widehat{\epsilon}_X$ has monotonicity property (m4). This fact can be used to reduce the search space of elementary conditions that can be created for each attribute [5].

Induced rules can be applied on a new set of objects. In such case, a rule covers every object that for each attribute $q_i \in P$ considered in the rule, either satisfies an elementary condition on q_i, or has missing value. Ambiguity of classification suggestions given by covering rules is resolved by the strategy described in [3].

4 Results of Computational Experiment and Discussion

The aim of the experiment was to compare the classifier that employs DRSA-mv_2 with other classifiers. The comparison was performed on 14 ordinal data sets characterized in Table 2 (most of the data sets were already used in [5]).

Data sets: `ERA`, `ESL`, `LEV`, and `SWD` come from [1]. Data sets: `denbosch` and `windsor` were taken from [6] and [16], respectively. Remaining data sets were taken from the UCI repository[1] and other public repositories (`fame`). For `windsor`, `cpu`, and `housing` data sets, decision attribute was discretized into four levels, containing equal number of objects. Each data set was transformed by uniform random introduction of a specified percentage of missing values, ranging from 5% to 50%.

Presented results were derived from 10-fold cross validation, repeated 5 times to get a better reproducibility. We compared the following classifiers. First one, VC-DomLEM-mv_2, employs VC-DomLEM adapted to induce rules in DRSA-mv_2 coupled with VC-DRSA (Section 3), implemented in jRS and jMAF

[1] See `http://www.ics.uci.edu/~mlearn/MLRepository.html`

Table 2. Characteristics of data sets

Data set	#Objects	#Condition attributes	#Classes
australian	690	14	2
balance	625	4	3
breast-cancer	286	7	2
breast-w	699	9	2
car	1296	6	4
cpu	209	6	4
fame	1328	10	5
denbosch	119	8	2
ERA	1000	4	9
ESL	488	4	9
housing	506	13	4
LEV	1000	4	5
SWD	1000	10	4
windsor	546	10	4

frameworks[2]. Another classifier, VC-DomLEM-smv, employs a non-invasive transformation of data (similar to the one presented in [2]) that substitutes missing values, followed by application of VC-DRSA, and induction of rules by the original VC-DomLEM [5]. In the transformation, each $q_i \in P$ with missing values is substituted by four attributes – q_i^1, q_i^2, for which an "increasing" monotonic relationship between evaluations of objects on a condition attribute and their class labels is assumed (i.e., the greater the attribute value, the better the class label is expected to be), and q_i^3, q_i^4, for which a "decreasing" monotonic relationship between evaluations of objects on a condition attribute and their class labels is assumed (i.e., the smaller the attribute value, the better the class label is expected to be). Then, for each $y \in U : q_i(y) \neq *$, evaluation $q_i^j(y)$, $j = 1, \ldots, 4$, is taken to be equal to $q_i(y)$, and for each $z \in U : q_i(z) = *$, evaluations $q_i^1(y), q_i^4(y)$ are taken to be equal to the maximum value in V_{q_i}, while evaluations $q_i^2(y), q_i^3(y)$ are taken to be equal to the minimum value in V_{q_i}. Moreover, we considered two other ordinal classifiers that preserve monotonicity constraints: Ordinal Learning Model (OLM), and Ordinal Stochastic Dominance Learner (OSDL). As they cannot handle missing values directly, we substituted missing values by means or modes. In general, it is not always the case that ordinal classifiers have better predictive accuracy than non-ordinal ones (as the former are biased by monotonicity constraints). Therefore, we also used some non-ordinal classifiers: Naive Bayes, Support Vector Machine (SVM) with linear kernel and default complexity. Additionally, Ripper and C4.5 classifiers were used with default settings. Each of the non-ordinal classifiers was able to handle missing values directly.

Table 3 contains values of Mean Absolute Error (MAE) and its standard deviation for data sets with 5% of missing values. For each data set we calculated ranks of MAE (presented in brackets; the smaller the rank, the better). Last row of Table 3 shows average ranks obtained by the classifiers.

We applied Friedman test [8] to compare the classifiers. The null-hypothesis was that all of them perform equally well. The result of test (p-value below 0.0001) and differences in average ranks allow us to conclude that there is a

[2] See `http://www.cs.put.poznan.pl/jblaszczynski/Site/jRS.html`

Table 3. MAE resulting from repeated 10-fold cross validation for 5% of missing values

Data set	VC-DomLEM -mv_2	VC-DomLEM -smv	Naive Bayes	SVM	Ripper	C4.5	OLM	OSDL
australian	0.1962 (4.5) ±0.001966	0.1962 (4.5) ±0.001966	0.2252 (6) ±0.002688	0.1704 (3) ±0.003735	0.1632 (2) ±0.004815	0.151 (1) ±0.004798	0.3783 (8) ±0.00698	0.3609 (7) ±0.00485
balance	0.2698 (3.5) ±0.007409	0.2698 (3.5) ±0.007409	0.1674 (1) ±0.007409	0.2205 (2) ±0.01468	0.3408 (5) ±0.01127	0.3558 (6) ±0.01097	0.6966 (7) ±0.02344	0.7427 (8) ±0.005935
breast-cancer	0.2552 (1.5) ±0.006255	0.3028 (7) ±0.01119	0.2664 (3) ±0.009486	0.3294 (8) ±0.01334	0.2972 (6) ±0.00383	0.2552 (1.5) ±0.007334	0.2944 (5) ±0.001399	0.2713 (4) ±0.006485
breast-w	0.05179 (5) ±0.003186	0.05293 (7) ±0.002023	0.04006 (2) ±0	0.03262 (1) ±0.001668	0.04578 (4) ±0.003731	0.05207 (6) ±0.002803	0.1854 (8) ±0.007919	0.04177 (3) ±0.001071
car	0.1426 (2) ±0.002036	0.1685 (4) ±0.006594	0.1997 (7) ±0.006716	0.1437 (3) ±0.004065	0.2514 (8) ±0.006591	0.1701 (5) ±0.002105	0.1801 (6) ±0.005909	0.1418 (1) ±0.006977
cpu	0.1158 (1.5) ±0.01108	0.1158 (1.5) ±0.01108	0.1837 (4) ±0.02021	0.4459 (7) ±0.01613	0.2727 (5) ±0.04312	0.1349 (3) ±0.01022	0.5971 (8) ±0.02484	0.3885 (6) ±0.01108
fame	0.4178 (3) ±0.00924	0.4849 (6) ±0.003898	0.4821 (5) ±0.003414	0.3696 (1) ±0.003064	0.4318 (4) ±0.01078	0.4054 (2) ±0.003551	1.707 (7) ±0.01062	1.741 (8) ±0.004354
denbosch	0.1513 (3) ±0.01406	0.1714 (4) ±0.01465	0.1361 (1) ±0.01946	0.2084 (7) ±0.01794	0.1815 (5) ±0.02636	0.1849 (6) ±0.01188	0.3126 (8) ±0.008234	0.1479 (2) ±0.004117
ERA	1.367 (2) ±0.01209	1.482 (5) ±0.005528	1.409 (3) ±0.006693	1.365 (1) ±0.01174	1.66 (7) ±0.006841	1.412 (4) ±0.01976	1.690 (8) ±0.01695	1.594 (6) ±0.01465
ESL	0.4484 (3.5) ±0.009904	0.4484 (3.5) ±0.009904	0.3754 (1) ±0.008636	0.4639 (5) ±0.002780	0.5008 (6) ±0.01663	0.3939 (2) ±0.007601	1.084 (8) ±0.01404	0.566 (7) ±0.008923
housing	0.3585 (1) ±0.01177	0.4194 (4) ±0.01826	0.5091 (6) ±0.002015	0.3708 (2) ±0.006324	0.4486 (5) ±0.02781	0.4047 (3) ±0.01337	0.9166 (7) ±0.01307	1.062 (8) ±0.009952
LEV	0.538 (6) ±0.001414	0.5568 (7) ±0.005269	0.4834 (4) ±0.002653	0.4762 (3) ±0.004622	0.4586 (2) ±0.00826	0.4434 (1) ±0.00857	0.707 (8) ±0.002366	0.5116 (5) ±0.005276
SWD	0.4702 (3) ±0.01007	0.502 (6) ±0.007376	0.4864 (5) ±0.002417	0.4506 (1) ±0.003382	0.476 (4) ±0.007563	0.458 (2) ±0.007376	0.7708 (8) ±0.01165	0.506 (7) ±0.00827
windsor	0.5473 (1) ±0.005838	0.6714 (5) ±0.01261	0.5579 (2) ±0.003397	0.6048 (4) ±0.01244	0.7278 (8) ±0.02491	0.6912 (7) ±0.02045	0.6894 (6) ±0.008715	0.5916 (3) ±0.004486
average rank	2.89	4.86	3.57	3.43	5.07	3.54	7.29	5.36

significant difference between the classifiers. VC-DomLEM-mv_2 is the best one, followed by SVM, C4.5, and Naive Bayes. Other ordinal classifiers and Ripper perform significantly worse. However, VC-DomLEM-smv outperforms OLM, OSDL, and Ripper.

In Figure 1, we present average ranks of MAE for different percentages of missing values (for each percentage, the differences between average ranks were significant). The best methods to handle missing values are: VC-DomLEM-mv_2, SVM, Naive Bayes, and – for lower amounts of missing values – also C4.5. VC-DomLEM-mv_2 and SVM obtain rather stable ranks regardless of the amount of missing values. VC-DomLEM-mv_2 performs best for smaller amounts of missing values. SVM is better for higher amounts. However, Naive Bayes is the best classifier when the amount of introduced missing values is 20% or higher. OLM, OSDL, and Ripper are clearly worse than the other compared methods.

We recommend to use VC-DomLEM-mv_2 when dealing with ordinal classification problems with monotonicity constraints that include low amount of missing values. For problems characterized by larger amount of missing values ($\geq 20\%$), one could consider Naive Bayes or SVM.

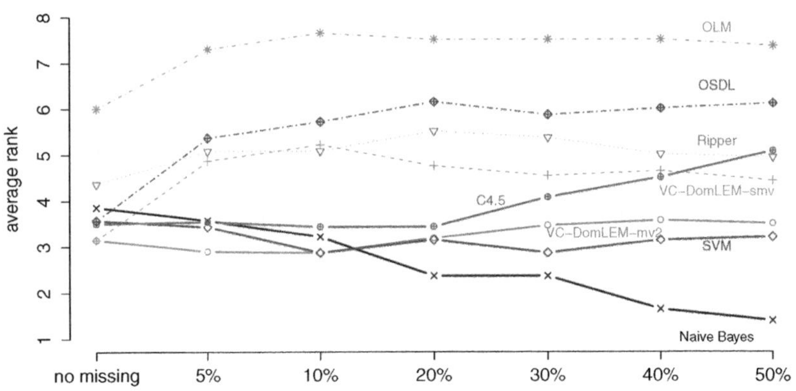

Fig. 1. Avg. rank of MAE for different percentages of missing values

Acknowledgment. The 1st and 3rd author acknowledge financial support from the Poznań University of Technology, grant no. 91-516/DS-MLODA KADRA. The 2nd author acknowledges financial support from the Polish National Science Centre, grant no. N N519 441939.

References

1. Ben-David, A.: Monotonicity maintenance in information-theoretic machine learning algorithms. Machine Learning 19(1), 29–43 (1995)
2. Błaszczyński, J., Greco, S., Słowiński, R.: Inductive discovery of laws using monotonic rules. Engineering Applications of Artif. Intelligence 25, 284–294 (2012)
3. Błaszczyński, J., Greco, S., Słowiński, R.: Multi-criteria classification – a new scheme for application of dominance-based decision rules. European Journal of Operational Research 181(3), 1030–1044 (2007)
4. Błaszczyński, J., Greco, S., Słowiński, R., Szeląg, M.: Monotonic variable consistency rough set approaches. Int. J. of Approx. Reasoning 50(7), 979–999 (2009)
5. Błaszczyński, J., Słowiński, R., Szeląg, M.: Sequential covering rule induction algorithm for variable consistency rough set approaches. Information Sciences 181, 987–1002 (2011)
6. Daniels, H., Kamp, B.: Applications of mlp networks to bond rating and house pricing. Neural Computation and Applications 8, 226–234 (1999)
7. Dembczyński, K., Greco, S., Słowiński, R.: Rough set approach to multiple criteria classification with imprecise evaluations and assignments. European Journal of Operational Research 198(2), 626–636 (2009)
8. Demsar, J.: Statistical comparisons of classifiers over multiple data sets. Journal of Machine Learning Research 7, 1–30 (2006)
9. Greco, S., Matarazzo, B., Słowiński, R.: Handling Missing Values in Rough Set Analysis of Multi-attribute and Multi-criteria Decision Problems. In: Zhong, N., Skowron, A., Ohsuga, S. (eds.) RSFDGrC 1999. LNCS (LNAI), vol. 1711, pp. 146–157. Springer, Heidelberg (1999)

10. Greco, S., Matarazzo, B., Słowiński, R.: Dealing with missing data in rough set analysis of multi-attribute and multi-criteria decision problems. In: Zanakis, S., et al. (eds.) Decision Making: Recent Developments and Worldwide Applications, pp. 295–316. Kluwer, Dordrecht (2000)
11. Greco, S., Matarazzo, B., Słowiński, R.: Rough sets theory for multicriteria decision analysis. European Journal of Operational Research 129(1), 1–47 (2001)
12. Greco, S., Matarazzo, B., Słowiński, R.: Granular computing for reasoning about ordered data: the dominance-based rough set approach. In: Pedrycz, W., et al. (eds.) Handbook of Granular Computing, ch.15, Wiley, Chichester (2008)
13. Grzymała-Busse, J.W., Hu, M.: A Comparison of Several Approaches to Missing Attribute Values in Data Mining. In: Ziarko, W., Yao, Y. (eds.) RSCTC 2000. LNCS (LNAI), vol. 2005, pp. 378–385. Springer, Heidelberg (2001)
14. Grzymała-Busse, J.W.: Mining Incomplete Data—A Rough Set Approach. In: Yao, J., Ramanna, S., Wang, G., Suraj, Z. (eds.) RSKT 2011. LNCS, vol. 6954, pp. 1–7. Springer, Heidelberg (2011)
15. Hastie, T., Tibshirani, R., Friedman, J.: The Elements of Statistical Learning. Springer, Berlin (2009)
16. Koop, G.: Analysis of Economic Data. Wiley, Chichester (2000)
17. Kryszkiewicz, M.: Rough set approach to incomplete information systems. Information Sciences 112, 39–49 (1998)
18. Pawlak, Z.: Rough Sets. Theoretical Aspects of Reasoning about. Data. Kluwer Academic Publishers, Dordrecht (1991)
19. Słowiński, R., Greco, S., Matarazzo, B.: Rough sets in decision making. In: Meyers, R.A. (ed.) Encyclopedia of Complexity and Systems Science, pp. 7753–7786. Springer, New York (2009)
20. Stefanowski, J., Tsoukias, A.: Incomplete information tables and rough classification. Computational Intelligence 17, 545–566 (2002)

Interval Valued Fuzzy Rough Set Model on Two Different Universes and Its Application

Hai-Long Yang

College of Mathematics and Information Science,
Shaanxi Normal University, Xi'an 710062, China
yanghailong@snnu.edu.cn

Abstract. Based on the interval valued fuzzy compatible relation, the interval valued fuzzy rough set model on two different universes is presented. Some properties of the interval valued fuzzy rough set model are discussed. Finally, an example is applied to illustrate the application of the interval valued fuzzy rough set model presented in this paper.

Keywords: Interval valued fuzzy sets, Rough sets, Interval valued fuzzy compatible relations, Interval valued fuzzy rough sets.

1 Introduction

Rough set theory developed by Pawlak [1], is a formal tool for representing and processing information in database. In the classical Pawlak rough set model [1], an equivalence relation is a key and primitive notion in the construction of an approximation space. This equivalence relation, however, seems to be a very restrictive condition that may limit the application domain of the rough set model. Many proposals have been put forward for generalizing and interpreting rough sets. Dubois and Prade [2] first introduced the concepts of rough fuzzy set and fuzzy rough set by combining fuzzy sets [3] and rough sets. Majority of researches about rough sets are based on one universe [4]. Nowadays, more and more efforts have been made [5-12] based on two different universes.

Recently, in [4], Sun and Ma studied the fuzzy rough set on two different universes based on a fuzzy compatible relation. Sun and Ma illustrated the application of the fuzzy rough set model on two different universes in clinical diagnosis systems [13]. However, in cases where the degrees of membership are interval values rather that single values, we need to use the interval valued fuzzy rough set model to make decision. The purpose of the present paper is to establish the interval valued fuzzy rough set model based on the interval valued fuzzy compatible relation and apply this model to clinical diagnosis systems.

2 Preliminaries

2.1 Interval Valued Fuzzy Sets

Definition 2.1 [14]. An interval valued fuzzy set A on a universe U is a mapping $A : U \longrightarrow Int([0,1])$, where $Int([0,1])$ stands for the set of all closed subintervals of $[0,1]$, the set of all interval valued fuzzy sets on U is denoted by $IVF(U)$.

J.T. Yao et al. (Eds.): RSCTC 2012, LNAI 7413, pp. 66–72, 2012.

Suppose that $A \in IVF(U)$, $\forall x \in U$, $A(x) = [\mu_A^-(x), \mu_A^+(x)]$ is called the degree of membership an element x to A. $\mu_A^-(x)$ and $\mu_A^+(x)$ are referred to as the lower and upper degrees of membership x to A where $0 \leq \mu_A^-(x) \leq \mu_A^+(x) \leq 1$. We can also denote A by $A = \{[x, \mu_A^-(x), \mu_A^+(x)] \mid x \in U\}$.

Definition 2.2 [15]. An interval valued fuzzy subset R of $U \times V$ is called an interval valued fuzzy relation from U to V, We call R an interval valued fuzzy relation on U if $U = V$. We denote the family of all interval valued fuzzy relations from U to V by

$IVF(U \times V)$.

$\forall R, S \in IVF(U \times V)$,

$R \subseteq S$ iff $\mu_R^-(x,y) \leq \mu_S^-(x,y)$ and $\mu_R^+(x,y) \leq \mu_S^+(x,y)$ for all $x, y \in U$;

$R \supseteq S$ iff $S \subseteq R$;

$R = S$ iff $R \subseteq S$ and $S \subseteq R$;

$R \cup S = \{(x,y), [\max(\mu_R^-(x,y), \mu_S^-(x,y)), \max(\mu_R^+(x,y), \mu_S^+(x,y))] \mid (x,y) \in U \times V\}$;

$R \cap S = \{(x,y), [\min(\mu_R^-(x,y), \mu_S^-(x,y)), \min(\mu_R^+(x,y), \mu_S^+(x,y))] \mid (x,y) \in U \times V\}$.

Definition 2.3 [15]. Let $R \in IVF(U \times U)$, then

(1) R is reflexive, if $\mu_R^-(u,u) = 1$ and $\mu_R^+(u,u) = 1$ for any $u \in U$;
(2) R is symmetric, if $\mu_R^-(u_1,u_2) = \mu_R^-(u_2,u_1)$ and $\mu_R^+(u_1,u_2) = \mu_R^+(u_2,u_1)$ for any $u_1, u_2 \in U$;
(3) R is transitive, if $\bigvee_{y \in U}(R(x,y) \wedge R(y,z)) \leq R(x,z)$ for any $x, y, z \in U$.

$\forall R \in IVF(U \times U)$, if R is reflexive and symmetric, then R is called an interval valued fuzzy similarity relation on U; if R is reflexive, symmetric, and transitive, then R is called an interval valued fuzzy equivalence relation on U.

2.2 Rough Sets

Let U be a non-empty finite universe, R be an equivalence relation on U. We use U/R to denote the family of all equivalence classes of R (or classifications of U), and $[x]_R$ to denote an equivalence class of R containing the element $x \in U$. The pair (U, R) is called an approximation space. For any $X \subseteq U$, we can define the lower and upper approximations of X [1] as follows:

$$\underline{R}(X) = \{x \in U \mid [x]_R \subseteq X\},\ \overline{R}(X) = \{x \in U \mid [x]_R \cap X \neq \emptyset\}.$$

The pair $(\underline{R}X, \overline{R}X)$ is referred to as the rough set of X. The rough set $(\underline{R}X, \overline{R}X)$ gives rise to a description of X under the present knowledge, i.e., the classification of U.

Furthermore, the positive region, negative region, and boundary region of X about the approximation space (U, R) are defined as follows, respectively:

$\mathrm{pos}(X) = \underline{R}(X)$, $\mathrm{neg}(X) = \sim \overline{R}(X)$, $\mathrm{bn}(X) = \overline{R}(X) - \underline{R}(X)$,

where $\sim X$ stands for the complementation of the set X, i.e., $U - X$.

3 Interval Valued Fuzzy Rough Set Model on Two Different Universes

First we present some basic notions.

Definition 3.1. Let U, V be two non-empty finite universes. R is an interval valued fuzzy relation from U to V, $\forall \alpha, \beta \in (0,1]$, we define the interval valued fuzzy compatible relation $R_{(\alpha,\beta)}$ from U to V as follows:

$$R_{(\alpha,\beta)}(u) = \{v \in V \mid \mu_R^-(u,v) \geq \alpha \text{ and } \mu_R^+(u,v) \geq \beta\}, \tag{3.1}$$

$$\text{i.e., } R_{(\alpha,\beta)} = \{(u,v) \in U \times V \mid v \in R_{(\alpha,\beta)}(u)\}. \tag{3.2}$$

Obviously, Definition 3.1 is an interval valued fuzzy extension of the fuzzy compatible relation (see Definition 3.3 in [4]). α and β can be viewed as the given least thresholds on the lower level values and the upper level values, respectively. For real-life applications of decision making, the thresholds α and β are usually in advance chosen by decision makers.

Definition 3.2. Let U, V be two non-empty finite universes, $R_{(\alpha,\beta)}$ be an interval valued fuzzy compatible relation from U to V, where $\alpha, \beta \in (0,1]$. For any X ($X \subseteq V$), we define the lower and upper approximations of X w.r.t. $R_{(\alpha,\beta)}$ as:

$$\underline{R_{(\alpha,\beta)}}(X) = \{u \in U \mid R_{(\alpha,\beta)}(u) \subseteq X\}, \tag{3.3}$$

$$\overline{R_{(\alpha,\beta)}}(X) = \{u \in U \mid R_{(\alpha,\beta)}(u) \cap X \neq \emptyset\}. \tag{3.4}$$

The pair $(\underline{R_{(\alpha,\beta)}}(X), \overline{R_{(\alpha,\beta)}}(X))$ is called the interval valued fuzzy rough set of X based on the interval valued fuzzy compatible relation $R_{(\alpha,\beta)}$.

Furthermore, the positive region $\mathrm{pos}_{R_{(\alpha,\beta)}}(X)$, negative region $\mathrm{neg}_{R_{(\alpha,\beta)}}(X)$, and boundary region $\mathrm{bn}_{R_{(\alpha,\beta)}}(X)$ of X w.r.t. $R_{(\alpha,\beta)}$ are define as: $\mathrm{pos}_{R_{(\alpha,\beta)}}(X) = \underline{R_{(\alpha,\beta)}}(X)$, $\mathrm{neg}_{R_{(\alpha,\beta)}}(X) = U - \overline{R_{(\alpha,\beta)}}(X)$, $\mathrm{bn}_{R_{(\alpha,\beta)}}(X) = \overline{R_{(\alpha,\beta)}}(X) - \underline{R_{(\alpha,\beta)}}(X)$.

Remark 3.1. If R is a fuzzy relation from U to V and $\beta = \alpha$, then the interval valued fuzzy rough set is degenerated to be the fuzzy rough set in [4].

Remark 3.2. If $U = V$, then $R_{(\alpha,\beta)}$ is a crisp binary relation on U. Therefore, $R_{(\alpha,\beta)}$ will be a crisp similarity relation on U if R is an interval valued fuzzy similarity relation U, and $R_{(\alpha,\beta)}$ will be an equivalence relation on U if R is an interval valued fuzzy equivalence relation on U. Thus, if R is an interval valued fuzzy equivalence relation on U, then the interval valued fuzzy rough set will be degenerated to be the classical Pawlak rough set.

Theorem 3.1. Let U, V be two non-empty finite universes and $R_{(\alpha,\beta)}$ be an interval valued fuzzy compatible relation from U to V, where $\alpha, \beta \in (0,1]$. For any $X, Y \subseteq V$. Then the lower approximation operator $\underline{R_{(\alpha,\beta)}}$ and the upper approximation operator $\overline{R_{(\alpha,\beta)}}$ have the following properties:

(1) $\underline{R_{(\alpha,\beta)}}(X) \subseteq \overline{R_{(\alpha,\beta)}}(X)$, $\overline{R_{(\alpha,\beta)}}(\emptyset) = \emptyset$, $\underline{R_{(\alpha,\beta)}}(V) = U$;

(2) $\underline{R_{(\alpha,\beta)}}(X \cap Y) = \underline{R_{(\alpha,\beta)}}(X) \cap \underline{R_{(\alpha,\beta)}}(Y)$, $\overline{R_{(\alpha,\beta)}}(X \cup Y) = \overline{R_{(\alpha,\beta)}}(X) \cup \overline{R_{(\alpha,\beta)}}(Y)$;

(3) $\underline{R_{(\alpha,\beta)}}(X \cup Y) \supseteq \underline{R_{(\alpha,\beta)}}(X) \cup \underline{R_{(\alpha,\beta)}}(Y)$, $\overline{R_{(\alpha,\beta)}}(X \cap Y) \subseteq \overline{R_{(\alpha,\beta)}}(X) \cap \overline{R_{(\alpha,\beta)}}(Y)$;

(4) If $X \subseteq Y$, then $\underline{R_{(\alpha,\beta)}}(X) \subseteq \underline{R_{(\alpha,\beta)}}(Y)$ and $\overline{R_{(\alpha,\beta)}}(X) \subseteq \overline{R_{(\alpha,\beta)}}(Y)$;

(5) $\underline{R_{(\alpha,\beta)}}(X) = \sim \overline{R_{(\alpha,\beta)}}(\sim X)$, $\overline{R_{(\alpha,\beta)}}(X) = \sim \underline{R_{(\alpha,\beta)}}(\sim X)$.

Proof. The proof is straightforward from Definition 3.2.

Remark 3.3. In general, $\underline{R_{(\alpha,\beta)}}(\emptyset) \neq \emptyset$ and $\overline{R_{(\alpha,\beta)}}(V) \neq U$. For example, let $U = \{u_1, u_2\}$, $V = \{v_1, v_2\}$, the interval valued fuzzy relation R from U to V is defined as $R = \{[(u_1, v_1), 0.2, 0.6], [(u_1, v_2), 0.4, 0.5], [(u_2, v_1), 0.5, 0.8], [(u_2, v_2), 0.1, 0.1]\}$. Take $\alpha = 0.5$ and $\beta = 0.7$, then we have $\underline{R_{(\alpha,\beta)}}(\emptyset) = \{u_1\} \neq \emptyset$ and $\overline{R_{(\alpha,\beta)}}(V) = \{u_2\} \neq U$.

Theorem 3.2. Let U, V be two non-empty finite universes, $R_{(\alpha_1,\beta_1)}$ and $R_{(\alpha_2,\beta_2)}$ be two interval valued fuzzy compatible relations from U to V, where $\alpha_1, \beta_1, \alpha_2, \beta_2 \in (0, 1]$. If $\alpha_1 \leq \alpha_2$ and $\beta_1 \leq \beta_2$, then we have

(1) $\underline{R_{(\alpha_1,\beta_1)}}(X) \subseteq \underline{R_{(\alpha_2,\beta_2)}}(X)$;

(2) $\overline{R_{(\alpha_2,\beta_2)}}(X) \subseteq \overline{R_{(\alpha_1,\beta_1)}}(X)$.

Proof. The proof is analogous to Theorem 3.2 in [8].

Theorem 3.3. Let R, S be two interval valued fuzzy relations from U to V. For any $X \subseteq V$, $\alpha, \beta \in (0, 1]$. If $R \subseteq S$, then

(1) $\underline{S_{(\alpha,\beta)}}(X) \subseteq \underline{R_{(\alpha,\beta)}}(X)$;

(2) $\overline{R_{(\alpha,\beta)}}(X) \subseteq \overline{S_{(\alpha,\beta)}}(X)$.

Proof. The proof is analogous to Theorem 3.4 in [8].

Theorem 3.4. Let U, V be two non-empty finite universes, where $U = \{u_1, u_2, \cdots, u_n\}$, and R be a interval valued fuzzy relation from U to V. If the interval valued fuzzy compatible relation $R_{(\alpha,\beta)}$ satisfies the included relation of $R_{(\alpha,\beta)}(u_1) \subseteq R_{(\alpha,\beta)}(u_2) \subseteq \cdots \subseteq R_{(\alpha,\beta)}(u_n)$, where $\alpha, \beta \in (0, 1]$. Then for any $X \subseteq V$, we have the following:

(1) If $u_i \in \underline{R_{(\alpha,\beta)}}(X)$, then $u_1, u_2, \cdots, u_{i-1} \in \underline{R_{(\alpha,\beta)}}(X)$ $(i \in \{2, 3, \cdots, n\})$;

(2) If $u_i \in \overline{R_{(\alpha,\beta)}}(X)$, then $u_2, u_3, \cdots, u_{i+1} \in \overline{R_{(\alpha,\beta)}}(X)$ $(i \in \{1, 2, \cdots, n-1\})$.

Proof. The proof is similar to Theorem 3.5 in [4].

Definition 3.3. Let U, V be two non-empty finite universes, $R_{(\alpha,\beta)}$ be an interval valued fuzzy compatible relation from U to V, where $\alpha, \beta \in (0, 1]$. For any $X \subseteq V$ $(X \neq \emptyset)$, the approximate precision $\rho_{R_{(\alpha,\beta)}}(X)$ of X w.r.t. $R_{(\alpha,\beta)}$ is defined as:

$$\rho_{R_{(\alpha,\beta)}}(X) = \frac{|\underline{R_{(\alpha,\beta)}}(X)|}{|\overline{R_{(\alpha,\beta)}}(X)|}, \tag{3.5}$$

where $|X|$ denotes the cardinality of the set X. Let $\mu_{R_{(\alpha,\beta)}}(X) = 1 - \rho_{R_{(\alpha,\beta)}}(X)$, and $\mu_{R_{(\alpha,\beta)}}(X)$ is called the rough degree of X w.r.t. $R_{(\alpha,\beta)}$.

Theorem 3.5. Let U, V be two non-empty finite universes and $R_{(\alpha,\beta)}$ be an interval valued fuzzy compatible relation from U to V, where $\alpha, \beta \in (0,1]$. For any $X, Y \subseteq V$ $(X \neq \emptyset, Y \neq \emptyset)$. Then the rough degree and the approximate precision of the sets X, Y, $X \cup Y$, and $X \cap Y$ are satisfied the following relations:

(1) $\mu_{R_{(\alpha,\beta)}}(X \cup Y)|\overline{R_{(\alpha,\beta)}}(X) \cup \overline{R_{(\alpha,\beta)}}(Y)| \leq \mu_{R_{(\alpha,\beta)}}(X)|\overline{R_{(\alpha,\beta)}}(X)| + \mu_{R_{(\alpha,\beta)}}(Y)$ $|\overline{R_{(\alpha,\beta)}}(Y)| - \mu_{R_{(\alpha,\beta)}}(X \cap Y)|\overline{R_{(\alpha,\beta)}}(X) \cap \overline{R_{(\alpha,\beta)}}(Y)|$;
(2) $\rho_{R_{(\alpha,\beta)}}(X \cup Y)|\overline{R_{(\alpha,\beta)}}(X) \cup \overline{R_{(\alpha,\beta)}}(Y)| \geq \rho_{R_{(\alpha,\beta)}}(X)|\overline{R_{(\alpha,\beta)}}(X)| + \rho_{R_{(\alpha,\beta)}}(Y)$ $|\overline{R_{(\alpha,\beta)}}(Y)| - \rho_{R_{(\alpha,\beta)}}(X \cap Y)|\overline{R_{(\alpha,\beta)}}(X) \cap \overline{R_{(\alpha,\beta)}}(Y)|$.

Proof. The proof is analogous to Theorem 3.3 in [4].

Remark 3.4. The approximate precision and the rough degree of X may be used to help decision-makers to make decision.

4 The Application of the Interval Valued Fuzzy Rough Set Model on Two Different Universes

Let U and V denote the set of sufferers and the set of symptoms, respectively. The thresholds α and β are in advance chosen by decision-makers. In general, one disease always has several basic symptoms. For any subset X of V, X denotes a certain disease which has basic symptoms $\{v_i\}$ $(v_i \in V)$. Given a certain sufferer u, if he belongs to the set $\underline{R_{(\alpha,\beta)}}(X)$ and $R_{(\alpha,\beta)}(u) \neq \emptyset$, i.e., he must suffer the disease X, then the sufferer need remedy immediately. If he belongs to the set $\overline{R_{(\alpha,\beta)}}(X) - \underline{R_{(\alpha,\beta)}}(X)$, i.e., the set $\mathrm{bn}_{R_{(\alpha,\beta)}}(X)$, then he may suffer the disease X or not, and he will be on the second choice by the doctor since he is not diagnosed according to these symptoms. If he belongs to the set $\mathrm{neg}_{R_{(\alpha,\beta)}}(X)$, then he does not suffer the disease X and he does not need the treatment.

Example 4.1. Let $U = \{u_1, u_2, u_3, u_4\}$ be the set of sufferers. $V = \{v_1, v_2, v_3, v_4\}$ is the set of symptoms. Suppose that the degrees of membership for every sufferer $u_i \in U$ w.r.t. the symptom $v_j \in V$ are given in the following table:

Table 1. Tabular representation of $R \in IVF(U \times V)$

$U \diagdown V$	v_1	v_2	v_3	v_4
u_1	$[0.74, 0.90]$	$[0.25, 0.57]$	$[0.17, 0.36]$	$[1, 1]$
u_2	$[0.62, 0.80]$	$[0.45, 0.68]$	$[0.87, 0.95]$	$[0.45, 0.55]$
u_3	$[0.53, 0.66]$	$[1, 1]$	$[0.24, 0.77]$	$[0.18, 0.75]$
u_4	$[0.12, 0.29]$	$[0.77, 0.90]$	$[0.43, 0.49]$	$[0.69, 1]$

Let $X = \{v_1, v_2\} \subseteq V$ denote a certain disease.

Case 1. Take $\alpha = 0.5$ and $\beta = 0.8$. By Definition 3.1, we can obtain

$R_{(0.5,0.8)}(u_1) = \{v_1, v_4\}$, $R_{(0.5,0.8)}(u_2) = \{v_1, v_3\}$, $R_{(0.5,0.8)}(u_3) = \{v_2\}$, $R_{(0.5,0.8)}(u_4) = \{v_2, v_4\}$.

Then, we can calculate the lower approximation, the upper approximation, and the negative region of X as follows, respectively:

$$\underline{R_{(0.5,0.8)}}(X) = \{u_3\},\ \overline{R_{(0.5,0.8)}}(X) = U,\ \mathrm{neg}_{R_{(0.5,0.8)}}(X) = \emptyset.$$

By Definition 3.3, we have $\rho_{R_{(0.5,0.8)}}(X) = \frac{1}{4}$, $\mu_{R_{(0.5,0.8)}}(X) = \frac{3}{4}$.

Based on the above analysis, we can obtain the following conclusions:

(1) The sufferer u_3 must suffer the disease X and he needs the treatment immediately.
(2) We do not assure if the sufferers u_1, u_2, and u_4 suffer the disease X according to these symptoms. And the decision of the doctor will be on second choice for them.
(3) None of the sufferers does not suffer the disease X.

Case 2. Take $\alpha = 0.7$ and $\beta = 0.9$. By Definition 3.1, we can obtain

$R_{(0.7,0.9)}(u_1) = \{v_1, v_4\}$, $R_{(0.7,0.9)}(u_2) = \{v_3\}$, $R_{(0.7,0.9)}(u_3) = \{v_2\}$, $R_{(0.7,0.9)}(u_4) = \{v_2\}$.

Then, we have

$$\underline{R_{(0.7,0.9)}}(X) = \{u_3, u_4\},\ \overline{R_{(0.7,0.9)}}(X) = \{u_1, u_3, u_4\},\ \mathrm{neg}_{R_{(0.7,0.9)}}(X) = \{u_2\}.$$

By Definition 3.3, $\rho_{R_{(0.7,0.9)}}(X) = \frac{2}{3}$, $\mu_{R_{(0.7,0.9)}}(X) = \frac{1}{3}$.

Based on the above analysis, we can obtain the following conclusions:

(1) The sufferers u_3 and u_4 must suffer the disease X and they need the treatment immediately.
(2) We do not assure if the sufferer u_1 suffers the disease X according to these symptoms. And the decision of the doctor will be on second choice for him.
(3) The sufferer u_2 does not suffer the disease X.

Remark 4.1. By Example 4.1, the method above is actually an adjustable method. Many decision making problems are essentially humanistic and thus subjective in nature; hence for decision making there actually does not exist a unique or uniform criterion. Thus, the adjustable feature makes the model more appropriate for many real world applications. Besides, the model presented in this paper also permits to control the risk of the misdiagnose in practice.

References

1. Pawlak, Z.: Rough sets. International Journal of Computer and Information Sciences 11, 341–356 (1982)
2. Dubois, D., Prade, H.: Rough fuzzy sets and fuzzy rough sets. International Journal of General System 17(2-3), 191–209 (1990)
3. Zadeh, L.A.: Fuzzy sets. Information and Control 8, 338–353 (1965)
4. Sun, B.Z., Ma, W.M.: Fuzzy rough set model on two different universes and its application. Applied Mathematical Modelling 35, 1798–1809 (2011)
5. Li, T.-J., Zhang, W.-X.: Rough fuzzy approximations on two universes of discourse. Information Sciences 178(3), 892–906 (2008)
6. Pei, D.W., Xu, Z.-B.: Rough Set Models on Two universes. International Journal of General System 33(5), 569–581 (2004)
7. Yan, R., Zheng, J., Liu, J., Zhai, Y.: Research on the model of rough set over dual-universes. Knowledge-Based Systems 23(8), 817–822 (2010)
8. Yang, H.-L., Li, S.-G., Guo, Z.-L., Ma, C.-H.: Transformation of bipolar fuzzy rough set models. Knowledge-Based Systems 27, 60–68 (2012)
9. Yao, Y.Y., Wong, S.K.M., Wang, L.S.: A non-numeric approach to uncertain reasoning. International Journal of General System 23, 343–359 (1995)
10. Liu, G.L.: Rough set theory based on two universal sets and its applications. Knowledge-Based Systems 23(2), 110–111 (2010)
11. Zhang, W.X., Wu, W.Z.: The rough set model based on the random set (I). Journal of Xi'an Jiaotong University 34(12), 15–47 (1998) (in Chinese)
12. Wu, Z.J., Qin, K.Y., Qiao, Q.X.: L-fuzzy rough sets on double domains. Computer Engineering and Applications 43(5), 10–11 (2007) (in Chinese)
13. Tsumcto, S.: Automated extraction of medical expert system rules from clinical databases based on rough set theory. Information Sciences 112, 67–84 (1998)
14. Gorzaaczany, M.B.: A method of inference in approximate reasoning based on interval valued fuzzy sets. Fuzzy Sets and Systems 21, 1–17 (1987)
15. Bustince, H., Burillo, P.: Mathematical analysis of interval-valued fuzzy relations: application to approximate reasoning. Fuzzy Sets and Systems 113, 205–219 (2000)

Maximal Characteristic Sets and Neighborhoods Approach to Incomplete Information Systems

Chien-Chung Chan

Department of Computer Science, University of Akron,
Akron, OH, 44325-4003, USA
chan@uakron.edu

Abstract. Data with missing values are represented as incomplete information systems in rough sets approaches. There are two possible interpretations of missing values as "do not care" or "lost" values. Many existing works considered only the former case. The use of characteristic sets to deal with both cases was first introduced by Grzymala-Busse. In this paper, we introduce a refinement of characteristic set approach to incomplete information systems, and we show that it can improve approximation accuracy similar to the improvement obtained by applying the techniques of maximal consistent blocks and binary neighborhood systems approaches to dealing with "do not care" missing values. Additionally, subset and concept based approximations are introduced for binary neighborhood systems in order to preserve upper approximations.

Keywords: Rough sets, Incomplete information systems, Characteristic sets, Maximal characteristic sets, Binary neighborhood systems.

1 Introduction

Extensions to the classical rough set theory [1-3] have been introduced for processing data with missing values represented as incomplete information systems [4-9]. It has been pointed out in [6] that missing values can be interpreted as "lost values" or "do not care". Many works considered only the "do not care" missing value [4], [5], [10].

For the case of "do not care" missing values, it has been shown that better approximation accuracy can be obtained by representing incomplete information systems using the technique of maximal consistent blocks [5], which could further be refined by using binary neighborhood systems [10]. However, these results are not applicable for the case of "lost" missing values. An approach to deal with both cases was introduced in [6] based on the concept of characteristic sets.

In this paper, we introduce the concept of maximal characteristic sets and binary neighborhood systems consisting of these sets as elementary sets for incomplete information systems. The maximal consistent blocks approach is based on similarity relations that are reflexive and symmetric. The maximal characteristic sets are derived from reflexive characteristic relations. The refinement of one characteristic set into a family of maximal characteristic sets are represented as a binary neighborhood system. As indicated in [6], [11], there are

J.T. Yao et al. (Eds.): RSCTC 2012, LNAI 7413, pp. 73–82, 2012.

three different ways to define approximations, namely, singleton, subset or concept based approximations. Additionally, it has been shown that upper approximations obtained by singleton approximations may not be definable [6]. Since approximations of binary neighborhood systems as defined in [10], [11] are singleton based approximations, we introduce the use of subset and concept based approximations for binary neighborhood systems in order to preserve the upper approximations, which is similar to the concept of global approximations for incomplete data introduced in [8].

The paper is organized as follows. In section two, we review the related concepts of representing incomplete information systems. Section three introduces the concept of maximal characteristic sets and shows how to derive binary neighborhood systems. Examples are used to show that better approximation accuracy can be obtained by the use of maximal characteristic sets. Finally, we present our concluding remarks in section four.

2 Related Concepts

In rough set theory, information of objects is represented by an information system $S = (U, A)$, where U is a nonempty finite set of objects and A is a nonempty finite set of attributes such that $a : U \to V_a$, for any attribute $a \in A$ with V_a as the domain of a. Each nonempty subset $B \subseteq A$ defines an *indiscernibility relation* $IND(B)$ on U as follows. For any $x, y \in U$, $(x, y) \in IND(B)$ if and only if $a(x) = a(y)$ for all $a \in B$. The indiscernibility relation $IND(B)$ is an equivalence relation, and equivalence classes of $IND(B)$ are called *elementary sets* of B, denoted by $[x]_B$. Elementary sets can be computed as intersections of blocks of attribute-value pairs. For any $x \in U$, we have $[x]_B = \cap\{[(a, v)] : a \in B, a(x) = v\}$.

Let $X \subseteq U$ and $B \subseteq A$, the B-lower and B-upper approximations of X by elementary sets of $IND(B)$ are defined as $\underline{B}X = \{x \in U : [x]_B \subseteq X\} = \cup\{[x]_B : x \in U \text{ and } [x]_B \subseteq X\}$ and $\overline{B}X = \{x \in U : [x]_B \cap X \neq \emptyset\} = \cup\{[x]_B : x \in U \text{ and } [x]_B \cap X \neq \emptyset\}$. The accuracy of approximation of X by B is defined as $|\underline{B}X|/|\overline{B}X|$, where $|Y|$ denotes the cardinality of set Y.

When objects of an information system with all attribute values specified, it is called a *complete information system*; otherwise, it is called an *incomplete information system.* There are two interpretations of missing attribute values in incomplete information systems [6]: (1) when a missing value could be any value of the domain, hence irrelevant, it is called "do not care" missing value and is denoted by '*'; (2) when a missing value is some specific value, but it is lost. It is called a "lost" value and denoted by '?'.

Decision tables are special cases of information systems applied in many applications where the attribute set is $A \cup \{d\}$, the attribute d is called *decision attribute*, and other attributes are called *condition attributes.* A decision table with missing values is called *incomplete decision table.* An example of incomplete decision table is shown in Table 1 where cases in $U = \{1, 2, 3, 4, 5, 6, 7\}$, the set of condition attributes is $A = \{$PACKER, KEYS ADDED, API CALLS, DLLs, UNIQUE STRINGS$\}$, and the set of decision attribute is $d = \{$Malware$\}$.

Table 1. An incomplete decision table with "do not care" and "lost" missing values

Case	PACKER	KEYS ADDED	API CALLS	DLLs	UNIQUE STRINGS	Maleware
1	yes	2	2	1	yes	no
2	yes	2	1	2	yes	no
3	no	2	1	2	no	no
4	*	2	1	2	no	yes
5	yes	1	1	2	no	yes
6	yes	2	1	2	*	yes
7	?	2	1	2	no	yes

2.1 Similarity Relations, Similarity Classes, and Maximal Consistent Blocks

To deal with "do not care" missing values, one way is to extend an equivalence relation to a similarity relation [4]. Let $IS = (U, A)$ be an incomplete information system and $B \subseteq A$, the similarity relation defined on U by B is given as

$$SIM(B) = \{(x, y) \in U \times U : a \in B, a(x) = a(y) or a(x) =' *' or a(y) =' *'\}. \quad (1)$$

For $x \in U$, the similarity class $S_B(x)$ of $SIM(B)$ containing x is defined as

$$S_B(x) = \{y \in U : (x, y) \in SIM(B)\}. \quad (2)$$

Similarity classes are elementary sets. The similarity relation $SIM(B)$ is reflexive and symmetric, and it is a tolerance relation. Let $X \subseteq U$ and $B \subseteq A$, the B-lower and B-upper approximations of X are defined as

$$\underline{T}_B(X) = \{x \in U : S_B(x) \subseteq X\} = \{x \in X : S_B(x) \subseteq X\} \quad (3)$$

and

$$\overline{T}_B(X) = \{x \in U : S_B(x) \cap X \neq \emptyset\} = \cup\{S_B(x) : x \in X\}. \quad (4)$$

The similarity class $S_B(x)$ of $SIM(B)$ can be further refined by the concept of maximal consistent blocks introduced in [5]. A subset $Y \subseteq U$ is a *consistent block* of B, if the similarity relation $SIM(B)$ restricted to Y is transitive, i.e., any two objects in Y are similar. A subset $Y \subseteq U$ is a *maximal consistent block* when all its proper subsets are not consistent blocks.

Let $M(B)$ denote the set of all maximal consistent blocks determined by B. Let $X \subseteq U$ and $B \subseteq A$, the B-lower and B-upper approximations of X are defined as

$$\underline{M}_B(X) = \cup\{Y \in M(B) : Y \subseteq X\} \quad (5)$$

and

$$\overline{M}_B(X) = \cup\{Y \in M(B) : Y \cap X \neq \emptyset\}. \quad (6)$$

It has been shown in [5] that maximal consistent blocks can improve approximation accuracy, since $\underline{T}_B(X) \subseteq \underline{M}_B(X)$ and $\overline{T}_B(X) = \overline{M}_B(X)$.

Table 2. An incomplete decision table with do not care missing values only

Car	Price	Mileage	Size	MaxSpeed	d
1	high	low	full	low	good
2	low	*	full	low	good
3	*	*	compact	high	poor
4	high	*	full	high	good
5	*	*	full	high	excellent
6	low	high	full	*	good

Example 1. Let us consider the incomplete decision table shown in Table 2. For simplicity and comparison purpose, it is taken from [5]. Let $B = A$ be the set of all condition attributes {Price, Mileage, Size, MaxSpeed}, and d is the decision attribute. Follow the idea introduced in [6], let objects with attribute value '*' be included in all blocks. Then, we have the following blocks of attribute-value pairs:

[(Price, high)] = {1, 3, 4, 5},
[(Price, low)] = {2, 3, 5, 6},
[(Mileage, high)] = {2, 3, 4, 5, 6},
[(Mileage, low)] = {1, 2, 3, 4, 5},
[(Size, full)] = {1, 2, 4, 5, 6},
[(Size, compact)] = {3},
[(MaxSpeed, high)] = {3, 4, 5, 6},
[(MaxSpeed, low)] = {1, 2, 6}.

We have the following similarity classes

$S_B(1) = [(Price, high)] \cap [(Mileage, low)] \cap [(Size, full)] \cap [(MaxSpeed, low)]$
$= \{1,3,4,5\} \cap \{1,2,3,4,5\} \cap \{1,2,4,5,6\} \cap \{1,2,6\}$
$= \{1\}$,
$S_B(2) = [(Price, low)] \cap [(Mileage, *)] \cap [(Size, full)] \cap [(MaxSpeed, low)]$
$= \{2,3,5,6\} \cap \{1,2,3,4,5,6\} \cap \{1,2,4,5,6\} \cap \{1,2,6\}$
$= \{2,6\}$,
$S_B(3) = [(Price, *)] \cap [(Mileage, *)] \cap [(Size, compact)] \cap [(MaxSpeed, high)]$
$= \{1,2,3,4,5,6\} \cap \{1,2,3,4,5,6\} \cap \{3\} \cap \{3,4,5,6\}$
$= \{3\}$,
$S_B(4) = [(Price, high)] \cap [(Mileage, *)] \cap [(Size, full)] \cap [(MaxSpeed, high)]$
$= \{1,3,4,5\} \cap \{1,2,3,4,5,6\} \cap \{1,2,4,5,6\} \cap \{3,4,5,6\}$
$= \{4,5\}$,
$S_B(5) = [(Price, *)] \cap [(Mileage, *)] \cap [(Size, full)] \cap [(MaxSpeed, high)]$
$= \{1,2,3,4,5,6\} \cap \{1,2,3,4,5,6\} \cap \{1,2,4,5,6\} \cap \{3,4,5,6\}$
$= \{4,5,6\}$,
$S_B(6) = [(Price, low)] \cap [(Mileage, high)] \cap [(Size, full)] \cap [(MaxSpeed, *)]$
$= \{2,3,5,6\} \cap \{2,3,4,5,6\} \cap \{1,2,4,5,6\} \cap \{1,2,3,4,5,6\}$
$= \{2,5,6\}$.

The similarity class $S_B(5) = \{4, 5, 6\}$ is not a maximal consistent block, since the pair (4, 6) is not in $SIM(B)$. The subsets $\{5, 6\}$ and $\{4, 5\}$ of $S_B(5)$ are maximal consistent blocks. Similarly, $S_B(6) = \{2, 5, 6\}$ is not maximal, since the pair (2, 5) is not in $SIM(B)$, but its subsets {2, 6} and {5, 6} are maximal. Therefore, from Table 2, the set $M(B)$ of all maximal consistent blocks determined by B is {{1}, {2, 6}, {3}, {4, 5}, {5, 6}}.

For the decision class $X = [(d, good)] = \{1, 2, 4, 6\}$, the B-lower and B-upper approximations of X based on the similarity classes are

$$\underline{T}_B(X) = \{1, 2\} \text{ and } \overline{T}_B(X) = \{1, 2, 4, 5, 6\}.$$

The B-lower and B-upper approximations of X based on the maximal consistent blocks are

$$\underline{M}_B(X) = \{1, 2, 6\} \text{ and } \overline{M}_B(X) = \{1, 2, 4, 5, 6\}.$$

It is clear that maximal consistent block based approximation has better approximation accuracy based on Table 2.

2.2 Binary Neighborhood Systems, Maximal Consistent Block Neighborhoods

The use of binary neighborhood systems to study approximations of database and knowledge-based systems was introduced in [12]. A binary neighborhood system is a Binary Granular Data Model, also referred as the 4th Grc Model, which is a pair (U, β) where $\beta = \{R^1, R^2, \ldots\}$ is a family of binary relations on U [13]. For each $x \in U$, the binary neighborhoods of x are derived from $\beta = \{R^1, R^2, \ldots\}$ as

$$N_i(x) = \{y \in U : (x, y) \in R^i\}, i = 1, 2, \ldots \tag{7}$$

The binary neighborhood system of x is the collection of binary neighborhoods of x:

$$NS(x) = \{N_1(x), N_2(x), \ldots\}. \tag{8}$$

The binary neighborhood system of U is

$$NS(U) = \{NS(x) : x \in U\}. \tag{9}$$

Let $X \subseteq U$, the lower and upper approximations of X are defined as [13]:

$$\underline{NS}(X) = \{x \in U : \exists N(x) \in NS(x) s.t. N(x) \neq \emptyset and N(x) \subseteq X\} \tag{10}$$

and

$$\overline{NS}(X) = \{x \in U : \forall N(x) \in NS(x) s.t. N(x) \cap X \neq \emptyset\}. \tag{11}$$

The representation of maximal consistent blocks as binary neighborhood system was shown in [10]. Let $M(B)$ be a set of all maximal consistent blocks determined by B, the corresponding neighborhoods of $x \in U$, called *maximal consistent block neighborhoods*, is defined as

$$MNS_B(x) = \{Y \in M(B) : x \in Y\}. \tag{12}$$

For $X \subseteq U$, the B-lower and B-upper approximations of X in a maximal consistent blocks neighborhood system are defined as [9]:

$$\underline{MNS}_B(X) = \{x \in U : \exists Y \in MNS_B(x) s.t. Y \neq \emptyset and Y \subseteq X\} and \tag{13}$$

$$\overline{MNS}_B(X) = \{x \in U : \forall Y \in MNS_B(x) s.t. Y \cap X \neq \emptyset\}. \tag{14}$$

It is shown that $\underline{MNS}_B(X) = \underline{M}_B(X)$ and $\overline{MNS}_B(X) \subseteq \overline{M}_B(X)$, thus maximal consistent blocks neighborhood systems provide better approximation accuracy than maximal consistent blocks of similarity classes [10].

Example 2. Consider Table 2 used in Example 1. For each $x \in U$, we have the following binary neighborhood systems derived from the set of all maximal consistent blocks $M(B) = \{\{1\}, \{2, 6\}, \{3\}, \{4, 5\}, \{5, 6\}\}$.

$MNS_B(1) = \{\{1\}\}$,
$MNS_B(2) = \{\{2, 6\}\}$,
$MNS_B(3) = \{\{3\}\}$,
$MNS_B(4) = \{\{4, 5\}\}$,
$MNS_B(5) = \{\{4, 5\}, \{5, 6\}\}$,
$MNS_B(6) = \{\{2, 6\}, \{5, 6\}\}$.

For the decision class $X = [(d, good)] = \{1, 2, 4, 6\}$, the B-lower and B-upper approximations of X based on the binary neighborhood systems are

$$\underline{MNS}_B(X) = \{1, 2, 6\} = \underline{M}_B(X) \text{ and } \overline{MNS}_B(X) = \{1, 2, 4, 5, 6\} = \overline{M}_B(X).$$

For the decision class $X = [(d, excellent)] = \{5\}$, the B-lower and B-upper approximations of X based on maximal consistent blocks are

$$\underline{M}_B(X) = \emptyset \text{ and } \overline{M}_B = \{4, 5, 6\}.$$

The B-lower and B-upper approximations of X based on the binary neighborhood systems are

$$\underline{MNS}_B(X) = \emptyset \text{ and } \overline{MNS}_B(X) = \{4, 5\}.$$

Thus, the binary neighborhood systems provide a smaller B-upper approximation of X.

2.3 Characteristic Relations and Characteristic Sets

Incomplete information systems with both "do not care" and "lost" missing values were first considered in [6] where elementary sets are represented by characteristic sets computed from intersections of blocks of attribute-value pairs. The basic idea is that objects with "do not care" missing value denoted by '*' are included in all blocks for every possible value of an attribute and objects with "lost" missing value denoted by '?' are excluded in all blocks. It is further assumed that at least one attribute-value pair is specified for each object.

For $x \in U$, the characteristic set of x is defined as

$$K_B(x) = \cap\{[x]_a : a \in B\}$$

where $[x]_a = \{y \in U : a(x) = a(y) \neq' ?' or a(y) =' *'\}$.

The characteristic relation $K(B)$ is a reflexive relation on U defined for $x, y \in U$ as:

$$(x, y) \in K(B) \text{ if and only if } y \in K_B(x).$$

Example 3. Let us consider the incomplete decision table shown in Table 1, and let $B = A$ be the set of all condition attributes. Then, we have the following blocks of attribute-value pairs:

$[(PACKER, yes)] = \{1, 2, 4, 5, 6\}$,
$[(PACKER, no)] = \{3, 4\}$,
$[(PACKER, *)] = \{1, 2, 3, 4, 5, 6\}$,
$[(KEYSADDED, 1)] = \{5\}$,
$[(KEYSADDED, 2)] = \{1, 2, 3, 4, 6, 7\}$,
$[(APICALLS, 1)] = \{1\}$,
$[(APICALLS, 2)] = \{2, 3, 4, 5, 6, 7\}$,
$[(DLLs, 1)] = \{1\}$,
$[(DLLs, 2)] = \{2, 3, 4, 5, 6, 7\}$,
$[(UNIQUESTRINGS, yes)] = \{1, 2, 6\}$,
$[(UNIQUESTRINGS, no)] = \{3, 4, 5, 6, 7\}$,
$[(UNIQUESTRINGS, *)] = \{1, 2, 3, 4, 5, 6, 7\}$.

We have the following characteristic sets computed from Table 1:

$K_B(1) = \{1, 2, 4, 5, 6\} \cap \{1, 2, 3, 4, 6, 7\} \cap \{1\} \cap \{1\} \cap \{1, 2, 6\} = \{1\}$,
$K_B(2) = \{2, 6\}$,
$K_B(3) = \{3, 4\}$,
$K_B(4) = \{3, 4, 6\}$,
$K_B(5) = \{5\}$,
$K_B(6) = \{2, 4, 6\}$,
$K_B(7) = \{3, 4, 6, 7\}$.

3 Neighborhood Systems of Maximal Characteristic Sets

In the following, we introduce the representation of characteristic sets as neighborhood systems of maximal characteristic sets. Additionally, we introduce a modified definition for approximations of sets based on neighborhood systems. As pointed out in [6], there are three different ways to define approximations of sets based on singleton, subset, or concept. As shown in [6] that upper approximations of singleton-based approximations may not be B-definable. The maximal consistent block binary neighborhood systems given in Section 2.1 use singleton based approximations, and they need to be modified. Lower and upper approximations that are globally definable were introduced in [8], where a union of characteristic sets $K_B(x)$, for $x \in X \subseteq U$ is called a *B-globally definable set.* For a concept X, the *B-global lower* approximation of X is defined as

$$G\underline{B}X = \cup\{K_B(x) : x \in X and K_B(x) \subseteq X\}, \tag{15}$$

and the *B-global upper* approximation of X is a set with the minimal cardinality containing X and is defined as

$$G\overline{B}X = \cup\{K_B(x) : \exists Y \subseteq U, x \in Y and K_B(x) \cap X \neq \emptyset\}. \tag{16}$$

3.1 Maximal Characteristic Sets and Neighborhoods

Let $K(B)$ be a characteristic relation and $K_B(x)$ be a characteristic set derived from $K(B)$. We say that $Y \subseteq K_B(x)$ is a *maximal characteristic set* if and only if Y is a maximal subset of $K_B(x)$ such that for any $x, y \in Y, (x, y) or (y, x) \in K(B)$. For $x \in U$, we derive a *maximal characteristic neighborhood system* from each characteristic set $K_B(x)$ consisting of its maximal characteristic subsets, and it is defined as:

$$NS(K_B(x)) = \{Y \subseteq K_B(x) : Y \text{ is a maximal characteristic set of } K_B(x)\}.$$

Example 4. From Example 3, each characteristic set is represented as a neighborhood system of maximal characteristic sets as follows.

$$\begin{array}{ll}
K_A(1) = \{1\}, & NS((K)_A(1)) = \{\{1\}\}, \\
K_A(2) = \{2, 6\}, & NS((K)_A(2)) = \{\{2, 6\}\}, \\
K_A(3) = \{3, 4\}, & NS((K)_A(3)) = \{\{3, 4\}\}, \\
K_A(4) = \{3, 4, 6\}, & NS((K)_A(4)) = \{\{3, 4\}, \{4, 6\}\}, \\
K_A(5) = \{5\}, & NS((K)_A(5)) = \{\{5\}\}, \\
K_A(6) = \{2, 4, 6\}, & NS((K)_A(6)) = \{\{2, 6\}, \{4, 6\}\}, \\
K_A(7) = \{3, 4, 6, 7\} & NS((K)_A(7)) = \{\{3, 4, 7\}, \{4, 6, 7\}\}.
\end{array}$$

3.2 Approximations Based on Maximal Characteristic Neighborhoods

Follow the singleton B-lower and B-upper approximations of X introduced in [10], [13], we have the B-lower and B-upper approximations of $X \subseteq U$, in a maximal characteristic neighborhood system as

$$\begin{array}{r}
\underline{B}X = \{x \in U : \exists Y \in NS((K_B(x)) s.t. Y \neq \emptyset and Y \subseteq X\} \text{ and} \\
\overline{B}X = \{x \in U : \forall Y \in NS((K_B(x)) s.t. Y \cap X \neq \emptyset\}.
\end{array}$$

In general, the upper approximations of singleton based approximations are not B-definable as shown in the following. For the decision in Table 1, apply the singleton A-lower and A-upper approximations of $[(d, no)] = \{1, 2, 3\}$, we have

$$\underline{A}\{1,2,3\} = \{1\} \text{ and } \overline{A}\{1,2,3\} = \{1,2,3\}.$$

Note that $\overline{A}\{1,2,3\} = \{1,2,3\}$ is not A-definable based on definitions (10) and (11) given in Section 2.2.

Let $X \subseteq U$, we define the subset B-lower and B-upper approximations of X as

$$\underline{B}X = \cup\{Y \in NS((K_B(x)) : x \in U, Y \subseteq X\} \tag{17}$$

and

$$\overline{B}X = \cup\{Y \in NS((K_B(x)) : x \in U, Y \cap X \neq \emptyset\}. \tag{18}$$

We define the concept B-lower and B-upper approximations of X as

$$\underline{B}X = \cup\{Y \in NS((K_B(x)) : x \in X, Y \subseteq X\} \tag{19}$$

and

$$\overline{B}X = \cup\{Y \in NS((K_B(x)) : x \in X, Y \cap X \neq \emptyset\}. \tag{20}$$

Apply the subset A-lower and A-upper approximations for the decision in Table 1, we have

$$\begin{aligned} \underline{A}\{1,2,3\} &= \{1\}, \\ \overline{A}\{1,2,3\} &= \{1,2,3,4,6,7\}, \\ \underline{A}\{4,5,6,7\} &= \{4,5,6\}, \\ \overline{A}\{4,5,6,7\} &= \{2,3,4,5,6,7\}. \end{aligned}$$

The concept A-lower and A-upper approximations are

$$\begin{aligned} \underline{A}\{1,2,3\} &= \{1\}, \\ \overline{A}\{1,2,3\} &= \{1,2,3,4,6\}, \\ \underline{A}\{4,5,6,7\} &= \{4,5,6\}, \\ \overline{A}\{4,5,6,7\} &= \{2,3,4,5,6,7\}. \end{aligned}$$

If we use the characteristic sets instead of maximal characteristic set neighborhoods to compute the approximations. The subset A-lower and A-upper approximations are

$$\begin{aligned} \underline{A}\{1,2,3\} &= \{1\}, \\ \overline{A}\{1,2,3\} &= \{1,2,3,4,6,7\}, \\ \underline{A}\{4,5,6,7\} &= \{5\}, \\ \overline{A}\{4,5,6,7\} &= \{2,3,4,5,6,7\}. \end{aligned}$$

The concept A-lower and A-upper approximations are

$$\begin{aligned} \underline{A}\{1,2,3\} &= \{1\}, \\ \overline{A}\{1,2,3\} &= \{1,2,3,4,6\}, \\ \underline{A}\{4,5,6,7\} &= \{5\}, \\ \overline{A}\{4,5,6,7\} &= \{2,3,4,5,6,7\}. \end{aligned}$$

It shows that the use of maximal characteristic neighborhoods provides a better approximation accuracy than using characteristic sets.

4 Conclusions

We have introduced maximal characteristic neighborhood system representation for incomplete decision tables containing both do not care and lost missing values. It is a refinement of characteristic sets, and hence, it improves approximation accuracy as shown in our examples. Additionally, we refined the singleton approximations of neighborhood systems to subset and concept approximations, which preserve the B-definability of upper approximations.

References

1. Pawlak, Z.: Rough sets: basic notion. International Journal of Computer and Information Science 11(15), 344–356 (1982)
2. Pawlak, Z.: Rough Sets and Decision Tables. In: Skowron, A. (ed.) SCT 1984. LNCS, vol. 208, pp. 186–196. Springer, Heidelberg (1985)
3. Pawlak, Z.: Rough Sets: Theoretical Aspects of Reasoning about Data. Kluwer Academic Publishing (1991)
4. Kryszkiewicz, M.: Rough set approach to incomplete information systems. Information Sciences 112, 39–49 (1998)
5. Leung, Y., Li, D.: Maximal consistent block techniques for rule acquisition in incomplete information systems. Information Sciences 153, 85–106 (2003)
6. Grzymała-Busse, J.W.: Characteristic Relations for Incomplete Data: A Generalization of the Indiscernibility Relation. In: Tsumoto, S., Słowiński, R., Komorowski, J., Grzymała-Busse, J.W. (eds.) RSCTC 2004. LNCS (LNAI), vol. 3066, pp. 244–253. Springer, Heidelberg (2004)
7. Grzymala-Busse, J.W.: On the Unknown Attribute Values in Learning from Examples. In: Raś, Z.W., Zemankova, M. (eds.) ISMIS 1991. LNCS (LNAI), vol. 542, pp. 368–377. Springer, Heidelberg (1991)
8. Grzymała-Busse, J.W., Rząsa, W.: Local and Global Approximations for Incomplete Data. In: Peters, J.F., Skowron, A. (eds.) Transactions on Rough Sets VIII. LNCS, vol. 5084, pp. 21–34. Springer, Heidelberg (2008)
9. Stefanowski, J., Tsoukiàs, A.: On the Extension of Rough Sets under Incomplete Information. In: Zhong, N., Skowron, A., Ohsuga, S. (eds.) RSFDGrC 1999. LNCS (LNAI), vol. 1711, pp. 73–82. Springer, Heidelberg (1999)
10. Yang, X., Zhang, M., Dou, H., Yang, J.: Neighborhood systems-based rough sets in in-complete information system. Knowledge-Based Systems 24, 858–867 (2011)
11. Yao, Y.Y.: Two views of the theory of rough sets in finite universes. International J. of Approximate Reasoning 15, 291–317 (1996)
12. Lin, T.Y.: Neighborhood systems and approximation in database and knowledge base systems. In: Proceedings of the Fourth International Symposium on Methodologies of Intelligent Systems, Poster Session, October 12-15, pp. 75–86 (1989)
13. Lin, T.Y.: Granular computing I: the concept of granulation and its formal model. International Journal of Granular Computing, Rough Sets and Intelligent Systems 1, 21–42 (2009)

Probabilistic Fuzzy Rough Set Model over Two Universes

Bingzhen Sun[1,2], Weimin Ma[1,*], Haiyan Zhao[1,3], and Xinxin Wang[4]

[1] School of Economics and Management,
Tongji University, Shanghai 200092, P.R. China
[2] School of Traffic and Transportation,
Lanzhou Jiaotong University, Lanzhou 730070, P.R. China
[3] Vocational Education Department,
Shanghai University of Engineering Science, Shanghai, 200092, P.R. China
[4] School of Mathematics and Statics,
Longdong University, Qingyang, 745000, P.R. China
mawm@tongji.edu.cn, bzh--sun@163.com

Abstract. As a new generalization of Pawlak rough set, the theory and applications of rough set over two universes has brought the attention by many scholars in various areas. In this paper, we propose a new model of probabilistic fuzzy rough set by introducing the probability measure to the fuzzy compatibility approximation space over two universes. That is, the model defined in this paper included both of probabilistic rough set and fuzzy rough set over two universes. The probabilistic fuzzy rough lower and upper approximation operators of any subset were defined by the concept of the fuzzy compatible relation between two different universes. Since there has two parameters in the lower and upper approximations, we also give other definitions for probabilistic fuzzy rough set model under the framework of two universes with different combination of the parameters. Furthermore, we discuss the properties for the established model in detail and present several valuable conclusions. The results show that this model has more extensively applied fields.

Keywords: Probabilistic fuzzy rough set, fuzzy compatibility relation, Two universes.

1 Introduction

Many our traditional tools for formal modeling, reasoning and computing are crisp, deterministic and precise in character. However, most of practical problems within fields as diverse as economics, engineering, environment, social science, medical science involve data that contain uncertainties. We cannot successfully use traditional mathematical tools because of various types of uncertainties existing in these problems. There have been a great amount of research and applications in the literature concerning some special tools such as probability

* Corresponding author.

J.T. Yao et al. (Eds.): RSCTC 2012, LNAI 7413, pp. 83–93, 2012.

theory [22], fuzzy set theory [25], rough set theory [6 – 8], interval mathematics [3, 11, 12, 17], and etc.

Rough set [6] theory is a new mathematical tool to deal with vagueness and uncertainty. One of the main advantages of rough set theory is that it does not need any preliminary or additional information about data, such as probability distribution in statistics, basic probability assignment in the Dempster-Shafer theory, or grade of membership or the value of possibility in fuzzy set theory [9]. The standard rough set model is a qualitative model that defines three regions for approximating a subset of a universe of objects based on an equivalence relation on the universe. A lack of consideration of the degree of overlap between an equivalence class and the set motivates many researchers to study quantitative rough set models [18, 26]. Therefore, probabilistic approaches to rough sets are one of the most important and successful schools of quantitative rough sets [5].

As a non-numeric methods to represent and manage uncertainty contained in various information systems, rough set theory also has been regarded as an interval structures to manage uncertainty. In order to establish a unified framework for treating uncertain information, a general notion of interval structure has already been proposed and discussed in detail [10, 15, 16, 21], and a generalized rough set model on two universes was proposed based on the so-called compatibility view. Subsequently, many results have been generated in the rough set theory over two universes [4]. In [24], Zhang and Wu defined a rough set model based on random set, in which, the lower approximation and the upper approximation were generalized to two universes. Subsequently, they proposed a general model of the interval-valued fuzzy rough set on two universes by integrating the rough set theory with the interval-valued fuzzy set theory according to the constructive and axiomatic approaches [23]. Based on the existed results, we present a definition of fuzzy rough set by a fuzzy compatibility relation between two universes [13]. Moreover, we study the probabilistic rough set over two universes by the constructive methods. We also study the relationship between probabilistic rough set and Bayesian decision making over two universes [2, 4, 5]. In this paper, we will present a new rough set model named probabilistic fuzzy rough set based on probabilistic rough set and fuzzy rough set over two different universes. Both of the models are our previous works. Though probabilistic fuzzy rough set over two universes is a generalization of the existed work, there are many differences among of the models.

In view of this opinion, the main objective of this paper was to study the basic theory for probabilistic fuzzy rough set model over two universes. We present various definitions for the probabilistic fuzzy rough set over two universes according to different parameters combination. Furthermore, we discuss the differences and relationships between the proposed model and the other existed rough set models. Meantime, we also establish several important properties for probabilistic fuzzy rough set model over two universes.

The remainder of this paper is organized as follows: Section 2 briefly introduces the basic concept which is needed in this paper such as rough set over two universes, fuzzy compatibility relation, probabilistic rough set over two universes.

Section 3 establishes the probabilistic fuzzy rough set model and also give various diversified definitions with the parameters combination. In section 4, we discuss the basic properties for the proposed model in Section 3. At last we conclude our research and set further research directions in Section 5.

2 Preliminaries

In this section, we review some basic concepts such as fuzzy compatibility relation, probabilistic rough set over two universes to be used in this paper.

2.1 Fuzzy Relation and Fuzzy Compatibility Relation

Throughout in this paper U, V denotes a non empty finite set unless stated otherwise.

Firstly, we give the fuzzy relation on two universes.

Let U, V be non-empty finite universes. A fuzzy sets $\widetilde{R} \in F(U \times V)$ of the universe $U \times V$(i.e., $\widetilde{R} : U \times V \longrightarrow [0,1]$) is called a fuzzy relation from U to V. In general, for any $u \in U, v \in V$, the value $\widetilde{R}(u,v)$ denotes the related degree of u and v [13].

If $U = V$, the fuzzy relation $\widetilde{R} \in F(U \times V)$ will be called a fuzzy relation on U.

Definition 2.1 [13]. Let $\widetilde{R} \in F(U \times U)$, then

(1) $\widetilde{R}$ is reflexive, if $\widetilde{R}(u,u) = 1$, for any $u \in U$;
(2) $\widetilde{R}$ is symmetric, if $\widetilde{R}(u_1,u_2) = \widetilde{R}(u_1,u_2)$, for any $u_1, u_2 \in U$,
(3) $\widetilde{R}$ is transitive, if $\widetilde{R} \circ \widetilde{R} \leq \widetilde{R}$.

Let $\widetilde{R} \in F(U \times U)$, if $\widetilde{R}$ is reflexive, symmetric, then $\widetilde{R}$ is called a fuzzy similarity relation on U, if $\widetilde{R}$ is reflexive, symmetric,and transitive, then $\widetilde{R}$ is called a fuzzy equivalence relation on U.

In the following, we present the fuzzy compatibility relation.

Definition 2.2 [13]. Let U, V be non-empty finite universes. $\widetilde{R}$ be a fuzzy relation from U to V, for any $\delta \in (0,1]$, we define the fuzzy compatibility relation $\widetilde{R}_\delta$ between the universe U and V as follows:

$$\widetilde{R}_\delta(u) = \{v \in V | \widetilde{R}(u,v) \geq \delta, \forall\ \delta \in (0,1], u \in U\}.$$

If $\delta = 1$, then $\widetilde{R}_\delta(u) = \{v \in V | \widetilde{R}(u,v) = 1, \forall\ u \in U\} = \{v \in V | uRv, \forall\ u \in U\}$ hold.

By the definition 2.2, we know that all the relationships of the elements between universe U and V are considered while the parameter δ takes all the value of $(0,1]$. As a result, we can obtain the expected aim of our by taking the reasonable threshold values δ according to the demanding of the different problems.

2.2 Rough Set Models over Two Universes

In this section, we will review the basic rough set models over two universes with some mainly results.

Definition 2.3 [10, 14]. Let U and V be two universes, and R be a binary relation from U to V, i.e. a subset of $U \times V$. R is said to be compatibility, or a compatibility relation, if for any $u \in U$; $v \in V$; there exist $t \in V$; $s \in U$ such that $(u, t), (s, v) \in R$.

Definition 2.4 [10, 14]. Let U and V be two universes, and R be a compatibility relation from U to V. The mapping $F : U \to 2^V, u \mapsto \{v \in V | (u, v) \in R\}$ is called the mapping induced by R.

Obviously, the above-defined binary relation R can uniquely determine the mapping F, and vice versa. Then the rough set over two universes is defined as follows:

Let U and V be two universes, and R be a compatibility relation from U to V. The ordered triple (U, V, R) is called a (two-universe) approximation space. The lower and upper approximations of $Y \subseteq V$ are, respectively, defined as follows [10, 15, 16, 21]:

$$\underline{apr}(Y) = \{x \in U | F(x) \subseteq Y\};$$
$$\overline{apr}(Y) = \{x \in U | F(x) \cap Y \neq \emptyset\}.$$

The ordered set-pair $(\underline{apr}(Y), \overline{apr}(Y))$ is called a generalized rough set, and the ordered operator-pair $(\underline{apr}, \overline{apr})$ is an interval structure.

In the following, we present fuzzy compatibility relation-based fuzzy rough set model and probabilistic rough set model over two universes [2, 4, 13].

Let U, V be two non-empty finite universes. $\widetilde{R}_\delta$ is the fuzzy compatibility relation from universe U to V. Then, $(U, V, \widetilde{R}_\delta)$ is called fuzzy compatibility approximation space over two universes.

Let U, V be non-empty finite universes, $\widetilde{R}_\delta$ be a fuzzy compatibility relation of the universe U and V. For any $X (X \subseteq V)$, we define the lower and upper approximations of X about $\widetilde{R}_\delta$ on the universe U and V as follows, respectively:

$$\underline{apr}_{\widetilde{R}_\delta}(X) = \{u \in U | \widetilde{R}_\delta(u) \subseteq X\},$$
$$\overline{apr}_{\widetilde{R}_\delta}(X) = \{u \in U | \widetilde{R}_\delta(u) \cap X \neq \emptyset\}.$$

Furthermore, we also define the positive region $pos_{\widetilde{R}_\delta}(X)$, negative region $neg_{\widetilde{R}_\delta}(X)$ and boundary region $bn_{\widetilde{R}_\delta}(X)$ of X about $\widetilde{R}_\delta$ on the universe U and V as follows, respectively:

$$pos_{\widetilde{R}_\delta}(X) = \underline{apr}_{\widetilde{R}_\delta}(X); \qquad neg_{\widetilde{R}_\delta}(X) = U - \overline{apr}_{\widetilde{R}_\delta}(X),$$
$$bn_{\widetilde{R}_\delta}(X) = \overline{apr}_{\widetilde{R}_\delta}(X) - \underline{apr}_{\widetilde{R}_\delta}(X).$$

This is the fuzzy rough set over two universes defined in reference [13].

Let U, V be two non-empty finite universes. $R \subseteq U \times V$ is the set-valued mapping from universe U to V. P is a probability measure defined on the σ

algebra formed by the image(*That is, the subset classes of the universe* V) of element $x(x \in U)$. Then, (U, V, R, P) is called probabilistic approximation space over two universes.

Let (U, V, R, P) be probabilistic approximation space over two universes. For any $0 \leq \beta < \alpha \leq 1$, $X \in 2^V$. Then the lower and upper approximations of X with parameter α and β as follows, respectively.

$$\underline{apr}_P^{\alpha}(X) = \{x \in U | P(X|R(x)) \geq \alpha\},$$
$$\overline{apr}_P^{\beta}(X) = \{x \in U | P(X|R(x)) > \beta\}.$$

Meanwhile, the positive region, boundary region and negative region of X in (U, V, R, P) with parameter α and β could be given as follows, respectively.

$$pos(X, \alpha) = \{x \in U | P(X|R(x)) \geq \alpha\},$$
$$bn(X, \alpha, \beta) = \{x \in U | \beta < P(X|R(x)) < \alpha\},$$
$$neg(X, \beta) = \{x \in U | P(X|R(x)) \leq \beta\} = U - \overline{apr}_P^{\beta}(X).$$

This is the probabilistic rough set over two universes defined in reference [2, 4].

3 Probabilistic Fuzzy Rough Set Model over Two Universes

In this section, we establish the concept of probabilistic fuzzy approximation operators based on the fuzzy compatible relation over two universes

First of all, we present the concept of probabilistic fuzzy compatibility approximation space over two universes.

Definition 3.1. Let U, V be two non-empty finite universes. $\widetilde{R}_\delta$ is the fuzzy compatibility relation from universe U to V. P is a probability measure defined on the σ algebra of the subset family of universe V. Then, $(U, V, \widetilde{R}_\delta, P)$ is called probabilistic fuzzy compatibility approximation space over two universes.

By the probabilistic fuzzy compatibility approximation space over two universes, we give the lower and upper approximations for any subset of universe V as follows.

Let $(U, V, \widetilde{R}_\delta, P)$ be probabilistic fuzzy compatibility approximation space over two universes. For any $0 \leq \beta < \alpha \leq 1$, $X \in 2^V$. Then the lower and upper approximations of X with parameters α and β are, respectively, defined as follows:

$$\underline{P}_{\widetilde{R}_\delta}^{\alpha}(X) = \{u \in U | P(X|\widetilde{R}_\delta(u)) \geq \alpha\},$$
$$\overline{P}_{\widetilde{R}_\delta}^{\beta}(X) = \{u \in U | P(X|\widetilde{R}_\delta(u)) > \beta\}.$$

The ordered set-pair $(\underline{P}_{\widetilde{R}_\delta}^{\alpha}(X), \overline{P}_{\widetilde{R}_\delta}^{\alpha}(X))$ is called probabilistic fuzzy rough set of two universes, $\underline{P}_{\widetilde{R}_\delta}^{\alpha}$ and $\overline{P}_{\widetilde{R}_\delta}^{\alpha}$ are the approximate operators from $P(V)$ to $P(U)$ ($P(\bullet)$ denotes the subset family of universe). We also call ordered operator-pair$(\underline{P}_{\widetilde{R}_\delta}^{\alpha}, \overline{P}_{\widetilde{R}_\delta}^{\alpha})$ is an interval structure.

Meanwhile, the positive region, boundary region and negative region of X in $(U,V,\widetilde{R}_\delta,P)$ with parameter α and β could be given as follows, respectively.

$$pos(X,\alpha)=\{u\in U|P(X|\widetilde{R}_\delta(u))\geq\alpha\},$$
$$bn(X,\alpha,\beta)=\{x\in U|\beta<P(X|\widetilde{R}_\delta(u))<\alpha\},$$
$$neg(X,\beta)=\{x\in U|P(X|\widetilde{R}_\delta(u))\leq\beta\}=U-\overline{P}^{\beta}_{\widetilde{R}_\delta}(X).$$

Remark 3.1. If we confine parameter $\alpha\in(0.5,1]$ and omit the parameter β, and re-define the lower and upper approximations of $X(X\in 2^V)$ with $(U,V,\widetilde{R}_\delta,P)$ are, respectively, as follows:

$$\underline{P}^{\alpha}_{\widetilde{R}_\delta}(X)=\{u\in U|P(X|\widetilde{R}_\delta(u))\geq\alpha\},$$
$$\overline{P}^{\beta}_{\widetilde{R}_\delta}(X)=\{u\in U|P(X|\widetilde{R}_\delta(u))>1-\alpha\}.$$

Then we obtain the variable probabilistic fuzzy rough set model over two universes.

Remark 3.2. If $\delta=1$, since that $\widetilde{R}_1(u)=\{v\in V|\widetilde{R}(u,v)=1,\forall u\in U\}=\{v\in V|(u,v)\in R,\forall\ u\in U\}=F(u)$, so the fuzzy compatibility relation $\widetilde{R}_\delta$ is degenerated the ordinary compatibility F on the universe U and V. Then we have the following relations:

$$\underline{P}^{\alpha}_{\widetilde{R}_\delta}(X)=\underline{apr}^{\alpha}_{P}(X)=\{u\in U|P(X|F(u))\geq\alpha\},$$
$$\overline{P}^{\beta}_{\widetilde{R}_\delta}(X)=\overline{apr}^{\beta}_{P}(X)=\{u\in U|P(X|F(u))>\beta\}.$$

This is the probabilistic rough set model between two different universes defined by Gong and Sun [2].

Remark 3.3. If $\delta=1$ and $\alpha=1,\beta=0$. By the Remark 3.2, we have the following results:

$$\underline{P}^{\alpha}_{\widetilde{R}_\delta}(X)=\underline{apr}(X)=\{u\in U|F(u)\subseteq X\},$$
$$\overline{P}^{\beta}_{\widetilde{R}_\delta}(X)=\overline{apr}(X)=\{u\in U|F(u)\cap X\neq\emptyset\}.$$

This is the rough set model between two different universes proposed by Yao et al. [21].

Remark 3.4. If $U=V$, it can be easily verified that $\widetilde{R}_\delta$ will be changed into a general binary relation of universe U. Then, for any $X\subseteq U$, we have the following results:

$$\underline{P}^{\alpha}_{\widetilde{R}_\delta}(X)=\underline{apr}^{\alpha}_{P}(X)=\{u\in U|P(X|R(u))\geq\alpha\},$$
$$\overline{P}^{\beta}_{\widetilde{R}_\delta}(X)=\overline{apr}^{\beta}_{P}(X)=\{u\in U|P(X|R(u))>\beta\}.$$

This is the probabilistic rough set based on general binary relation over the same universe [4].

The above results show the relationship between probabilistic fuzzy rough set over two universes and the existed rough set model. The results also illuminate the model proposed in this paper is more widely and broader application filed.

Actually, we also can obtain other generalized rough set model over two(one) universes by discuss various cases for the fuzzy compatibility relation $\widetilde{R}_\delta$. However, it does not deduce the fuzzy rough set over two universes from the definition of probabilistic fuzzy rough set over two universes. So, it shows that the rough set model proposed in this paper does not a directly generalization of the existed model but a new result. This could be the theory value that establishes the probabilistic fuzzy rough set over two universes.

Similar to the classical Pawlak rough set, we also define the uncertainty measure of probabilistic fuzzy rough set over two universes as the way of the Pawlak rough set in the following:

We call $\rho_{(\alpha,\beta)}(X) = \frac{|\underline{P}^{\alpha}_{\widetilde{R}_\delta}(X)|}{|\overline{P}^{\beta}_{\widetilde{R}_\delta}(X)|}$ the accuracy of approximation for subset $X(X \subseteq V)$ in probabilistic fuzzy compatibility approximation space.

Moreover, the approximated quality of lower and upper approximations are, respectively, define as follows:

$$\underline{q}(X) = \frac{|\underline{P}^{\alpha}_{\widetilde{R}_\delta}(X)|}{|U|} = P(\underline{P}^{\alpha}_{\widetilde{R}_\delta}(X)), \qquad \overline{q}(X) = \frac{|\overline{P}^{\beta}_{\widetilde{R}_\delta}(X)|}{|U|} = P(\overline{P}^{\beta}_{\widetilde{R}_\delta}(X)).$$

Furthermore, the relationship between the accuracy and quality of approximation can be expressed as follows: $\rho_{(\alpha,\beta)}(X) = \frac{\underline{q}(X)}{\overline{q}(X)}$.

Then, we call $\sigma_{(\alpha,\beta)}(X) = 1 - \rho_{(\alpha,\beta)}(X) = \frac{|bn(X,\alpha,\beta)|}{|\overline{P}^{\beta}_{\widetilde{R}_\delta}(X)|}$ the roughness for set X in probabilistic fuzzy compatibility approximation space.

Actually, there have the similar properties for the accuracy of approximation and roughness of the probabilistic fuzzy rough set and also can establish the relationship between the accuracy of approximation and roughness like the existed probabilistic rough set models over two universes [4].

Like the probabilistic rough set on the same universe, there are other forms for the definition of lower and upper approximations since it includes two parameters. So is true for probabilistic fuzzy rough set over two universes. In the following, we present the definitions for the probabilistic fuzzy rough lower and upper approximations respectively.

Definition 3.2. Let $(U, V, \widetilde{R}_\delta, P)$ be probabilistic fuzzy compatibility approximation space over two universes. For any $0 \leq \beta < \alpha \leq 1$, $X \in 2^V$. Then the other cases for lower and upper approximations of X with parameter α and β are, respectively, defined as follows:

Case 1. $\underline{P}^{\alpha}_{\widetilde{R}_\delta}(X) = \{u \in U | P(X|\widetilde{R}_\delta(u)) > \alpha\}$,
$\overline{P}^{\beta}_{\widetilde{R}_\delta}(X) = \{u \in U | P(X|\widetilde{R}_\delta(u)) \geq \beta\}$.

Case 2. $\underline{P}^{\alpha}_{\widetilde{R}_\delta}(X) = \{u \in U | P(X|\widetilde{R}_\delta(u)) > \alpha\}$,
$\overline{P}^{\beta}_{\widetilde{R}_\delta}(X) = \{u \in U | P(X|\widetilde{R}_\delta(u)) > \beta\}$.

Case 3. $\underline{P}^{\alpha}_{\widetilde{R}_\delta}(X) = \{u \in U | P(X|\widetilde{R}_\delta(u)) \geq \alpha\}$,
$\overline{P}^{\beta}_{\widetilde{R}_\delta}(X) = \{u \in U | P(X|\widetilde{R}_\delta(u)) \geq \beta\}$.

Though the above models have different definition form, all of them have similar properties. However, they could have different decision objects when it was applied to the management decision in practice.

4 Properties for Probabilistic Fuzzy Rough Set over Two Universes

In this section, we discuss the properties in detail for probabilistic fuzzy rough set under the framework of two universes.

Theorem 4.1. Let $(U, V, \widetilde{R}_\delta, P)$ be probabilistic fuzzy compatibility approximation space over two universes. For any $0 \le \beta < \alpha < 1, X, Y \in 2^V$. The lower and upper approximation operators satisfy the following properties.

(1) $\underline{P}^{\alpha}_{\widetilde{R}_\delta}(\emptyset) = \overline{P}^{\beta}_{\widetilde{R}_\delta}(\emptyset) = \emptyset, \qquad \underline{P}^{\alpha}_{\widetilde{R}_\delta}(V) = \overline{P}^{\beta}_{\widetilde{R}_\delta}(V) = U,$

(2) $\underline{P}^{\alpha}_{\widetilde{R}_\delta}(X) =\sim \overline{P}^{(1-\alpha)}_{\widetilde{R}_\delta}(\sim X), \qquad \overline{P}^{\beta}_{\widetilde{R}_\delta}(X) =\sim \underline{P}^{(1-\beta)}_{\widetilde{R}_\delta}(\sim X),$

(3) $\underline{P}^{\alpha}_{\widetilde{R}_\delta}(X \cap Y) \subseteq \underline{P}^{\alpha}_{\widetilde{R}_\delta}(X) \cap \underline{P}^{\alpha}_{\widetilde{R}_\delta}(Y)$, $\overline{P}^{\beta}_{\widetilde{R}_\delta}(X \cup Y) \supseteq \overline{P}^{\beta}_{\widetilde{R}_\delta}(X) \cup \overline{P}^{\beta}_{\widetilde{R}_\delta}(Y)$,

(4) $\underline{P}^{\alpha}_{\widetilde{R}_\delta}(X \cup Y) \supseteq \underline{P}^{\alpha}_{\widetilde{R}_\delta}(X) \cup \underline{P}^{\alpha}_{\widetilde{R}_\delta}(Y)$, $\overline{P}_{\widetilde{R}_\delta}(X \cap Y) \subseteq \overline{P}_{\widetilde{R}_\delta}(X) \cap \overline{P}_{\widetilde{R}_\delta}(Y)$,

(5) If $X \subseteq Y$. Then $\underline{P}^{\alpha}_{\widetilde{R}_\delta}(X) \subseteq \underline{P}^{\alpha}_{\widetilde{R}_\delta}(Y)$, $\overline{P}^{\beta}_{\widetilde{R}_\delta}(X) \subseteq \overline{P}^{\beta}_{\widetilde{R}_\delta}(Y)$,

(6) If $\alpha_1 \le \alpha_2, \beta_1 \le \beta_2$. Then $\underline{P}^{\alpha_2}_{\widetilde{R}_\delta}(X) \subseteq \underline{P}^{\alpha_1}_{\widetilde{R}_\delta}(x), \qquad \overline{P}^{\beta_2}_{\widetilde{R}_\delta}(X) \subseteq \overline{P}^{\beta_1}_{\widetilde{R}_\delta}(X)$.

Proof. Tt can be easily verified by the definition in Section 3.

By the relation (6) in Theorem 4.1, we know positive region will increase with parameter α decrease but negative region will increase with parameter β increase and boundary region will dwindle for two universes probability rough set for given threshold δ.

Theorem 4.2. Let $(U, V, \widetilde{R}_\delta, P)$ be probabilistic fuzzy compatibility approximation space over two universes. For any $0 < r < 1, \quad X \in 2^W$. The following relationships hold.

(1) $\lim\limits_{\alpha \to r^+} \underline{P}^{\alpha}_{\widetilde{R}_\delta}(X) = \bigcup\limits_{\alpha > r} \underline{P}^{\alpha}_{\widetilde{R}_\delta}(X) = \overline{P}^{r}_{\widetilde{R}_\delta}(X),$

(2) $\lim\limits_{\beta \to r^-} \overline{P}^{\beta}_{\widetilde{R}_\delta}(X) = \bigcap\limits_{\beta < r} \overline{P}^{\beta}_{\widetilde{R}_\delta}(X) = \underline{P}^{r}_{\widetilde{R}_\delta}(X).$

Proof. Tt is similar to the Theorem 3.2 of Ref.[4].

Theorem 4.3. Let $(U, V, \widetilde{R}_\delta, P)$ be probabilistic fuzzy compatibility approximation space over two universes. For any $0 < r < 1, \quad X \in 2^W$. The following relationships hold.

(1) $\lim\limits_{\alpha \to r^-} \underline{P}^{\alpha}_{\widetilde{R}_\delta}(X) = \bigcap\limits_{\alpha < r} \underline{P}^{\alpha}_{\widetilde{R}_\delta}(X) = \underline{P}^{r}_{\widetilde{R}_\delta}(X),$

(2) $\lim\limits_{\beta \to r^+} \overline{P}^{\beta}_{\widetilde{R}_\delta}(X) = \bigcup\limits_{\beta > r} \overline{P}^{\beta}_{\widetilde{R}_\delta}(X) = \overline{P}^{r}_{\widetilde{R}_\delta}(X).$

Proof. Tt is similar to the Theorem 3.3 of Ref.[4].

Theorem 4.2 and Theorem 4.3 show the continuous of the probabilistic fuzzy lower and upper approximation operators with parameters for the given threshold parameter δ. The results are similar to the probabilistic rough set over two universes.

However, the following conclusions could not hold which are similar to the fuzzy rough set over two universes [13] for the given parameters α and β.

Remark 4.1. Let $(U, V, \widetilde{R}_\delta, P)$ be probabilistic fuzzy compatibility approximation space over two universes. For any $\delta_1, \delta_2 \in (0,1]$, and they are satisfied the relation $\delta_1 \leq \delta_2$. For any $X(X \subseteq V)$, the following relations

(1) $\underline{P}^{\alpha}_{\widetilde{R}_{\delta_2}}(X) \subseteq \underline{P}^{\alpha}_{\widetilde{R}_{\delta_1}}(X)$,

(2) $\overline{P}^{\beta}_{\widetilde{R}_{\delta_2}}(X) \subseteq \overline{P}^{\beta}_{\widetilde{R}_{\delta_1}}(X))$.

could not hold.

Remark 4.2. Let $(U, V, \widetilde{R}_\delta, P)$ be probabilistic fuzzy compatibility approximation space over two universes. Let $\widetilde{R}, \widetilde{S}$ be two fuzzy binary relation between U and V and satisfy $\widetilde{R} \subseteq \widetilde{S}$. For any $X \subseteq V$, the following relations

(1) $\underline{P}^{\alpha}_{\widetilde{S}_\delta}(X) \subseteq \underline{P}^{\alpha}_{\widetilde{R}_\delta}(X)$,

(2) $\overline{P}^{\beta}_{\widetilde{R}_\delta}(X) \subseteq \overline{P}^{\beta}_{\widetilde{S}_\delta}(X)$,

could not hold.

The above results with the properties illustrate the relationships and differences between the model proposed in this paper and the existed rough set theory over two universes.

5 Conclusions

In this paper, we have developed a new concept of probabilistic fuzzy rough set over two universes by combing the fuzzy rough set with the probabilistic rough set over two universes. It is also can be viewed as a generalization of probabilistic rough set based on the fuzzy compatibility relation over two universes. The relationship between the probabilistic fuzzy rough set with the existing rough set over two universes were established. Furthermore, three interrelated definitions of the probabilistic fuzzy rough set over two universes were established in detail. In addition, we briefly discuss the properties for the proposed model and present several important conclusions. It would enrich the basic theory of the rough set over two universes and also provide a new model and tool to handle uncertainty and fuzziness contained in management decision for practice.

In this paper, we focus on the basic definition and the related properties for the probabilistic fuzzy rough set over two universes. Actually, the applications of any new rough set model or generalized rough set model could be as important as the theory research of themselves. So further work should consider the approaches to the management decision with probabilistic fuzzy rough set model.

Acknowledgements. The work was partly supported by the National Science Foundation of China (71161016, 71071113), a Foundation for the Author of National Excellent Doctoral Dissertation of PR China (200782), the Shuguang Plan of Shanghai Education Development Foundation and Shanghai Education Committee (08SG21), Shanghai Pujiang Program, and Shanghai Philosophical and Social Science Program (2010BZH003), the Fundamental Research Funds for the Central Universities.

References

1. Duda, R.O., Hart, P.E.: Pattern classification and scene analysis. Wiley, New York (1973)
2. Gong, Z.T., Sun, B.Z.: Probability rough sets model between different universes and its applications. In: Proc. International Conference on Machine Learning and Cybernetics, pp. 561–565. IEEE Press, China (2008)
3. Gong, Z.T., Sun, B.Z., Chen, D.G.: Rough set theory for the interval-valued fuzzy information systems. Information Science (178), 1986–1985 (2008)
4. Ma, W.M., Sun, B.Z.: Probabilistic rough set over two universes and rough entropy. International Journal of Approximate Reasoning (53), 608–619 (2012)
5. Ma, W.M., Sun, B.Z.: On relationship between probabilistic rough set and Bayesian risk decision over two universes. International Journal of General Systems 41(3), 225–245 (2012)
6. Pawlak, Z.: Rough sets. International Journal of Computer and Information Science (11), 341–356 (1982)
7. Pawlak, Z.: Rough Sets: Theoretical Aspects of Reasoning about Data. Kluwer Academic Publishers, Dordrecht (1991)
8. Pawlak, Z., Skowron, A.: Rudiments of rough sets. Information Science (177), 3–27 (2007)
9. Pawlak, Z., Grzymala-Busse, J.W., Slowinski, R.: Rough sets. Communications of the ACM (38), 88–95 (1995)
10. Pei, D.W., Xu, Z.B.: Rough set models on two universes. International Journal of General Systems 33(5), 569–581 (2004)
11. Skowron, A., Stepaniuk, J.: Tolerance approximation spaces. Fundamenta Information (27), 245–253 (1996)
12. Sun, B.Z., Gong, Z.T., Chen, D.G.: Fuzzy rough set for the interval-valued fuzzy information systems. Information Science (178), 2794–2815 (2008)
13. Sun, B.Z., Ma, W.M.: Fuzzy rough set model on two different universes and its application. Applied Mathematical Modelling (35), 1798–1809 (2011)
14. Shafer, G.: Belief functions and possibility measures. In: Bezdek, J.C. (ed.) Analysis of Fuzzy Information, vol. (1), pp. 51–84. CRC Press, Boca Raton (1987)
15. Wong, S.K.M., Wang, L.S., Yao, Y.Y.: Interval structures: a framework for representing uncertain information. In: Proceeding of 8th Conference on Uncertainty Artificial Intelligent, pp. 336–343 (1993)
16. Wong, S.K.M., Wang, L.S., Yao, Y.Y.: On modeling uncertainty with interval structures. Computer Intelligent (11), 406–426 (1993)
17. Wu, W.Z., Zhang, W.X.: Constructive and axiomatic approaches of fuzzy approximation operators. Information Science (159), 233–254 (2004)
18. Yao, Y.Y.: Probabilistic rough set approximations. International Journal of Approximate Reasoning 49(2), 255–271 (2008)

19. Yao, Y.Y.: Constructive and algebraic methods of the theory of rough sets. Information Science (109), 21–47 (1998)
20. Yao, Y.Y.: Relational interpretations of neighborhood operators and rough set approximation operators. Information Science (111), 239–259 (1998)
21. Yao, Y.Y., Wong, S.K.M., Wang, L.S.: A non-numeric approach to uncertain reasoning. International Journal of General Systems (23), 343–359 (1995)
22. Yan, J.A.: Theory of measures. Science Press, Beijing (1998)
23. Zhang, H.Y., Zhang, W.Z., Wu, W.Z.: On characterization of generalized interval-valued fuzzy rough sets on two universes of discourse. International Journal of Approximate Reasoning 51(1), 56–70 (2009)
24. Zhang, W.X., Wu, W.Z.: Rough set models based on random sets(I). Journal of Xi'an Jiaotong University (12), 75–79 (2000)
25. Zadeh, L.A.: Fuzzy sets. Information & Control (8), 338–353 (1965)
26. Ziarko, W.: Probabilistic approach to rough sets. International Journal of Approximate Reasoning 49(2), 272–284 (2008)

Rough Optimizations of Complex Expressions in Infobright's RDBMS

Dominik Ślęzak[1,2], Piotr Synak[2], Jakub Wróblewski[2], Janusz Borkowski[2], and Graham Toppin[3]

[1] Institute of Mathematics, University of Warsaw, ul. Banacha 2, 02-097 Warsaw, Poland
[2] Infobright Inc., Poland, ul. Krzywickiego 34, lok. 219, 02-078 Warsaw, Poland
[3] Infobright Inc., Canada, 47 Colborne St., Suite 403, Toronto, ON M5E1P8 Canada
{slezak,synak,jakubw,januszb,toppin}@infobright.com

Abstract. We discuss the importance of analytic SQL statements with complex expressions in the business intelligence and knowledge discovery applications. We report the recent improvements of the execution of complex expressions in the Infobright's RDBMS, which is based on the paradigms of columnar databases and adaptive rough computations over the granulated metadata layer.

Keywords: RDBMS, Analytic SQL, Complex Expressions, Rough Computing.

1 Introduction

An important trend in the IT industry relates to the RDBMS solutions specialized in database analytics, aimed at advanced reporting and ad-hoc querying against massive amounts of data. Infobright's technology[1] is an example of such a solution, optimized particularly with regard to the analysis and exploration of rapidly growing machine-generated data sets[2]. Infobright's approach to solving the underlying computational scalability problems is based on a specific application of rough computing [1] in combination with the principles of columnar databases [2]. An important ingredient here is a layer of fast heuristic algorithms that attempt to minimize and optimize the data access during the query execution [3]. There is an ongoing process of improving this layer by means of new ideas and implementations. This paper reports one of the areas of such improvements, dedicated to SQL statements with complex expressions.

We concentrate on complex expressions that can be interpreted as dynamically derived columns of the original or intermediately created tables involved into the analytic SQL statement execution. Such expressions may occur in the conditions (e.g.: `WHERE expression = ...`), aggregations (e.g.: `SELECT SUM(expression)`), groupings (e.g.: `SELECT expression as A ... GROUP BY A`) et cetera. They may take a form of date functions, arithmetic operations, conditional expressions et cetera

[1] `www.infobright.org, www.infobright.com`
[2] `en.wikipedia.org/wiki/machine-generated_data`

J.T. Yao et al. (Eds.): RSCTC 2012, LNAI 7413, pp. 94–99, 2012.

[4]. They may be also used to combine the results of the aggregate functions and they may involve or be involved in the correlated subqueries [5].

The queries with expressions often occur in business intelligence and decision making applications [6]. The string functions may be useful in the analysis of the abovementioned machine-generated data, where web logs, url addresses et cetera take a form of long varchar columns [7]. Expressions are also important in the knowledge discovery applications. Consider, e.g., the task of the feature extraction [8], where the SQL-based scripts can derive new useful attributes from a relational database [9]. As another example, consider the methods of SQL-based machine learning, such as the decision tree construction proposed in [10], which can be further extended by searching for the cuts on linear combinations of the original columns [11]. In all such cases, expressions represent the new dimensions of a decision model. The resulting model can be represented by expressions too, e.g.: boolean expressions defining the root-to-leaf paths in a tree, or arithmetic expressions defining the clusters of homogeneous rows.

In this paper, we show how to optimize the Infobright's process of resolving the analytic SQL statements with expressions. In Section 2, we recall the foundations of the discussed database architecture, emphasizing its relationships to rough approximations. In Section 3, we outline the drawbacks of our previous implementation of expressions. We follow up with an overview of possible improvements and the solution introduced in the latest version of our product. Section 4 concludes the paper.

2 Infobright's Architecture

Infobright's RDBMS is an example of a rough-columnar database engine. It combines the columnar data storage/processing [2] with the metadata layer containing the granulated tables used to minimize the data access intensities, basing on the principles of rough sets [1]. Rows of a granulated table (further called rough rows) correspond to some partition blocks of the original data rows. Columns of a granulated table (further called rough attributes) store statistics (further called rough values) describing the original data columns within particular blocks of rows. The rough values may contain information such as min/max values (interpreted specifically for different data types), a sum of values (or, e.g., a total length of strings), a number of nulls et cetera.

New rough values are computed after receiving each block of 2^{16} freshly loaded rows. The gathered rows correspond to a new rough row in the granulated table. The engine first decomposes rows onto particular columns' values and creates data packs – collections of 2^{16} values of each column. In other words, each data pack corresponds to a single block of rows and a single data column. Each data pack is represented by a rough value summarizing its statistics at a metadata level. Last but not least, each of data packs can be compressed independently. Figure 1 illustrates the overall framework of loading, storing and querying the data. As displayed, rough values can be applied to categorize some data packs as not requiring access with respect to the query conditions. However, rough values can assist also in resolving other parts of SQL clauses, such as aggregations, different forms of joins, correlated subqueries et cetera [3].

Sticking with the example of the `WHERE` clause, there are two cases when the given portion of data can be categorized as not requiring further access. The first case is to

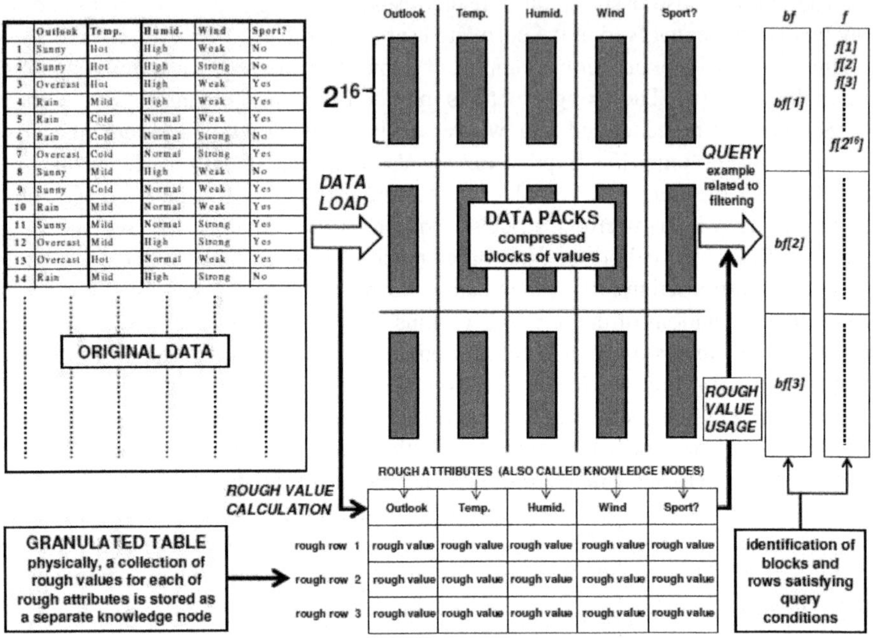

Fig. 1. Loading, storing and querying in Infobright's RDBMS. The metadata layer containing the granulated tables is often called the Infobright's knowledge grid. The contents of rough attributes are stored in so called knowledge nodes. During the query execution, all applicable knowledge nodes are put into memory. Rough values may be used, e.g., to qualify (disqualify) the blocks of rows that satisfy (do not satisfy) the given `WHERE` clause. Only the borderline blocks (precisely, their data packs corresponding to the columns used in the clause) need to be decompressed and examined row by row. The filtering results obtained at the level of blocks and single rows are passed as the arrays denoted by bf and f to the further stages of the query execution.

use rough values against `WHERE` clauses in order to eliminate the blocks that are for sure out of the scope of a given query [12]. The second case occurs when it is enough to use a given block's statistics. It may happen, e.g., when we are sure that all rows in a block satisfy query conditions and, therefore, some of its rough values can represent its contribution into the final result. In our research, we noted an analogy between the two above-discussed cases and the positive/negative regions defined within the theory of rough sets [1]. It helped us to employ various AI-based heuristics aimed at minimizing the amounts of data packs that need to be accessed, i.e., the data portions that cannot be categorized as positive/negative by using their corresponding rough values.

Infobright's performance depends on the quality of rough values and the algorithms that use them. It is worth adding that we assign rough values not only to physical data packs but also to intermediate structures generated during the query execution (e.g.: hash tables used in aggregations). Also, we dynamically produce rough values applicable at further execution stages. This will be better visible in the next section, where we explain why it is worth computing such new rough values for expressions.

3 New Implementation of Complex Expressions

The presented approach enables Infobright to gather positive feedback in the areas such as online analytics, financial services and telecommunications, with the largest customers approaching one petabyte of the processed data. Besides rough operations, the speed of analytic queries is assured by the already-mentioned advantages of columnar database design [2], as well as the adaptive query optimization [13] and parallel execution of data operations [14]. On the other hand, a challenge in front of technologies combining many architectural aspects is to assure stable performance, i.e., comparable execution speed of the SQL statements differing only in details.

One of our problems with regard to the above-mentioned performance stability used to refer to the way of processing complex expressions when compared to the original data columns. As already stated, expressions can occur in many places of the SQL syntax, including filters, attributes to be aggregated et cetera. On the other hand, for the users of the business intelligence and data mining tools, it is quite natural to expect that the speed of execution of the analytic queries does not decrease significantly when replacing a single column by an expression involving several columns. Using a simple example, the users realize that execution of the clause `SELECT X+Y as A ... GROUP BY A` takes longer than `SELECT X ... GROUP BY X`, but they will not accept an order of magnitude difference in the query execution time.

Prior to the recent 4.1 release of Infobright's RDBMS, the default way of processing expressions relied on the MySQL code. Basically, we treated expressions as additional data columns with dynamically derivable values and we internally used the MySQL structures to compute a given expression's value for each particular row. Interpretation of expressions as such additional columns may be actually compared to creation of features of information systems corresponding to the physical data tables or intermediate (although not necessarily materialized) tables resulting from joins, orderings, groupings et cetera, treated as dynamically derived information systems [15].

Such a strategy enabled us to quickly extend the query functionality onto all expressions implemented within MySQL. On the other hand, there were three issues: 1. Computation based on MySQL structures was slightly slower than expected; 2. The MySQL code turned out as not fully thread safe [16], which limited our abilities to parallelize some steps of the execution of SQL statements with expressions [17]; 3. There was no rough computation support for the dynamically derivable columns corresponding to expressions, i.e., using the above example of query `SELECT X+Y as A ... GROUP BY A`, the rough values for `A` did not exist, so they could not help in optimizing the access to data packs of the underlying data columns `X` and `Y`.

In order to solve the two first problems, we re-implemented the code for computing complex expressions by ourselves. Although time consuming, this development project was finished successfully, providing the functional coverage for a wide range of expressions useful in the database analytics. However, addressing the challenges only at the level of the row-by-row calculations was not sufficient to obtain the performance comparable to queries with no expressions. Therefore, some algorithms extracting rough values of complex expressions for particular blocks of rows or, more generally, portions of data involved in the query execution turned out to be necessary.

Table 1. Functions & operators optimized at the rough level in Infobright 4.1

Logical
=
<>, ! =
<=
<
>
>=
BETWEEN
IN
NOT, !
AND, &&
OR
XOR
NOT BETWEEN
COALESCE
NOT IN
ISNULL
CASE
IF
IFNULL
NULLIF
IS NULL
IS NOT NULL

Numerical
+
−
*
/
%
DIV
ABS
EXP
LOG10
LOG2
LOG
SQRT
FLOOR
BIN
OCT

String
LIKE
NOT LIKE
CONCAT
LENGTH
SUBSTR
SUBSTRING
INSTR
LOCATE
LEFT
MID
RIGHT
LOWER
LCASE
UPPER
UCASE

Others
INET_NTOA

Date/Time
CURDATE
CURRENT_DATE
CURRENT_TIME
DATE
DATEDIFF
DAY
HOUR
MONTH
YEAR
CURRENT_TIMESTAMP
CURTIME
DAYOFMONTH
DAYOFYEAR
LOCALTIME
LOCALTIMESTAMP
EXTRACT
MINUTE
NOW
QUARTER
SECOND
TIME
TO_DAYS

There are various potential ways of providing rough values of expressions to the Infobright's algorithms minimizing the data access. For instance, a popular research trend in the database industry is to use the statistics of the occurrence of expressions in the query logs. In our case, it would mean memorizing the knowledge nodes of some of dynamically derived columns corresponding to the occurred expressions and reusing them in the future queries. However, the question then arises how to maintain the right expressions, so the overall size of the metadata layer does not grow in an uncontrolled way. This is especially difficult to address in the case of ad-hoc query workloads, where each next statement may contain slightly different expressions.

We chose another solution, which is implementing at the rough level the selected group of popular types of expressions. Table 1 displays the functions and operators, which are supported in this way in the Infobright 4.1 release. This means that for a given block of rows or, in other words, its corresponding rough row in the granulated table the rough value of a new attribute corresponding to the complex expression is computed basing on the rough values of data columns involved in the expression's definition. Such dynamically created rough values are used by the query optimization and execution algorithms exactly in the same way as the rough values of the original data columns. The way of computing rough values for particular types of expressions turned out to be an interesting research task, closely related to the principles of granular and interval computing [18]. The tests conducted over real-world data sets prove that this strategy allows us to achieve the wanted stability of the query performance.

4 Conclusions

We presented the recently improved Infobright's implementation of SQL statements with expressions. It is based on computing the metadata statistics for attributes corresponding to expressions, so they can be processed similarly to the data columns.

References

1. Pawlak, Z., Skowron, A.: Rudiments of Rough Sets. Information Sciences 177(1), 3–27 (2007)
2. White, P., French, C.: Database System with Methodology for Storing a Database Table by Vertically Partitioning all Columns of the Table. US Patent 5,794,229 (1998)
3. Ślęzak, D., Eastwood, V.: Data Warehouse Technology by Infobright. In: Proc. of the 2009 ACM SIGMOD Int. Conf. on Management of Data (SIGMOD), pp. 841–846 (2009)
4. Pöss, M., Nambiar, R.O., Walrath, D.: Why You Should Run TPC-DS: A Workload Analysis. In: Proc. of the 33rd Int. Conf. on Very Large Data Bases (VLDB), pp. 1138–1149 (2007)
5. Synak, P.: Rough Set Approach to Optimisation of Subquery Execution in Infobright Data Warehouse. In: Proc. of the Int. Workshop on Soft Computing for Knowledge Technology (SCKT), Hanoi University of Technology (2008)
6. Davenport, T.H., Harris, J.G.: Competing on Analytics - The New Science of Winning. Harvard Business School Press (2007)
7. Ślęzak, D., Toppin, G.: Injecting Domain Knowledge into a Granular Database Engine: A Position Paper. In: Proc. of the 19th ACM Conf. on Information and Knowledge Management (CIKM), pp. 1913–1916 (2010)
8. Liu, H., Motoda, H. (eds.): Feature Extraction, Construction and Selection: A Data Mining Perspective. Kluwer Academic Publishers (1998)
9. Wróblewski, J.: Analyzing Relational Databases Using Rough Set Based Methods. In: Proc. of the 8th Int. Conf. on Information Processing and Management of Uncertainty (IPMU), Part I, pp. 256–262 (2000)
10. Nguyen, H.S., Nguyen, S.H.: Fast Split Selection Method and Its Application in Decision Tree Construction from Large Databases. International Journal of Hybrid Intelligent Systems 2(2), 149–160 (2005)
11. Nguyen, H.S.: From Optimal Hyperplanes to Optimal Decision Trees. Fundamenta Informaticae 34(1-2), 145–174 (1998)
12. Metzger, J.K., Zane, B.M., Hinshaw, F.D.: Limiting Scans of Loosely Ordered and/or Grouped Relations Using Nearly Ordered Maps. US Patent 6,973,452 (2005)
13. Deshpande, A., Ives, Z.G., Raman, V.: Adaptive Query Processing. Foundations and Trends in Databases 1(1), 1–140 (2007)
14. Hellerstein, J.M., Stonebraker, M., Hamilton, J.: Architecture of a Database System. Foundations and Trends in Databases 1(2), 141–259 (2007)
15. Pawlak, Z.: Information Systems - Theoretical Foundations. Information Systems 6, 205–218 (1981)
16. Oracle: Multithreaded Programming Guide (Beta). Online Documentation (2010)
17. Borkowski, J.: Performance Debugging of Parallel Compression on Multicore Machines. In: Wyrzykowski, R., Dongarra, J., Karczewski, K., Wasniewski, J. (eds.) PPAM 2009, Part II. LNCS, vol. 6068, pp. 82–91. Springer, Heidelberg (2010)
18. Kreinovich, V.: Towards Faster Estimation of Statistics and ODEs Under Interval, P-Box, and Fuzzy Uncertainty: From Interval Computations to Rough Set-Related Computations. In: Kuznetsov, S.O., Ślęzak, D., Hepting, D.H., Mirkin, B.G. (eds.) RSFDGrC 2011. LNCS (LNAI), vol. 6743, pp. 3–10. Springer, Heidelberg (2011)

Rough Sets Based Inequality Rule Learner for Knowledge Discovery

Yang Liu[1,2], Guohua Bai[2], Qinglei Zhou[1], and Elisabeth Rakus-Andersson[3]

[1] School of Information Engineering,
Zhengzhou University, Zhengzhou, Henan 450001, China
[2] School of Engineering, Blekinge Institute of Technology,
Karlskrona, Blekinge 371 79, Sweden
[3] School of Applied Mathematics, Blekinge Institute of Technology,
Karlskrona, Blekinge 371 79, Sweden

Abstract. Traditional rule learners employ equality relations between attributes and values to express decision rules. However, inequality relationships, as supplementary relations to equation, can make up a new function for complex knowledge acquisition. We firstly discuss an extended compensatory model of decision table, and examine how it can simultaneously express both equality and inequality relationships of attributes and values. In order to cope with large-scale compensatory decision table, we propose a scalable inequality rule leaner, which initially compresses the input spaces of attribute value pairs. Example and experimental results show that the proposed learner can generate compact rule sets that maintain higher classification accuracies than equality rule learners.

Keywords: Classification, machine learning, rough sets, rule induction, inequality rule.

1 Introduction

Knowledge acquisition methods have caused widespread concerns [1]. Knowledge discovered is often modeled by the relations between attributes and values. This line of research ideas is closely related to rule learner [2]. In the classical rough sets model, the upper or lower approximations are defined by elementary sets, i.e., the collections of objects that are taken the same value with respect to a set of attributes [3]. The relationship between attributes and values is equality. Both certain rules and possible rules can be generated from lower and upper approximations. Within knowledge acquisition systems, knowledge is usually expressed as IF-THEN production rule form. The conditional part of a rule is a conjunction of equalities over conditional attributes. Rules with this form of representation can be consequently embedded in classification module [4].

The main goal of rule learner is to improve the simplicity and prediction accuracy of mined rules [5]. Rules that are expressed by complex forms are hardly for people to comprehend. Therefore, it may lead to over-fit training data.

J.T. Yao et al. (Eds.): RSCTC 2012, LNAI 7413, pp. 100–105, 2012.

In order to improve the efficiency, rule learners are designed by "simplicity first" methodology to guarantee that the extracted rules are simple representation and at the same time they can maintain comparable classification accuracy on testing data [6]. Most of rule learners generate rules that express relationship between attributes and values. Numerous types of relationship can be used to expand rule form, such as equivalence relation, similarity relation and order relations [7]. The common characteristics of these models are their expression of the relationship between attributes and values. Therefore, these methods are particularly important to mine domain knowledge [8].

In first order logic, antecedents of a IF-THEN implication is a conjunction of positive or negative propositions. A negative proposition may require hundreds of its complement positive propositions to express its meanings. Unless data contains the complete information, mining only equality rule forms is hardly to generate a complete set of rules. Therefore, the efficiency of classification module will be affected negatively [2]. Cios et al. conducted exploratory rule induction method that generates production rules that exclusively use inequalities in the conditional part of generated rules, instead of the equalities generated by other learners. But the extracted rules are still more complicated [9]. In [10], we builded a series of extended models of decision table to express both equality and inequality relationships. Complexity analysis observed that the compensatory model exhibits square space complexity, which is bottleneck for the classical rough sets based rule learners. However, none of these works focuses on the framework of design effective inequality involved rule induction method.

2 Compensatory Model for Expressing Inequality Relations

The core task of knowledge acquisition is the expression of acquired knowledge. Most approaches directly build relationships of attributes and values. Yao et al. extended classical model for mining ordered relationships, i.e., $<, >, \leqslant, \geqslant$ [11]. In practice, inequality plays a very important role in decision analysis. For example, an image recognition task is to distinguish plates and non-plate problem. For simplicity, we assume that plates are circular. The traditional rule learner may produce rules represent the concept of non-plate as "IF Shape = rectangle THEN is not a plate", "IF shape = Star THEN is not a plate" and "IF shape = Round THEN is a plate". However, this learning procedure cannot learn the nature of knowledge. If a new shape of object is added to system, the robot will be difficult to recognize this object. Therefore, a classification rule contains inequality can solve above problem, such as "IF Shape $\neq$ round THEN is not a plate". When a triangle object is added, the inequality rule can classify it to a correct decision class. Therefore, inequality as a complement of equality can help learning incomplete data sets.

Definition 1. *Given a decision table* $S = (U, A = C \cup \{d\}, \{V_a | a \in A\}, \{f_a | a \in A\})$, *its compensatory model is* $S^* = (U, C^* \cup \{d\}, \{V^*_{p^v_a} | p^v_a \in C^*\} \cup \{V_d\}, \{f^*_{p^v_a} | p^v_a \in C^*\} \cup \{f_d)\}$, *Where:*

$$C^* = \{p_a^v | a \in C, v \in V_a\};$$

$$V^*_{p_a^v} \in \{0,1\} \text{ for } p_a^v \in C^*;$$

$$f^*_{p_a^v}(x) = \begin{cases} 1, & \text{if } f_a(x) = v, \\ 0, & \text{otherwise.} \end{cases}$$

By definition, the compensatory model of decision table is a binary valued decision table. Assume that $m = |U|$ and $n = |C|$, i.e., the decision table contains m objects and n conditional attributes. For any decision table, each grid can be considered as information content. As we know in fundamental information theory, encoding complexity of information X is $\log_2 |X|$ bits. Therefore, the encoding complexity of extended model can be calculated by its grids.

Theorem 1. *For any attribute $a \in C$, If $|V_a|$ is constant, then the asymptotic encoding complexity of compensatory model S^* is $O(nm)$.*

Proof. Because $|V_a|$ is constant, $c = max_{a \in C}|V_a|$ is also a constant. The compensatory model of decision table S^* has at most m rows and nc columns, and for each grid a bit of encoding complexity is needed. Therefore, encoding complexity is $O(nmc) = O(nm)$. ■

Theorem 1 shows that when the range of attribute values is constant relative to the size of objects, the compensatory model exhibits linear encoding complexity.

3 An Inequality Rule Learner

Let us denote B is the lower or upper approximation of a concept set from the training data. Rule learner firstly uses a procedure to compress input set. The reduced input set is used to mine rules so that it can reduce search space and further improve its efficiency on large-scale data.

1) The compression procedure

The input sets are optimized by a data reduction method before searching for a local covering. It is used the similar idea of [6]. The procedure is a data reduction method with linear time complexity, which is used to reduce searching space in the process of finding attribute-value pairs. The reduced set characterizes the common attribute-value information in original input set, and it additionally records the distribution of cases it generalized. Thus, the reduced set is called as a *generalized set* and elements in this set are called as *generalized cases*.

2) Inequality rule learner

Rule learner uses a different heuristic method for selecting candidate attribute-value pairs compared with algorithm LEM2. In order to check wether a specific generalized case is included in original case space U, we introduce a *consistent check function*, which is a boolean function that returns true if a given generalized case is included in a given block set.

Definition 2. *Let* $T = \{t|t = (a, v), a \in A, v \in V\}$ *be a set of attribute value pairs, the block of* T *is* $[T] = \{x|f(x, a) = v, x \in U\}$, C *is a conditional attributes set,* $g \in G$ *is a generalized case, the consistency check function is:*

$$Cons_C(g, T) = \begin{cases} \textbf{\textit{false}} \text{ } \textit{if } (\exists x \in [T])(\exists c \in C)(x_c \neq ? \wedge g_c \neq ? \wedge x_c \neq g_c), \\ \textbf{\textit{true}} \text{ } \textit{otherwise.} \end{cases} \quad (1)$$

Definition 3. *Given* $a \in C$ *as a conditional attribute,* $v \in V_a$ *is a value of the attribute,* $| V_a |$ *denotes the number of values of attribute* a*. The score function of the attribute value pair relative to generalized set G is defined as:*

$$Score_G(a, v) = \sum_{g \in G \wedge g_a = v} g_{num} * \mid V_a \mid . \quad (2)$$

where g_{num} *is the occurrence of the skeleton* g *in original case space* U*.*

The pseudocode of proposed rule learner QLEM2 presents below.

Algorithm: QLEM2 Rule Learner
Input: A non-empty generalized set G, conditional attribute set C.
Output: A local covering set $\mathbb{T}$.

$\mathbb{T} \Leftarrow \emptyset$;
while $G \neq \emptyset$ **do**
  $T \Leftarrow \emptyset$, $TmpG \Leftarrow G$;
  while $(T = \emptyset)$ **or** $([T] \nsubseteq B)$ **do**
    $t \Leftarrow \arg\max_{(a,v) \notin T} Score_G(a, v)$;
    $T \Leftarrow T \cup \{t\}$;
    $G \Leftarrow \{g \in G | Cons_C(g, T)\}$;
  end while
  for $\forall t \in T$ **do**
    if $[T \setminus \{t\}] \subseteq B$ **and** $T \setminus \{t\} \neq \emptyset$ **then**
      $T \Leftarrow T \setminus \{t\}$;
    end if
  end for
  $\mathbb{T} \Leftarrow \mathbb{T} \cup \{T\}$;
  $G \Leftarrow TmpG \setminus \{g \in TmpG | Cons_C(g, T)\}$;
end while
for $\forall T \in \mathbb{T}$ **do**
  if $\cup_{S \in \mathbb{T} \setminus \{T\}} [S] = B$ **then**
    $\mathbb{T} \Leftarrow \mathbb{T} \setminus \{T\}$;
  end if
end for
return $\mathbb{T}$;

Algorithm QLEM2 uses one threshold to reinforce the stopping condition. QLEM2 uses default value for the threshold, which is set to zero, unless a user specifies alternative value. The threshold is used to prune minimal complex for producing very specific or complex rule forms.

4 Experiment

Now we focus on comparing accuracy and rule complexity between equality rule learner and inequality one on a variety of realistic datasets. A detailed description of datasets is presented in Table 1. The datasets were obtained from the UCI machine learning repository [12]. The continuous attributes are firstly discretized by CAIM tool [13]. In the experiment, QLEM2 generates rules contain inequality relations from compensatory model of decision table. The threshold parameter is set to be 0.015. DataSqueezer [6] is a well-known equality based rule learner, whose goal is to gradually add the candidate attribute-value pair of rules to increase the generalization ability of rule sets.

Table 1. Description of data sets

Abbr.	data set	# cases	# classes	# attributes
tae	TA Evaluation	151	3	5
hea	StatLog heart disease	270	2	13
cle	Cleve database	303	2	14
bos	Boston hoursing	506	3	13
cra	Credit approval	690	2	15
aca	Australian credit approval	690	2	14
gec	StatLog German credit data	1000	2	20
hyp	Hypothyroid disease	3163	2	25

We repeat ten times 10-fold cross validation on each dataset. The results of classification accuracy and rule complexity are shown in Table 2. We only evaluate rules with confidence up to 0.01. In this table, "# rules" represents the number of rules extracted, and "L/R" means the average length per rule. The size of rule set and average length per rule are used as measurements of rule complexity. To summarize, QLEM2 is characterized by achieving higher classification accuracy than DataSqueezer, while it also generates smaller size of compact rule set.

Table 2. Comparison results of QLEM2 and equality rule learner

data sets	DataSqueezer			QLEM2		
	accuracy	# rules	L/R	accuracy	# rules	L/R
tae	55	21	2.5	59	19	2.5
hea	79	7	3.2	82	14	2.4
cle	46	16	3.0	61	18	2.8
bos	70	22	5.2	73	21	4.3
cra	73	27	3.6	83	25	4.5
aca	62	20	2.5	76	14	3.3
gec	80	134	3.3	81	63	4.6
hyp	95	15	4.2	95	14	4.5
Mean	70.0	32.8	3.4	76.3	23.5	3.6

5 Conclusion

This paper examines a compensatory model of decision table, which provide rule leaner a function of mining rules with inequalities. While the rule learner in LERS system generates rules directly from lower and upper approximations of a concept, the proposed learner QLEM2 is based on a reduced set and searches minimal complex from the optimized input set. Because the proposed learner targets at finding rules in the form of both equality and inequality relations, it is efficient to generate compact rule set. Experimental results show that the proposed learner obtains more concise rule sets without sacrificing classification accuracy than state of the art rule learner.

References

1. Skowron, A.: Extracting laws from decision tables: a rough set approach. Computational Intelligence 11(2), 371–388 (2007)
2. Ruckert, U., De Raedt, L.: An experimental evaluation of simplicity in rule learning. Artificial Intelligence 172(1), 19–28 (2008)
3. Grzymala-Busse, J., Rzasa, W.: A Local Version of the MLEM2 Algorithm for Rule Induction. Fundamenta Informaticae 100(1), 99–116 (2010)
4. Wang, G.: Rough Set Based Uncertain Knowledge Expressing and Processing. In: Kuznetsov, S.O., Ślęzak, D., Hepting, D.H., Mirkin, B.G. (eds.) RSFDGrC 2011. LNCS, vol. 6743, pp. 11–18. Springer, Heidelberg (2011)
5. Janssen, F., Furnkranz, J.: On the quest for optimal rule learning heuristics. Machine Learning 78(3), 343–379 (2010)
6. Kurgan, L.A., Cios, K.J., Dick, S.: Highly scalable and robust rule learner: Performance evaluation and comparison. IEEE Transactions On Systems Man and Cybernetics 36, 32–53 (2006)
7. Blaszczynski, J., Slowinski, R., Szelag, M.: Sequential covering rule induction algorithm for variable consistency rough set approaches. Information Sciences (2010)
8. Pawlak, Z., Skowron, A.: Rough sets and Boolean reasoning. Information Sciences 177(1), 41–73 (2007)
9. Cios, K.J., Kurgan, L.A.: CLIP4: Hybrid inductive machine learning algorithm that generates inequality rules. Information Sciences 163(1-3), 37–83 (2004)
10. Liu, Y., Bai, G., Feng, B.: On mining rules that involve inequalities from decision table. In: 7th IEEE International Conference on Cognitive Informatics, pp. 255–260. IEEE (2008)
11. Yao, Y., Zhou, B., Chen, Y.: Interpreting Low and High Order Rules: A Granular Computing Approach. In: Kryszkiewicz, M., Peters, J.F., Rybiński, H., Skowron, A. (eds.) RSEISP 2007. LNCS (LNAI), vol. 4585, pp. 371–380. Springer, Heidelberg (2007)
12. Frank, A., Asuncion, A.: UCI machine learning repository (2010), `http://archive.ics.uci.edu/ml`
13. Kurgan, L.A., Cios, K.J.: CAIM discretization algorithm. IEEE Transactions on Data and Knowldge Engineering 16(2), 145–153 (2004)

A Computer Aided Diagnosis System for Breast Cancer Using Support Vector Machine

Omar S. Soliman[1] and Aboul Ella Hassanien[2]

[1] Cairo University, Faculty of Computers and Information, Cairo, Egypt
Dr.omar.soliman@gmail.com

[2] Cairo University, Faculty of Computers and Information, Cairo, Egypt,
Scientific Research Group in Egypt
abo@egyptscience.net
www.egyptscience.net

Abstract. This article introduces a computer aided diagnosis scheme using support vector machine, in conjunction with moment-based feature extraction. An application of ultrasound breast cancer imaging has been chosen and computer aided diagnosis scheme have been applied to see their ability and accuracy to classify the breast cancer images into two outcomes: cancer or non-cancer. The introduced scheme starts with a preprocessing phase to enhance the quality of the input breast ultrasound images and to reduce speckle without destroying the important features of input ultrasound images for diagnosis. This is followed by performing the seeded-threshold growing region algorithm in order to identify the region of interest and to detect the boundary of the breast pattern. Then, moment-based features are extracted. Finally, a support vector machine classifier were employed to evaluate the ability of the lesion descriptors for discrimination of different regions of interest to determine whether they represent cancer or not. To evaluate the performance of presented scheme, we present tests on different breast ultrasound images. The experimental results obtained, show that the overall accuracy offered by the employed support vector machine was 98.1%, whereas classification ratio using neural network was 92.8%.

1 Introduction

The breast cancer screening tests performed on a regular basis play a crucial role in reducing the rate of mortality, especially among women ages 50 and older [17]. The screening tests include digital mammography, clinical breast examination, breast self-examination, or a combination of the above. Digital mammography refers to the application of digital system techniques on digital Mammograms. Currently, digital mammography is one of the most promising cancer control strategies since the cause of breast cancer is still unknown. Computer-assisted reading of medical images is a relatively new concept which has been developed during the last 10 years and which is growing into diagnostic radiology. Especially in mammography, image processing techniques and automated pattern recognition schemes is applied to assist radiologists in the interpretation of mammogram.

However due to low contrast of breast cancer ultrasound imaging; the automatic cancer segmentation is still a challenging task. According to the National Cancer Institute,

J.T. Yao et al. (Eds.): RSCTC 2012, LNAI 7413, pp. 106–115, 2012.

each year about 180,000 women in the United States develop breast cancer, and about 48,000 lose their lives to this disease. It is also reported that a woman's lifetime risk of developing breast cancer is 1 in 8 [10,3,17].

Ultrasound is a very useful complementary imaging technique which not only provides a different assessment of the lesion, but also allows detecting very small lesions and analyze dense breasts, Many algorithm were used to detect breast cancer by knowing the region of cancer and classify it to which type will be and if it normal or up normal [10,3,2]. However there are a lot of research works had been done for breast cancer diagnosis.

Cheng et al. [3] introduced an extentive reveiw of the automated breast cancer detection and classification using ultrasound images analysis, and their advanteges and disadvanteges were discussed. Also, Saeys et al. [1] showed that feature selection techniques had been applied in the feild of bioinformatics. Eadie et al. [2] presented a systematic reveiw of computer-assisted diagnosis in diagnostic cancer imaging; they disscused some evidence relating to the use of CAD with various cancers and imaging modalities, and investigates whether CAD provides a benefit to radiologists, with comparisons made between radiologists diagnosing images alone and results from CAD systems. Sree et al. [10] reveiwed various modalities used in breast cancer dediction includeing mammography, breast ultrasound, thermography, magnetic resonance imaging (MRI), positron emission tomography (PET), scintimammography, optical imaging, electrical impedance based imaging, and computed tomography (CT). Shu-Ting et al. [11] investigated diagnosing breast masses in digital mammographyusing feature selection and ensemble methods. They showed that the application of decision tree (DT), support vector machine(SVM) - sequential minimal optimization (SVM-SMO) and their ensembles to solve the breast cancer diagnostic problem were preformed well.

In [4] a detection of HER2 breast cancer biomarker using the opto-fluidic ring resonator (OFRR) biosensor is proposed. The OFRR is utilized for the rapid detection of breast cancer biomarker HER2 ECD. Their results showed that the OFRR is capable of rapidly detecting HER2 ECD in human serum at clinically relevant concentrations in approximately 30 min. An Automatic seeded region growing for color image segmentation is proposed in [5]. Zaim in [6] proposed an automatic segmentation schema of the prostate from ultrasound data using feature-based self organizing map. Kelly et al. [8] proposed a breast cancer detection using automated whole breast ultrasound and mammography in radiographically dense breast. And in [7] introduced a breast cancer detection approch : radiologists' performance using mammography with and without automated whole-breast ultrasound . it provides a demonstration that experienced breast radiologists can learn to interpret 2D AWBU quickly. So radiologists would significantly improve their cancer detection rates in dense-breasted women by adding AWBU to mammography. Tumen et al. [12] introduced components of the experimental framework that we use to explore how recognition accuracy changes with respect to factors including the choice of feature extraction parameters, feature selection, feature combination, and classifier fusion.Talebi et al. [13] developed a genetic active contour algorithm for medical ultrasound image segmentation. In [14] Huang et al. proposed schema for breast cancer dignoses and prediction that combine neural network classifier with entropy based feature selection. Miguel et al.[15] proposed a filtering, segmentation and feature extraction methodology in ultrasound evaluation of breast lesions. Mancas et. al. [16] applied a segmentation using a region-growing thresholding formedical image . Mat-Isa et al. [17] improved screening for cervical cancer using seeded region growing features extraction algorithm.

The aim of this paper is to develop a computer aided diagnosis system for breast cancer based on ultrasound image analysis using support vector machine to classify the suspicious regions of breast ultrasound images into different categories such as benign findings and malignancy. The rest of the paper is organized as follows: The proposed system including preprocessed, segmentation, feature extraction, classification phase are introduced in section 2. In section 3, experimental results and analysis are presented. Where section 4 is devoted to conclusions and further research.

2 The Proposed Computer Aided Diagnosis System for Breast Cancer

The computer aided diagnosis system for breast cancer proposed in this paper is composed of the following four fundamental building phases:

- **Pre-processing** In the first phase of the investigation, a pre-processing algorithms based on basics image processing filters are presented. It is adopted and used to improve the quality of the images and to make the segmentation and feature extraction phases more reliable.
- **Segmentation phase** In the second phase, seeded-threshold growing region algorithm have been used to detect cavity contours region of breast cancer.
- **Feature extraction based on moments** In the third phase, features have been extracted and represented in a database as vector values.
- **Classification using support vector machine:** The last phase is the classification and prediction of new objects, it is dependent on the support vector machine.

The general architecture of the proposed system are depicted in Fig. 1 and these four phases are described in detail in this section along with the steps involved and the characteristics feature for each phase.

2.1 Preprocessing Phase

The aim of preprocessing phase is to enhance the quality of the input breast ultrasound images, to reduce speckle without destroying the important features of input ultrasound images for diagnosis. Preprocessing does not increase image information content, but its main target is to improve images that contain an undesired distortions also it enhances some image features which is very important for feature extraction and classification process. The main steps of preprocessing algorithm is described in algorithm 1.

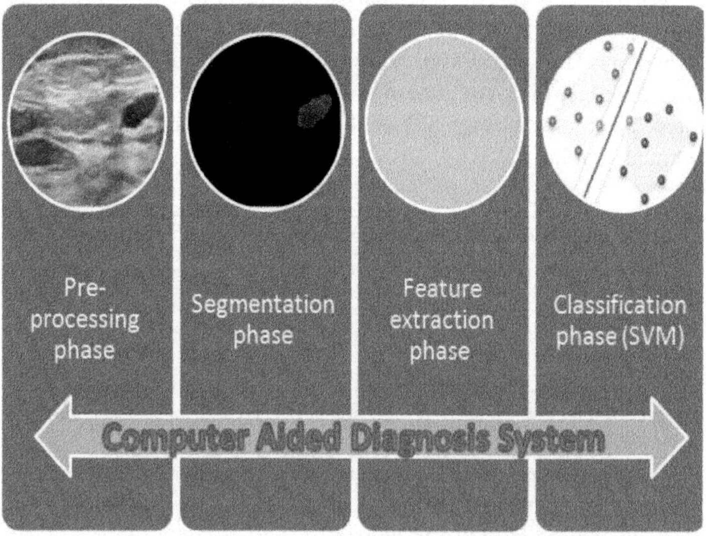

Fig. 1. Main phases of the proposed breast cancer diagnoses system

Algorithm 1. Breast ultrasound image preprocessing algorithm

Input ultrasound breast cancer image.
for each image **do**
  Calculate the average value of input image by dividing sum of all values of image on number of pixels.
  Set threshold equal to average value.
  for each pixel **do**
    if value of pixel < threshold **then**
      Set its output pixel = 0
    else
      Set its output pixel = the value of pixel
    end if
  end for
end for
Obtain the enhanced binary ultrasound breast cancer image.

2.2 Segmentation Phase: Seeded-Threshold Growing Region Algorithm

In this phase, the seeded-threshold growing region (STGR) algorithm is employed to detect cavity contours region of breast cancer by dividing the ultrasound breast image into non-overlapping regions, and separating objects (lesions) from the background. The boundaries of the lesions are delineated for feature extraction and make the region that contains the cancer only in another image. After the threshold to segment the breast image is determined, the region growing is started till detect all region of breast cancer by doing a boundary around the region. In the region growing, the pixel direction is determined based on neighboring information include horizontal, vertical and all directional. If the value of the Horizontal edge is higher than the value of the vertical edge, the pixel direction is Horizontal; if the value of the vertical edge is higher than the value of the horizontal edge, the pixel direction is vertical; otherwise, the pixel direction is all-directional. The main steps the STGR algorithm to segment breast ultrasound image is described in algorithm 2.

Algorithm 2. Breast ultrasound image segmentation using STGR algorithm

```
Input preprocessed ultrasound breast cancer image.
for each image do
    Calculate the average value of input image by dividing sum of all values of image on
    number of pixels.
    Set threshold (T) equal to average value.
    Define seed point
    for each pixel do
        Start region growing algorithm using neighborhood information direction.
        if (|seedpoint| − value of pixel) < T then
            Put the value of pixel in a vector.
            Detect the cancer part by doing boundary around this pixel.
        end if
    end for
    Delete the other parts of the breast.
end for
```

2.3 Feature Extraction Phase: Moment Algorithm

The aim of this phase is to extract the feature vector of breast cancer lesions that can accurately distinguish lesion, non-lesion, benign or malignant. The most effective features should be selected since the feature space could be very large and complex. One set of the useful object descriptors is based on the moments theory. The moment invariants are moment-based descriptors of planar shapes, which are invariant under translational, rotational, scaling, and reflection transformation [9]. The feature vector of each image is extracted using moment algorithm as described in algorithm 3 and these feature vectors are stored for purpose of classification.

2.4 Classification Phase: Support Vector Machine

In a last phase, the moment-based features were used for classification. These feature vectors were used to obtain high classification precision using SVM, which seeks the optimal separating hyperplane between two classes by focusing on the training samples that lie on the class boundaries while discarding other training samples effectively. The simplest classification problem that SVM could deal with is the linearly separable binary classification.

Algorithm 3. Moment algorithm

Input ultrasound breast cancer image.
for each ultrasound image **do**
  Compute the median of each image as:

$$X_m = \sum_{i=1}^{N}\sum_{j=1}^{N} i * X(i,j) / \sum_{i=1}^{N}\sum_{j=1}^{N} X(i,j) \tag{1}$$

$$Y_m = \sum_{i=1}^{N}\sum_{j=1}^{N} j * X(i,j) / \sum_{i=1}^{N}\sum_{j=1}^{N} X(i,j) \tag{2}$$

  Assign each image a matrix $A(p,q)$.
  Fill the matrix $A(p,q)$ using the following equation:

$$A(p,q) = \sum_{m=1}^{N}\sum_{n=1}^{N} X(m,n) * (m - X_m)^p * (n - X_n)^q \tag{3}$$

  Compute feature matrix $E(i,j)$ as:

$$E(i,j) = A(p,q)/A(1,1)^M \tag{4}$$

  where $M = (i+j+2)*(1/2)$
  Convert the feature matrix E(i,j) into feature vector V.
  Store the feature vector V.
end for

3 Experimental Results and Discussions

The proposed system is evaluated and tested using ultrasound of breast cancer images data set. In the preprocessing phase of ultrasound image, is enhanced and the noisy is removed and obtaining a smoothed ultrasound image. The input ultrasound breast cancer image is enhanced without any destroying of its important features by applying the preprocessing algorithm 1. Fig. 2 shows the original and preprocessed breast image. The preprocessed retina image is used as an input image segmentation and detection phases.

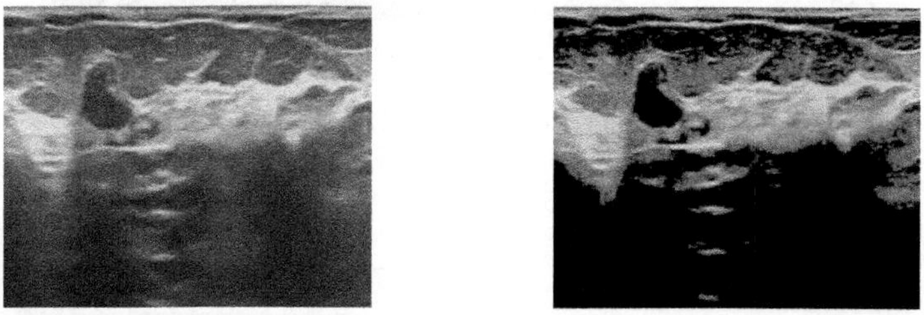

(a) Original breast cancer ultrasound image (b) Enhanced breast cancer ultrasound image

Fig. 2. Preprocessed breast cancer ultrasound image

In the segmentation phase the ultrasound image is segmented using the seeded-threshold growing region (STGR) algorithm, by dividing it into non-overlapping regions, and it separates the objects (lesions) from the background. The boundaries of the lesions are delineated for feature extraction and make the region that contain the cancer only in another image. Fig. 3 shows the enhanced breast cancer ultrasound image and the segmented image that contains only the region of lesions.

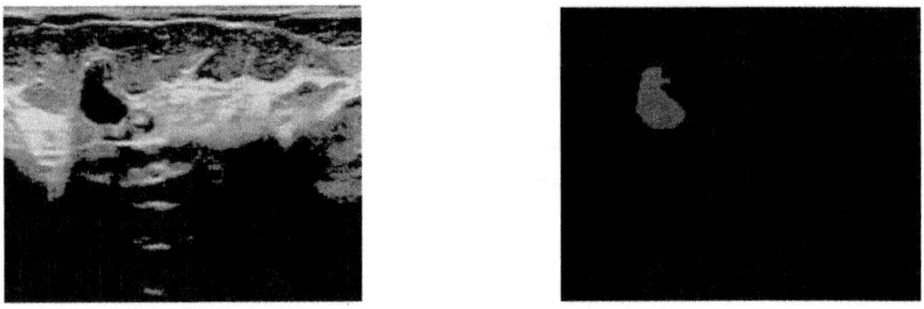

(a) Enhanced breast cancer ultrasound image (b) Segmented breast cancer ultrasound image

Fig. 3. Segmentation of breast cancer ultrasound image

The feature vectors of the segmented ultrasound breast cancer images with lesions region are extracted using the moment algorithm and stored for classification phase. In order to to classify the suspicious regions into different categories, such as benign findings and malignancy. The SVM algorithm is implemented to classify suspicious regions, and to detect weather these regions of ultrasound image benign or malignant cancers. The classification results of suspicious regions with malignant cancers are shown in Fig. 6 and Fig. 5. Where the lesions region benign tumor is shown in Fig. 4.

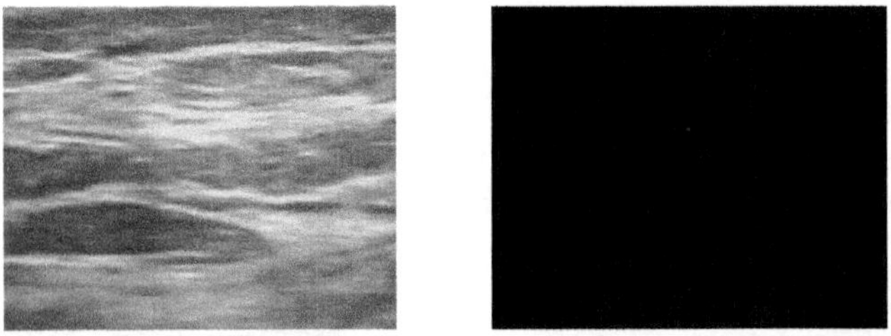

(a) Original breast ultrasound image (b) Normal breast ultrasound images

Fig. 4. Classification results of Normal breast ultrasound image

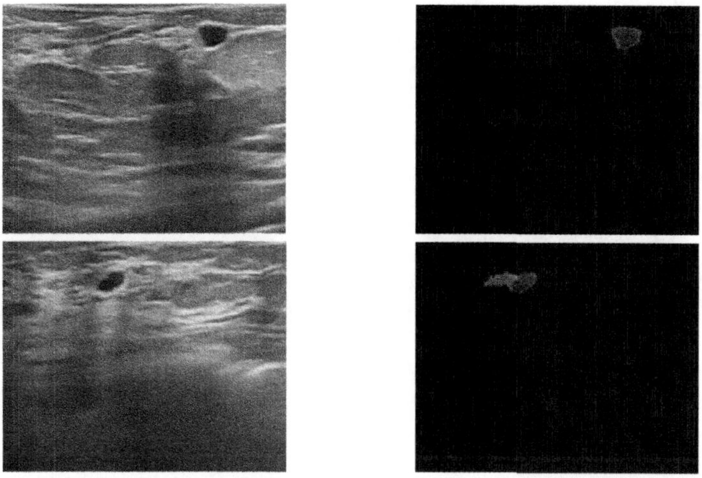

(a) Enhanced breast ultrasound images (b) Detected cancer in breast ultrasound images

Fig. 5. Classification results of detected breast cancer ultrasound image

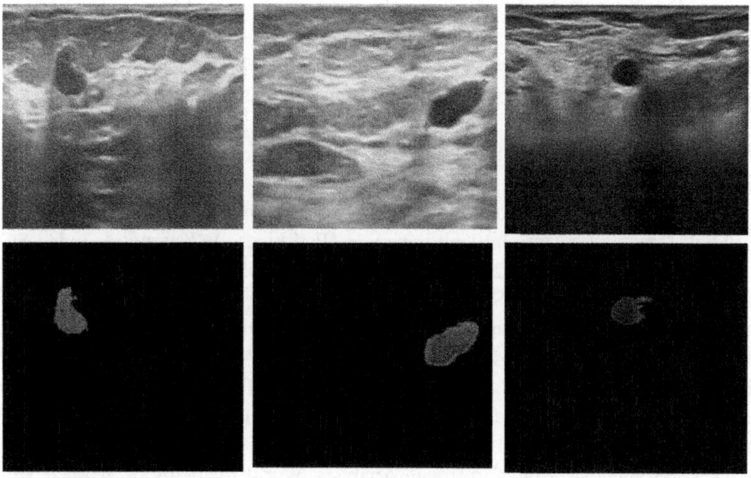

Fig. 6. Classification results of detected breast cancer ultrasound image: The upper row represent original while the lower represented the detected region

4 Conclusions and Further Research

This article introduces a computer aided diagnosis scheme using support vector machine, in conjunction with moment-based feature extraction and seeded-threshold growing region algorithm. An application of ultrasound breast cancer imaging has been chosen and computer aided diagnosis scheme have been applied to see their ability and accuracy to classify the breast cancer images into two outcomes: cancer or non-cancer. To evaluate the performance of our approach, we presented tests on different ultrasound breast images which show that the overall accuracy offered by the employed support vector machine was 98.1%, whereas classification ratio using neural network was 92.8%.

Our future work is to integrate the proposed system with resent advanced machine learning techniques to extract an accurate features vectors and classification of breast cancer diseases digeneses and identification as early as possible.

Acknowledgment. This work has been supported by Cairo University, project Bio-inspired Technology in Women Breast Cancer Classification, Prediction and Visualization.

References

1. Saeys, Y., Inza, I., Larraaga, P.: A review of feature selection techniques in bioinformatics. Journal of Bioinformatics 23(19), 2507–2517 (2007)
2. Eadie, L.H., Taylor, P., Gibson, A.P.: A systematic review of computer-assisted diagnosis in diagnostic cancer imaging. European Journal of Radiology (2011)
3. Cheng, H.D., Shan, J., Ju, W., Guo, Y., Zhang, L.: Automated breast cancer detection and classification using ultrasound images: a survey. Pattern Recogn. 43, 299–317 (2009)
4. Gohring, J.T., Dale, P.S., Fan, X.: Detection of HER2 breast cancer biomarker using the opto-fluidic ring resonator biosensor. Sensors and Actuators, B: Chemical 146(1), 226–230 (2010)
5. Shih, F.Y., Cheng, S.: Automatic seeded region growing for color image segmentation. Image and Vision Computing 23(10), 877–886 (2005)
6. Zaim, A.: Automatic Segmentation of the Prostate from Ultrasound Data Using Feature-Based Self Organizing Map. In: Kalviainen, H., Parkkinen, J., Kaarna, A. (eds.) SCIA 2005. LNCS, vol. 3540, pp. 1259–1265. Springer, Heidelberg (2005)
7. Kelly, K.M., Dean, J., Lee, S.-J., Comulada, W.S.: Breast cancer detection: radiologists' performance using mammography with and without automated whole-breast ultrasound. Eur. J. of Radiology 20, 2557–2564 (2010)
8. Kelly, K.M., Dean, J., Lee, S.-J., Scott Comulada, W.: Breast cancer detection using automated whole breast ultrasound and mammography in radiographically dense breasts. Eur. J. of Radiology 20, 734–742 (2010)
9. Flusser, J.: Moment Invariants in Image Analysis. World Academy of Science. Engineering and Technology 11, 376–381 (2005)
10. Sree, S.V., Ng, E.Y.-K., Acharya, R.U., Faust, O.: Breast imaging: A survey. World J. Clinical Oncology 2(4), 171–178 (2011)
11. Luo, S.-T., Cheng, B.-W.: Diagnosing Breast Masses in Digital MammographyUsing Feature Selection and Ensemble Methods. J. Medical System (2010)
12. Tumen, R.S., Acer, M.E., Sezgin, T.M.: Feature extraction and classifier combination for image-based sketch recognition. In: Proc. SBIM 2010, pp. 63–70 (2010)
13. Talebi, M., Ayatollahi, A., Kermani, A.: Medical ultrasound image segmentation using genetic active contour. J. Biomedical Science and Engineering 4, 105–109 (2011)
14. Huang, M.-L., Hung, Y.-H., Chen, W.-Y.: Neural Network Classifier with Entropy Based Feature Selection on Breast Cancer Diagnosis. J. Medical System 34, 865–873 (2010)

15. Alemn-Flores, M., Alemn-Flores, P., Lvarez-Len, L., Fuentes-Pavn, R., Santana-Montesdeoca, J.M.: Filtering, Segmentation and Feature Extraction in Ultrasound Evaluation of Breast Lesions. In: Bildverarbeitung fr die Medizin, pp. 168–172 (2008)
16. Mancas, M., Gosselin, B., Macq, B.: Segmentation using a region-growing thresholding. In: Image Processing: Algorithms and Systems, pp. 388–398 (2006)
17. Mat-Isa, N.A., Mashor, M.Y., Othman, N.H.: Seeded Region Growing Features Extraction Algorithm; Its Potential Use in Improving Screening for Cervical Cancer. International Journal of The Computer, the Internet and Management 13(1), 61–70 (2005)

A New Method of the Accurate Eye Corner Location*

Yong Yang and Yunxia Lu

Institute of Computer Science & Technology,
Chongqing University of Posts and Telecommunications,
Chongqing, 400065, P.R. China
yangyong@cqupt.edu.cn, 297086784@qq.com

Abstract. Eye corner location is a hot research topic in recent years. A novel eye corner Location method is proposed in the paper. Firstly, a haar features face detection based on adaboost is used to detect the face in an image. Secondly, a step of accurate eye corner location is proposed, which consists of rough eye location, contour extraction ellipse fitting, corner detection and eye corner location. Rough eye location is used for reducing the search range in the image. Then contour extraction based on ellipse fitting is taken. In the following, the curvature scale space(CSS) corner detection operator is used for corners detection. At last, the inner and outer eye corners can be determined according to statistics result of frequency distribution of the corner points projection. The proposed method is proved to be an effective and robust method according to the result of comparative experiments.

Keywords: contour extraction, ellipse fitting, corner detection, eye corner location.

1 Introduction

Eyes are the most important characters in the face. Research on location of eyes is a very hot issue in recent years. There are lots of applications related with eye location. For example, it is very important part for monitoring fatigue drivers based on eye location. Nowadays, the main approaches of eye location and detection adopt the pupil and iris [1–6]. However, there is another method which is more stable and based on eye location instead pupil and iris. It just because the eye corner location is seldom affected by the direction of gaze and the eye states. Furthermore, corner is a kind of significant local features of object in digital image. It can be taken as a feature of image while it can decrease the amount of

* Part of this work is supported by National Natural Science Foundation of China (No. 61075019), Scientific Research Foundation of Chongqing Municipal Education Commission (No. KJ110522, KJ110512), Chongqing Key Lab of Computer Network and Communication Technology Foundation (No. CY-CNCL-2009-02), Natural Science Foundation of Chongqing University of Posts and Telecommunications(No. A2009-26, No. JK-Y-2010002).

J.T. Yao et al. (Eds.): RSCTC 2012, LNAI 7413, pp. 116–125, 2012.

information of the image. Therefore, it is often used in pattern recognition and image analysis. On the other hand, eye corner location is much more difficult than iris location since it is located in skin and influenced by wrinkles, dark circles, swells and cosmetic for eye corner location, and there is no such unique gray scale character of iris.

There are lots of research works on eye location. J.Song used the binary edge and intersity information to locate the eyes[7]. Zhang et al used the knowledge of facial characteristics to extract the eye corners, which through the binarization of the characteristic block[8]. Xu et al presented an angle model, in that model an angle of the eyelid was made, which is based on the semantic features[9].

Based on the aforementioned methods, a new eye corner location method is proposed to avoid the influence from the varying light, wrinkles, dark circles, etc. The proposed method consists of two phases: face detection and accurate eye corner location. Firstly, a haar features face detection based on adaboost is used to detect the face in image. Secondly, a step of accurate eye corner location is proposed, which consists of rough eye location, contour extraction ellipse fitting, corner detection and eye corner location. Rough eye location is used for reducing the search range in image. Then contour extraction based on ellipse fitting is taken. In the following, the CSS corner detection operator is used for corners detection. At last, the inner and outer eye corner can be determined according to statistics result of frequency distribution of the corner points projection. The proposed method is proved to be an effective and robust method according to the result of comparative experiments. The structure of the following paragraph is listed as follows. Face detection is introduced in section 2, proposed method of accurate eye corner location is introduced in section 3, experiments and discussion is introduced in section 4. In the end, there is a conclusion and future work.

2 Face Detection

It must be extensively and inaccurately if eye location is taken directly on a full image. Therefore, a step of face detection should be taken to reduce the searching range. It is shown that combing haar features and adaboost method would be more efficiency and more accuracy compared with other face detection methods [2]. The cascade haar classifier is a classical classifier for face detection, and it is proved to be a good choice for frontal face detection [10]. In this paper, the cascade haar classifier is used for detecting face, and developed based on VC6.0 and OpenCV.

The principle of face detection used in this work is described as follows. Firstly, cascade haar classifier is used to search where the face rectangles are. A series of rectangle regions would be returned. Based on multiple searching, the overlap area on a different scale of sliding window would be detected. The overlap regions which pass through the cascade classifier would be collected to combining a mean rectangular.In the end, the face area can be gotten. It means that the final face location is the average of the multiple detection results.

3 Accurate Eye Corner Location

3.1 Rough Eye Location

According to the cognitive principle of human face, geometric constraints are applied to localize eyes. That is to say, that eyes are only searched in the top half of a face. The eyes in frontal face is not less than 1/2 of the detection region of face, therefore, the precise area of eye is restricted in 1/8 to 1/2 of upper part of face and 1/8 to 7/8 of left part of face in order to reduce the unnecessary interference detection and operation consumption. As shown in Fig. 1.

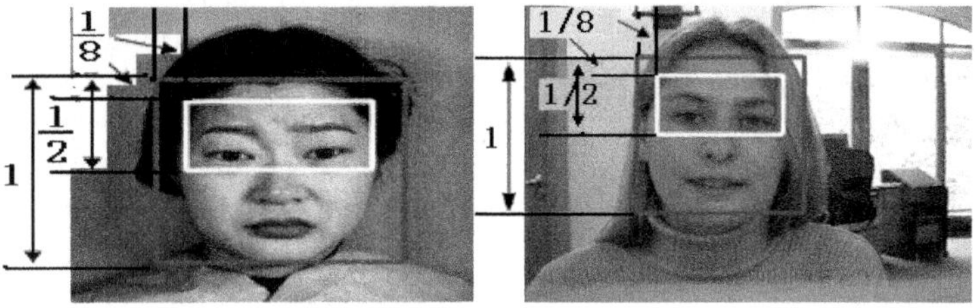

Fig. 1. The Process Schema for Rough Eye Location

3.2 Extraction Eye Contour Based on Ellipse Fitting

The eyelids of human are arcuate, and the veins of skin in eye corner are multi-textures and density. It is a strong edge, and the contour points we obtained from eye corner are closely spaced. Therefore, the ellipse fitting method can be adopted here to match the eyes, and ensure outline points and elliptic curve are roughly superposed. It will separate the outline points from some interferential pouch or wrinkles. As a result, eye contour can be accurately described and the eye corner can be orientated. Since this method is more accurate than hough transform to represent eye corner and is used in this work.

The basic operation of ellipse fitting is described as follows.

Firstly, the outline of a binary image can be extracted and stored by cvFind-Contours function of OpenCV. Secondly, position of a sequence of points can be gotten from the curve. As for a curve containing a series of points, it can be fit by ellipse fitting if it contains no less than 6 points. Because, there are 6 coefficients for complete binary indefinite quadratic equations, so the sequence of points is no less than 6. The ellipse fitting used in this paper is based on Euclidean distance and the least square method [11].

The principle of ellipse fitting is introduced as follows. An Orthogonal distance d_i is defined as a distance between the random point $X_i = (x_i, y_i)$ and the point of tangency $X_t = (x_{ti}, y_{ti})$ on the ellipse $Q(x, y)$ (Fig. 2). The connection of two points and the tangents which pass through the point X_t is orthogonal. Given n points $X_i\,(1, \ldots, n, n \geq 6)$, the minimum value of the following function (2) is the ellipse fitting based on the least square method.

$$F(p) = \sum_{i=1}^{n} d_i^2 \tag{1}$$

The ellipse equation is the following function.

$$Q(x, y) = A(x - x_0)^2 + B(x - x_0)(y - y_0) + C(y - y_0)^2 - 1 = 0 \tag{2}$$

The point of $X_t = (x_{ti}, y_{ti})$ is satisfied with the following two requirements:

$$A(x_t - x_0)^2 + B(x_t - x_0)(y_t - y_0) + C(y_t - y_0)^2 - 1 = 0 \tag{3}$$

$$(y - y_t)\frac{\partial}{\partial x}Q(x_t, y_t) = (x - x_t)\frac{\partial}{\partial y}Q(x_t, y_t) \tag{4}$$

Make$\Delta x = x_t - x_0$,$\Delta y = y_t - y_0$ through calculation and reduction we can get the orthogonal distance as the following function expresses.

$$d = \sqrt{(x - x_0 - \Delta x)^2 + (y - y_0 - \Delta y)^2} \tag{5}$$

Complicated calculations have to be taken here, and four solutions can be obtained. There is only one real solution satisfying *F(p)* to be minimum. Ellipse obtained in this way is the best one.

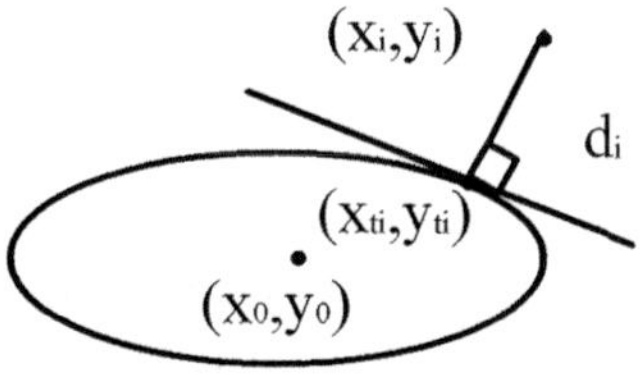

Fig. 2. Orthogonal Distance of a Point to a Conic

3.3 Corner Detection and Eye Corner Location

The eye corner is the intersection area of the upper and lower eyelids [7] which is usually influenced by furrow and cosmetic. The results would be some deviation if detection method based on gray level projection is used directly. Usually, the corner detection method based on the edge profile would be more accurate to get the candidate corner points of eyes compared with method based on gray level projection. In this paper, one of the methods, a corner detection method based on the CSS [12] is used for detecting the corner of curve.

Firstly, from the CSS corner detection we get the the candidate points. Among the candidate points, there are real eye corner points, eyebrows and other interference corners. In order to find eye corner point accurately, a method based on statistics result of frequency distribution of the corner points projection is proposed in this paper. It is based on the distinguish ability of statistics result of frequency distribution. (Fig. 3, Fig. 4) Then the eye corners can be gotten according to the statistics result of frequency distribution of the corner points projection. The basic operation is described as follows.

Obtains these n candidate points through the CSS corner detection. Coordinate these n candidate points on y axis ($y_1 \leq y_2 \leq \ldots \leq y_n$), notes for $hk(x, y)(k = 1, 2, \ldots, n)$, for each y definition $h(x, y)$, a distribution function of projection frequency is used to calculate the projection number $N(Y = y_j)$.

$$N(Y = y_j) = \sum_i |h(x_i, y_j)|, (i = 1, 2, \cdots l; j = 1, 2, \cdots d) \tag{6}$$

Here, l and d is the length and width of image of accurate eye area.

For the purpose of separating the candidate points of eyes from eyebrows, a new variable Δ is defined. Δ denotes pixels of width of template and can be used to separat the candidate points of eyes from eyebrows. It is defined as follows.

$$\Delta = \left\lceil \lambda \times d/l \times \frac{d \cdot l}{s} \right\rceil = \left\lceil \lambda \cdot \frac{d}{l} \right\rceil \tag{7}$$

In (11), l, d and s are the length, width and area of accurate eye area image respectively, and λ is the ratio of images in width and height. In this work, λ is determined on 50 images with coarse position of eye and the image size varying from 60 * 35 to 170 * 90 pixels in BioID[13]. It is determined based on eyes corner accuracy varying with different Δ value (take 5, 6, 7... 20 pixels). And that value is the Δ value when it got the highest accuracy in each of eye corner point Locating image. The λ values in the eye area image are uniform distribution in near function $y = 7 \times \log((x - 25) \times 2)$ while aspect ratio of the image is a certain percentage. So $\lambda = 7 \times \log((l - 25) \times 2)$. When the image is $l \times d = 120 \times 60$, $\lambda = 30$, the effect is the best, and Δ is 15, and the template pixels for Positioning eyes and eyebrows width is 15. Along to the y axis, when meeting the first corner point began to calculation with the templates width. Corner for statistics in Δ is the following function.

$$N_\Delta = \sum_j \sum_i |h(x_i, y_j)|, (i = 1, 2, \cdots l; j = k, \cdots k + \Delta) \tag{8}$$

The inner and outer human eyes are made up of four eye corners, so $N_\Delta \geq 4$. When $N_\Delta < 4$, the point k is defined as a interference point and so repetition of an operation from $k+1$. Repetition of an operation formula (8) from $k+1$ only after meeting $N_\Delta \geq 4$. The final Δ area is the eyes corner area. (See Fig. 4)

In the fitting image corner area of Δ eye, since the two outer eye corners are located in both ends of the x axis, the two outer eye corners are $h(x_{min}, y_{\Delta eye})$ and $h(x_{max}, y_{\Delta eye})$.

Based on the morphological characteristics of eyes, two eye corners are located in between two outer eye corners and near the center line $\Phi = (x_{\min,\Delta eye} + x_{\max,\Delta eye})/2$ of the two outer eye corners, and the two outer corners located on both sides of the Φ and the distance of the two to Φ almost equal.

$$x_{inner1} < \Phi < x_{inner2} \tag{9}$$

$$\left| \frac{x_{inner1} + x_{inner2}}{2} - \Phi \right| \leq 5 \tag{10}$$

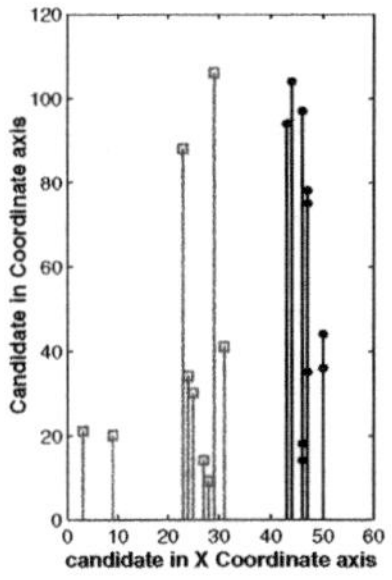

Fig. 3. The Corners Projection on y Axis

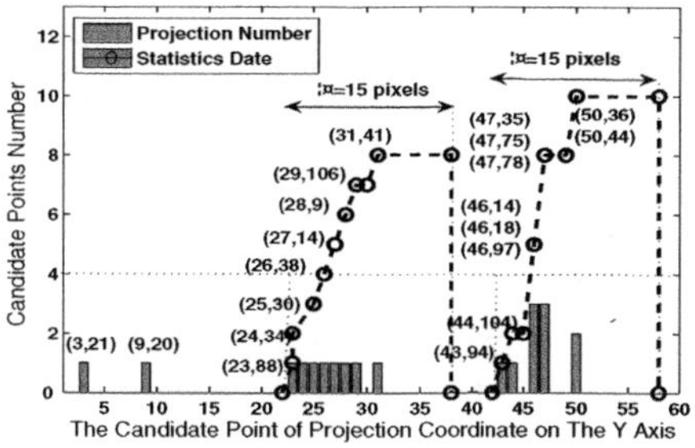

Fig. 4. The Frequency Distributions and the Frequency Statistical of Corners Projection

Began to calculation from the left Φ nearby, the two inner eye corners are first points meet the conditions. That is to say the inner eye corners are $h(x_{inner1}, y_{\Delta eye})$ and $h(x_{inner2}, y_{\Delta eye})$.

4 Experiments and Discussion

The wrinkle and pouch will influence the result of eye corner location when there is a kind of expression. Furthermore, light will affect the result also. To test the effectiveness of the proposed method, 100 images random selected respectively from two datasets, JAFFE [14] and BioID, are used in the experiments. Based on the method of rough eye location introduced in 3.1, image size between 100*50 to 130*75 pixels is gotten for the following experiments.

On the other hand, another two methods are selected for compared experiments, the first is SNS (Structured Neighborhood Similarity) introduced in [9], the second is eye corner detection based on the mean threshold segmentation of binary features introduced in [8] (abbreviate as MTSBF in the following).

4.1 Comparative Experiment on Accuracy

The first experiment is for the purpose of testing the accuracy of the proposed method compared with the other two methods in eye corner location. Here, the same 100 images are used for the 3 methods, and location results can be gotten automatically. On the other hand, another location result indicated the real eye corner is also gotten by hand. It can be seen correct location if it is within 5 pixels when we compared the two results, otherwise, it is a wrong location [8]. A ratio indicated accuracy of each method can be gotten and listed as follows.

From table 1, we can find that the proposed method can get the highest location rate. Therefore, the effectiveness of the proposed method can be proved. Moreover, since light is varied in images of the BioID dataset, the proposed method can get the highest location rate among the three methods, the robustness of the proposed method can be shown. Furthermore, we found that location

Table 1. The Detection Rate of Inner and Outer Eye Corners

	SNS		MTSBF		Proposed Method	
	Inner eye corners	Outer eye corners	Inner eye corners	Outer eye corners	Inner eye corners	Outer eye corners
JAFFE	78%	89%	86%	92%	95%	99%
BioID	65%	81%	72%	84%	73%	92%

rate of outer eye corners are higher than inner eye corners in the three experiments. That is to say, outer eye corners can be more accurate location compared with inner eye corners. This phenomenon is due to the contour points (feature point) of outer eye corners are more than the points of inner eye corners.

4.2 Case Study

To test the accuracy of the proposed method further, a case study is taken. 6 images among the 100 images in both datasets are randomly selected in the compared experiments. the location result and a location ratio are be listed as follows. Fig. 5, Fig. 6, Fig. 7, and Table 2 are results of JAFFE. Fig. 8, Fig. 9,

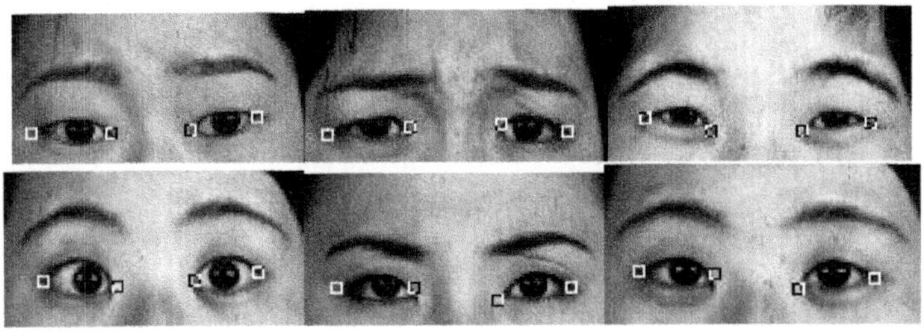

Fig. 5. Inner and Outer Eye Corners location of the Proposed Method

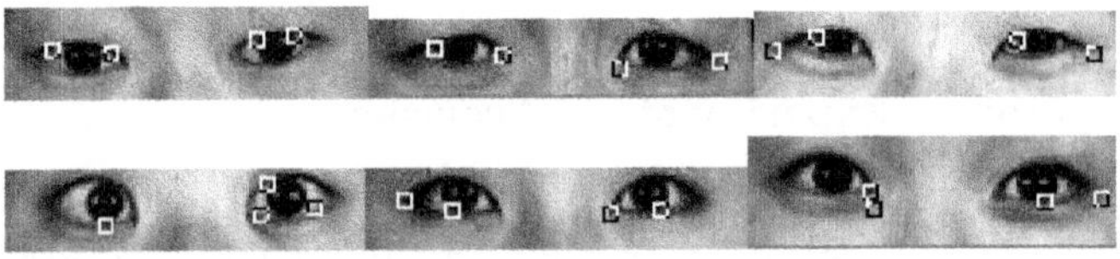

Fig. 6. Inner and Outer Eye Corners location of SNS

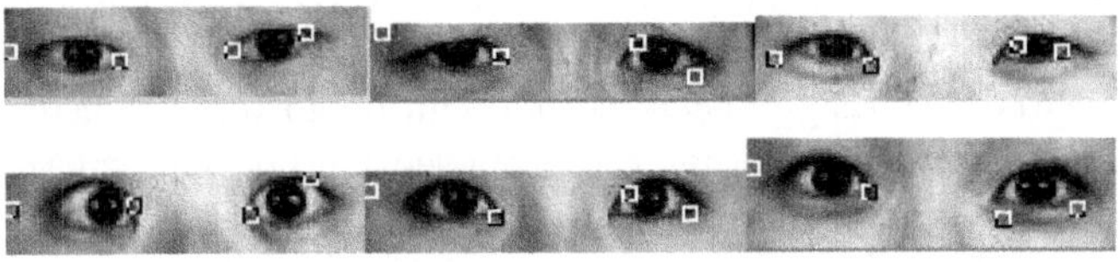

Fig. 7. Inner and Outer Eye Corners location of MTSBF

Table 2. The Detection Rate of Eye Corners of Different Image in JAFFE

JAFFE	SNS		MTSBF		Proposed Method	
	Inner eye corners	Outer eye corners	Inner eye corners	Outer eye corners	Inner eye corners	Outer eye corners
Img 1	100%	100%	100%	100%	100%	100%
Img 2	100%	50%	100%	50%	100%	100%
Img 3	50%	100%	100%	100%	100%	100%
Img 4	50%	50%	100%	0	100%	100%
Img 5	50%	50%	50%	50%	100%	100%
Img 6	50%	50%	100%	50%	100%	100%

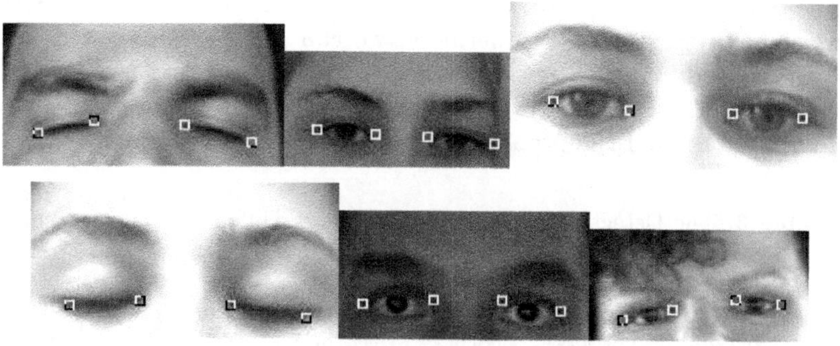

Fig. 8. Inner and Outer eye Corners location of the Proposed Method

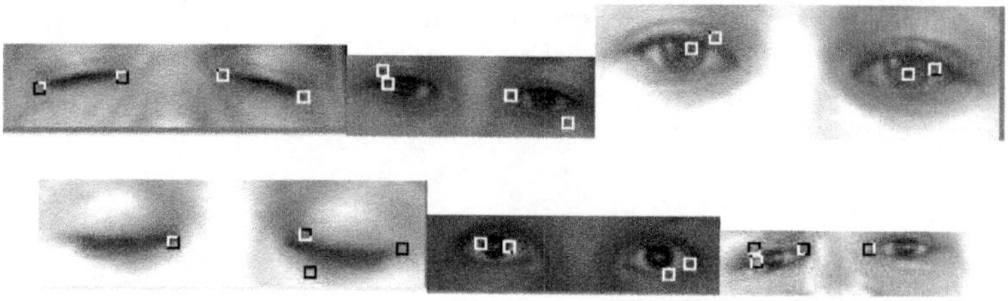

Fig. 9. Inner and Outer Eye Corners location of SNS

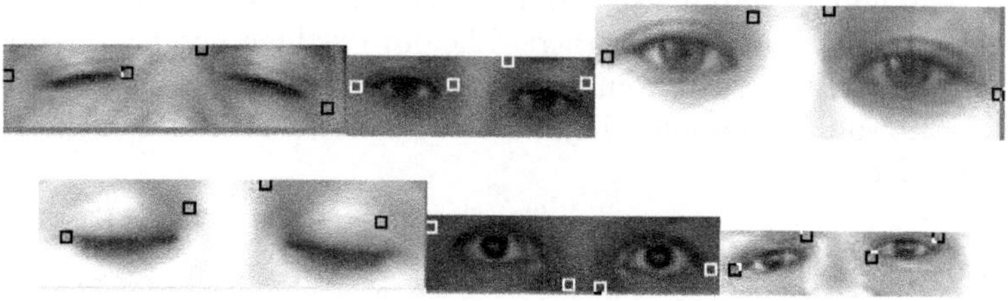

Fig. 10. Inner and Outer Eye Corners location of MTSBF

Fig. 10, and Tab. 3 are results of BioID. When we compare the proposed method with the other two methods, we can found that the proposed method is more accurate. There are three cases. Firstly, all the three methods can correctly location, such as inner eye corners in image No.1 of JAFFE. Although all the three methods can correctly location, we can found the proposed method are more accurate among the three results. Secondly, only the proposed method is correctly location among the three methods, such as inner eye corners in image No.2 of BioID. When we compared the three results, we can found the two outer eye corners can be correctly located, whereas SNS miss the left side inner eye corner, and MTSBF miss the right side inner eye corner. Thirdly, there are wrong results in all the three methods, such as inner eye corners in image No.5 of BioID. When we compare the three results, we can found the proposed method miss the right eye corner, SNS miss the right eye corner also, and MTSBF miss the two inner eye corners. Furthermore, we can found even though both the proposed method and SNS miss the right side inner eye corner, but the result of the proposed method approach to the correct result. Based on these in depth analysis, we can found the proposed method is more accurate and robust.

Table 3. The Detection Rate of Eye Corners of Different Image in BioID

BioID	SNS		MTSBF		Proposed Method	
	Inner eye corners	Outer eye corners	Inner eye corners	Outer eye corners	Inner eye corners	Outer eye corners
Img 1	100%	100%	50%	50%	100%	100%
Img 2	50%	100%	50%	100%	100%	100%
Img 3	0	50%	0	100%	100%	100%
Img 4	100%	50%	50%	0	100%	100%
Img 5	50%	100%	0	50%	50%	100%
Img 6	100%	50%	50%	100%	100%	100%

5 Conclusion

In the paper a novel eye corner location method based on the statistics result of frequency distribution of the corner points projection is proposed. This method takes advantage of contour extraction, ellipse fitting and corner detection to extract candidate points of eye corners. The proposed method is proved to be an accurate and robust method even if the proposed used in the images with varied light and emotion. In the following, different face would be researched further, such as side face.

References

1. Moein, L., AmirMohamad, S.Y.: Corner Sharpening with Modified Harris Corner Detection to Localize Eyes in Facial Images. In: IEEE International Conference on International Symposium ELMAR 2010, vol. 9, pp. 27–31 (2010)

2. Paul, V., Michael, J.: Robust Real-time Object Detection. International Journal of Computer Vision 57(2), 137–154 (2004)
3. Lihui, C., Christos, G., Yang, M.Y.: Reduced Complexity Eye Detector for Colour Images using Harris Corners, Color Heuristics and Edge Maps. IEEE PRIME 11, 245–248 (2006)
4. Stylianos, A., Nikolaos, N., Ioannis, P.: Detection of Facial Characteristics based on Edge Information. In: Proceedings of VISAPP (2) 2007, pp. 247–252 (2007)
5. Zhihua, Z., Xin, G.: Projection Functions for Eye Detection. Pattern Recognition 37(5), 1–10 (2004)
6. Peng, W., Qiang, J.: Multi-View Face and Eye Detection Using Discriminant Features. Computer Vision and Image Understanding 105(2), 99–111 (2007)
7. Song, J., Chi, Z.: A robust eye diction method using combined binary edge and intensity information. Pattern Recognition 39, 1110–1125 (2006)
8. Zhigang, Z., Mingquan, Z., Guohua, G.: Research on Automatic Localization of Facial Key Features. Computer Engineering and Applications 43(21), 197–198 (2007)
9. Cui, X., Ying, Z., Zengfu, W.: Semantic Feature Extraction for Accurate Eye Corner Detection. In: IEEE International Conference on Digital Object Identifier, pp. 1–4 (2008)
10. Rainer, L., Jochen, M.: An Extended Set of Haar-like Features for Rapid Object Detection. In: IEEE ICIP, pp. 900–903 (2002)
11. Zhengyou, Z.: Parameter Estimation Techniques: A Tutorial with Application to Conic Fitting. INRIA 10(2676), 59–76 (1995)
12. Xiaochen, H., Nelson, Y.: Corner Detector Based on Global and Local Curvature Properties. Optical Engineering 47(5), 057008-1–057008-12 (2007)
13. Jesorsky, O., Kirchberg, K.J., Frischholz, R.W.: Robust Face Detection Using the Hausdorff Distance. In: Bigun, J., Smeraldi, F. (eds.) AVBPA 2001. LNCS, vol. 2091, pp. 90–95. Springer, Heidelberg (2001)
14. Michael, L., Shigeru, A., Miyuki, K.: The Japanese Female Facial Expression (JAFFE) Database (1998), http://www.mis.atr.co.jp/~mlyons/jaffe.html

A Hybrid Approach of Verbal Decision Analysis and Machine Learning

Isabelle Tamanini[1], Plácido Rogério Pinheiro[1], and Cícero Nogueira dos Santos[2]

[1] University of Fortaleza (UNIFOR), Graduate Program in Applied Computer Sciences, Av. Washington Soares, 1321 - Bl J Sl 30 - 60.811-905, Fortaleza, Brazil
isabelle.tamanini@gmail.com, placido@unifor.br
[2] IBM Research - Brazil, Rio de Janeiro - RJ
cicerons@br.ibm.com

Abstract. The elicitation of preferences is the most costly process of the ZAPROS III-*i* considering the human-computer interaction. We intend to integrate the method and decision trees in order to improve this process. As a study case, the data from a battery of tests of patients with possible diagnosis of Alzheimer's disease will be used to structure a decision tree based on the characteristics that play the main role on the diagnosis, and a preference's scale will be establish through the analysis of the resulting tree. Then, this scale will be loaded into the ZAPROS method in order to rank-order the involved tests.

Keywords: Verbal Decision Analysis, ZAPROS III-*i*, Decision Trees, Diagnosis of the Alzheimer's Disease.

1 Introduction

The Verbal Decision Analysis (VDA), which is the focus of this work, is based on the multi-criteria problem solving through its qualitative analysis. The ZAPROS methodology, which belong to the VDA framework, aims at ranking multi-criteria alternatives, and it needs the Decision Maker (DM) to compare all the values of criteria involved in the problem so that a decision rule can be structured. This makes the stage the costliest process from the DM's point of view. Thus, the aim of this work is to integrate the ZAPROS III-*i* method and decision trees in order to minimize the human participation in the preferences elicitation process, and increase the granularity of the problem without decreasing the method's performance.

A hybrid model to determine which questionnaire from a battery of tests would identify faster a possible case of the Alzheimer's disease will be presented. This way, the social contribution of this work is to determine which test from a battery would present the highest probability of detecting the disease on its early stages, evaluating the characteristics that play main role on the diagnosis. Some works involving multi-criteria and the early diagnosis of diseases have already been developed as in: [3], which was developed based on previous models [2][9], applies multi-criteria in the health area and a model validated by the data provided on the battery of CERAD [5]; and also in [10][11][12].

J.T. Yao et al. (Eds.): RSCTC 2012, LNAI 7413, pp. 126–131, 2012.

2 ZAPROS III-*i* Method

The Verbal Decision Analysis (VDA) framework is structured on the acknowledgment that most of the decision making problems can be verbally described.

The ZAPROS III-*i* method belongs to the VDA framework and it will be used in this work. The method is structured in three well-defined main stages: *Problem Formulation*, *Elicitation of Preferences* and *Comparison of Alternatives*, as proposed in the main version of the ZAPROS method [7], and it aims at ranking multi-criteria alternatives in scenarios involving a rather small set of criteria and criteria values, and a great number of alternatives.

Regarding the *Formal Statement of the Problem*, the methodology follows the same problem formulation proposed in [7]:

Given:

1) K = 1, 2,..., N, representing a set of N criteria;
2) n_q represents the number of possible values in the scale of q-th criterion, $(q \in K)$; for the ill-structured problems, as in this case, usually $n_q \leq 4$;
3) $X_q = \{x_{iq}\}$ represents a set of values to the q-th criterion, and this set is the scale of this criterion; $|X_q| = n_q \, (q \in K)$;
4) $Y = X_1 * X_2 * ... * X_N$ represents a set of vectors y_i, in such a way that: $y_i = (y_{i1}, y_{i2}, ..., y_{iN})$, and $y_i \in Y$, $y_{iq} \in X_q$ and $P = |Y|$, where $|Y| = \prod_{i=1}^{i=N} n_i$.
5) $A = \{a_i\} \in Y$, i=1,2,...,t, where the set of t vectors represents the description of the real alternatives.

Required: The ranks of set A alternatives based on the DM's preferences.

In the *Elicitation of Preferences* stage, the scale of preferences for quality variations (Joint Scale of Quality Variations - JSQV) is defined. The elicitation of preferences follows the order of steps proposed in [13]. This structure is the same proposed in [7], however, the substages that compare values of criteria against the first and the second reference situations[1] were put together in just one substage. The consolidation of these substages reflects on an optimization of the process, reducing the number of questions presented to the DM on 50%.

The *Comparison of Alternatives* process starts after the decision rule is obtained. It follows same structure proposed in [7], with a modification in the comparison of pairs of the alternatives' substage according to the one proposed in [8] aiming at reducing the number of incomparability cases. Fig. 1 shows the structure of the comparison of the alternatives process. The complete functioning of the process is exposed in [13]. A tool was proposed in [14] in order to facilitate the decision making process and perform it consistently, observing its complexity and aiming at making it accessible.

[1] The *First Reference Situation* is represented by an alternative that has the best evaluations for each criterion of the problem. In the same way, the *Second Reference Situation* can be defined as the alternative that has the worst evaluations for each criterion of the problem.

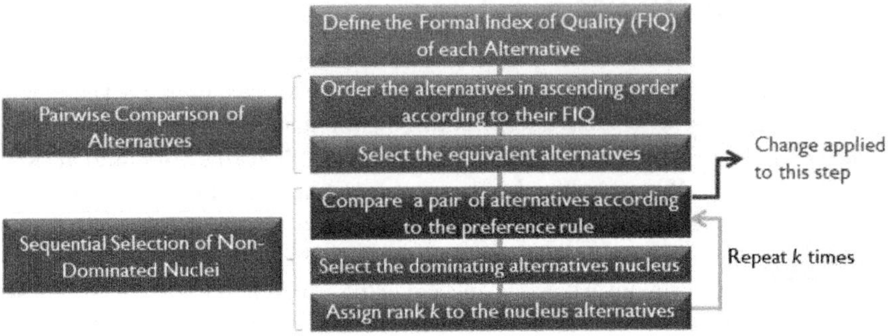

Fig. 1. Comparison of alternatives process

3 Decision Trees

The process of extracting knowledge of a bit amount of data, either automated or semi-automated, is a great necessity nowadays [4]. The data mining task represents one stage of this process. There are four types of machine learning [16]: Classification Learning, Associative Learning, Clustering, Numerical Prediction. Decision trees is a hierarchical model [1], based on the supervised classification learning technique, and it aims at classifying a new sample based on a set of classified samples. The final classification model is based on *if-then* structures, starting from the most informative attribute to the less one.

The Waikato Environment for Knowledge Analysis (WEKA) [15] will be used to structure the decision tree in this approach. Structured on Java, it contains a collection machine learning algorithms for data mining tasks. Its first version was created on 1993, and it is an open source software issued under the GNU General Public License, thus, it is possible to study and modify the source code.

4 The Diagnosis of the Alzheimer's Disease

Researchers agree that the advances in the medical area have significant importance on the increase of life expectancy. Along with this fact, there is a major increase in the number of health problems among the elderly. The Alzheimer's disease is difficult to be diagnosed, since the initial symptoms are subtle and they progress slowly until they are clear and irreversible. According to studies conducted by the Alzheimer's Association [6], the Alzheimer's disease is one of the costliest diseases, second only to cancer and cardiovascular diseases, and it is expensive for the long-term care system [5], since a great part of the population might suffer from dementia and, from this group, only a small number of patients may be capable of affording their own care.

With the purpose simplifying the diagnosing process this disease, this study aims at establishing an order of relevance of the tests to be applied to a patient in order to get to the diagnosis faster. To do this, we sought to choose the most important tests in the diagnosis of Alzheimer's disease, using the battery of the

Consortium to Establish a Registry for Alzheimer's Disease (CERAD) [5]. This battery has been chosen because it encompasses all the steps of the diagnosis and it is used all over the world.

5 A Hybrid Model to Determine the Main Characteristics to Diagnose the Alzheimer's Disease

With the aim of determining which questionnaire would analyze the most important indicators that a patient might have the Alzheimer's disease, a hybrid model was structured using decision trees and the ZAPROS III-*i* method. The model is based on all the characteristics analyzed by two tests of the CERAD's neuropathological battery: Clinical History and Neuropathological Diagnosis.

The criteria established were defined based on each question from the questionnaires, so that we would have one criteria for each question on the test. Regarding the CERAD data, only the results of the tests of patients that had already died and on which the necropsy has been done were selected (122 cases), because it is known that necropsy is essential for validating the clinical diagnosis of dementing diseases. After gathering the data from these patients, the dataset was filtered so that the duplicated cases and the ones with a considerable amount of missing information would be removed. This way, the decision tree was built considering 52 cases. The criteria selected were: A. Other Dementia, B. Parkinson's Disease, C. Heart Disease, D. Hypertension, E. Stroke or Def TIA, F. Seizures, G. Thyroid, H. Diabetes, I. Alcoholism, J. Drug Intoxication, K. Severe Head Injury, L. Byes2 Deficiency, M. Affective Disorder (Depression), N. Hemorrhage, O. Infarct only. Fig. 2 shows the criteria considered on the evaluation and the resulting tree.

Based on the analysis of the decision tree, a scale of preferences was built, such that the most informative criteria would be preferable to the others and the ones that were not selected to be a decision tree node were considered irrelevant based on the dataset provided. Thus, the preferences scale obtained was the following: Seizures $\prec$ Hemorrhage $\prec$ Diabetes $\prec$ Parkinson's Disease $\prec$ Affective Disorder $\prec$ Other Dementia $\equiv$ Heart Disease $\equiv$ Hypertension $\equiv$ Stroke of Def. TIA $\equiv$ Thyroid $\equiv$ Alcoholism $\equiv$ Drug Intoxication $\equiv$ Severe Head Injury $\equiv$ Byes2 Deficiency $\equiv$ Infarct. Before, this analysis of the most informative criteria would be made by a decision maker, and with the application of a decision tree algorithm, we were able to automate and construct a preferences rule based on the dataset provided.

Then, the alternatives were formulated identifying which facts would be identified by each questionnaire. The questionnaires were represented as criteria values following the rule: if it considers the criteria on its questionings, then its criteria value will be 1 for that characteristic, otherwise, it will be 2. Thus, the questionnaire Clinical History was be described as A1B1C1D1E1F1G1H1I1J1K1L1 M1N2O2, and, the questionnaire Neuropathological Diagnosis, as A1B1C2D2E1 F2G2H2I2J2K2L2M2N1O1. Then, the criteria set and their values, the preferences scale and the alternatives were loaded into the Aranaú Tool, and, after

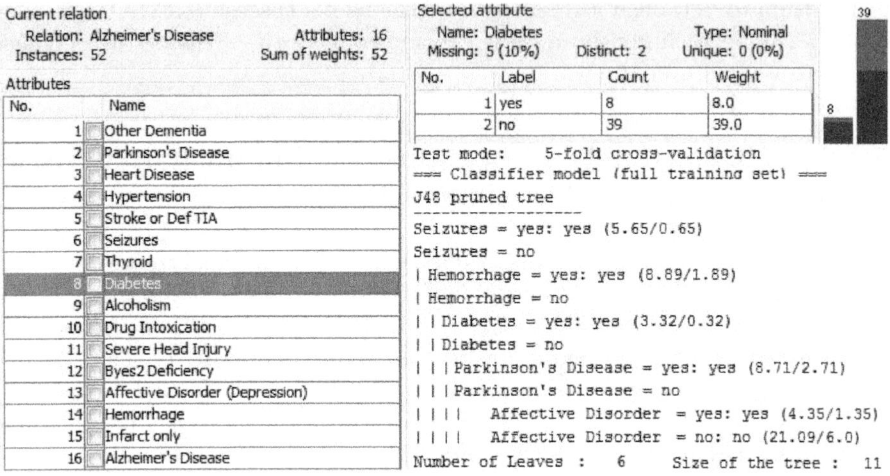

Fig. 2. Criteria analyzed and the resulting decision tree

it, it was shown that the Clinical History questionnaire is the more likely to determine a diagnosis of the Alzheimer's disease.

6 Conclusions, Future Works and Acknowledgments

With the aim of reducing the time demanded by the elicitation of preferences process and in order to reduce the human participation in the process, this work presented an hybrid model to support the decision making process using decision trees and the ZAPROS III-*i* method. Since the human participation in the preferences elicitation process was reduced when combining the two methodologies, the amount of criteria and values of criteria of ZAPROS III-*i* method could be increased: the preferences were learned by the decision tree algorithm and based on a dataset from CERAD's battery of tests. After the hierarchical structure was built, we were able to determine which attributes had the greatest informative index and these were selected as dominants against the others on the scale of preferences. Then, this scale was used to determine which test from CERAD's battery would lead to an early diagnosis of the Alzheimer's disease.

We intend to involve more questionnaires from the CERAD's battery of tests on the model, such that a complete analysis of the characteristics of each questionnaire would be made. Also, the interaction of the ZAPROS III-*i* method with other machine learning techniques will be investigated.

The authors are thankful to the Consortium to Establish a Registry for Alzheimer's Disease for making available the data used in this study.

References

1. Alpaydin, E.: Introduction to Machine Learning, 2nd edn. The MIT Press, Massachusetts (2010)
2. de Castro, A.K.A., Pinheiro, P.R., Pinheiro, M.C.D.: A Hybrid Model for Aiding in Decision Making for the Neuropsychological Diagnosis of Alzheimer's Disease. In: Chan, C.-C., Grzymala-Busse, J.W., Ziarko, W.P. (eds.) RSCTC 2008. LNCS (LNAI), vol. 5306, pp. 495–504. Springer, Heidelberg (2008)
3. Castro, A.K.A., Pinheiro, P.R., Pinheiro, M.C.D., Tamanini, I.: Towards the Applied Hybrid Model in Decision Making: A Neuropsychological Diagnosis of Alzheimer's Disease Study Case. International Journal of Computational Intelligence Systems 4, 89–99 (2011)
4. Fayyad, U., Piatetsky-Shapiro, G., Smyth, P.: From Data Mining to Knowledge Discovery in Databases. AI Magazine 17(3), 37–54 (1996)
5. Fillenbaum, G.G., van Belle, G., Morris, J.C., Mohs, R., Mirra, S., Davis, P., Tariot, P., Silverman, J., Clark, C., Welsh-Bohmer, K.: Consortium to Establish a Registry for Alzheimer's Disease (CERAD): The first twenty years. Alzheimer's & Dementia 4(2), 96–109 (2008)
6. Koppel, R.: Alzheimer's Disease: The Costs to U.S. Businesses in 2002, Alzheimer's Association (2002)
7. Larichev, O.: Ranking Multicriteria Alternatives: The Method ZAPROS III. European Journal of Operational Research 131(3), 550–558 (2001)
8. Moshkovich, H., Mechitov, A., Olson, D.: Ordinal, Judgments in Multiattribute Decision Analysis. European Journal of Operational Research 137(3), 625–641 (2002)
9. Pinheiro, P.R., Castro, A.K.A., Pinheiro, M.C.D.: Multicriteria Model Applied in the Diagnosis of Alzheimer's Disease: A Bayesian Network. In: 11th IEEE International Conference on Computational Science and Engineering, vol. 1, pp. 15–22 (2008)
10. Tamanini, I., Castro, A.K.A., Pinheiro, P.R., Pinheiro, M.C.D.: Verbal Decision Analysis Applied on the Optimization of Alzheimer's Disease Diagnosis: A Study Case Based on Neuroimaging. Special Issue: Software Tools and Algorithms for Biological Systems, Advances in Experimental Medicine and Biology 696(7), 555–564 (2010)
11. Tamanini, I., Pinheiro, P.R., Pinheiro, M.C.D.: The Neuropathological Diagnosis of the Alzheimer"s Disease under the Consideration of Verbal Decision Analysis Methods. In: Yu, J., Greco, S., Lingras, P., Wang, G., Skowron, A. (eds.) RSKT 2010. LNCS (LNAI), vol. 6401, pp. 427–432. Springer, Heidelberg (2010)
12. Tamanini, I., Pinheiro, P.R., Pinheiro, M.C.D.: The Choice of Neuropathological Questionnaires to Diagnose the Alzheimer"s Disease Based on Verbal Decision Analysis Methods. In: Zhu, R., Zhang, Y., Liu, B., Liu, C. (eds.) ICICA 2010. LNCS, vol. 6377, pp. 549–556. Springer, Heidelberg (2010)
13. Tamanini, I.: Improving the ZAPROS Method Considering the Incomparability Cases. Master Thesis – Graduate Program in Applied Computer Sciences, University of Fortaleza. Available in UNIFOR Digital Library (2010)
14. Tamanini, I., Pinheiro, P.R.: Pinheiro: Reducing Incomparability in Multicriteria Decision Analysis; An Extension of the ZAPROS Method. Pesquisa Operacional 31(2), 251–270 (2011)
15. Hall, M., Frank, E., Holmes, G., Pfahringer, B., Reutemann, P., Witten, I.: The WEKA Data Mining Software: An Update. SIGKDD Explorations 11(1) (2009)
16. Witten, I., Frank, E.: Data Mining: Practical Machine Learning Tools and Techniques. Elsevier, San Francisco (2005)

An Efficient Data Integrity Verification Method Supporting Multi-granular Operation with Commutative Hash*

Long Chen, Banglan Liu, and Wei Song

Institute of Computer Forensics,
Chongqing University of Posts and Telecommunications,
Chongqing, 400065, P.R. China
chenlong@cqupt.edu.cn, liubanglan@yahoo.cn

Abstract. Cloud computing has been envisioned as the next-generation architecture of IT enterprise. However the security problem of Cloud computing hinder itself from large-scale application. One major challenge is to verify the integrity of data at untrusted server under the condition of supporting public auditability and dynamic data operation. This paper analyzed the problems of existing schemes, introduced commutative hash into Merkle hash tree, aggregated basic block into subblock and block, constructed a kind of hierarchical Merkle hash tree by using the concept of hierarchical structure in granular computing and proposed an improved scheme to offer effective, multi-granular dynamic operation. We analyzed the efficiency and security of the protocol. The analysis result showed that the protocol is efficient and security.

Keywords: cloud storage, data integrity, commutative hash, multi-granular.

1 Introduction

Cloud computing provides an extensible environment for growing requirement of data storage and computing. Cloud computing relieves the burden of managing and maintaining data. But, if such an important service is vulnerable to attack, or the cloud storage service provider is untrusted, it will cause an irretrievable loss to users. For example, the storage service provider experienced a data failure occasionally, they may hide the message of data error from the clients for the benefit of their own. What is more, for saving money and storage space, the service provider might neglect to keep or deliberately delete rarely accessed data files which belong to an ordinary client [1]. In addition to this, considering

* This work is supported by Natural Science Foundation Project of CQ CSTC of P. R. China (No.cstc2011jjA40031), Science & Technology Research Program of the Municipal Education Committee of Chongqing of P.R.China (No.KJ110505), and Open Program of Key Lab of Computer Network & Communication of Chongqing of P.R.China (No. JK-Y-2010003).

J.T. Yao et al. (Eds.): RSCTC 2012, LNAI 7413, pp. 132–141, 2012.

the large volume of the outsourced data and the user's constrained of computing resource and capability, it is different from concern on isolating errors and efficiency of verification while directly verifying data integrity locally[2,3]. The client will not keep any data copy, and the original data will not be processed locally while executing integrity verification. From the above, the key issue of the problem can be generalized as how the users can find an efficient way to perform periodical integrity verifications without the local copy of data files [1].

However, tranditional data integrity verification schemes are based on hash and signature, they can not verify outsourced data without a local copy. In addition, it's impractical to download the whole data for integrity verification, for the huge I/O and communication cost, especially when data volume is big. So it's expensive to execute remote integrity verification, cloud storage users have to resort to TPA(Third Party Auditor) who has professional knowledge and capability for helping them to audit outsourced data periodically. Besides, for protecting secret data from leaking to the third party, the process of verification should have a capability of privacy-preserving.

1.1 Granular Computing

Granular computing includes all the theory, methods and techniques about granularity [4]. The two basic problem in Granular computing are granulation and the computing of granules. Granulation involves the construction of the three basic components, granules, granulated views and hierarchies[5]. Granular computing emphasizes on the effective use of multiple levels of granularity. The granularity of granules and levels enables us to construct a hierarchical structure called a hierarchy. The three tasks of GrC are: constructing granular structures, working within a particular level of the structure, and switching between levels [6].

1.2 Design Goals

This paper analyses the problem of [1], which we call it Wang's integrity verification scheme, or Wang's scheme for short. Inspired by Granular computing theory, through aggregate basic block into subblock and block, and construct a kind of hierarchical Merkle hash tree, we designed an integrity verification scheme that it not only achieves Wang's goal, but also achieves the following purposes: (1)Multi-granular operation: to support different dynamic operation on different granularity of file block; (2) Lightweight: to allow TPA to perform verification with minimum communication overhead; (3) Privacy-preserving: to ensure that TPA can't derive user's data content from the information collected during the verification process.

2 Proposed Scheme

2.1 Notation and Preliminaries

Bilinear Map. A bilinear map is a map $e : G \times G \rightarrow G_T$, where G is a Gap Diffie-Hellman (GDH) group and G_T is another multiplicative cyclic group of prime order p with the following properties: (1)Computable: there exists an efficiently computable algorithm for computing e; (2)Bilinear: for all $h_1, h_2 \in G$ and $a, b \in Z_p$, $e(h_1^a, h_2^b) = e(h_1, h_2)^{ab}$; (3) Non-degenerate: $e(g, g) \neq 1$, where g is a generator of G [7].

One-Way Accumulators. A hash function h is commutative if $h(x, y) = h(y, x)$, for all x and y. A hash function is collision-free if, given (a, b), it is difficult to compute a pair (c, d) such that $h(a, b) = h(c, d)$, while $((a, b) \neq (c, d)$ and $(a, b) \neq (d, c)$ [8]. A family of one-way accumulators is a family of one-way hash functions each of which is quasi-commutative and collision-free. one-way accumulators h_A ensure that if one starts with an initial value $x \in X$, and a set of values $y_1, y_2, \cdots, y_m \in Y$, then the accumulated hash $z = h_A(h_A(h_A(\cdots h_A(h_A(h_A(x, y_1), y_2), y_3), \cdots, y_{m-2}), y_{m-1}), y_m)$ would be unchanged if the order of the y_i were permuted [9], and if a new element was added to one-way accumulators, generate a new accumulated hash value, it is dependent with the order of elements. We use h representative of cryptographic hash functions, h_A representative of commutative hash functions.

Merkle Tree(Merkle Hash Tree, MHT). A Merkle Hash Tree is a kind of tree structure. The leaves of Merkle Tree are the hashes of authentic data values. The inner nodes of MHT are the hashes of the data concatenation of their child nodes. It is used to verify the integrity of one data block or a few data blocks quickly. It is only need to process one or several paths from leaves to the root of MHT while executing integrity verification. So it is intended to efficiently and securely prove that a group of elements are unaltered and undamaged.

2.2 File Decomposition and Aggregation

In the view of granular computing, granulation is a process of constructing a problem solving space, the method of granulation include decomposing rough granules to fine granules or aggregating fine granules to rough granules. Wang's scheme decomposed a file into several basic block, each leave node in Merkle Tree representive of one basic block, so the height of Wang's Merkle Tree is too high to manage data blocks. What's worse, while executing dynamic operation, it must operate basic blocks one by one.

In our scheme, the file is decomposed to I blocks, $F = (m_1, m_2, \cdots, m_I)$, then, each block m_i will be decomposed into J subblocks, $F = (m_{1,1}, m_{1,2}, \cdots, m_{I,J})$, each subblock $m_{i,j}$ will continuously be decomposed into K basic blocks at the last, so the file F can be represented as the collection of basic blocks, that is $F = (m_{1,1,1}, m_{1,1,2}, \cdots, m_{1,1,K}, m_{1,2,1}, \cdots, m_{I,J,K})$. From the above three kinds of definition of file F, the file could be considered as a 3-demension structure, compared to Wang's 1-demension file structrue. That is to say, our scheme decompose rough granules to fine granules in different granularity.

Hierarchical Merkle Hash Tree. The main target of granular computing model is to solving problem in different level of granularity. Granulation is essentially a hierarchical structure, the granules in the same level always have the same properties and functions. In order to simplify the process of integrity verification. We construct a hierarchical structure using granular computing method.

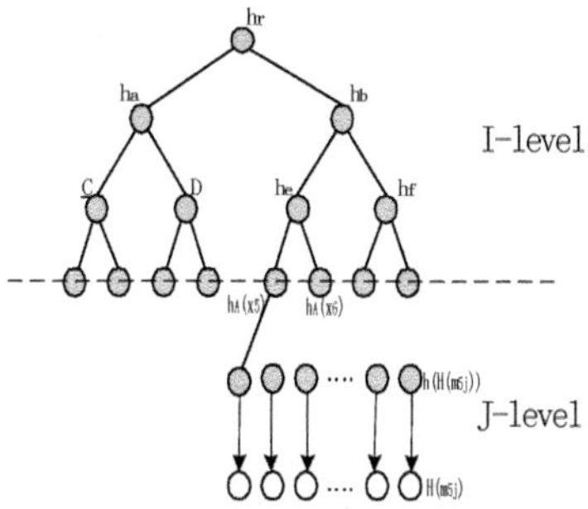

Fig. 1. Structure of Hierarchical Merkle Hash Tree

Then we introduce commutative hash into Merkle Hash Tree. The structure of Hierarchical Merkle Hash Tree is shown as Fig. 1, the tree has two level: I-level and J-level. We use $H(m_{i,j})$ as the tag for subblock $m_{i,j}$ (hash function H is viewed as a random oracle [1]). One leaf node of J-level denote the hash of $H(m_{i,j})$, that is $h(H(m_{i,j}))$. The number of leaf node in J-level cannot more than J, J-level nodes are stored as a sequenced structure. $h_A(x_i)$ denote the accumulate hash of $h(H(m_{i,j}))$ in i-th block, $1 \leq j \leq J$, the leaf node of I-level is $h_A(x_i)$, the structure of I-level are similar to Wang's scheme. As shown in Fig. 1: $h_A(x_5) = h_A(x, h(H(m_{5,1})), h(H(m_{5,2})), \cdots, h(H(m_{5,J}))), x \in X$. The introduction of one-way accumulator (commutative hash) reduces the communication cost.

In our scheme, the prover sends auxiliary authenticate information (AAI) to the auditor, we denote it by $\Omega_{i,j}, (i \in I, j \in I_2)$ or $\Omega_i, (i \in I)$. We define Ω_i as the nodes those are the sibling on the path from the leaves $h(H(m_i))$ in I-level to the root node R. $z_{i,j}$ is the accumulated hash of all J-level leaf nodes except $h(H(m_{i,j}))$. $\Omega_{i,j}$ includes both Ω_i and $z_{i,j}$.

2.3 Verification Protocol

The process of integrity verification is the process of computing on granules. The procedure of our verification protocol is executed as follows: Setup: By invoking $KeyGen()$, the client generates its public key and private key. The data file F is pre-processed by running $SigGen()$, and produce the homomorphic authenticators together with metadata. Let $I_1 = \{1, 2, \cdots, I\}$, $I_2 = \{1, 2, \cdots, J\}$, $I_3 = \{1, 2, \cdots, K\}$.

$KeyGen()$: The client generates a random signing key pair $\{spk, ssk\}$, Choose a random α from Z_p, and compute $v = g^{\alpha}$, then, picks a random K-element subset $\{u_k\}_{1 \leq k \leq K}, w_k = u_k^{\alpha}$, the private key $sk = (\alpha, ssk)$, the public key $pk = (\nu, spk, \{w_k\}_{k \in I_3}, g, \{u_k\}_{k \in I_3})$.

$SigGen()$: for the given file F, firstly, compute the file tag for F, $t = name||u_1||u_2|| \cdots ||u_k||SSig_{ssk}(name||u_1||u_2|| \cdots ||u_k)$, then the client computes signature $\sigma_{i,j}$ for each subblock $m_{i,j}$ as $\sigma_{i,j} = (\omega_{i,j} \cdot \prod_{k=1}^{K} u_k^{m_{i,j,k}})^{\alpha}$,

$\omega_{i,j} = h(H(m_{i,j})||f(j))$, $f(j)$ is a random function. We denote the set of signature by $\Phi = \{\sigma_{i,j}\}_{1\leq i\leq I, 1\leq j\leq J}$. The client then generates a root R based on the structure of our hierarchical Merkle Hash Tree. the client signs the root R with the private key α: $sig_{sk}(H(R)) \leftarrow (H(R))^{\alpha}$. At last, the client sends $\{F, t, \Phi, sig_{sk}(H(R))\}$ to the server and deletes $\{F, \Phi, sig_{sk}(H(R))\}$ from its local storage.

The TPA or client can verify the integrity of the outsourced data by challenging the server. First, the TPA use private key *spk* to verify the signature on t, if the verification fails, return $FALSE$ to reject; otherwise, recover u_k. the TPA (verifier) picks a random c-element subset $I_c = \{s_1, s_2, \cdots, s_c\}$ from set I_1, we assume that $s_1 \leq s_2 \leq, \cdots, \leq s_c$, TPA uses element i in set I_c as the serial number of block, and randomly select element j as the serial number of subblock, then, randomly select an element $v_{i,j\ (i\in I_c, j\in I_2)}$ from Z_p, thus, the message *chal* is generated. The message *chal* specifies the positions of the blocks to be checked in this challenge phase. Then, the verifier sends the *chal* to the prover(server).

GenProof(): After receiving the challenge $chal\{(i, j, v_{i,j})\}_{s_1\leq i\leq s_c, j\in I_2}$, the server will run *GenProof*() to generate response proof of the correctness of stored data. In detail, the server randomly selects a element o from z_p, and computes $Q_k = (w_k)^o = (u_k^{\alpha})^o$, $Q_k \in G, k \in I_3$, computes $\mu'_k = \sum_{i=s_1, j\in I_2}^{i=s_c, j\in I_2} v_{i,j} m_{i,j,k}$, $k \in I_3, \mu_k = \mu'_k + oh(Q_k) \in Z_p$, aggregation signatures will also be computed: $\sigma = \prod_{i=s_1, j\in I_2}^{i=s_c, j\in I_2} (\sigma_{i,j})^{v_{i,j}}$, the prover will also provides the verifier with a small amount of auxiliary information AAI $\{\Omega_{i,j}\}_{i\in I_c, j\in I_2}$ or $\{\Omega_i\}_{i\in I_c}$, at the last, the server respond TPA with proof $P = \{\{\mu_k\}_{k\in\{1,2,\cdots,K\}}, \sigma, \{Q_k\}_{k\in\{1,2,\cdots,K\}}, \{H(m_{i,j}), \Omega_{i,j}\}_{i\in I_c, j\in I_2}, sig_{sk}(H(R))\}$.

VerifyProof(): once receiving the response from the prover, TPA runs *VerifyProof*(), generates root R using $\{H(m_{i,j}), \Omega_{i,j}\}_{s_1\leq i\leq s_c, j\in I_2}$, and verify its integrity by checking $e(sig_{sk}(H(R)), g)\underline{?}e(H(R), g^{\alpha})$. If the verification fails, the verifier rejects by return $FALSE$. Otherwise, the verifier checks if equation 1 holds:

$$e(\sigma \cdot \prod_{k=1}^{K} Q_k^{h(Q_k)}, g)\underline{?}e(\prod_{i=s_1, j\in I_2}^{i=s_c, j\in I_2} (\omega_{i,j})^{v_{i,j}} \cdot \prod_{k=1}^{k} u_k^{\mu_k}, v) \tag{1}$$

The correctness of the verification equation 1 can be elaborated as follows:

$$\begin{aligned}
&e(\sigma \cdot \prod_{k=1}^{K} Q_k^{h(Q_k)}, g) \\
&= e(\prod_{i=s_1, j\in I_2}^{i=s_c, j\in I_2} (\sigma_{i,j})^{v_{i,j}} \cdot \prod_{k=1}^{K} (u_k^{\alpha\cdot o})^{h(Q_k)}, g) \\
&= e(\prod_{i=s_1, j\in I_2}^{i=s_c, j\in I_2} ((\omega_{i,j}) \cdot \prod_{k=1}^{K} u_k^{m_{i,j,k}})^{\alpha})^{v_{i,j}} \cdot \prod_{k=1}^{K} (u_k^{\alpha\cdot o})^{h(Q_k)}, g)
\end{aligned}$$

$$= e([\prod_{i=s_1,j\in I_2}^{i=s_c,j\in I_2} (\omega_{i,j})^{v_{i,j}} \cdot \prod_{i=s_1,j\in I_2}^{i=s_c,j\in I_2} \prod_{k=1}^{K} (u_k^{m_{i,j,k}})^{v_{i,j}}]^\alpha \cdot [\prod_{k=1}^{K} u_k^{oh(Q_k)}]^\alpha, g)$$
$$= e(\prod_{i=s_1,j\in I_2}^{i=s_c,j\in I_2} (\omega_{i,j})^{v_{i,j}} \cdot \prod_{i=s_1,j\in I_2}^{i=s_c,j\in I_2} (\omega_{i,j})^{v_{i,j}} \cdot \prod_{k=1}^{K} (u_k^{\mu_k'}) \cdot \prod_{k=1}^{K} u_k^{oh(Q_k)}, g^\alpha)$$
$$= e(\prod_{i=s_1,j\in I_2}^{i=s_c,j\in I_2} (\omega_{i,j})^{v_{i,j}} \cdot \prod_{k=1}^{K} u_k^{\mu_k}, v)$$

Apparently, Q_k did not affect the result of integrity verification. If user don't worry about data leakage to a third party. Then, the prover (server) only need to send $\{\{\mu_k'\}_{k\in\{1,2,\cdots,K\}}, \sigma, \{H(m_{i,j}), \Omega_{i,j}\}_{i\in I_c, j\in I_2}, sig_{sk}(H(R))\}$ to the verifier as the proof of the correctness and integrity of stored data, the communications cost is reduced without privacy preserving.

3 Dynamic Operation

3.1 Data Modification

Firstly, we introduce the operation of data modification. Suppose the user want to replace m_i with m^*, according to different size of block granule, the protocol will be introduced respectively:

(1)modify block: at first, the client compute corresponding signatures of the new data block(compute signatures of its subblocks as a set), then, he sends an update request $update = (M, i, m^*, \sigma^*)$ to the server, where, M denote modification operation, i denote the i-th block, while receiving the request, the server executes $ExecUpdate(F, \Phi, update)$, the process is shown in detail as follows: 1) replaces the original block m_i with the new one m^*; 2)replaces signature σ with σ^*; 3) replace hash of tag of subblocks which contained by m_i with hash of tag of subblocks which contained by m^*, that is, replace $h(H(m_{i,j}))$ with $h(H(m_{i,j}{}^*)), j \in I_2$; 4)computes the accumulated hash of $h(H(m_{i,j}{}^*))$, generate $h_A(x_i{}^*)$, 5)replace $h_A(x_i)$ with $h_A(x_i{}^*)$; 6)recomputes the MHT, generate a new root node R^*, Finally, the server generates a proof of this operation and response it to the client(TPA), $P_{update} = (\Omega_i, h_A(x_i), Sig_{sk}(H(R)), R^*)$, where Ω_i is the AAI for authentication of m_i.

(2)modify subblock: at first, the client compute corresponding signatures of the new data block σ^*, then, he sends an update request $update = (M, i, j, m^*, \sigma^*)$ to the server, j denote the j-th subblock in i-th block, upon receiving the request, the server run $ExecUpdate(F, \Phi, update)$, replace $h(H(m_{i,j}))$ with $h(H(m_{i,j}{}^*))$, generate $h_A(x_i{}^*)$, replace $h_A(x_i)$ with $h_A(x_i{}^*)$, recompute the MHT, generate a new root node R^*, Finally, the server generate a proof of this operation and response it to the client(TPA), $Pupdate = (\Omega_{i,j}, h(H(m_{i,j})), Sig_{sk}(H(R)), R^*)$, where $\Omega_{i,j}$ is the AAI for authentication of $m_{i,j}$, and $z_{i,j}$ is included.

Then the server send the proof to the client. After receiving the proof of modification operation from server. the client execute integrity verification for dynamic operation. First, The client check if the node is trusted, if not, output $FALSE$, else generate the root R using $\{\Omega_i, h_A(x_i)\}$ or $\{\Omega_{i,j}, h(H(m_{i,j}))\}$

and authentic R by check $e(sig_{sk}(H(R)), g) \stackrel{?}{=} e(H(R), g^{\alpha})$, if the equation is not true, then return $FALSE$, otherwise the client can compute new root R_{new} by $\{\Omega_{i,j}, h_A(x_i{}^*)\}$ or $\{\Omega_{i,j}, h(H(m_{i,j}{}^*))\}$, and authenticate server complete Modification operation by check whether R_{new} equals to R^*, if it is not true, return $FALSE$, otherwise, return $TURE$, then client signs the new root R^*, and send it to the server. In the end, executes the default integrity verification. if its output is $TURE$, Modification operation is completed, deletes $sig_{sk}(H(R^*)), P_{update}$ and m^* from its local storage.

3.2 Data Insertion

insert block: Suppose user want to insert m^* after m_i, the procedure of protocol is similar to that of data modification operation. At the first, the client compute corresponding signatures of the new data block σ^*(compute signatures of its subblocks as a set), then, he sends an update request $update = (I, i, m^*, \sigma^*)$ to the server, where, I denotes insertion operation. After receiving the request, the server executes $ExecUpdate(F, \Phi, update)$, the process is shown in detail as follows: stores m^* and its signature, compute hash value of tag of subblocks which contained by m_i, that is $h(H(m_{i,j}))$, adds leaf nodes $h(H(m_{i,j}))$ to J-level of MHT, compute the accumulated hash $h_A(x_i{}^*)$, find $h_A(x_i)$ in I-level, insert $h_A(x_i{}^*)$ after I-level leaf node $h_A(x_i)$, add an internal node C to the original tree in I-level, where $h_C = h(h(H(m_i))||h(H(m^*)))$, recomputed the hash of nodes related to the path, generates a new root node R^*. After generates a new root node R^*, the server generates a proof of this operation and response it to the client(TPA), $P_{update} = (\Omega_i, h_A(x_i), sig_{sk}(H(R)), R^*)$. After receiving the proof of modification operation from server, the client audit the proof, the procedure is similar to modification operation.

This scheme does not support dynamic operation of inserting subblock. And data deletion is the opposite of data insertion. there is something special that if there is not any J-level node in i-th block, then deletes the i-th block. The details of the protocol procedures are thus omitted here.

4 Security Analysis

We mainly evaluate the storage correctness and privacy-preserving. All proofs are derived based on probabilistic, it could be envision as security if it has a high probability security assurance.

(1) Storage Correctness Guarantee: the cloud servers can not offer valid response if they are not faithfully storing the data. If the cloud server passes the integrity verification, then it must indeed possess of the specified data. The procedure of proof consists of three steps: Firstly, we show that there is no such a malicious server that can forge a valid response $\{\{\mu_k\}_{k\in\{1,2,\cdots,K\}}, \sigma,$ $\{Q_k\}_{k\in\{1,2,\cdots,K\}}, \{H(m_{i,j}), \Omega_{i,j}\}_{i\in I_c, j\in I_2}, sig_{sk}(H(R))\}$ to pass the verification equation 1. Value Q_k, which was used for privacy-preserving, will not affect the result of the equation, because of the hardness of discrete-log and the

commutativity of modular exponentiation in pairing. Next, if the response $\{\{\mu_k\}_{k\in\{1,2,\ldots,K\}}, \sigma, \{Q_k\}_{k\in\{1,2,\cdots,K\}}, \{h(H(m_{i,j})), \Omega_{i,j}\}_{i\in I_c, j\in I_2}, sig_{sk}(H(R))\}$ is valid, where $\mu_k = \mu'_k + oh(Q_k) \in Z_p, Q_k = (w_k)^o = (u_k^\alpha)^o$, then μ'_k must also be valid. it can be deduced from the determinism of discrete exponentiation and the collision-free property of hash function. In the end, the validity of μ'_k implies the correctness of $\{m_{i,j,k}\}_{i\in I_c, j\in I_2, k\in I_3}$ where $\mu'_k = \sum_{i=s_1, j\in I_2}^{i=s_c, j\in I_2} v_{i,j} m_{i,j,k}$. Therefore, the correctness of those specific blocks is ensured. In conclusion, the storage correctness of our scheme is guaranteed.

(2) Privacy Preserving Guarantee: We must make sure that TPA can't derive users' data content by collecting information during verifying process. If TPA can derive μ'_k, then $\{m_{i,j,k}\}_{i\in I_c, j\in I_2, k\in I_3}$ can be easily computed by solving a group of linear equations when TPA have collected enough combinations of the same blocks. we will prove the Privacy-Preserving Guaranting in two steps: Firstly, there is not any information on μ'_k can be derived from μ_k, for μ_k is blinded by o as $\mu_k = \mu'_k + oh(Q_k) \in Z_p Q_k = (w_k)^o = (u_k^\alpha)^o$, where o is a random value selected by cloud server and it is unknown to TPA, because of the hardness of discrete-log assumption, the value o is still hidden against TPA, so, μ_k guarantees the privacy of μ'_k. Secondly, there isn't any information about μ'_k can be derived from σ, therefore, the correctness of specific sampled blocks are ensured. All above sums up to guarantee the privacy of user's data. In conclusion, the privacy of user's data in our scheme is guaranteed.

5 Performance Analysis

We assume that if 1% basic block errored, and the probability of detect errors is 99%, we need to authenticate 460 data blocks. In our experiment, we select 1GB file as experiment data, select 460 sample data blocks. First, we respectively computed and analysed the communication cost of different size of blocks, we respectively compared and analysed the experiment result of different size of data blocks and subblocks, at the last on condition of a set of optimum parameters, we done experiment on different size of files, and compare our experiment result with Wang's scheme.

Fig. 2 shows the communication cost at different granularity of file block in our scheme, one curve denote a block contain 8, 16, 32, 64, 128 subblock, respectively. Fig. 3 show when one block contains 64 subblocks, namely J=64, the comparison of communication cost at different block size between our scheme and Wang's(based on RSA), From Fig. 3, it can be observed that the communication cost of our scheme is better than Wang's, especially with the growing of block, it's more obvious that our scheme is better in communication cost. while 512KB as the block size(contain 64 subblocks), our scheme achieve the optimal value in communication cost. Fig. 4 shows that, at the above-mentioned optimal value, for different size of file, the communication of our scheme is obviously better than Wang's. What's more, with the grown of file size, our scheme is still better.

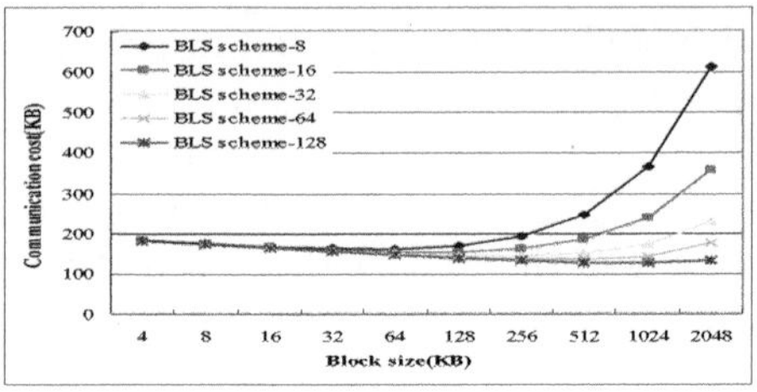

Fig. 2. the comparison of communication cost at different granularity of file block

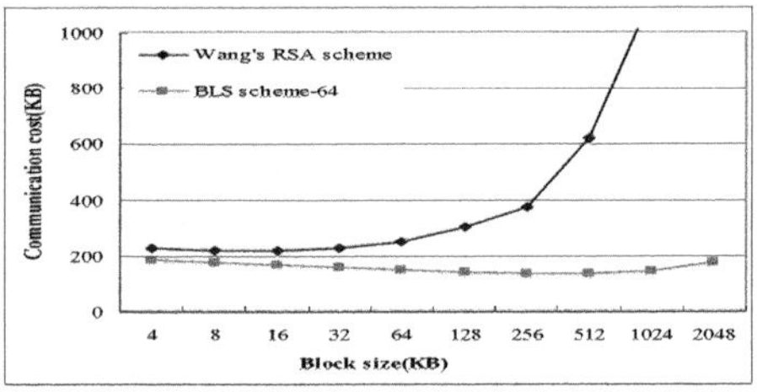

Fig. 3. the comparison of communication cost at different block size between our scheme and Wang's

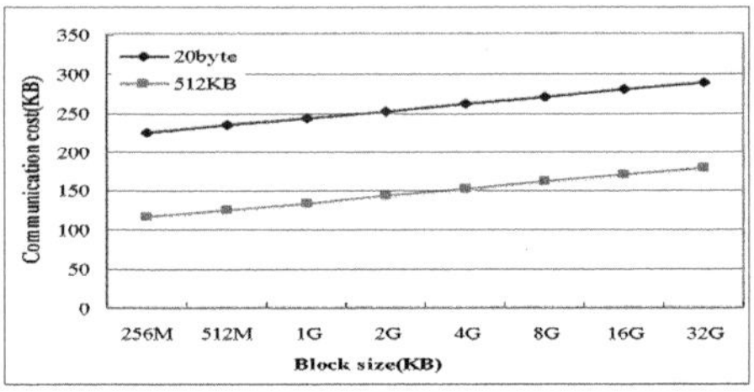

Fig. 4. the comparison of communication cost at different file size between our scheme and Wang's

6 Conclusion

This paper proposes a remote data integrity verification scheme on the basis of Wang's scheme. Compared to the state-of-the-art, our scheme supports dynamic operation at block and subblock, reduces the communication cost and achieves privacy-preserving.

References

1. Qian, W., Cong, W., Kui, R., Wenjing, L., Jin, L.: Enabling Public Auditability and Data Dynamics for Storage Security in Cloud Computing. IEEE Transactions on Parallel and Distributed Systems 22(5), 847–859 (2011)

2. Long, C., Guoyin, W.: An Efficient Integrity Check Method for Fine-Grained Data over Galois Field. Chinese Journal of Computers 34(5), 847–855 (2011)
3. Long, C., Guoyin, W.: An integrity check method for fine-grained data. Journal of Software 20(4), 902–909 (2009)
4. Guoyin, W., Qinghua, Z., Jun, H.: An overview of granular computing. CAAI Transactions on Intelligence Systems 6(2), 8–25 (2007)
5. Tsauyoung, L.: Granular computing: practices, theories, and future directions. In: Encyclopedia of Complexity and Systems Science, pp. 4339–4355 (2009)
6. Yiyu, Y.: Granular Computing: Past, Present and Future. In: 2008 IEEE International Conference on Granular Computing, vol. 8, pp. 80–85 (2008)
7. Cong, W., Qian, W., Kui, R., Wenjing, L.: Privacy-preserving public auditing for data storage security in cloud computing. In: InfoCom2010. IEEE (2010)
8. Goodrich, M., Tamassia, R., Schwerin, A.: Implementation of an authenticated dictionary with skip lists and communitative hashing. In: Proceedings of DARPA DISCEX II (June 2001)
9. Benaloh, J., de Mare, M.: One-Way Accumulators: A Decentralized Alternative to Digital Signatures. In: Helleseth, T. (ed.) EUROCRYPT 1993. LNCS, vol. 765, pp. 274–285. Springer, Heidelberg (1994)
10. Zhuo, H., Sheng, Z., Nenghai, Y.: Privacy-Preserving Remote Data Integrity Checking Protocol with Data Dynamics and Public Verifiability. IEEE 23(9), 1432–1437 (2011)
11. Giuseppe, A., Randal, B., Reza, C.: Provable data possession at untrusted stores. In: De Capitani di Vimercati, S., Syverson, P. (eds.) Proceedings of CCS 2007, vol. 10, pp. 598–609. ACM Press (2007)
12. Sravan, K.R., Ashutosh, S.: Data Integrity Proofs in Cloud Storage. In: Third InternationalConference on Communication Systems and Networks (COMSNETS), vol. 1, pp. 1–4 (2011)
13. Chris Erway, C., Alptekin, K., Charalampos, P., Roberto, T.: Dynamic Provable Data Possession. In: Proc. of CCS 2009, pp. 213–222. ACM, Chicago (2009)

Density-Based Method for Clustering and Visualization of Complex Data

Tomasz Xięski, Agnieszka Nowak-Brzezińska, and Alicja Wakulicz-Deja

University of Silesia, Institute of Computer Science,
ul. Będzińska 29, 41-200 Sosnowiec, Poland
{tomasz.xieski,agnieszka.nowak,alicja.wakulicz-deja}@us.edu.pl

Abstract. In this paper the topic of clustering and visualization of the data structure is discussed. Authors review currently found in literature algorithmic solutions ([3], [5]) that deal with clustering large volumes of data, focusing on their disadvantages and problems. What is more the authors introduce and analyze a density-based algorithm OPTICS (Ordering Points To Identify the Clustering Structure) as a method for clustering a real-world dataset about the functioning of transceivers of a cellular phone operator located in Poland. This algorithm is also presented as an relatively easy way for visualization of the data's inner structure, relationships and hierarchies. The whole analysis is performed as a comparison to the well-known and described DBSCAN algorithm.

Keywords: cluster visualization, clustering, OPTICS, DBSCAN.

1 Introduction

The problem of choosing the proper clustering algorithm with regard to complex data is not a trivial task [10]. Such an algorithm should have the smallest possible computational complexity, high resistance to the presence of noise in data and allow to discover clusters of arbitrary shapes. As shown in [8], [9], most of these requirements are fulfilled by a density DBSCAN (Density-Based Spatial Clustering of Applications with Noise) algorithm. Its strength lies in the naturally perceived definition of a cluster as a densely-packed area containing similar objects. Unfortunately, although this assumption seems logical and allowing for a clear definition of the boundaries of groups, in practice it may be also responsible for one of two main disadvantages of the algorithm – it is not able to detect a cluster hierarchy if such naturally exists in the source data. Furthermore, results of this algorithm (quality of the created clusters) depend on the proper selection of initial parameters (Eps – radius of the considered neighborhood and MinPts – the number of objects in a group). The choice of these parameters (such as presented in [8]), which consists of carrying out at least a few clusterings and selecting the best one (and its initial parameters), is time consuming and therefore not always possible to apply, especially in the context of analysis of large and complex databases. That is why the authors decided to investigate and implement a different algorithm based on the idea of

J.T. Yao et al. (Eds.): RSCTC 2012, LNAI 7413, pp. 142–149, 2012.

density – OPTICS (Ordering Points To Identify the Clustering Structure) [1]. In contrast to its original, it does not generate a clustering of objects, but only an ordering based on the reachability-distance. However, the generated ordering can be used to partition the dataset into groups (for any fixed neighborhood radius) [1]. In addition, the obtained ordering of objects allows the identification and visualization of the internal structure and relationships in data.

2 Approaches Used to Cluster Large Volumes of Data

The amount of collected data is constantly growing, and hence the term "large datasets" also changes its meaning – until recently it referred to several thousand samples, but today there are processed millions or even billions of data. In the domain literature [3], [5], [7] there are found six main approaches to the problem of clustering large volumes of complex data.

The first of them – *data sampling* – involves selecting a random sample of objects from the dataset and using a cluster analysis' algorithm only on the selected sample. Then it is possible to assign the remaining objects to the already generated representative clusters. This method is the essence of the popular partitional algorithm called CLARA (Clustering for Large Applications). Unfortunately, using only selected samples instead of the entire dataset often results in a clustering which is far from optimal [6]. What is more, the final effect can be different each time the algorithm is run, depending on the methodology of choosing samples.

Another technique used to reduce memory consumption is the *discretization* of data. Generally, two types of discretization are used: *static* – where the set of rules and classes to which one should assign data objects is apriori known, and *dynamic* – where a clustering algorithm is applied taking into account only one attribute and then classes are determined on the basis of the created groups. The biggest problem connected with data discretization is to determine the optimal number of intervals. Even when using dynamic discretization the generated division may be far from ideal.

Some scientific publications [3], [5] describe the *divide and conquer* method. The whole dataset is stored in a larger auxiliary memory (e.g. on the hard disk) and it is divided into smaller portions. Each of these portions are being subjected to the process of cluster analysis separately. At the end of this method, the aggregate results from all clusterings are presented to the user. The CURE (Clustering Using Representatives) algorithm is a popular representative of the divide and conquer approach. It operates on a random sample of data, which is further divided into smaller portions. Then, each portion is subjected to a hierarchical algorithm, to determine the representatives of the clusters. Other objects of the dataset are assigned to clusters based on their degree of correlation with the formed representatives. Unfortunately, it is not possible to apply the divide and conquer method in all cases – some algorithms (such as Hierarchical Agglomerative Clustering) need to operate on the whole dataset. In addition, depending on how much and what portions of the data will be chosen, the final clustering can be of better or worse quality. Not without significance is also the homogeneity of the dataset – it should be uniform to get the best results.

Another solution to the problem of clustering large, complex volumes of data is to use an *incremental algorithm* (if such an implementation is available). The basic assumption used in this approach is the ability to analyze each object from the set independently. In the main memory there are usually stored only the representatives of clusters, and each object is correlated with the existing representatives, to determine its belonging to a group. BIRCH (Balanced Iterative Reducing and Clustering Using Hierarchies) is an incremental, hierarchical algorithm, often used in the clustering task. It gives good results when discovering spherical clusters of similar size, but such a situation (in the context of complex data analysis) rarely takes place. It is also an example of an algorithm (called order-dependent) in which the processing sequence of objects has a huge impact on the final results. Unfortunately, a large group of incremental algorithms can be characterized by this feature [5].

If the main problem is the high computational complexity of an algorithm, *parallelism* can be used. There are generally two ways to do this: through the use of multiple central processing units (usually multi-core) or by taking advantage of the power of a graphics processing unit (GPU), which is specialized in performing floating point operations. However, this implies the need for defining and allocating tasks to individual computing units. This could include a division of the dataset into parts (as in the divide and conquer method), so that each unit carries out a clustering of its data portion, or the allocation of tasks resulting from the construction of the cluster analysis' algorithm. Parallelization of the algorithm is unfortunately not always possible. Furthermore, there are some speed considerations, which have to be taken into account when dealing with parallelization. If one uses mainly the CPU for this task, then the greatest impact on the calculation speed will have the number of available processors (and cores) as well as the frequency and method of communication between threads. Using the computational units included on the graphics card, one can get much higher speeds (e.g. in [4] there was achieved an acceleration of 10 – 200), but the process of optimizing code for a given family of graphics cards is much more complicated than for the CPU. Moreover, if the algorithm requires frequent data exchange between computational units, then the benefit from its parallelization is usually restricted. It is also worth mentioning that graphics cards (from different companies) have their own unique programming libraries, which leads to a large attachment to a particular hardware platform [4].

3 Comparison of OPTICS and DBSCAN

During the process of clustering complex data, apart from problems related to the processing of large amounts of data, one must also take into account its internal structure. Relationships contained in real, multidimensional data can create hierarchies and have a heterogeneous nature. That is why the approach to cluster data regarding cellular telephony with the use of the DBSCAN algorithm (described by authors in [8], [9]) highlighted two major problems of this technique.

The first problem is the selection of appropriate initial parameters (Eps, MinPts). Too large values of the neighborhood radius (Eps), may cause that two naturally existing small clusters could be interpreted by the algorithm as one bigger. Small values of this parameter will not be good, if the dataset will consist of sparsely located objects. The second input parameter (MinPts) has the greatest impact on the number of objects in groups, and thus the size of the outliers' group – the greater its value, the potentially more objects can be classified as informational noise. The simplest approach to solving this problem, is to perform a number of clusterings with different values of initial parameters, assess their quality, and select the best result. Unfortunately, just the clustering process by itself is a time-consuming task.

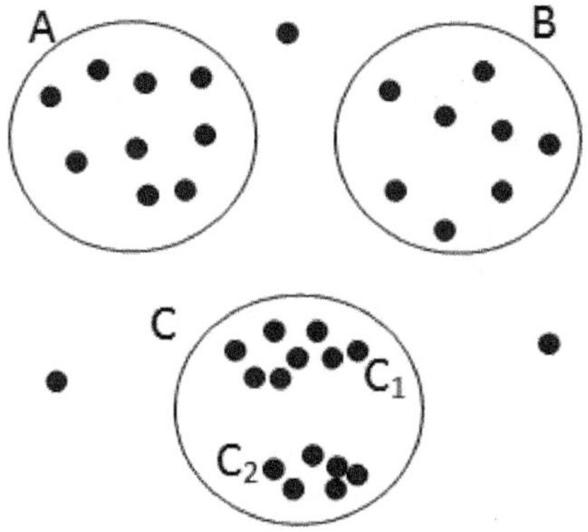

Fig. 1. The presence of a hierarchy of clusters

Another disadvantage of the DBSCAN algorithm (directly related to the way it works) is the inability to detect a cluster hierarchy. This situation is shown in figure 1. For the presented dataset it is not possible to detect all four clusters (labeled A, B, C1, C2), using a constant value for the neighborhood radius. The DBSCAN algorithm would generate a partitioning, consisting of groups A, B, C or groups C1, C2, depending on the chosen parameters values. In the latter case, the actual clusters A and B would be treated as informational noise. Using this fact, one could modify the DBSCAN algorithm, to create clusters of different densities at the same time – it would use different values for the neighborhood radius. However, to obtain a consistent result, it would be necessary to maintain a certain order in which objects would be processed. One should always choose the object, which is density-reachable at the smallest possible neighborhood radius (Eps), so that clusters with the highest density were discovered first. The OPTICS algorithm works in a similar way (taking into account all possible values of the neighborhood radius to a set limit), except that it does not assign labels to objects belonging to a specific group. Instead, it stores the order in which objects were processed, and information that can be used to identify individual objects belonging to a specific group. This information consists of two parameters (calculated for each object from the set) – a so-called *core-distance* and a *reachability-distance*.

The *core-distance* (of an object p from the dataset) is the smallest distance between p and an object within its Eps-neighborhood[1], such that p would be classified as a core object. Otherwise (if p can not be regarded as a core object), the *core-distance* is undefined. The *reachability-distance* of an object p (with regard to object q) is the smallest distance, such that p is directly density-reachable from q, if q is a core object. In that case, the *reachability-distance* of p can not be smaller than the core-distance of q, because for smaller values p is not density-reachable from q. However, if q is not a core object, the reachability-distance of p (with regard to q) is undefined.

Taking into account the presented definitions, one can specify a simplified flowchart of the OPTICS algorithm. It is as follows:

1. Select an object from the dataset.
2. Determine the Eps-neighborhood of the currently analyzed object.
3. Calculate the core-distance of the analyzed object.
4. If the core-distance is undefined, go to step one.
5. Calculate the reachability-distances of objects contained in the neighborhood of the currently analyzed (relative to this object), and then sort the objects in ascending order according to their reachability-distance.
6. Continue this process (from the first point) until all of the objects from the dataset have been analyzed.
7. Print or store the order in which objects were processed in a dataset, together with the values of core and reachability-distances.

Based on the presented flowchart, one can see strong similarities to the DBSCAN algorithm. For both methods the most important element is to determine the Eps-neighborhood. The DBSCAN algorithm on its basis, determines the membership of each object to a particular cluster, while OPTICS generates a specific arrangement of objects. This arrangement can be used not only to assign object to groups, but also to visualize the data's structure. Considering the similarities between these algorithms, it can be stated, that the execution time of OPTICS (and in consequence also its computational complexity) does not differ drastically from its original execution time. Experiments conducted in [1] have shown, that the execution time of the OPTICS algorithm is equal to 1.6 times of DBSCAN's execution time. This is not a surprising result because, as noted earlier, the computational complexity of both methods is highly dependent on the process of determining the Eps-neighborhood, which must be performed for each object from the dataset. If there are no data indexes used, to obtain the answer to a query regarding the neighborhood of an object, a scan through the entire database would have to be performed. In that case, the computational complexity of the OPTICS algorithm would be $O(n^2)$. The situation is considerably

[1] Detailed definitions of Eps-neighborhood, density-reachability and related terms are described in [1], [2], [7].

improved, if one uses an index based on tree structures (called an tree-based spatial index), because the average computational complexity of the whole algorithm is reduced to $O(n \log n)$ [1].

4 OPTICS as a Visualization Method

Visualization of the dataset's structure can be very useful, if the analyst wants to know the general outline of the data at a high level of abstraction. Then, attention is paid to the presence of a hierarchy, or the level of consistency between the created groups. Important may be also the fact, if there exist certain dominant groups (of very high cardinality in relation to the entire dataset), or if the nature of this structure is more homogeneous. Very often, based only on the generated clustering, an analyst (especially in the context of large, complex datasets) may not be able to explain or interpret correctly the obtained results. That is why more and more emphasis is placed on visualization tools to assist in the process of analysis and interpretation of clustering results.

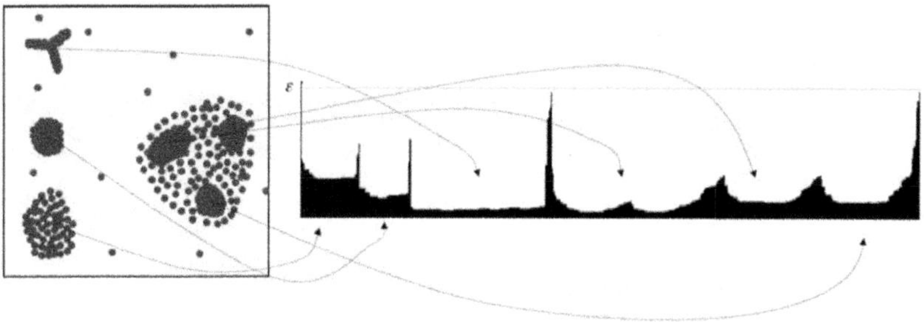

Fig. 2. Example of a reachability plot

To see the detailed structure of the dataset, one has to create a plot of the reachability-distance of every object, according to the ordering generated by the OPTICS algorithm. This bar plot is called a reachability-plot. A sample reachability-plot for artificially generated two-dimensional data is presented in figure 2. Clusters on the plot are represented by valleys. The more narrow the valley is, the less objects are included in a particular cluster. However, the smaller the reachability-distance value, the more dense the cluster is (more coherent).

Based on the reachability-plot, one can easily detect the existence of a cluster hierarchy. Such hierarchy is presented in the plot (figure 2) as a series of small valleys contained in one deeper. When identifying potential cases of cluster containment, one should pay particular attention that the valley should be very shallow. The low value of the reachability-distance means that another object is very close to the previous one in a given order, which for shallow valleys implies that, two clusters are located very near. In addition, if these shallow valleys are located within another very deep, with a high degree of probability we are dealing with a situation, where a number of small, cohesive clusters, is contained in one bigger (of a much smaller density). Unfortunately for the real-world complex datasets, the identification of a hierarchy of clusters based only on the reachability-plot is far more difficult. Figure 3 shows the reachability-plot for real-world complex data, regarding the operation of one of cellular network's transceivers[2]. This plot however is truncated to the first 32000 objects (structured according to the result of the OPTICS algorithm), due to the readability aspects and limitations of the MS Excel software, which does not allow to visualize a greater number of two-dimensional data in a bar plot (in one serie).

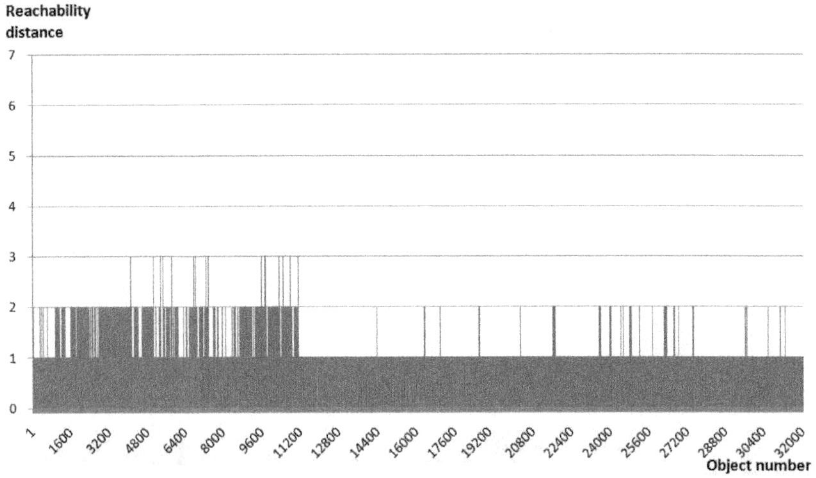

Fig. 3. Reachability plot for real-world data

The presented in figure 3 plot shows that already in its initial areas (between objects numbered 8000 to about 10,000), one can observe the presence of a hierarchy of clusters – several smaller clusters are potentially contained in a bigger one. These relationships are not as easily identifiable as for the artificial dataset (shown in figure 2).

Moreover, despite of a significant reduction in the number of objects shown in the plot – the entire dataset consists of about 143000 objects – its overall readability is not high. It is therefore necessary, to develop other ways to create a reachability-plot, in order to present the entire dataset in an understandable and human readable form. The used measure of similarity (distance) between objects has also a big impact on the readability of the generated plot. In this case, the measure of similarity was defined as the number of common features. This implies that, if objects close to each other differ only in the values of two features, it is seen as a relatively large change on the plot. For this reason, in the future there should be performed tests using other distance measures.

[2] The analyzed dataset is described in detail in [9].

5 Summary

The aim of this paper was to present currently found in literature solutions for clustering large volumes of data, with particular emphasis on their shortcomings and problems. Another key objective of this study was to analyze and discuss the OPTICS algorithm, as a method of visualizing the structure of large complex datasets. A particular aspect of this analysis was to compare the density-based DBSCAN algorithm with OPTICS. This comparison showed that the analyzed algorithms have many similarities with each other, even though they may be used in completely different purposes. The theoretical considerations are supported by applying the OPTICS algorithm to a real-world dataset concerning cellular telephony, in order to visualize and interpret the structure of this set.

References

1. Ankerst, M., Breunig, M.M., Kriegel, H.P., Sander, J.: Optics: Ordering points to identify the clustering structure. In: Proceedings of ACM SIGMOD International Conference on Management of Data, Philadelphia, USA (1999)
2. Ester, M., Kriegel, H.P., Sander, J., Xu, X.: A density-based algorithm for discovering clusters in large spatial databases with noise. In: Proceedings of 2nd International Conference on Knowledge Discovery and Data Mining, USA (1996)
3. Jain, A.K., Murty, M.N., Flynn, P.J.: Data Clustering: A Review. ACM Computing Surveys 31(3) (1999)
4. Böm, C., Noll, R., Plant, C., Wackersreuther, B.: Density-based Clustering using Graphics Processors. In: Proceeding of the 18th ACM Conference on Information and Knowledge Management, USA (2009)
5. Berry, M.W., Browne, M.: Lecture notes in Data Mining. World Scientific Publishing Co. Pte. Ltd., Singapur (2009)
6. Tufféry, S.: Data Mining and Statistics for Decision Making. Wiley & Sons Ltd., UK (2011)
7. Han, J., Kamber, M., Pei, J.: Data Mining. Concepts and Techniques. Elsevier Inc., USA (2012)
8. Wakulicz-Deja, A., Nowak-Brzezińska, A., Xięski, T.: Efficiency of Complex Data Clustering. In: Yao, J., Ramanna, S., Wang, G., Suraj, Z. (eds.) RSKT 2011. LNCS, vol. 6954, pp. 636–641. Springer, Heidelberg (2011)
9. Xięski, T.: Clustering complex data. In: Wakulicz-Deja, A. (ed.) Decision Support Systems. Institute of Computer Science of the University of Silesia (2011) (in Polish)
10. Nowak-Brzezińska, A., Jach, T., Xięski, T.: Choice of the clustering algorithm and the efficiency of finding documents. Studia informatica, Scientific Papers of Silesian Technical University 31(2A(89)), 147–162 (2010) (in Polish)
11. Xu, R., Wunsch, D.: Clustering. IEEE Press Series on Computational Intelligence, USA (2008)

Inference Processes Using Incomplete Knowledge in Decision Support Systems – Chosen Aspects

Agnieszka Nowak-Brzezińska, Tomasz Jach, and Alicja Wakulicz-Deja

Institute of Computer Science, University of Silesia,
Będzinska 39, 41–200 Sosnowiec, Poland
{agnieszka.nowak,tomasz.jach,alicja.wakulicz-deja}@us.edu.pl

Abstract. The authors propose to use cluster analysis techniques (particularly clustering) to speed-up the process of finding rules to be activated in complex decision support systems with incomplete knowledge. The authors also wish to inference within such decision support systems using rules, of which premises are not fully covered by the facts. The AHC or mAHC algorithm is used. The authors adapted Salton's most promising path method with own modifications for a fast look-up of the rules.

Keywords: knowledge bases, cluster analysis, clustering, decision support systems, incomplete knowledge, inference, AHC.

1 Introduction

Currently developed knowledge bases try to support human experts in the process of solving decision problems. The complexity of these bases rapidly increases, the best example here would be medical data and knowledge bases. The inference within these is completely non-trivial, because modern knowledge bases often consist of thousands of rules.

Under the classical definition of Decision Support System the authors mean the combination of knowledge base and inference algorithms. Both rely on rules, in which every one of it consists of two parts: decisional and conditional. Formally, the Decision Support System with structures added by the authors is given by:

$DSS =< R, A, V, F_{sim}, Tree >$ where:
$R = \{r_1, \cdots, r_n\}$ – set of rules with Horn's forms,
$A = \{a_1, \cdots, a_m\}$ – where $A = C \cup D$ (condition and decision attributes),
V – nonempty, finite set of values of attributes; $V = \cup_{a \in A} V_a$
V_a – the domain of attribute a; $F_{sim} : X \times X \rightarrow R|[0 \cdots 1]$,
$dec : R \rightarrow V_{dec}$, where $V_{dec} = \{d_1, \cdots, d_m\}$,
$Tree = \{w_1, \cdots, w_{2n-1}\} = \cup_{i=1}^{2n-1} w_i$ (or $Tree = \{w_1, \cdots, w_k\} = \cup_{i=1}^{k} w_i$ where $k \leq 2n-1$).

Using these, it can be said that each rule $r \in R$ (set of rules in DSS) is considered to be an conjunction of attribute-value pairs (noted further as descriptors). Additionally, each rule is marked with specific value of decision attribute ($d \in V_{dec}$).

J.T. Yao et al. (Eds.): RSCTC 2012, LNAI 7413, pp. 150–155, 2012.

To sum things up: $r_i = (a_1 = v_1) \cdot (a_2 = v_2) \cdot \ldots \cdot (a_m = v_m) \Rightarrow d_j$, where $m \leq card(A)$. The increasing number of attributes, connected with the rapid increase of the number of samples on basis of which rules are generated, makes efficient inference algorithms in complex data structures essential to the quality of results.

However, the number of rules and the size of attribute set are not the only aspects of proper inference. In real life situations, it is hardly possible to obtain full consistency of knowledge base. The inconsistency is understood by the authors both as the situation, where the same conjunction of conditional attributes and their respective values lead to different decisions[1] and when at least one rule's in knowledge base condition's are not fully satisfied by the facts.

In order to address this problem, various methods can be used. The authors of this paper propose the cluster analysis approach to cluster similar rules and to identify those which can be activated during the inference process.

Let us consider the following example:

```
R1: (attr4=8600)    & (attr8=177) & (attr1=152) =>(class=2)
R2: (attr4=8600)    &    (attr1=151) =>(class=2)
R3: (attr4=8600)    &    (attr7=30)   =>(class=2)
Facts:
(attr4=8600),          (attr7=40),              (attr1=152)
```

The classical decision support system will not activate any of the rules because in neither of them the conditions are fully satisfied. The closest to be fully satisfied is R1 rule, therefore proposed system will activate it, but flags it as uncertain. This methods allows the user to fine-tune the *precision* of inference process: to balance between accurate but limited inference and approximate but giving more potentially useful information.

1.1 Search Using Hierarchy Structure

The AHC algorithm generates the complete rules' tree[1]. On the other hand, the mAHC algorithm stops completing the process (the difference can be seen on Figure 1). This property can be used to speed-up the process of searching relevant rules by comparing user's query to the representatives of clusters, rather than to the rules themselves. On each level, one must compare the query to the left and right branch and choose the path, which is more promising. Formally, by d_i authors mean the descriptors set, f is the similarity function between two rules and k_i, l_i are the nodes being merged. Using these notations each cluster w_i can be defined as: $w_i = (d_i, f, k_i, l_i)$, where $d_i = \{d_1, \ldots, d_m\}, f : R \times R \rightarrow R|[0 \cdots 1]$.

The idea of the most promising path was firstly stated in Salton's SMART system[2], which was the great inspiration for the authors when creating the proposed system. This approach starts the look-up process from the root of the structure comparing the left and right branch using the f function to determine

[1] Where $((a_1 = v_1) \cdot (a_2 = v_2) \cdot \cdots \cdot (a_n = v_n) \Rightarrow d_1) \wedge ((a_1 = v_1) \cdot (a_2 = v_2) \cdot \cdots \cdot (a_m = v_m) \Rightarrow d_2)$.

which one is the most probable to have relevant rules. The operation progresses until the leaves level is reached.

In order to implement the most promising path method, the authors must have also taken into consideration the method of computing the similarity of the query to particular nodes (the f function). The preliminary research about representatives was published in previous works [3], therefore here we are going to discuss only the differences and improvements which evolved from then.

The first method which comes in mind, so-called *descriptors coverage*, computes the number of descriptors occurring both in the question, as well as in the individual nodes according to the formula: $f_d(k,l) = card(d_k \cap d_l)$ where d_k and d_l are the sets consisting of descriptors of nodes l and k respectively. Unfortunately, this method boosts the value of those nodes, which have a large number of repeating descriptors, often common for a vast majority of rules in the system. However, when one has to deal with the incomplete knowledge, the information about common attributes can be vital for proper distinguishing the clusters.

The second approach, called *attributes coverage*, takes into consideration only the number of common attributes, regardless of their values: $f_a(k,l) = card(a_k \cap a_l)$ where a_k and a_l denotes the attributes' set of the $k-th$ and $l-th$ cluster respectively. As it was stated before, this approach addresses the problem of multiple, common descriptors which disturb the proper similarity computing. In another words the situation when clusters' representatives consist of many commonly occurring descriptors is undesirable because of the lack of proper distinction between them.

During the preliminary studies, authors combined the above methods into one called *hybrid coverage*: $f_h(k,l) = card(d_k \cap d_l) \cdot C_1 + card(a_k \cap a_l) \cdot C_2; \quad (C_1 + C_2 = 1) \wedge C_1 > 0, C_2 > 0$.

The authors suggest that the hybrid coverage will benefit both from the advantages of attribute and descriptors coverages. The scaling factors C_1 and C_2 are used to fine-tune the influence of both of the mentioned coverages. During the experiments two opposite set of values were chosen: one which greatly favors the descriptors part, and the other boosts the attribute part.

To clear things up, the authors propose the following example. Given two nodes: $k : d_k = \{(A = 1), (A = 1), (A = 2), (B = 1), (B = 1), (C = 1)\}$ $l : d_l = \{(A = 2), (A = 2), (B = 1), (B = 1), (B = 1), (C = 1)\}$ and a query: $Q : (A = 2) \cdot (C = 1)$ the following factors can be computed:

- $f_d(k,Q) = 2; f_d(l,Q) = 3 \quad f_a(k,Q) = 4; f_a(l,Q) = 3$
- If $C_1 = 0,75$ and $C_2 = 0,25$, then $f_{h1}(k,Q) = 2,5; f_{h1}(l,Q) = 3$
- If $C_1 = 0,25$ and $C_2 = 0,75$, then $f_{h2}(k,Q) = 3,5; f_{h2}(l,Q) = 3$

2 Computational Experiments

In order to compare the proposed solutions, the authors implemented two hierarchical clustering algorithms: AHC (which uses the complete hierarchical tree of rules) and mAHC (using the authors' method of choosing the optimal number of clusters). The difference can be schematically seen on Figure 1. The results of these experiments are shown in Figure 2. For four databases from Machine Learning Repository (Wine, Lymphography, Spect, Balance) the authors conducted both clustering algorithms assuming every observation from those databases as the rule in knowledge base. The process of preparing the data for the clustering is explained in detail in authors' previous paper [3]. On each case, 10 random queries were chosen (the query was in fact one randomly chosen rule from knowledge base). *Recall* and *precision* values were computed and the average from those 10 queries was computed.

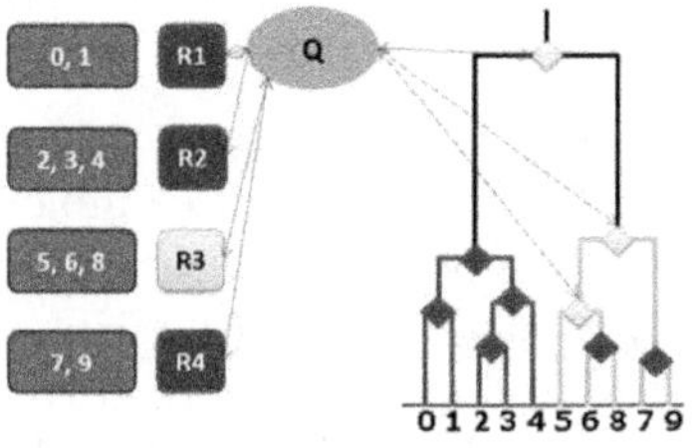

Fig. 1. Search using mAHC (left) and AHC (right)

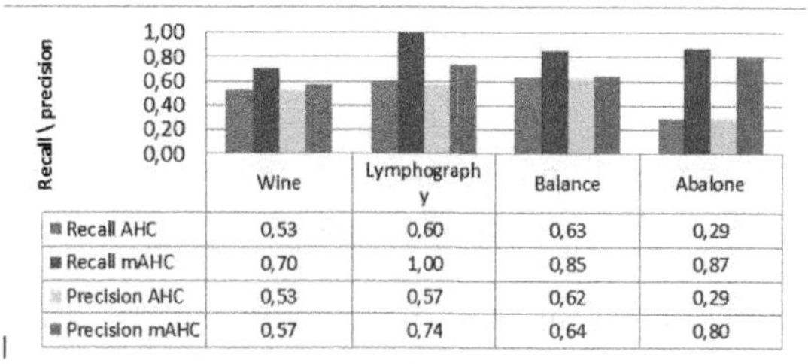

	Wine	Lymphography	Balance	Abalone
Recall AHC	0,53	0,60	0,63	0,29
Recall mAHC	0,70	1,00	0,85	0,87
Precision AHC	0,53	0,57	0,62	0,29
Precision mAHC	0,57	0,74	0,64	0,80

Fig. 2. The quality of hierarchical and structural search

2.1 The Most Promising Path

In order to practically verify the results, the experiments were conducted (this works are the basis of currently developed DSS to inference in complex knowledge bases with uncertain knowledge). Firstly, it was assumed that currently analyzed rule becomes the query to the system. To the complete system, computed by different combinations of the most promising path method and cluster joining criteria, query containing all of the descriptors of currently analyzed rule was submitted. The answer given was saved as the goal answer. Following, that particular rule was deleted from the knowledge base and the process of forming clusters was repeated. Again, the system was queried and the given answer was being analyzed along with the one saved in the previous step. *Recall* and *precision* was computed both to the goal answer (assumed to be the optimal answer) and to the submitted query (if the system has found the proper answer).

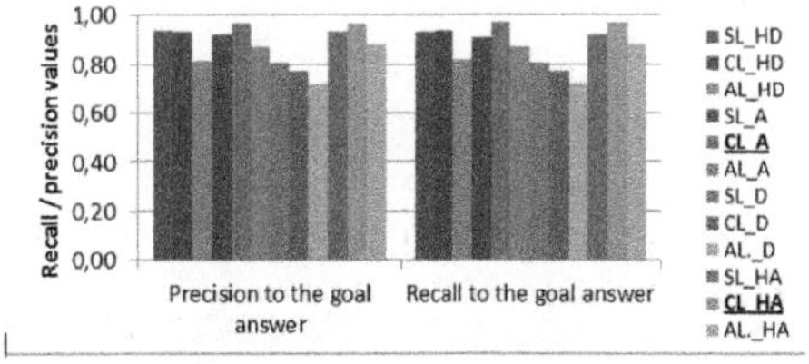

Fig. 3. Experiments involving the most promissing path

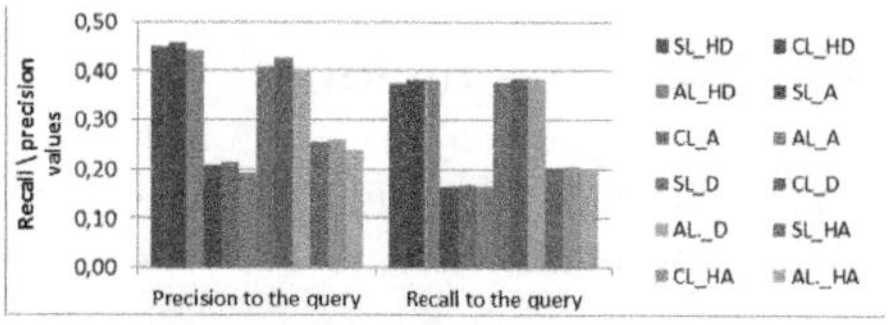

Fig. 4. The results of computational experiments

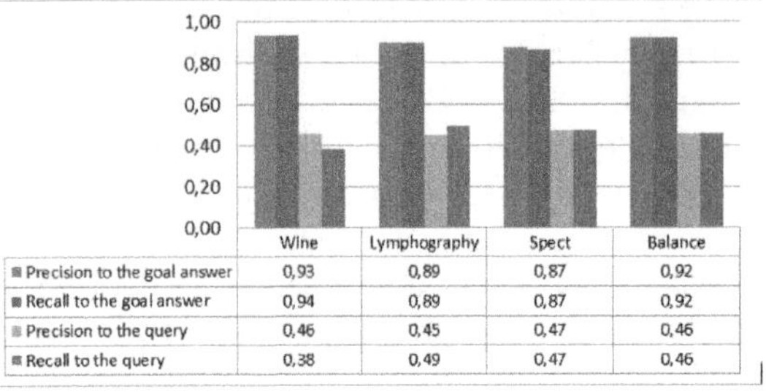

	Wine	Lymphography	Spect	Balance
Precision to the goal answer	0,93	0,89	0,87	0,92
Recall to the goal answer	0,94	0,89	0,87	0,92
Precision to the query	0,46	0,45	0,47	0,46
Recall to the query	0,38	0,49	0,47	0,46

Fig. 5. The results of hybrid method for chosen knowledge bases

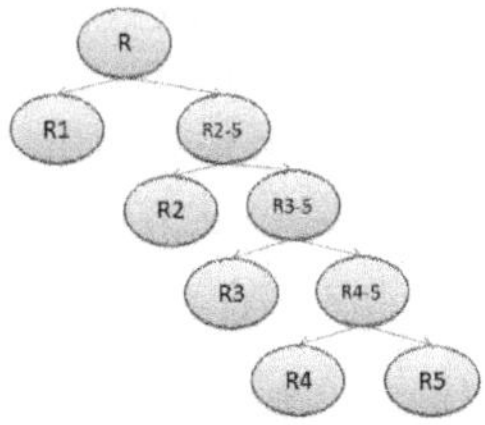

Fig. 6. Chaining the clusters in the AHC tree

Figures 3 and 4 share the same marks: SL - Single Linkage, CL - Complete Linkage, AL - Average Linkage, HD - the hybrid version of the most promising path coverage having the parameter C_1 significantly smaller than C_2 (descriptors more important than the attributes), HA - the same, but C_1 was far more greater than C_2 (on the contrary: attributes more important than descriptors), A - attribute coverage, D - descriptor coverage.

It seems obvious, that the best results were achieved when using the CL joining criterion. Both *recall* and *precision* to the goal answer values were more or less on the same level with the slight favor of HA and A methods. It could be believed to be the confirmation of the authors' assumptions about a better distinction of the clusters using the information about common attributes.

In the second part of the experiments *precision* and *recall* values to the submitted query were computed for the limited system. By doing this, the authors wished to investigate if the proposed system is able to compensate the incompleteness of the knowledge[2].

The figure 4 clearly shows the superiority of the proposed hybrid coverage method, especially the one with the significant boost for the descriptors. Regardless of the method of joining the clusters, overall quality of the results was a few times better than using other coverage methods.

For further investigations, the authors chose the complete linkage method along with the hybrid coverage with descriptors' boost. The same methodology was used to the test on different knowledge bases. The results are shown on Figure 5. The preliminary results from the tuning parameters phase were confirmed for all the databases analyzed by the authors.

3 The Conclusions

The authors of the study came across a serious problem with a tendency for the clusters to chain (Figure 6). Due to the fact of a relatively brief description of

[2] However, one has to keep in mind, that because of the removing of the rule, which is the optimal answer for the query (limited system) the maximal values of the quality parameters can not be achieved.

each rule, and their small distinguishability between each other, often leads to impaired uniformity dendrogram (during one of the experiments in one of the subtrees at every level we had only one rule, and the second - others). After analyzing the situation, the authors pointed out a disturbing fact of the poor quality of the distinguishability matrix built at the beginning of the algorithm. For example, in the Abalone base, there were 7138531 cells in the similarity matrix, where the entire database had only 43 different values of similarity factors. Further research will aim to eliminate this phenomenon.

The authors were able to improve Salton's most promising path method of searching the rules. In future works the authors will focus on further investigating distance measures and other ways to further distinguish the rules in order to create better quality clusters. The method of certainty factors CF[4] is also considered as the next approach for the correct modeling of uncertainty and inference.

References

1. Kaufman, L., Rousseeuw, P.J.: Finding Groups in Data: An Introduction to Cluster Analysis. Wiley, New York (1990)
2. Salton, G.: Automatic Information Organization and Retreival. McGraw-Hill, New York (1975)
3. Wakulicz-Deja, A., Nowak-Brzezińska, A., Jach, T.: Inference Processes in Decision Support Systems with Incomplete Knowledge. In: Yao, J., Ramanna, S., Wang, G., Suraj, Z. (eds.) RSKT 2011. LNCS, vol. 6954, pp. 616–625. Springer, Heidelberg (2011)
4. Simiński, R., Nowak-Brzezińska, A., Jach, T., Xięski, T.: Towards a Practical Approach to Discover Internal Dependencies in Rule-Based Knowledge Bases. In: Yao, J., Ramanna, S., Wang, G., Suraj, Z. (eds.) RSKT 2011. LNCS, vol. 6954, pp. 232–237. Springer, Heidelberg (2011)
5. Jain, A., Dubes, R.: Algorithms for clustering data. Prentice Hall (1988)
6. Koronacki, J., Ćwik, J.: Statystyczne systemy uczące się. Exit, Warszawa (2008)
7. Frank, A., Asuncion, A.: UCI Machine Learning Repository. UC, SoIaCS, Irvine, CA (2010), `http://archive.ics.uci.edu/ml`
8. Myatt, G.: Making Sense of Data. A Practical Guide to Exploratory Data Analysis and Data Mining. John Wiley and Sons, Inc., New Jersey (2007)
9. Kumar, V., Tan, P., Steinbach, M.: Introduction to Data Mining. Addison-Wesley (2006)
10. Pawlak, Z.: Rough set approach to knowledge-based decision suport. European Journal of Operational Research, 48–57 (1997)

Interactive Document Indexing Method Based on Explicit Semantic Analysis*

Andrzej Janusz[1], Wojciech Świeboda[1], Adam Krasuski[1,2], and Hung Son Nguyen[1]

[1] Faculty of Mathematics, Informatics and Mechanics, The University of Warsaw, Banacha 2, 02-097, Warsaw Poland

[2] Chair of Computer Science, The Main School of Fire Service, Słowackiego 52/54, 01-629 Warsaw Poland

Abstract. In this article we propose a general framework incorporating semantic indexing and search of texts within scientific document repositories. In our approach, a semantic interpreter, which can be seen as a tool for automatic tagging of textual data, is interactively updated based on feedback from the users, in order to improve quality of the tags that it produces. In our experiments, we index our document corpus using the Explicit Semantic Analysis (ESA) method. In this algorithm, an external knowledge base is used to measure relatedness between words and concepts, and those assessments are utilized to assign meaningful concepts to given texts. In the paper, we explain how the weights expressing relations between particular words and concepts can be improved by interaction with users or by employment of expert knowledge. We also present some results of experiments on a document corpus acquired from the PubMed Central repository to show feasibility of our approach.

Keywords: Semantic Search, Interactive Learning, Explicit Semantic Analysis, PubMed, MeSH.

1 Introduction

The main idea of a keyword search is to look for texts (documents) that contain one or more words specified by a user. Then, using a dedicated ranking algorithm, relevance of the matching documents to the user query is predicted and the results are served as an ordered list. In contrast, semantic search engines try to improve the search accuracy by understanding both, the user's information need and the contextual meaning of texts, which are then intelligently associated.

* This work is partially supported by the National Centre for Research and Development (NCBiR) under Grant No. SP/I/1/77065/10 by the Strategic scientific research and experimental development program: "Interdisciplinary System for Interactive Scientific and Scientific-Technical Information" and grants from Ministry of Science and Higher Education of the Republic of Poland N N516 077837.

J.T. Yao et al. (Eds.): RSCTC 2012, LNAI 7413, pp. 156–165, 2012.

From the data processing point of view, the semantic search engine may be divided into three main components: semantic text representation module, interpretation and representation of the user query, and intelligent matching algorithm. The scope of the first two modules may be categorized as a semantic data representation. In opposite to the keyword search, the semantic data representation, and thus the semantic indexes, can not be calculated once and then utilized by intelligent matching algorithms. The text representation, as well as a query interpretation should be assessed with respect to the type of the users' group, a context of the words in the query and many others factors.

The better part of current search engines is based on a combination of textual keyword search and sophisticated document ranking methods. Only a few processes search queries, analysing both, a query and documents' content with respect to their meaning, and return the semantically relevant search results [1]. However, even this approach becomes insufficient. The process of information retrieval needs to be made intelligently in order to help users in relevant information. The key role in this process is recognition of the users' information needs and collecting feedback about the search effectiveness. The gathered information should be utilized to improve search algorithms and forge better responses to user requirements. Those challenges are in the scope of studies on adaptive search engines which interact with experts (users) and operate in a semantic representation space.

SONCA (Search based on ONtologies and Compound Analytics) platform [2] is developed at the Faculty of Mathematics, Informatics and Mechanics of University of Warsaw. It is a part of SYNAT project focusing on development of Interdisciplinary System for Interactive Scientific and Scientific-Technical Information (`www.synat.pl`). SONCA is a framework whose aim is to extend the functionality of search engines by more efficient search of relevant documents, intelligent extraction and synthesis of information, as well as more advanced interaction between users and knowledge sources.

Within the SYNAT project, some successful methods for semantic text representation and indexing have already been developed [3,4]. In this article we present an adaptive semantic search model which can be treated as a step forward a truly semantic search engine. The model interactively calculates the semantic text representation with respect to the user feedback. At the current development phase we utilize the user feedback explicitly, i.e. we work with a set of documents labelled with concepts by experts. However, in the future we plane to extend the use-case to interaction with implicit user feedback, such as user navigation patterns or a specific query context.

The rest of the paper is structured as follows: Section 2 contains basic information about Explicit Semantic Analysis method and its applications in the semantic indexing and semantic search. An interactive learning model and an adaptive method for improving the semantic text representation is proposed in Section 3. Section 4 describes an experiment conducted on a corpus of documents from PubMed Central Open Subset [5] and presents its result. Section 5 presents some concluding remarks and a plan for future works.

2 Semantic Indexing and Searching Using ESA

Explicit Semantic Analysis (ESA) proposed in [6] is a method for automatic tagging of textual data with predefined concepts. It utilizes natural language definitions of concepts from an external knowledge base, such as an encyclopaedia or an ontology, which are matched against documents to find the best associations. Such definitions are regarded as a regular collection of texts, with each description treated as a separate document.

In ESA, the semantic relatedness between concepts and documents is computed two-fold. First, after the initial processing (stemming, stop words removal, identification of terms), the corpus and the concept definitions are converted to the *bag-of-words* representation. Each of the unique terms in the texts is given a weight expressing its association strength. Assume that after the initial processing of a corpus consisting of M documents, $D = \{T_1, \ldots, T_M\}$, there have been identified N unique terms (e.g. words, stems, n-grams) $w_1, \ldots, w_N$. Any text T_i in the corpus D can be represented by a vector $\langle v_1, \ldots, v_N \rangle \in \mathbb{R}_+^N$, where each coordinate v_j expresses a value of some relatedness measure for j-th term in vocabulary (w_j), relative to this document. The most common measure used to calculate v_j is the *tf-idf* (term frequency-inverse document frequency) index (see [7]) defined as:

$$v_j = tf_{i,j} \times idf_j = \frac{n_{i,j}}{\sum_{k=1}^{N} n_{i,k}} \times \log\left(\frac{M}{|\{i : n_{i,j} \neq 0\}|}\right), \tag{1}$$

where $n_{i,j}$ is the number of occurrences of the term w_j in the document T_i.

Next, the bag-of-words representation of concept definitions is transformed into an inverted index that maps words into lists of K concepts described in a knowledge base. The inverted index is used as a semantic interpreter. Given a text from a corpus, it iterates over words from the text, retrieves the corresponding entries and merges them into a weighted vector of concepts that represents the given text.

Let $W_i = \langle v_1, \ldots, v_j, \ldots, v_N \rangle$ be a bag-of-words representation of an input text T_i, where v_j is the tf-idf index of w_j described in (1). Let $inv_{j,k}$ be an inverted index entry for w_j. It quantifies the strength of association of the term w_j with a knowledge base concept c_k, $k \in \{1, \ldots, K\}$. For convenience, all the weights $inv_{j,k}$ can be arranged in a sparse matrix structure with N rows and K columns, denoted by INV, such that $INV[j,k] = inv_{j,k}$ for any pair (j,k). The new vector representation of T_i will be denoted by $U_i = \langle u_1, \ldots, u_K \rangle$ where:

$$u_k = \sum_{j : w_j \in T_i} v_j \times inv_{j,k} = W_i * INV[\cdot, k]. \tag{2}$$

In the above equation $*$ is the standard dot product and $INV[\cdot, k]$ indicates k-th column of the sparse matrix INV. This new representation will be called a *bag-of-concepts* of a text T_i.

For practical reasons it may also be useful to represent documents only by the most relevant concepts. In such a case, the association weights can be used to rank the concepts and to select only the top concepts from the ranked list. One can also apply some more sophisticated methods that involve utilization of internal relations in the knowledge base (e.g. for semantic clustering of concepts and assigning only the most representative ones to the documents).

The original purpose of Explicit Semantic Analysis was to provide means for computing semantic relatedness between texts. However, an intermediate result – weighted assignments of concepts to documents (induced by the term-concept weight matrix) may be naturally utilized in document retrieval as a semantic index [8,9]. A user (an expert) may query a document retrieval engine for documents matching a given concept. If the concepts are already assigned to documents, this problem is conceptually trivial. However such a situation is relatively rare, since employment of experts who could manually labelled documents from a huge repository is expensive. On the other hand, utilization of an automatic tagging method, such as ESA, allows to infer labelling of previously untagged documents.

In the presented study, our main goal is to balance these two approaches (manually labelling and automatic tagging): we start with a default ESA method [6] and we update weights incorporated by this model by considering expert feedback. We describe the model and the algorithm which updates weights provided by ESA in an batch mode. We stress however, that essentially the same procedure can be used in an on-line fashion, analogically to stochastic updates in neural networks: the algorithm would process a text repository document by document and update underlying weights accordingly.

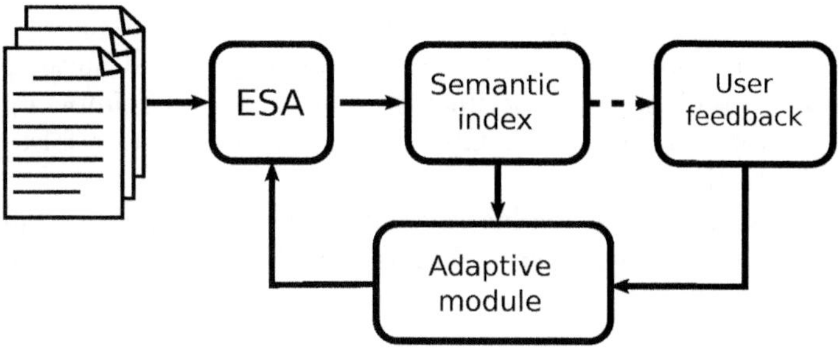

Fig. 1. A scheme of the interactive learning model

3 Learning Model

The model structure imposed by ESA can be interpreted as a one layer neural network [10] with N input nodes corresponding to terms and K output nodes corresponding to concepts. Initially, weights $inv_{j,k}$ for the j-th word ($j \in \{1, \ldots, N\}$) and the k-th concept ($k \in \{1, \ldots, K\}$) are assigned based on the *bag-of-words* representation of the k-th concept. A set of words occurring in a definition of the concept c_k will be denoted by $MeSH_k$. In our case study, this representation is computed using natural language annotations from the MeSH ontology [11].

We impose two constraints on the weight refinement procedure. The first one is that the descriptions of concepts available in a knowledge base determine the network structure (i.e. we restrict the weights $inv_{j,k}$ equal zero to remain zero throughout the updates – we do not construct any new connections in the network). The second constraint is that the particular updates are multiplicative rather than additive (thus, $inv_{j,k} \geq 0$, which provides a more transparent interpretation of weights that address several, otherwise unrelated topics).

For a document T_i, represented by a set of terms T_i', the network is initialized with inputs equal to term frequency coefficients of the terms from T_i'. Each raw output produced by the network is interpreted as the vector of concept association strengths to the document T_i (see equation 2). Next, the network selects a set of C concepts with the highest association strengths, denoted by $Top_C(T_i)$. We assume that an expert provides feedback for document T_i (i.e., labels it with a set of concepts $Exp(T_i) = \{c_1, \ldots, c_{n_i}\}$ or marks whether this document was relevant to his/her search), thus we are able to define a supervised learning problem.

Since the model resembles a neural network, we propose to adopt the idea of weight propagation algorithm from multi-layer neural networks for the problem at hand. For this purpose, we need to define a global model error and a local error for each concept in the output layer.

For the document T_i we define a set of false positives $FP_i = Top_C(T_i) \setminus Exp(T_i)$ and a set of false negatives $FN_i = Exp(T_i) \setminus Top_C(T_i)$. Within the set $Top_C(T_i)$ we also identify concepts which were truly relevant for experts, i.e. the set of true positives $TP_i = Top_C(T_i) \cap Exp(T_i)$. Our global error measures of interest are measures appropriate for document retrieval – namely, *recall*[1] and F_1-*score*[2] (for the top C concepts) [12]. By analogy to the back-propagation algorithm, in our preliminary experiments we utilized a very simple update rule, namely back-propagating +1 (for false negatives) and −1 (for false positives) proportionally to the inputs.

Figure 2 illustrates an exemplary output of the tagging algorithm. We sort concepts indicated by experts according to $ESA(T_i, c_k)$ (i.e., their strength of

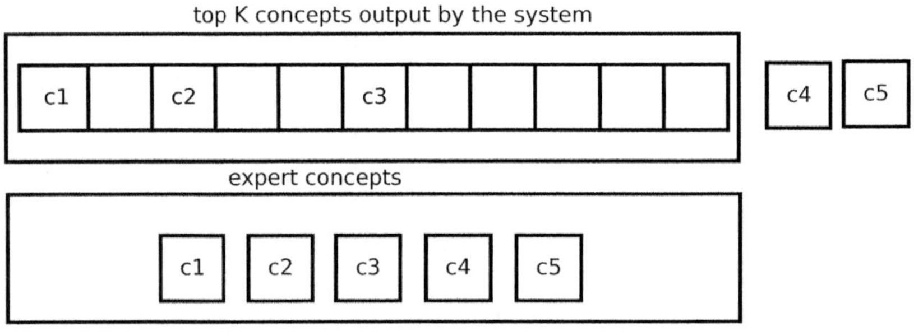

Fig. 2. Illustration of the weight updating algorithm for a single document T_i

[1] http://en.wikipedia.org/wiki/Precision_and_recall
[2] http://en.wikipedia.org/wiki/F1_score

association to the document computed using the ESA method). The false negatives are those expert concepts that fall outside the top C concepts assigned by ESA (i.e. c_4 and c_5). The set of false positives consists of non-expert concepts returned within the top C by the indexer (they correspond to the blank boxes on the illustration). The true positives are the expert concepts that are within the set of top C concepts (i.e. c_1, c_2 and c_3). While only the false positives and the false negatives contribute to model error, our update procedure also reinforces positive feedback for the true positives.

Sparse matrices ΔINV_i for each document are averaged and applied to matrix INV at the end of the loop over all documents, i.e., for each pair of a word and a concept (j, k):

$$INV^{new}[j, k] = INV^{old}[j, k] \times \left(1 + \frac{1}{M[j, k]} \times \sum_{i=1}^{M} \Delta INV_i[j, k]\right)$$

Algorithm 1 shows the update procedure for a single document.

Algorithm 1. An algorithm for computing a sparse update matrix ΔINV_i for a single document.

---

1 **begin**
2   Initiate an empty matrix ΔINV_i with N rows and K columns
3   **for** $c_k \in (Top_C(T_i) \cup Exp(T_i))$ **do**
4         $tmpWords = T_i' \cap MeSH_k$
5         $tmpNorm = \sum\limits_{w_j \in tmpWords} tf_{doc}(j, i)$
6   **if** $c_k \in FP_i$ **then**
7   **for** $w_j \in tmpWords$ **do**
8                 $\Delta INV_i[j, k] = -\frac{tf_{doc}(j,i)}{tmpNorm}$
9                 $M[j, k] = M[j, k] + 1$
10   **end**
11   **end**
12   **else**
13   **if** $c_k \in FN_i \cup TP_i$ **then**
14   **for** $w_j \in tmpWords$ **do**
15                     $\Delta INV_i[j, k] = +\frac{tf_{doc}(j,i)}{tmpNorm}$
16                     $M[j, k] = M[j, k] + 1$
17   **end**
18   **end**
19   **end**
20   **end**
21   **return** ΔINV_i
22 **end**

---

4 Experiments

We have conducted several experiments to verify feasibility of our approach. In the experiments we utilized a text corpus consisting of roughly 38.000 documents from the PubMed Central Open Subset repository [5].

As the external knowledge base we used the MeSH ontology [11], which is also employed by PubMed to index articles and to facilitate search through its resources. We adapted the ESA method to enable tagging documents from our corpus with the MeSH concepts (also known in MeSH terminology as terms or headings). In MeSH, each heading is accompanied by a natural language description prepared by domain experts, which we first process using text mining tools in order to determine the structure of our model (i.e. relations between words and concepts) and the initial values of weights.

Additionally, each document in the corpus was labelled by experts from PubMed with a set of MeSH concepts. We treat those tags as kind of user feedback and we utilize it for improving the word-concept associations in the learning model described in Section 3. We are also using the tags for the evaluation purpose. We split the corpus of documents into a training and a test set. In the experiments, we train our model on the first one and then verify its performance on the second. As the quality measures for each test document we used the F_1-*score* and *recall* for the top $C = 30$ concepts. In the following experiments, we will report the averaged quality measures for all documents from the test set.

We implemented and tested a prototype of our model in R System [13]. The average results obtained when we were training the model on 20.000 randomly selected documents (see Fig. 3) turned out to be very promising. On the test set, we observed a significant, by ≈ 107%, improvement of performance over the

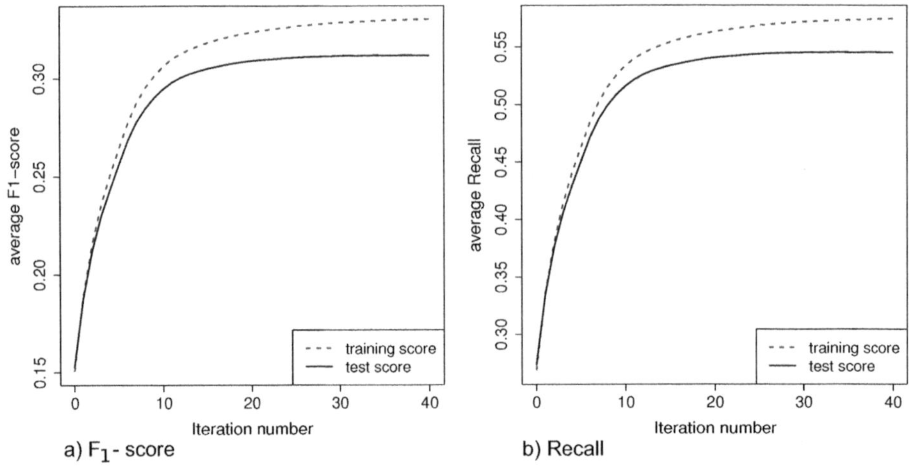

Fig. 3. F_1-score and recall curves for the training and test sets

classical ESA (the iteration 0 in the both plots). The most notable is a very high recall (≈ 0.54). This is a remarkable score, since the total number of possible concepts in our experiment is ≈ 26.000 and only 30 of them were assigned to any single document.

We also investigated stability of the weight updating procedure. The Figure 4a shows recall of the model for different sizes of the training set. It illustrates the diminishing returns in terms of recall, resulting from using an increasing number of observations. It suggests that using 20.000 documents for training may be sufficient (i.e. the solution cannot be improved much within this model merely by including additional observations). Moreover, it can be noticed that the difference between the performance of the training and test sets decreases with an increasing number of documents used for learning.

Figure 4b shows the average number of iterations required to train the model for different sizes of the training corpus. It is interesting to observe, that if we assume a reasonable stopping criterion, saying that we terminate learning when the improvement on the training set does not exceed 0.001, then the learning accuracy varies more or less at the same level regardless of the number of training documents. Additionally, for such a simple stopping criterion, the results on the test set consistently remains close to the maximum.

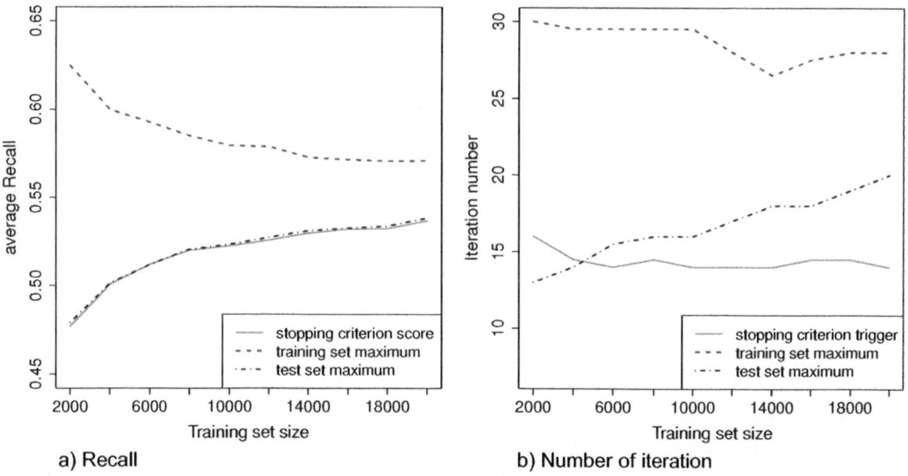

Fig. 4. The recall of the model on the test set (Fig. 4a) and the number of iterations required to train the model for different sizes of the training set (Fig. 4b)

5 Conclusions and Future Work

In this article we proposed a model which combines Explicit Semantic Analysis and expert feedback. The presented preliminary results show that the proposed approach significantly improves the baseline result provided by ESA.

We have shown that the network structure imposed by ESA is sparse enough to lead to convenient calculations, yet rich enough to further improve over the baseline. Questions we wish to address in the future are how this particular network structure compares to a randomly chosen structure of similar sparsity, and how it can be further augmented. In our future research we also aim to explore other weight update rules and other improvements to the proposed approach. In particular, we wish to investigate the problem of online model updating that utilizes user feedback in a form of click-through data.

Finally, in our research we are interested in methods that can utilize the improved representation of documents in tasks such as unsupervised clustering or topical classification of scientific articles. For instance, we have organized a data mining competition[3] whose aim was to verify whether the automatically generated associations between investigated texts and concepts are useful for predicting their topical classification [14]. Although the results of this challenge are already encouraging, we want to check if the improved semantic tagging will have a positive impact on the accuracy of predictions.

We are planning to design more efficient and scalable adaptive versions of ESA that can be implemented in Semantic Search Engines like SONCA system. The experimental results presented in Figure 4 are showing that the proposed solution does not require a big number of annotated training samples and long computation time to obtain a satisfactory accuracy.

Within the SONCA engine we are also developing interfaces which would facilitate search results visualization by grouping semantically similar documents. In this task the enhanced document representation provided by ESA can be used not only for construction of a appropriate similarity measure (as, for example, in [15]), but also for labelling the resulting clusters. Such semantic labels are meaningful for people and can make our system more comprehensive for future users.

References

1. Fazzinga, B., Gianforme, G., Gottlob, G., Lukasiewicz, T.: Semantic web search based on ontological conjunctive queries. Web Semantics: Science, Services and Agents on the World Wide Web (2011)
2. Nguyen, L.A., Nguyen, H.S.: On Designing the SONCA System. In: Bembenik, R., Skonieczny, Ł., Rybiński, H., Niezgodka, M. (eds.) Intelligent Tools for Building a Scient. Info. Plat. SCI, vol. 390, pp. 9–35. Springer, Heidelberg (2012)
3. Ślęzak, D., Janusz, A., Świeboda, W., Nguyen, H.S., Bazan, J.G., Skowron, A.: Semantic Analytics of PubMed Content. In: Holzinger, A., Simonic, K.-M. (eds.) USAB 2011. LNCS, vol. 7058, pp. 63–74. Springer, Heidelberg (2011)
4. Szczuka, M., Janusz, A., Herba, K.: Clustering of Rough Set Related Documents with Use of Knowledge from DBpedia. In: Yao, J., Ramanna, S., Wang, G., Suraj, Z. (eds.) RSKT 2011. LNCS (LNAI), vol. 6954, pp. 394–403. Springer, Heidelberg (2011)

[3] JRS'2012 Data Mining Competition: Topical Classification of Biomedical Research Papers `http://tunedit.org/challenge/JRS12Contest`.

5. Roberts, R.J.: PubMed Central: The GenBank of the published literature. Proceedings of the National Academy of Sciences of the United States of America 98(2), 381–382 (2001)
6. Gabrilovich, E., Markovitch, S.: Computing semantic relatedness using wikipedia-based explicit semantic analysis. In: Proc. of the 20th Int. Joint Conf. on Artificial Intelligence, Hyderabad, India, pp. 1606–1611 (2007)
7. Manning, C., Raghavan, P., Schütze, H.: Introduction to information retrieval, 2008. Online edition (2007)
8. Hliaoutakis, A., Varelas, G., Voutsakis, E., Petrakis, E.G.M., Milios, E.: Information retrieval by semantic similarity. Int. Journal on Semantic Web and Information Systems (IJSWIS). Special Issue of Multimedia Semantics 3(3), 55–73 (2006)
9. Rinaldi, A.M.: An ontology-driven approach for semantic information retrieval on the web. ACM Trans. Internet Technol. 9, 1–24 (2009)
10. Mitchell, T.M.: Machine Learning. McGraw Hill series in computer science. McGraw-Hill (1997)
11. United States National Library of Medicine: Introduction to MeSH - 2011 (2011), `http://www.nlm.nih.gov/mesh/introduction.html`
12. Feldman, R., Sanger, J. (eds.): The Text Mining Handbook. Cambridge University Press (2007)
13. R Development Core Team: R: A Language and Environment for Statistical Computing. R Foundation for Statistical Computing, Vienna, Austria (2008)
14. Janusz, A., Nguyen, H.S., Ślęzak, D., Stawicki, S., Krasuski, A.: JRS'2012 Data Mining Competition: Topical Classification of Biomedical Research Papers. In: Yan, J.T., et al. (eds.) RSCTC 2012. LNCS (LNAI), vol. 7413, pp. 422–431. Springer, Heidelberg (2012)
15. Janusz, A., Ślęzak, D., Nguyen, H.S.: Unsupervised similarity learning from textual data. Fundamenta Informaticae (2012)

Microblog Topic Detection Based on LDA Model and Single-Pass Clustering*

Bo Huang, Yan Yang, Amjad Mahmood, and Hongjun Wang

School of Information Science & Technology, Southwest Jiaotong University, Chengdu, 610031, P.R China
Key Lab of Cloud Computing and Intelligent Technology, Chengdu, Sichuan Province, 610031, P.R China
huangbo1582@163.com, {yyang,wanghongjun}@swjtu.edu.cn, amjad.pu@gmail.com

Abstract. Microblogging is a recent social phenomenon of Web2.0 technology, having applications in many domains. It is another form of social media, recognized as Real-Time Web Publishing, which has won an impressive audience acceptance and surprisingly changed online expression and interaction for millions of users.It is observed that clustering by topic can be very helpful for the quick retrieval of desired information. We propose a novel topic detection technique that permits to retrieve in real-time the most emergent topics expressed by the community. Traditional text mining techniques have no special considerations for short and sparse microblog data. Keeping in view these special characteristics of data, we adopt Single-pass Clustering technique by using Latent Dirichlet Allocation (LDA) Model in place of traditional VSM model, to extract the hidden microblog topics information. Experiments on actual dataset results showed that the proposed method decreased the probabilities of miss and false alarm, as well as reduced the normalized detection cost.

Keywords: Microblog,topic detection, LDA model,Single-pass clustering.

1 Introduction

Microblogging has become a primary channel by which people not only share information, but also search for information. It fills a gap between blogging and instant messaging, allowing people to publish short messages on the web about what they are currently doing. First Microblog was launched by Evan William in 2006. According to Twitter, there were 175 million registered users by the end of 2010. This rapid adoption has generated interest in gathering information from microblogging about real time news and opinions on specific topics. This interest, in turn, has led to a proliferation of microblog search services from

* This work is partially supported by the National Science Foundation of China (Nos. 61170111 , 61003142 and 61152001) and the Fundamental Research Funds for the Central Universities (No. SWJTU11ZT08).

J.T. Yao et al. (Eds.): RSCTC 2012, LNAI 7413, pp. 166–171, 2012.

both microblogging service providers (like Twitter) and general purpose search engines (like Bing and Google).However compared with traditional document retrieval and web search, microblog search is still in its infancy.

In a typical microblog search scenario using twitter, around 1500 tweets that contains the query terms, will be returned, ranked by their creation time. Although, other presentation formats are also available (e.g ordering results by author popularity, or by hyperlinks referenced), presentation formats optimized for topic monitoring are not yet widely available. The goal of this paper is to explore the potential for topic organization of microblog search results.

This is a challenging problem because microblog posts are short and sparse, so traditional topical clustering technique based on lexical overlap is necessarily weak. We use single - pass clustering method with Latent Dirichlet Allocation (LDA) Model instead of traditional VSM model[1]. The experimental results has proved the effectiveness of LDA model over VSM.

The rest of this paper is organized as follows: Section 2 presents the current state of topic detection. Section 3 explains the Latent Dirichlet Allocation (LDA) model and the MCMC method with Gibbs sampling for LDA. Section 4 covers the methodology of Single-pass clustering algorithm. Sectiona 5 describes the experiments and analysis of results. Finally, section 6 discusses the conclusion and future work.

2 Related Work

Yang et al. [2] investigates the use and extension of text retrieval and clustering techniques for event detection using hierarchical and non-hierarchical document clustering algorithm. They found that resulting clustering hierarchies are highly informative for retrospective detection of previously unidentified events. Trieschnigg and Kraaij[3] proposed an incremental hierarchical clustering algorithm. They take a sample from the corpus to build a hierarchical cluster structure, then optimize the resulting binary tree for the minimal cost metric, finally assign the remaining documents from the corpus to clusters in the structure obtained from the sample. Papka and Allen [4] detect topic by using a Single-pass clustering algorithm and a novel thresholding model. This model incorporates the properties of events as major component, but the priori report sparse will lead to the topic model is not accurate. Finally, explored that the probabilities of miss alarm and false alarm may increase with the Single-Pass Clustering. Cataldi et al. [5] proposed the new hot topic detection methods based on the relationship between the timing and the social evaluation Twitter. In an appropriate period of time, if a topic has been widely detected, but before this rarely occurs, then you can think that this topic is the new hot topic at this particular moment. Phuvipadawat and Murate [6] put forward a collection of breaking news on Twitter, He designed a program called "Hotstream" to provide users breaking news.

In the topic detection process, building Model is a basic challenge. The vector space model(VSM) is the most common model. For the short and sparse microbolgging text, VSM (using words or terms as characters) cannot perform accurate calculation of the text Similarity. In order to reduce the date scarcity and make it more topic-focused, we propose the LDA model [6] to the data modeling, extracting the hidden microblog topics information. High-dimensional sparse text vector is mapped to low-dimensional hidden topic space, combined with the classic single-pass clustering algorithm for text clustering different topic.

3 The Method of Microblog Text Modeling

3.1 Latent Dirichlet Allocation

Latent Dirichlet Allocation (LDA)[7] is a generative probabilistic model for a corpus of discrete data. It models the words in documents under the "bag-of-words" assumption, ignoring the orders of the words. Following this "exchangeability", the distribution of the words would be independent and identically distributed under some given conditioned of parameters. This conditionally independence allows us to build a hierarchical Bayesian model for a corpus of documents and words. This process can be described graphically as shown in Fig.1.

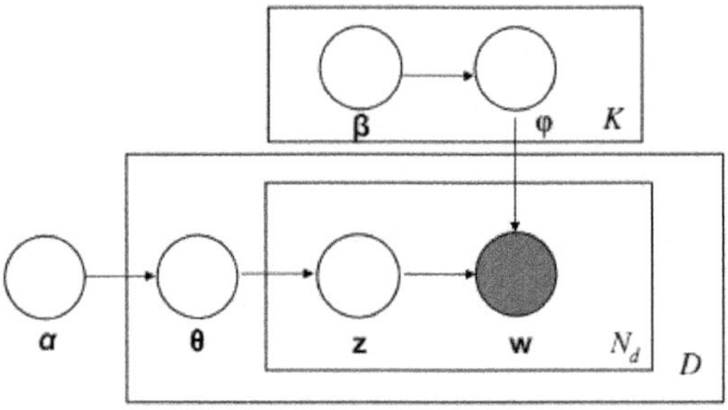

Fig. 1. Graphical model representation of LDA

For each document d in the corpus, the LDA model first picks a multinomial distribution $\theta_d = [\theta_{d1}...\theta_{dk}]^T$ from the Dirichlet distribution $\alpha_d = [\alpha_{d1}...\alpha_{dk}]^T$, and then the model assigns a topic $z_{id} = k$ to the ith word in the document according to the multinomial distribution θ_d. Given the topic $z_{id} = k$,the model then pick a word w_{id} from the vocabulary of V words according to the multinomoial distribution $[\phi_{k1}...\phi_{kV}]^T$ which is generated from the Dirichlet distribution $[\beta_{k1}...\beta_{kV}]^T$ for each topic k.

Markov Chain Monte Carlo (MCMC)[8] is a general method to obtain samples from complex distribution. We have to construct a Markov chain that is irreducible, a periodic, and reversible in order to make the chain have a unique stationary distribution. Such properties are guaranteed if we apply the Gibbs sampling for the state transitions [9].The algorithm is detailed as follows:

We first consider the joint distribution of z and w

$$p(w, z|\alpha, \beta) = \int_\theta \int_\phi p(w, z, \theta, \phi|\alpha, \beta) d\theta d\phi \tag{1}$$

Given the joint distribution of w and z under LDA, we can compute the conditional probability for the Gibbs sampler by

$$p(z_{id} = k|z^{\neg id}, x, \alpha, \beta) = \frac{p(z^{\neg id}, z_{id} = k|x, \alpha, \beta)}{\sum_{k'=1}^{K} p(z^{\neg id}, z_{id} = k'|x, \alpha, \beta)} \tag{2}$$

After the Markov chain reach the stationary distribution, we can start drawing samples from the chain. As shown in [8], given a sampled z, we can estimate the values of the other latent variables by

$$\theta_{dk} = \frac{\alpha_k + n_{dk}}{\alpha + n_d}, \phi_{kv} = \frac{\beta_{kv} + n_{kv}}{\beta_k + n_k} \tag{3}$$

where the counts are obtained from the assignment z. The above two equations are derived by computing the expectance of the Dirichlet distribution in the posterior form.

4 Topic Detection by Single-Pass Clustering

As a result of Gibbs sampling for LDA, θ is a $d * k$ matrix,where d is the total number of microblog texts, k is the number of latent topics. Matrix element value indicates the probability of each text data set to generate implicit topic, can also be seen as the document-topic vectors.

The proposed Single-pass clustering[3] algorithm is as following:

For each document d in the sequence loop;
  Find a cluster c that maximizes $\cos(c, d)$;
    If $\cos(c, d) > t$ then
        Include d in c;
    Else create a new cluster whose only document is d;
  End loop.

5 Experiment and Results

5.1 Evaluation Criteria

Detection performance is characterized in terms of the probability of Miss and False alarm errors (P_{Miss} and P_{FA}). These error probabilities are then combined into a single detection cost C_{Det}, by assigning costs to miss and false alarm errors[10]:

$$(C_{Det}) = C_{Miss} \cdot P_{Miss} \cdot P_{target} + C_{FA} \cdot P_{FA} \cdot P_{non-target} \tag{4}$$

According to the TDT standards,we set $C_{Miss} = 1.0, C_{FA} = 0.1, P_{target} = 0.02$.

Because these values vary with the application, C_{Det} will be normalized so that $(C_{Det})_{Norm}$ can be no less than one without extracting information from the source data. This is done as follows:

$$(C_{Det})_{Norm} = \frac{C_{Det}}{min(C_{Miss} \cdot P_{target}, C_{FA} \cdot P_{non-target})} \quad (5)$$

The $(C_{Det})_{Norm}$ is smaller,the quality of topic detection is better.

5.2 Dataset

We collected 108122 texts of Sina-microblog August 2011 by Web crawler. All data covering 957 topics discussed by the microbolg usrers. Before the experiment, data preprocessed by ICTCLAS Segmentation system.

5.3 Results

After fixing different similarity threshold t, the Single-pass clustering based on VSM model and the Single-Pass clustering based on LDA model are executed. The corresponding experimental results are shown in Table 1 and 2. Followings observations are found in the results:

1. With increasing similarity threshold t, the missing rate increases gradually, and the fault detection rate decreases gradually, consuming function is down then up.
2. The Single-pass Clustering based on LDA model can reduce the P_{Miss} ,P_{FA} and$(C_{Det})_{Norm}$ to improve the topic detection accuracy.

Table 1. The results of Single-pass based on VSM model

t	0.001	0.002	0.003	0.005	0.008	0.01	0.02	0.05	0.08
P_{Miss}	0.2145	0.2386	0.3258	0.3322	0.3664	0.4615	0.4962	0.5138	0.5567
P_{FA}	0.3360	0.3196	0.2862	0.2635	0.1179	0.1128	0.1012	0.1073	0.0943
$(C_{Det})_{Norm}$	0.0372	0.0361	0.0346	0.0325	0.0189	0.0203	0.0216	0.0249	0.0313

Table 2. The results of Single-pass based on LDA model

t	0.01	0.02	0.03	0.05	0.08	0.1	0.2	0.3	0.5
P_{Miss}	0.0110	0.0126	0.0531	0.0639	0.1128	0.1532	0.2704	0.3001	0.3552
P_{FA}	0.1538	0.1464	0.1052	0.0841	0.0533	0.0094	0.0002	0.0000	0.0000
$(C_{Det})_{Norm}$	0.0152	0.0146	0.0133	0.0095	0.0074	0.0040	0.0054	0.0060	0.0071

6 Conclusion

Considering the in-built characteristics of large-scale and high-sparse microblog data, we purposed the Single-Pass Clustering algorithm based on LDA model to solve the data sparseness problem faced by the traditional VSM. The experimental results show that the algorithm could decrease the probabilities of Miss Alarm and False Alarm, and finally reducing the normalized detection cost. Future research will optimize the LDA model, and consider the real-time processing of larger data.

References

1. Salton, G., Wong, A., Yang, C.S.: A vector space model for automatic indexing. Communications of the ACM 18, 613–620 (1995)
2. Yang, Y., Pierce, T., Carbonell, J.: A study on Retro-spective and On-Line Event detection. In: Proceedings of the 21st Annual International ACM SIGIR Conference on Research and Development in Information Retrieval, USA, pp. 28–36 (1998)
3. Trieschnigg, D., Kraaij, W.: TNO hierarchical topic detection report at TDT 2004. In: The 7th Topic Detection and Tracking Conf. (2004)
4. Papka, R., Allan, J.: On Line New Event Detection using Single Pass Clustering. UMass Computer Science (1998)
5. Cataldi, L., Caro, D., Schifanella, C.: Emerging Topic Detection on Twitter based on Temporal and Social Terms Evaluation. In: MDMKDD 2010 Proceedings of the Tenth International Workshop on Multimedia Data Mining, Washington, pp. 1–10 (2010)
6. Phuvipadawat, S., Murata, T.: Breaking News Detection and Tracking in Twitter. In: 2010 IEEE/WIC/ACM International Conference on Web Intelligence and Intelligent Agent Technology (WI-IAT), Toronto, pp. 120–123 (2010)
7. Blei, D., Ng, A., Jordan, M., et al.: Latent dirichlet allocation. Journal of Machine Learning Research 3, 993–1022 (2003)
8. Stuart, G., Donald, G.: Stochastic relaxation gibbs distributions and the bayesian restoration of images. IEEE Transactions on Pattern Analysis and Machine Intelligence 6, 7212–7411 (1984)
9. Griffiths, T.L., Steyvers, M.: Finding scientific topics. Proceedings of the National Academy of Science 101, 5228–5235 (2004)
10. The Linguistic Data Consortium.: The 2004 Topic Detection and Tracking. Task Definition and Evaluation Plan (2004), `http://www.itl.nist.gov/iad/mig/tests/tdt/2004/TDT04.Eval.Plan.v1.2.compare.1.1c`

Named Entity Matching in Publication Databases*
A Case Study of PubMed in SONCA

Marcin Szczuka, Paweł Betliński, and Kamil Herba

Institute of Mathematics, The University of Warsaw
Banacha 2, 02-097 Warsaw, Poland
{szczuka,pbetl}@mimuw.edu.pl, k.herba@students.mimuw.edu.pl

Abstract. We present a case study in approximate data matching for a database system that contains information about scientific publications. The approximate matching process is meant to identify whether several records in the database are in fact repeated instances of the same real-world object. In our case study we are concerned with matching instances of objects such as XML documents, persons' names, affiliations, journal names, and so on. The particular data we are dealing with is a representation of the PubMed Central document corpus within the data warehouse that is a part of the SONCA system. SONCA system is being developed as one of components of the general scientific information platform SYNAT.

Keywords: Text mining, approximate matching, document grouping, data cleaning, data matching, similarity function, record linkage, record matching, duplicate detection, object matching, entity resolution, data warehousing, granulation.

1 Introduction

Approximate data matching is a central problem in several data management processes, such as data integration, data cleaning, approximate queries, faceted search, similarity search and so on. As it is explained in the very useful survey [4], matching is the process of bringing together data from different, and sometimes heterogeneous, data sources and comparing them in order to find out whether they represent the same real-world object. Since data come from different sources, it is to be expected, that they differ from each other in the way they are represented. This is a complex problem, since it is not trivial to assert that two heterogeneous data instances represent the same object of the reality.

* This work was supported by the grant N N516 077837 from the Ministry of Science and Higher Education of the Republic of Poland, the Polish National Science Centre grant 2011/01/B/ST6/03867 and by the Polish National Centre for Research and Development (NCBiR) under Grant No. SP/I/1/77065/10 in frame of the strategic scientific research and experimental development program: "Interdisciplinary System for Interactive Scientific and Scientific-Technical Information".

J.T. Yao et al. (Eds.): RSCTC 2012, LNAI 7413, pp. 172–179, 2012.

Heterogeneity can happen in data structure as well as in data value. Therefore, a data matching process must be able to analyze both structure and data value.

In our particular case study we are dealing with the problem of matching so called *named entities* within a database (a data warehouse) derived from the collection of scientific articles. These articles are a part of PubMed Central (PMC) - the National Library of Medicine's digital archive of full-text journal literature (see [2]). The named entities in discourse are publications (documents), persons (authors, editors, etc.), affiliations (institution names and locations), publishers, journals, book series, to name just a few. The original documents from PubMed Central are first converted to a dedicated XML format (nXML - see [10]) and stored in the local repository. Then, each document is parsed and loaded to a dedicated data warehouse - the SONCA Analytic Index Server. The parser decomposes each document into meaningful parts and creates (loads) data instances in the database. The important feature of the parser is that it processes the references contained in each document and creates database instance for each title, name, affiliation, journal name, etc. This process contributes to creation of multiple instances for many named entities. One can say, that repeated instances for a given object come from different data sources. In order to deal with this situation it is necessary to perform matching procedure.

The problem of matching named entities of various kind in PubMed database has been discussed before (see [7,15]). Our approach is different from previous ones in the way we make use of PMC data. For us, the documents from PMC are just one example of possible document sources (repositories). Hence, we do not have PubMed database handy to verify the matching. All information we are given is contained in the relational database structure that results from processing (parsing) texts from PMC. In other words, even if a named entity (document, person, institution) exists in the original PubMed database, we may never be aware of it, if it does not appear in any of the documents that our system has processed. This is completely understandable, given the fact that our overall goal is not to re-create PubMed from a sample of documents, but to devise a method that deals with various types of document corpora that we may encounter in the future.

The fact that we derive both instances and (matched) objects directly from scientific texts creates some problems that are not common to other publication databases. In particular, the problem of matching persons' names in databases such as DBLP [13] or Arnetminer [16] has been discussed before. However, the problem we are tackling in our matching attempts is somewhat different. Since our instances come from processing of documents, the vast majority of names that we encounter is derived from citations. Therefore, most of authors only have initials of given name(s). That creates mach higher inconsistency in data. Therefore, we cannot rely on the name only and have to use secondary (constructed) features to perform matching.

The article is organized as follows. Section 2 describes the framework we work in, the SONCA system. Then we describe how our data set is represented in SONCA data warehouse (Section 2.1) and what the matching process entails in this case. Section 3 contains description of the actual state of the data warehouse we are dealing with and explanation of results of initial matching attempts. We finish with conclusions and directions for further work in Section 4.

2 SYNAT Platform and SONCA System

The SYNAT project (abbreviation of Polish "**SY**stem **NA**uki i **T**echniki", see [3]) is a large, national R&D program of Polish government aimed at establishment of a unified network platform for storing and serving digital information in widely understood areas of science and technology. The project is composed of nearly 50 modules developed by research teams at 16 leading research institutions in Poland.[1]

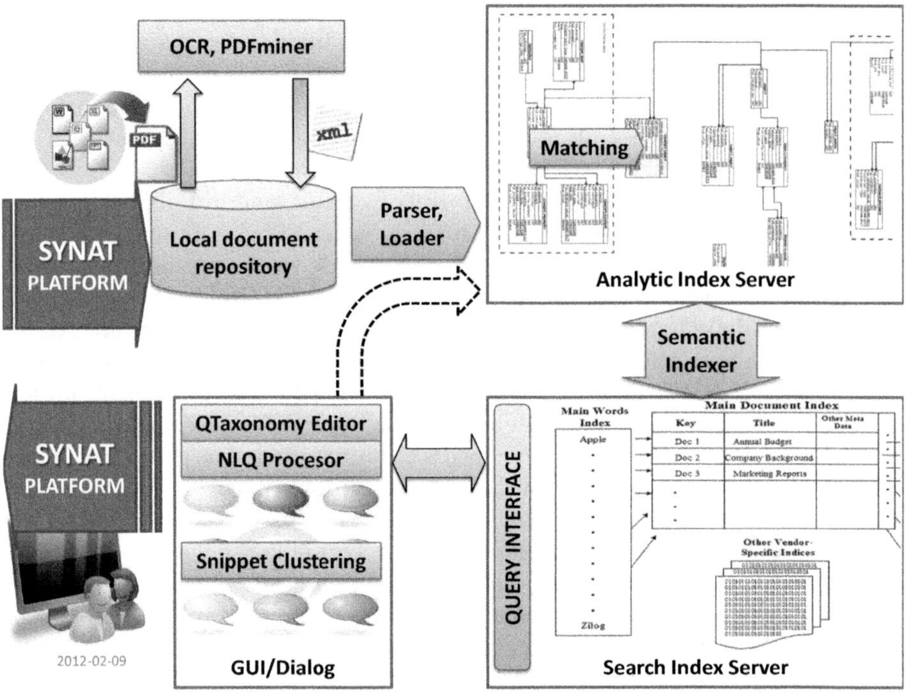

Fig. 1. The overview of SONCA

Within the framework of a larger (SYNAT) project we want to design and implement a solution that will make it possible for a user to search within repositories of scientific information (articles, patents, biographical notes, etc.) using their semantic content. Our prospective system for doing that is called SONCA (abbreviation for **S**earch based on **ON**tologies and **C**ompound **A**nalytics, see [8,11,12]).

[1] `http://www.synat.pl`

Ultimately, SONCA should be capable of answering the user query by listing and presenting the resources (documents, Web pages, et cetera) that correspond to it *semantically*. In other words, the system should have some *understanding* of the intention of the query and of the contents of documents stored in the repository as well as the ability to retrieve relevant information with high efficacy. The system should be able to use various knowledge bases related to the investigated areas of science. It should also allow for independent sources of information about the analyzed objects, such as, e.g., information about scientists who may be identified as the stored articles' authors.

The requirement for matching in SONCA warehouse model emerges quite naturally when we proceed with construction of the relational data schema aimed at efficient storage and querying of parsed scientific articles, as well as entities corresponding to authors, institutions, references, scientific concepts, and so on (see upper-right part of Figure 1). Another important requirement for the proposed model is the ability to answer a query about all possible entities that may be interesting for users in a well organized and efficient manner. To achieve that the process of instance matching, tantamount to generation of objects, has to be performed.

2.1 SONCA Analytic Data Warehouse

For the purpose of this study we will describe the structure of SONCA Analytic Index Server (data warehouse) only briefly. For details one may refer to [8]. The internal architecture of the database is implemented with use of EAV/CR (Entity-Attribute-Value with Classes and Relationships - see [9]) model. SONCA's warehouse structure is organized into three layers, of which we are concerned with upper two: instance and object. In order to keep the size of database at bay and execute the queries efficiently we have employed the Infobrigt's RDBMS engine [6].

The parser, which processes the documents from local repository, creates an instance (record) in the generic (instance related) part of the database for every type of entity that was identified within document. Entities (instances) are stored in the data table(s) with preservation of information about structure of they relations in the original document. The instances may be either very simple, like a single word or a number, or quite complex, like an instance representing a publication with all the underlying sub-instances like: title, publication year, publisher etc.

The result of matching is stored in special table on the database server. This table stores the binding information for instances (records in original data) and objects. In the particular case of matching instances to create objects in the SONCA data warehouse we sometimes have to perform two steps:

1. Instance matching. We have to decide if the two or more instances are in fact representing the very same object. Then we decide whether this object is already in the database or not. In the former case a new object is being created, in the latter we add information about new matching instances and modify object properties, if necessary (see point 2 below).

2. Object creation/modification. If the just matched instance(s) require creation of new object or if an instance matched with existing object carries new, previously unavailable piece of information, then we need to create/modify the objects. In the EAV/CR model that we use this is equivalent to creating and/or modifying the values of object's properties (attributes). This is sometimes a complicated task in itself.

The two steps mentioned above can be represented in the framework of Granular Computing. They correspond to creation of granules and derivation of granule representations. For more on that see [14].

3 Matching of PubMed Papers in SONCA Database

In the actual R&D work with matching in SONCA we have already progressed beyond what is described below. The case study explains the finished part of our investigations and presents those results that have already been positively verified. In order to preserve clarity and to save some space we present some key numbers about our experiments in the Table 1 below.

Table 1. Key information about matching experiment in SONCA

Feature	**Value**
No. of PMC documents processed	75,844
No. of instances of all types	433,597,303
No. of instances of type 'publication'	2,623,849
No. and percentage of 'publication' instances from citations	2,547,870 (97%)
No. of created objects of type 'publication'	1,195,878
No. and percentage of matched instances of type 'publication'	1,874,816 (71%)
No. and percentage of 'publication' objects matching one instance	884,858 (74%)
Highest number of instances matching a single 'publication' object	243
Average consistency for properties of 'publication' object (excl. `title`*)*	97.78%
Initial percentage of objects with inconsistency for property `title`	0.02%
Number of unique `<name,surname> pairs`	1,814,071
Number of unique instances of type 'affiliation'	387,602

The first matching process that we have launched was aimed at identifying and grouping together instances that represent the same publication. Following the categorization defined in [4], this step entails both content-based and structure-based matching. In case of matching documents for which at least one instance corresponds to the document that was originally stored in the local repository the task becomes a relatively easy one. Not only are the property values for this instance highly consistent across all instances, but we also have detailed information how they were created, as we possess the full technical specification of the algorithms used by parser/loader during their creation. The only noticeable problems we had with instances representing "real" documents occurred when the document they originated from was damaged or violated formatting rules. This, however was just a handful of cases (below 0.5%).

For the vast majority (97%) of instances of type 'publication' we had to process (match) them using the little information that was contained in references found in full-text documents. For these instances we needed to first check if something that matches instance have been already assigned an object and then match property values, if they existed, or create new ones. With 'publication' instances derived (parsed) from references we usually started from checking if any of the highly discriminative (unique) properties, such as: PubMed ID [10], DOI[2] or ISBN/ISSN[3]. Then we used such property value to check if it has already been associated with any object existing in our database. If no unique number was present we moved to secondary, less reliable properties. Fortunately, out of 16 properties that we have considered only three – `title`, `lastpage` and `publish_year` – showed significant inconsistency. As it turned out, most of inconsistencies in `publish_year` (year of publication) and `lastpage` fields were easy to identify and correct. Thanks to that parser/loader algorithm has subsequently been improved and the current version produces no inconsistencies in `publish_year`. To compare tiles we have used methods and experience from [5]. It was no longer possible to establish similarity of texts (titles) using only SQL queries. We had to extract data from the database and process using NLP methods. As it is quite common for references, the extracted titles were modifications, sometimes quite extensive, of the original full titles. We had to account for common practices such as shortening, usage of abbreviations, omission of parts of sentences and so on. The methods we have used made it possible to assign a unique and correct title to nearly all objects of type 'publication'.

Albeit the "success rate" in case of matching publications is just 71% we consider it a success, given how little useful and consistent information can be retrieved from some references. Unfortunately, when it comes to matching other types of named entities, especially persons and institutions (affiliations), we find ourselves in dire straits. In case of matching persons' names we are faced with difficult task resulting for the way these names enter our database. As vast majority of persons are identified as authors or editors mentioned in references, we usually do not have their full given names. Just a quick look at any list of references, even the one at the end of this paper, reveals the severity of the problem. In fact, as we have found out, the content based matching in this case is inefficient. There is simply not enough information in two or three words that constitute person's name, especially when only initials of given names are available. Therefore, we want construct additional attributes that make it possible to either discern or group names. The first step is to use the co-author relationship graph, a bit like it was done in DBLP [13] and Arnetminer [16]. By comparing co-author graphs for two instances we can decide if they represent the same named entity (object) or not. Also, we can make use of the valuable information contained in the citation graph. As we have large part of citation

[2] `http://www.doi.org/`

[3] `http://www.isbn-international.org/`

graph for matched publications stored in our data warehouse, we can use this graph to construct a filter that will eliminate the instances that are irrelevant for a current object and leave for further, more detailed examination only the instances that have some relationship to the object in discourse.

4 Conclusions and Further Work

Matching in the analytical part of SONCA database architecture is the work-in-progress. We constantly improve all aspects of the entire system, matching included. This case study spins around experiments made with relatively early version of SONCA components (local repository, data warehouse and parser/loader). The current experimental environment has been much improved, partly thanks to observations and conclusions drawn from the initial matching attempts.

As mentioned in Section 3 matching of named entities other than publications (documents) in the SONCA data warehouse is at early stage. The level of complication of matching task in case of objects such as persons, institutions, conferences, and so on, is not yet fully recognized. We will face this task in the immediate future. We expect that a significant involvement of text mining tools that make use of intelligent methods such as inductive learning, approximate reasoning, granular computing, as well as rough-set-based approaches, will make it possible to achieve observable improvement. We hope that use of intelligent methods for finding (constructing) secondary feature using citation and co-author graphs will bring the breakthrough on this front. At the same time we strive to keep the highest possible portion of computations within the SQL-based database. Since the data we are dealing with is massive, most of typical tools that work on "flat" data table is only capable to work on tiny samples of the entire data collection. Taking the data out of RDBMS is frequently too costly. on the other hand, the SQL and its extensions provide a fairly limited arsenal of text processing/mining methods.

As stated in the introduction, our approach differs from typical ones in the way we use our information (document) source. Inasmuch as we expect the data in the final SYNAT platform to be heterogenous and quite varied, we have to prepare for (almost) everything. In order to field test all methods (including matching) in SONCA we have already started initial works with another large source of publication data - the ACM's Digital Library [1]. We also expect to be able to use catalogs and selected resources of the Polish National Library[4], which will add another facet to the project, as we will have to develop methods to cope with texts in Polish.

References

1. Association for Computing Machinery: The Digital Library: the ACM Guide to Computing Literature. WWW Page (2012), `http://librarians.acm.org/acm-guide-computing-literature`

[4] `http://alpha.bn.org.pl/screens/libinfo.html`

2. Beck, J., Sequeira, E.: PubMed Central (PMC): An archive for literature from life sciences journals. In: McEntyre, J., Ostell, J. (eds.) The NCBI Handbook, ch. 9. National Center for Biotechnology Information, Bethesda (2003), http://www.ncbi.nlm.nih.gov/books/NBK21087/
3. Bembenik, R., Skonieczny, Ł., Rybiński, H., Niezgódka, M. (eds.): Intelligent Tools for Building a Scientific Information Platform. SCI, vol. 390. Springer, Heidelberg (2012)
4. Dorneles, C.F., Gonçalves, R., dos Santos Mello, R.: Approximate data instance matching: a survey. Knowl. Inf. Syst. 27(1), 1–21 (2011)
5. Herba, K.: Semantic recognition and tagging of scientific articles. Master's thesis, Faculty of Mathematics, Informatics, and Mechanics, The University of Warsaw, Warsaw, Poland (2011) (in Polish)
6. Infobright, Inc.: Infobright Enterprise Edition (IEE). WWW Page (2012), http://infobright.com
7. Jonnalagadda, S., Topham, P.: Nemo: Extraction and normalization of organization names from pubmed affiliation strings. Journal of Biomedical Discovery and Collaboration 5, 50–75 (2010)
8. Kowalski, M., Ślęzak, D., Stencel, K., Pardel, P., Grzegorowski, M., Kijowski, M.: Rdbms model for scientific articles analytics. In: Bembenik, et al. [3], ch. 4, pp. 49–60
9. Nadkarni, P.: The EAV/CR model of data representation. Tech. rep., Center for Medical Informatics, Yale University School of Medicine (2000), http://ycmi.med.yale.edu/nadkarni/eav_cr_frame.html
10. National Center for Biotechnology Information: Archiving and Interchange Tag Set (2008), http://dtd.nlm.nih.gov/archiving/
11. Nguyen, A.L., Nguyen, H.S.: On designing the sonca system. In: Bembenik, et al. [3], ch. 2, pp. 9–35
12. Nguyen, H.S., Ślęzak, D., Skowron, A., Bazan, J.: Semantic search and analytics over large repository of scientific articles. In: Bembenik, et al. [3], ch. 1, pp. 1–8
13. Reuther, P., Walter, B., Ley, M., Weber, A., Klink, S.: Managing the Quality of Person Names in DBLP. In: Gonzalo, J., Thanos, C., Verdejo, M.F., Carrasco, R.C. (eds.) ECDL 2006. LNCS, vol. 4172, pp. 508–511. Springer, Heidelberg (2006)
14. Szczuka, M., Ślęzak, D.: Representation and Evaluation of Granular Systems. In: Watada, J., Watanabe, T., Phillips-Wren, G., Howlett, R.J., Jain, L.C. (eds.) Intelligent Decision Technologies. SIST, vol. 15, pp. 287–296. Springer, Heidelberg (2012)
15. Tsai, R.T.H., Sung, C.L., Dai, H.J., Hung, H.C., Sung, T.Y., Hsu, W.L.: NERBio: using selected word conjunctions, term normalization, and global patterns to improve biomedical named entity recognition. BMC Bioinformatics 7(S-5) (2006)
16. Zhang, D., Tang, J., Li, J.Z., Wang, K.: A constraint-based probabilistic framework for name disambiguation. In: Silva, M.J., Laender, A.H.F., Baeza-Yates, R.A., McGuinness, D.L., Olstad, B., Olsen, Ø.H., Falcão, A.O. (eds.) CIKM, pp. 1019–1022. ACM (2007)

On the Homogeneous Ensembling with Balanced Random Sets and Boosting

Vladimir Nikulin

Department of Mathematical Methods in Economy,
Vyatka State University, Kirov, Russia
vnikulin.uq@gmail.com

Abstract. Ensembles are often capable of greater prediction accuracy than any of their individual members. As a consequence of the diversity between individual base-learners, an ensemble will not suffer from overfitting. On the other hand, in many cases we are dealing with imbalanced data and a classifier which was built using all data has tendency to ignore minority class. As a solution to the problem, we propose to consider a large number of relatively small and balanced subsets where representatives from the both patterns are to be selected randomly. Using different pre-processing technique combined with available background knowledge, which may have subjective treatment, we can generate many secondary databases for training. The relevance of those databases maybe tested with five folds cross-validation (CV5). Further, we can use CV5-results to optimise blending structure. Note that it is appropriate to use different software for CV5 evaluation and for the computation of the final solution. Our model was tested online during an International Carvana data mining Contest on the Kaggle platform. This Contest was highly popular and attracted 582 actively participating teams, where our team was awarded 2nd prize.

Keywords: ensembling, blending, decision trees, boosting, neural nets, cross validation, classification.

1 Introduction

In line with an ensembling theory, we are interested to generate a variety of high quality solutions for the problem, and there are two main directions how to do that. One very popular way is associated with 1) the usage of the different software in application to the same database, or 2) we can apply the same software (named classifier C1) to different databases. Based on our experience, the second direction is a more preferable in the case of Carvana data.

The next fundamental question is how to link (or how to blend) many different solutions in a most optimal way [1]. The answer is a quite straightforward: it appears to be natural to use cross-validation (CV) with fixed design matrix as a criterion for blending, where the number of 5 folds seems to be quite sufficient. However, implementation of the CV5 may represent significant computational

J.T. Yao et al. (Eds.): RSCTC 2012, LNAI 7413, pp. 180–189, 2012.

problem, particularly, in the case if we are dealing with imbalanced data, and have to construct the final solution as a homogeneous ensemble of many base learners, each of which is a function of the randomly selected balanced subset. Note that the main target of the CV is to compare different databases, where the quality of any particular solution may not necessarily be high.

We can implement here very important principle of invariance, which maybe described very briefly as follows. Suppose, we have another classifier, named C2, which is much faster compared to the C1. Note that the quality of the classification (or quality of the patterns separation) by the C2 maybe much poorer compared to the C1, but it is not essential here. Validity of the hypothesis of invariance is a subject of the fundamental importance. According to this hypothesis, the scaling in the quality of performance between C1 and C2 is about the same around all the secondary databases. Based on our experience with Carvana data, the hypothesis of invariance is true. Therefore, we can use C2 to conduct all the necessary experiments with CV5 in application to 10-20 secondary databases. After collection of the CV5 experimental results with fixed design matrix, we can optimise weighting coefficients for blending. Then, we can recompute solutions for the selected secondary databases with C1, and apply blending coefficients in order to calculate the final solution. In this particular project we used GBM in R as C1, and Neural Nets (NNs) in CLOP, Matlab, as C2.

2 Data Pre-processing

Carvana database[1] includes two parts 1) training with 72983 samples, where 8976 are positive (that means problematic), and all the other samples are negative (that means normal); 2) testing with 48707 samples (unlabelled).

The list of 36 original features is given in Table 1, where 3 features (index=0) were excluded from further consideration. Remaining 33 features were divided into 4 parts:

1) numerical (15 features in total including target variable);
2) textual (14 features in total);
3) categorical (3 features in total);
4) PurchDate.

Remark 1. Any missing values were replaced by "-1". "PurchDate" values were transferred to four integer values: 1) year (2009 or 2010); 2) month; 3) day of the week; and 4) day of the month.

2.1 Textual data

Using special software, written in Perl, we created list of all text-units for any feature, and counted the numbers of their occurrences in the training database.

[1] `http://www.kaggle.com`

Subject to the sufficient level (see, column Δ in Table 1) any particular text-unit was given sequential positive index, or zero index, which means infrequent (insignificant) value.

Table 1. List of 36 original features, where index=0 means that the feature was excluded from modelling (3 in total); index=1 - numerical feature (15 in total including target variable); index=2 - date (one feature); index=3 - textual features (14 features in total); index=4 - categorical features (3 features in total)

N	Field Name	Type	Index	Δ
1	RefID	NA	0	
2	IsBadBuy	target	1	
3	PurchDate	date	2	
4	Auction	txt	3	100
5	VehYear	year	1	
6	VehicleAge	num	1	
7	Make	txt	3	20
8	Model	txt	3	40
9	Trim	txt	3	39
10	SubModel	txt	3	20
11	Color	txt	3	50
12	Transmission	txt	3	100
13	WheelTypeID	cat	4	
14	WheelType	txt	3	100
15	VehOdo	num	1	
16	Nationality	txt	3	100
17	Size	txt	3	100
18	TopThreeAmericanName	txt	3	100
19	MMRAcquisitionAuctionAveragePrice	num	1	
20	MMRAcquisitionAuctionCleanPrice	num	1	
21	MMRAcquisitionRetailAveragePrice	num	1	
22	MMRAcquisitonRetailCleanPrice	num	1	
23	MMRCurrentAuctionAveragePrice	num	1	
24	MMRCurrentAuctionCleanPrice	num	1	
25	MMRCurrentRetailAveragePrice	num	1	
26	MMRCurrentRetailCleanPrice	num	1	
27	PRIMEUNIT	txt	3	50
28	AcquisitionType	NA	0	
29	AUCGUART	txt	3	50
30	KickDate	NA	0	
31	BYRNO	cat	4	
32	VNZIP	cat	4	
33	VNST	txt	3	20
34	VehBCost	num	1	
35	IsOnlineSale	num	1	
36	WarrantyCost	num	1	

As a consequence of the above pre-processing transformation/treatment, we produced two completely numerical matrices (for training and for testing) with 35 features each and without any missing values.

3 Synthetic Features

Let us consider 4 features: VehicleAge, VehOdo, VehBCost and WarrantyCost, where the first one is discrete, and the others are continuous.

Continuous features maybe investigated using method of the moving averages, applied to the sorted (according to the selected feature) vector of the target variable. We have found that the "get kicked" probability is an increasing function of VehOdo (V_{15}, see Table 1) and WarrantyCost (V_{36}), and decreasing function of VehBCost (or of any other Cost-related variable). Based on this observation, we can consider the following structure for the new (synthetic) variable:

$$f_{\text{new}} = \frac{V_{23}}{(1 + C_1 V_{36})(C_2 + V_{15} + C_3 V_6)}, \tag{1}$$

where non-negative parameters $C_i, i = 1, \ldots, 3$, were selected (optimised using specially designed software written in Matlab) in order to maximise diversity of the moving average corresponding to (1), see Figure 1(d).

Two sets of the coefficients C are given in Table 2. Additionally, we used third synthetic variable:

$$f_{\text{new}}^{(3)} = \frac{V_{23} + C_4 V_{34}}{(C_5 + V_{36})}, \tag{2}$$

where $C_4 = 1.49, C_5 = 173$.

Table 2. Two sets of coefficients for synthetic variables

C_1	C_2	C_3
0	267	14354
9.8	333	9229

Remark 2. Equation (1) represents just an example. In general terms, definition of the new synthetic variable may include many multipliers. For example, in the case of a very popular Credit Contest on the Kaggle platform we used 13 multipliers, corresponding to the different original variables. After optimisation of the coefficients, we can split the whole new variable by considering sub-products of 2, 3, 4,..., components. As a consequence, we shall create many new synthetic variables, which cannot be replaced by only one variables as a product of all components.

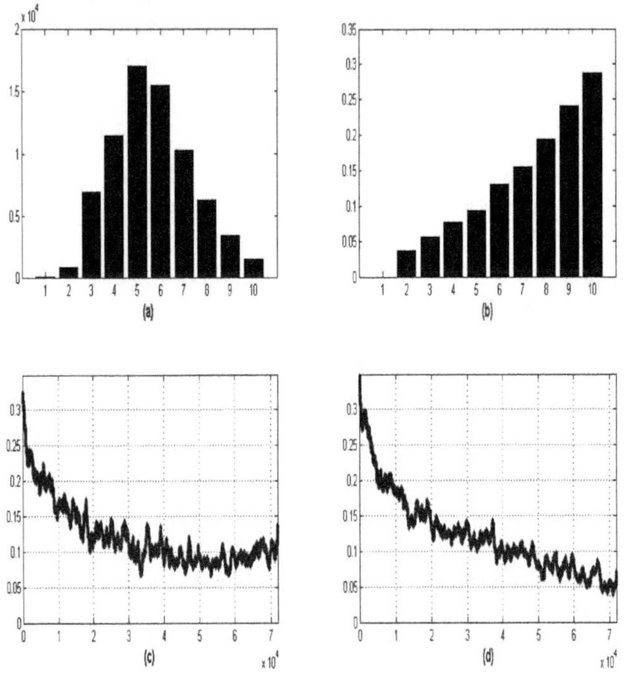

Fig. 1. (a) Numbers of occurrences for VehicleAge; (b) empirical probabilities for VehicleAge (ten values in total); (c) moving averages for VehBCost; (d) moving averages for the synthetic feature (see Section 3)

3.1 On the Comparison between Different Cost-Variables

In the most early stages of the Contest, we had noticed that the differences between Costs are much more informative compared to the Cost-variables themselves. Using database with 37 variables (=35-1+3), as described above, where we excluded VehYear variable (as it replicates VehicleAge) and replaced Cost-variables (NN19-26) by their differences with VehBCost (8 differences in total), we were able to achieve public score on the LeaderBoard 0.26023 in the terms of Gini Index.

As a next step, we decided to replace above 8 Cost-differences by all the 36 Cost-differences (the number of all combinations from 9 by 2):

1) for $i = 1, \ldots, 8$,
2) for $j = i + 1, \ldots, 9$,
3) $r_{ij} = 1 - \frac{X_i+1}{X_j+1}$
4) end;
5) end;

where by $X_j, j = 1, \ldots, 9$, we denote Costs/Prices (see, features NN19-26, 34 in Table 1).

New database includes 65 features. With this database, we observed a very significant improvement: public score on the LeaderBoard 0.26608.

4 Homogeneous Ensembling with Balances Random Sets

In many cases, ensembles have significantly better prediction accuracy compared to their individual members [2]. As a consequence of the diversity between individual base-learners, an ensemble will not suffer from overfitting. On the other hand, in many cases we are dealing with imbalanced data and a classifier which was built using all data has tendency to ignore minority class. As a solution to the problem, we propose to consider a large number of relatively small and balanced subsets where representatives from the both patterns are to be selected randomly [3].

4.1 On the Boosting Principles Applied to the Selection of the Balanced Subsets

In our previous publications [3], [4], we used a sequence of balanced subsets, which were selected from the training set independently (one subset was completely independent from the next, and so on). However, it appears to be logical if we shall apply here principles of boosting [5] based on the latest solution (known, also, as base-learner).

Principle of Complexity in Application to the Labelled Data. Let us describe the proposed boosting model in more details. Suppose that $y_t \in \{0,1\}$ is the target variable and $s_t^{(\alpha)} \in [0,\ldots,1]$ is the training solution corresponding to the sample t, α is a sequential index of the balanced random subset. Then, we shall select sample t as a prospective to be included in the following balanced subset $\alpha+1$ subject to the following conditions

$$\xi \leq \xi_1 \; if \; y_t = 1; \tag{3a}$$

$$\xi \leq \xi_2, \; otherwise, \tag{3b}$$

where

$$\xi_i = c_{i1} + (c_{i2} + c_{i3} \cdot \phi) \cdot w(y_t, s_t^{(\alpha)}), i = 1,\ldots,2, \tag{4}$$

$$w(y,s) = |y - s|^{\beta}, \tag{5}$$

where ξ and $\phi \in [0,\ldots,1]$ are standard uniform random variables, $\beta > 0$ and $c_{ij} > 0, i = 1,\ldots,2, j = 1,\ldots,3$, are regulation parameters. For example, we can select $\beta = 0.35$, and the recommended values for coefficients c are given in Table 3.

Table 3. Recommended values for the matrix of coefficients $c_{ij} > 0, i = 1, \ldots, 2$, $j = 1, \ldots, 3$.

0.25	0.4	0.35
0.12	0.08	0.06

Remark 3. We can see that values in the first row of Table 3, which correspond to the minority class are much bigger compared to the second row, which corresponds to the majority class. As a direct consequence, selection according to (3a) and (3b) will create relatively balanced subset. However, we considered selection in accordance with (3a) and (3b) as just a preliminary. After that, we conducted final adjustment to ensure that the relation between positives and negatives is exactly as required.

Remark 4. The function (5) in (4) represents a very important boosting multiplier to ensure that "difficult" samples will be given higher probability to be selected.

Principle of Simplicity in Application to the Unlabelled Data. We can extend selection (3a) and (3b) to the unlabelled test set. However, there is a fundamental difference between treatment of labelled and unlabelled data. In the case of labelled data, we shall be selecting more complex samples, but in the case of unlabelled data, we shall be selecting simpler samples with stronger indication regarding their classification in accordance with available training solution. We used this semi-supervised approach in application to the Credit Challenge on the Kaggle platform, where the data were stronger imbalanced and the quality of classification according to the AUC was significantly higher compared to the Carvana Challenge.

5 Some Other Ways to Construct Secondary Training Datasets

In Section 3.1 we introduced 36 ratios $r_{ij}, i = 1, \ldots, 8, j = i + 1, \ldots, 9$. Clearly, all those ratios have different importance, and we have found that the following 4 relations are the most influential: 1) $\{19, 23\}$; 2) $\{20, 24\}$; 3) $\{21, 25\}$ and 4) $\{22, 26\}$, where indexes of the involved features are given in Table 1. The next in line were 8 pairs: NN19-26 against N34.

Firstly, we decided to apply another formula to compare different Costs:

$$q_{ij} = \log \frac{X_i + \Delta}{X_j + \Delta}, \tag{6}$$

where we used value $\Delta = 100$ as a smoothing parameter.

Further, we added to the model all possible sums of the first 4 the most influential relations, plus an indicator whether or not all 8 involved Costs are available. In total, we calculated the block $\mathcal{B}$ of $C_4^2+C_4^3+C_4^4+1=12$ additional features compared to the previous database with 65 features, which is described in Section 3.1.

With this database (77 features) we observed LeaderBoard public score of 0.26810.

5.1 The Best Single Model

Compared to the previous database with 77 features, we removed feature N35 (see Table 1), also, we removed the indicator of the presence of eight Cost features.

As a very important innovation, we introduces a new definition, which represents a more advanced development compared to (6):

$$\lambda_{ij} = \frac{q_i - q_j}{q_i + q_j}, \tag{7}$$

where

$$q_i = \log \frac{X_i + \Delta}{X_0 + \Delta}, i = 1, \ldots, 8,$$

X_0 is feature N34 (VehBCost).

Using above definition (7), we computed 4 new features corresponding to the pairs: 1) $\{19, 23\}$; 2) $\{20, 24\}$; 3) $\{21, 25\}$ and 4) $\{22, 26\}$. Plus, we re-computed 11 features from the block $\mathcal{B}$. Consequently, new database included 79 features (= 77-2+4), and we observed LeaderBoard score 0.26867.

Remark 5. In order to reduce overfitting, we are interested to increase random factor in the model. It appears to be logical to split all the 79 features into 4-5 blocks, where importance of the features within any particular block is about the same. As it was discussed in Section 4, the final classifier represents an average of the base-learners, each of which is based on the randomly selected balanced subset. Importantly to note that the features in the model were, also, selected randomly from any particular block, based on our assessment of how important this block is.

6 Blending of the Different Databases with Neural Nets

In the above section we described three secondary databases with 65, 77 and 79 features. In fact, we created about 20 databases with up to 142 features.

As a next step, we can test any particular database using cross-validation as a standard tool for validation of the classification model, where five folds appears to be quite sufficient. Suppose, we are using the same design (known, also, as splitting) matrix for CV in application to all databases. Then, after computation of the CV-solutions for different databases, we consider performance of the linear combinations of those solutions (known, also, as blend).

After optimisation of the non-negative weighting coefficients, we can compute test-solutions for the selected datasets, which correspond to the sufficiently large weighting coefficients, and compute blend of those particular solutions for a final submission.

An implementation of the above scheme may require a lot of computational time taking into account the fact that any single solution represents a homogeneous ensemble of 100-200 base-learners (each of which corresponds to the randomly selected balanced subset).

6.1 Principle of Invariance

In order to reduce significantly the computational costs, we can implement a principle of invariance. In accordance with this principle, the quality of the CV-solutions are not important in an absolute scale. In contrast, important are relations between different CV-solutions.

We conducted CV5 with Neural Nets function from the Matlab-based CLOP package[2], which is significantly faster compared to the GBM function in R. Evaluation of one particular database with 200 balanced random subsets took about 4 hours time. The following CV5 results were observed: 0.270632 (F79), 0.266423 (F77) and 0.26154 (F65), where numbers in the brackets indicate numbers of the features in the corresponding database.

Remark 6. It is interesting to note that there were several other databases with CV5 result better than 0.26154. However, inclusion of those databases in the blend produced worse result.

The best solution (in both public and private) was a blend of F79, F77 and F65 solutions (used GBM package in R) with the following weighting coefficients: $\{\frac{100}{170}, \frac{55}{170}, \frac{15}{170}\}$. It produces Gini score 0.26885 in public (4th out of 582 participating teams), and 0.26655 in private (3rd place in the Contest).

7 Concluding Remarks

Selection bias [6] or overfitting represents a very important and challenging problem. As it was noticed in [7], if the improvement of a quantitative criterion such as the error rate is the main contribution of a paper, the superiority of a new algorithms should always be demonstrated on independent validation data. In this sense, an importance of the data mining contests is unquestionable. The rapid popularity growth of the data mining challenges [8] demonstrates with confidence that it is the best known way to evaluate different models and systems. Based on our own experience, cross-validation (CV) maybe easily overfit as a consequence of the intensive experiments. Further developments such as nested CV [9] are computationally too expensive [7], and should not be used

[2] `http://clopinet.com/CLOP/`

until it is absolutely necessary, because nested CV may generate secondary serious problems as a result of 1) the dealing with an intense computations, and 2) very complex software (and, consequently, high level of probability to make some mistakes) used for the implementation of the nested CV. Moreover, we do believe that in most of the cases scientific results produced with the nested CV are not reproducible (in the sense of an absolutely fresh data, which were not used prior).

Generally, we are satisfied with our results, and consider blending model applied to the different databases as a main innovation proposed in this paper. Note, that using conceptually similar method as described in this paper, we were able to achieve 9th place out of 970 actively participating teams in another data mining Contest, named "Credit", which was, also, based on the Kaggle platform.

References

[1] Koren, Y.: The BellKor Solution to the Netflix Grand Prize, Wikipedia, 10 pages (2009)
[2] Wang, W.: Some fundamental issues in ensemble methods. In: World Congress on Computational Intelligence, Hong Kong, pp. 2244–2251. IEEE (2008)
[3] Nikulin, V.: Classification of imbalanced data with random sets and mean-variance filtering. International Journal of Data Warehousing and Mining 4(2), 63–78 (2008)
[4] Nikulin, V., McLachlan, G.: Classification of imbalanced data with balanced random sets. Journal of Machine Learning Research, Workshop and Conference Proceedings 7, 89–100 (2009)
[5] Freund, Y., Schapire, R.: A decision-theoretic generalization of on-line learning and an application to boosting. Journal of Computer and System Sciences 55, 119–139 (1997)
[6] Heckerman, J.: Sample selection bias as a specification error. Econometrica 47(1), 153–161 (1979)
[7] Jelizarow, M., Guillemot, V., Tenenhaus, A., Strimmer, K., Boulesteix, A.-L.: Over-optimism in bioinformatics: an illustration. Bioinformatics 26(16), 1990–1998 (2010)
[8] Carpenter, J.: the best analyst win. Science 331, 698–699 (2011)
[9] Cudeck, R., Browne, M.: Cross-validation of covariance structures. Multivariate Behavioral Research 18(2), 147–167 (1983)

On Cost and Uncertainty of Decision Trees

Igor Chikalov, Shahid Hussain, and Mikhail Moshkov

Mathematical and Computer Sciences & Engineering Division,
King Abdullah University of Science and Technology,
Thuwal 23955-6900, Saudi Arabia
{igor.chikalov,shahid.hussain,mikhail.moshkov}@kaust.edu.sa

Abstract. This paper describes a new tool for the study of relationships between the cost (depth, average depth, number of nodes, etc.) and uncertainty of decision trees, which is closely connected with accuracy of trees. In addition to the algorithm the paper also presents the experimental results of application of our algorithm on some of the datasets acquired from UCI ML Repository [1].

Keywords: decision trees, cost functions, uncertainty measure.

1 Introduction

Decision trees are widely used as predictors, as a way of knowledge representation, and as algorithms for problem solving. There uses require optimizing decision trees for certain cost functions such as the number of misclassifications, depth/average depth, and number of nodes. That is minimizing one of these cost functions yields more accurate, faster, or more understandable decision trees (respectively).

We consider the concept of approximate decision trees using an uncertainty measure. This uncertainty, $R(T)$, for a decision table T is equal to the number of unordered pairs of rows in the decision table T, labeled with different decisions. The uncertainty measure $R(T)$ allows us to define the notion of α-decision trees. That is, for a fixed nonnegative integer α, an α-decision tree for T localizes a given row in a subtable T' of T such that $R(T') \leq \alpha$. The value α can also be considered as uncertainty of the this tree. The parameter α is also connected with accuracy of trees.

The aim of this paper is to study the relationships between cost and uncertainty of decision trees. For a given cost function ψ, decision table T, and a nonnegative integer α, we find the minimum cost of α-decision tree for T relative to ψ. To this end we have designed an algorithm based on dynamic programming approach [2,3,4] and integrated into a software tool called Dagger [5]. We also performed various experimental results to the datasets acquired from UCI ML Repository [1] (we show results for three datasets for illustrations purposes).

The presented algorithm and its implementation in the software tool Dagger together with similar algorithms devised by the authors (see for example [6]) can be useful for investigations in Rough Sets [7,8] where decision trees are used as classifiers [9].

J.T. Yao et al. (Eds.): RSCTC 2012, LNAI 7413, pp. 190–197, 2012.

This paper is organized into five sections. In Sect. 2 basic notions are discussed. Section 3 is devoted to the consideration of relationships between cost and uncertainty of decision trees. In Sect. 4, results of experiments are presented followed by conclusions in Sect. 5.

2 Basic Notions

In this section, we consider main notions connected with decision tables and decision trees.

2.1 Decision Tables and Decision Trees

A *decision table* T is a rectangular table with m columns labeled with conditional attributes $f_1, \ldots, f_m$. The entries of the table T are nonnegative integers as the value of attributes $f_1, \ldots, f_m$. Rows of the table are pairwise different and each row is labeled with a nonnegative integer describing the value of the decision attribute d. We denote by $E(T)$ the set of attributes (columns of the table T), each of which contains different values. For $f_i \in E(T)$, let $E(T, f_i)$ be the set of values from the column f_i.

Let $f_{i_1}, \ldots, f_{i_t} \in \{f_1, \ldots, f_m\}$ and $a_1, \ldots, a_t$ be nonnegative integers. We denote by $T(f_{i_1}, a_1) \ldots (f_{i_t}, a_t)$ the subtable of the table T, which consists of such and only such rows of T that at the intersection with columns $f_{i_1}, \ldots, f_{i_t}$ have numbers $a_1, \ldots, a_t$, respectively. Such nonempty tables (including the table T) will be called *separable subtables* of the table T. For a subtable Θ of the table T we will denote by $R(\Theta)$ the number of unordered pairs of rows that are labeled with different decisions. Later we will interpret the value $R(\Theta)$ as the *uncertainty* of the table Θ. A minimum decision value which is attached to the maximum number of rows in a nonempty subtable Θ will be called the *most common decision* for Θ.

A *decision tree* Γ *over* the table T is a finite directed tree with a root in which each terminal node is labeled with a decision. Each nonterminal node is labeled with a conditional attribute, and for each nonterminal node, the outgoing edges are labeled with pairwise different nonnegative integers. Let v be an arbitrary node of Γ. We now define a subtable $T(v)$ of the table T. If v is the root then $T(v) = T$. Let v be a node of Γ that is not the root, nodes in the path from the root to v be labeled with attributes $f_{i_1}, \ldots, f_{i_t}$, and edges in this path be labeled with values $a_1, \ldots, a_t$, respectively. Then $T(v) = T(f_{i_1}, a_1) \ldots (f_{i_t}, a_t)$.

Let α be a nonnegative integer. We will say that Γ is an *α-decision tree for* T if any node v of Γ satisfies the following conditions:

- If $R(T(v)) \leq \alpha$ then v is a terminal node labeled with the most common decision for $T(v)$.
- Otherwise, v is labeled with an attribute $f_i \in E(T(v))$ and, if $E(T(v), f_i) = \{a_1, \ldots, a_t\}$, then t edges leave node v, and these edges are labeled with $a_1, \ldots, a_t$ respectively.

For any α, an α-decision tree for T is called a *decision tree for* T.

2.2 Representation of Sets of Decision Trees

Consider an algorithm for construction of a graph $\Delta(T)$, which represents the set of all decision trees for the table T. Nodes of this graph are some separable subtables of the table T. During each step we process one node and mark it with the symbol *. We start with the graph that consists of one node T and finish when all nodes of the graph are processed.

Let the algorithm has already performed p steps. We now describe the step number $(p+1)$. If all nodes are processed then the work of the algorithm is finished, and the resulting graph is $\Delta(T)$. Otherwise, choose a node (table) Θ that has not been processed yet. Let b be the most common decision for Θ. If $R(\Theta) = 0$, label the considered node with b, mark it with symbol * and proceed to the step number $(p+2)$. If $R(\Theta) > 0$, then for each $f_i \in E(\Theta)$ draw a bundle of edges from the node Θ (this bundle of edges will be called *f_i-bundle*). Let $E(\Theta, f_i) = \{a_1, \ldots, a_t\}$. Then draw t edges from Θ and label these edges with pairs $(f_i, a_1), \ldots, (f_i, a_t)$ respectively. These edges enter into nodes $\Theta(f_i, a_1), \ldots, \Theta(f_i, a_t)$. If some of the nodes $\Theta(f_i, a_1), \ldots, \Theta(f_i, a_t)$ are not present in the graph then add these nodes to the graph. Mark the node Θ with the symbol * and proceed to the step number $(p+2)$.

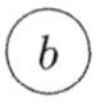

Fig. 1. Trivial decision tree

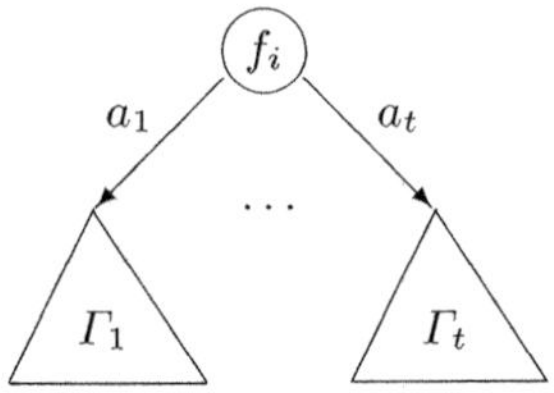

Fig. 2. Aggregated decision tree

Now for each node Θ of the graph $\Delta(T)$, we describe the set of decision trees corresponding to the node Θ. We will move from terminal nodes, which are labeled with numbers, to the node T. Let Θ be a node, which is labeled with a number b. Then the only trivial decision tree depicted in Fig. 1 corresponds to the node Θ.

Let Θ be a nonterminal node (table) then there is a number of bundles of edges starting in Θ. We consider an arbitrary bundle and describe the set of decision trees corresponding to this bundle. Let the considered bundle be an f_i-bundle where $f_i \in (\Theta)$ and $E(\Theta, f_i) = \{a_1, \ldots, a_t\}$. Let $\Gamma_1, \ldots, \Gamma_t$ be decision trees from sets corresponding to the nodes $\Theta(f_i, a_1), \ldots, \Theta(f_i, a_t)$. Then the decision tree depicted in Fig. 2 belongs to the set of decision trees, which correspond to this bundle. All such decision trees belong to the considered set, and this set does not contain any other decision trees. Then the set of decision trees corresponding to the node Θ coincides with the union of sets of decision trees corresponding to the bundles starting in Θ and the set containing one decision tree depicted in Fig. 1, where b is the most common decision for Θ. We denote by $D(\Theta)$ the set of decision trees corresponding to the node Θ.

The following proposition shows that the graph $\Delta(T)$ can represent all decision trees for the table T.

Proposition 1. *Let T be a decision table and Θ a node in the graph $\Delta(T)$. Then the set $D(\Theta)$ coincides with the set of all decision trees for the table Θ.*

2.3 Cost Functions

We will consider cost functions which are given in the following way: values of considered cost function ψ, which are nonnegative numbers, are defined by induction on pairs (T, Γ), where T is a decision table and Γ is an α-decision tree for T. Let Γ be an α-decision tree represented in Fig. 1. Then $\psi(T, \Gamma) = \psi^0$, where ψ^0 is a nonnegative number. Let Γ be an α-decision tree depicted in Fig. 2. Then $\psi(T, \Gamma) = F(N(T), \psi(T(f_i, a_1), \Gamma_1), \ldots, \psi(T(f_i, a_t), \Gamma_t))$. Here $N(T)$ is the number of rows in the table T, and $F(n, \psi_1, \psi_2, \ldots)$ is an operator which transforms the considered tuple of nonnegative numbers into a nonnegative number. Note that the number of variables $\psi_1, \psi_2, \ldots$ is not bounded from above.

The considered cost function will be called *monotone* if for any natural t, any nonnegative numbers $a, c_1, \ldots, c_t, d_1, \ldots, d_t$ and the inequalities $c_1 \leq d_1, \ldots, c_t \leq d_t$ the inequality $F(a, c_1, \ldots, c_t) \leq F(a, d_1, \ldots, d_t)$ follows. We will say that ψ is *bounded from below* if, $\psi(T, \Gamma) \geq \psi^0$ for any decision table T and any α-decision tree Γ for T.

Now we take a closer view of some monotone cost functions, which are bounded from below.

Number of nodes: $\psi(T, \Gamma)$ is the number of nodes in α-decision tree Γ. For this cost function $\psi^0 = 1$ and $F(n, \psi_1, \psi_2, \ldots, \psi_t) = 1 + \sum_{i=1}^{t} \psi_i$.

Number of nonterminal nodes: $\psi(T, \Gamma)$ is the number of nonterminal nodes in α-decision tree Γ. For this cost function $\psi^0 = 0$ and $F(n, \psi_1, \psi_2, \ldots, \psi_t) = 1 + \sum_{i=1}^{t} \psi_i$.

Number of terminal nodes: $\psi(T, \Gamma)$ is the number of terminal nodes in α-decision tree Γ. For this cost function $\psi^0 = 1$ and $F(n, \psi_1, \psi_2, \ldots, \psi_t) = \sum_{i=1}^{t} \psi_i$.

Depth: $\psi(T, \Gamma)$ is the maximum length of a path from the root to a terminal node of Γ. For this cost function $\psi^0 = 0$ and $F(n, \psi_1, \psi_2, \ldots, \psi_t) = 1 + \max\{\psi_1, \ldots, \psi_t\}$.

Total path length: for an arbitrary row $\bar{\delta}$ of the table T we denote by $l(\bar{\delta})$ the length of the path from the root to a terminal node v of Γ such that $\bar{\delta}$ is in $T(v)$. Then $\psi(T, \Gamma) = \sum_{\bar{\delta}} l(\bar{\delta})$, where we take the sum on all rows $\bar{\delta}$ of the table T. For this cost function $\psi^0 = 0$ and $F(n, \psi_1, \psi_2, \ldots, \psi_t) = n + \sum_{i=1}^{t} \psi_i$. Note that the *average depth* of Γ is equal to the total path length divided by $N(T)$.

3 Relationship between Cost and Uncertainty

Let T be a decision table with m columns labeled with $f_1, \ldots, f_m$ and ψ be a monotone and bounded from below cost function. The main aim of this paper is to provide an algorithm to compute the function $\mathcal{F}_{\psi,T}$, which is defined on the set $\{0, \ldots, R(T)\}$. For any $\alpha \in \{0, \ldots, R(T)\}$, the value of $\mathcal{F}_{\psi,T}(\alpha)$ is equal to the minimum cost of an α-decision tree for T, relative to the cost function ψ. This function can be represented by the tuple

$$(\mathcal{F}_{\psi,T}(0), \ldots, \mathcal{F}_{\psi,T}(R(T))) .$$

3.1 Computing the Function $\mathcal{F}_{\psi,T}$

Now for each node Θ of the graph $\Delta(T)$ we compute the function $\mathcal{F}_\Theta = \mathcal{F}_{\psi,\Theta}$ (we compute the $R(\Theta)$-tuple describing this function).

A node of $\Delta(T)$ is called *terminal* if there are no edges leaving this node. We will move from the terminal nodes, which are labeled with numbers, to the node T.

Let Θ be a terminal node of $\Delta(T)$ which is labeled with a number b that is the most common decision for Θ. We know that $R(\Theta) = 0$. Therefore, b is a common decision for Θ, and the decision tree depicted in Fig. 1 is the only 0-decision tree for Θ. Since $R(\Theta) = 0$, we should consider only one value of α – the value 0. It's clear that the minimum cost of 0-decision tree for Θ is equal to ψ^0. Thus, the function $\mathcal{F}_\Theta$ can be described by the tuple (ψ^0).

Let Θ be a nonterminal node of $\Delta(T)$ then it means that $R(\Theta) > 0$. Let $\alpha \in \{0, \ldots, R(\Theta)\}$. We need to find the value $\mathcal{F}_\Theta(\alpha)$, which is the minimum cost relative to ψ of an α-decision tree for Θ. Since $R(\Theta) > 0$, the root of any α-decision tree for Θ is labeled with an attribute from $E(\Theta)$. For any $f_i \in E(\Theta)$, we denote by $\mathcal{F}_\Theta(\alpha, f_i)$ the minimum cost relative to ψ of an α-decision tree for Θ such that the root of this tree is labeled with f_i. It is clear that

$$\mathcal{F}_\Theta(\alpha) = \min\{\mathcal{F}_\Theta(\alpha, f_i) : f_i \in E(\Theta)\}. \tag{1}$$

Let $f_i \in E(\Theta)$ and $E(\Theta, f_i) = \{a_1, \ldots, a_t\}$. Then any α-decision tree Γ for Θ with the attribute f_i attached to the root can be represented in the form depicted in Fig. 2, where $\Gamma_1, \ldots, \Gamma_t$ are α-decision trees for $\Theta(f_i, a_1), \ldots, \Theta(f_i, a_t)$. Since ψ is a monotone cost function, the tree Γ will have the minimum cost if the costs of trees $\Gamma_1, \ldots, \Gamma_t$ are minimum. Therefore,

$$\mathcal{F}_\Theta(\alpha, f_i) = F(N(\Theta), \mathcal{F}_{\Theta(f_i,a_1)}(\alpha), \ldots, \mathcal{F}_{\Theta(f_i,a_t)}(\alpha)). \tag{2}$$

If for some j, $1 \le j \le t$, we have $\alpha > R(\Theta(f_i, a_j))$ then $\mathcal{F}_{\Theta(f_i,a_j)}(\alpha) = \psi^0$, since the decision tree depicted in Fig. 1, where b is the most common decision for $\Theta(f_i, a_j)$, is an α-decision tree for $\Theta(f_i, a_j)$. The cost of this tree is ψ^0. Since ψ is a cost function, which is bounded from below, the cost of any α-decision tree for $\Theta(f_i, a_j)$ is at least ψ^0.

The formulas (1) and (2) allow us to find the value of $\mathcal{F}_{\Theta}(\alpha)$ if we know the values of $\mathcal{F}_{\Theta(f_i,a_j)}(\alpha)$, where $f_i \in E(\Theta)$ and $a_j \in E(\Theta, f_i)$. When we reach to the node T we will obtain the function $\mathcal{F}_T = \mathcal{F}_{\psi,T}$.

Figure 3 is the directed acyclic graph (DAG) as an illustration of working of the considered algorithm.

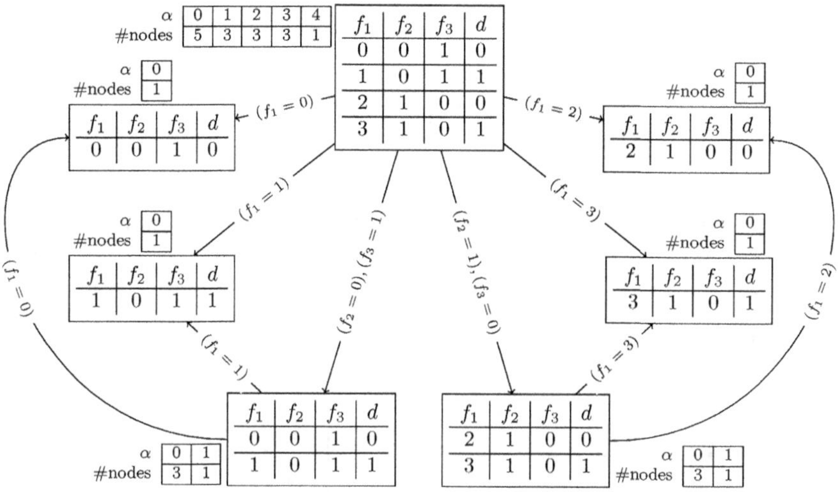

Fig. 3. DAG with relationships between uncertainty and number of nodes

4 Experimental Results

We consider now results of some experiments with decision tables from UCI ML Repository [1] based on software system DAGGER [4,5]. The resulting plots are depicted in Figs. 4, 5, and 6.

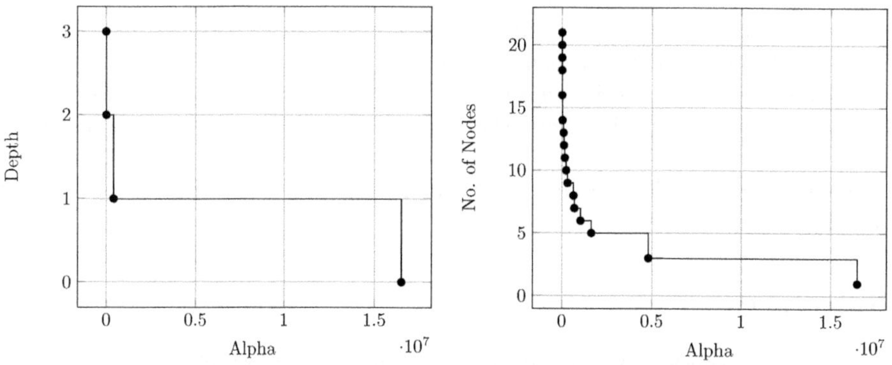

Fig. 4. MUSHROOM dataset (22 attributes and 8125 rows)

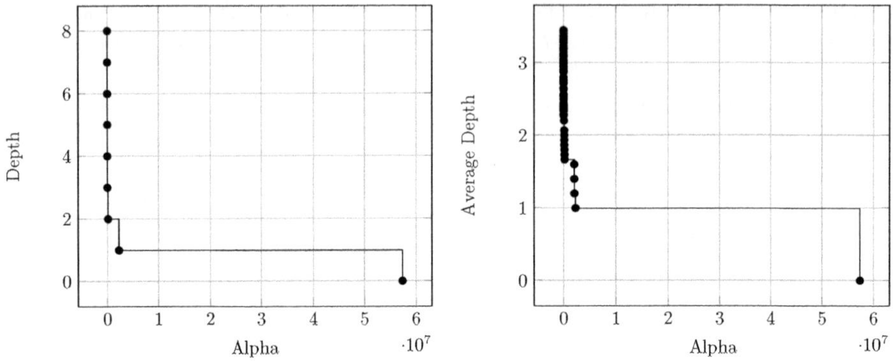

Fig. 5. NURSERY dataset: (8 attributes and 12960 rows)

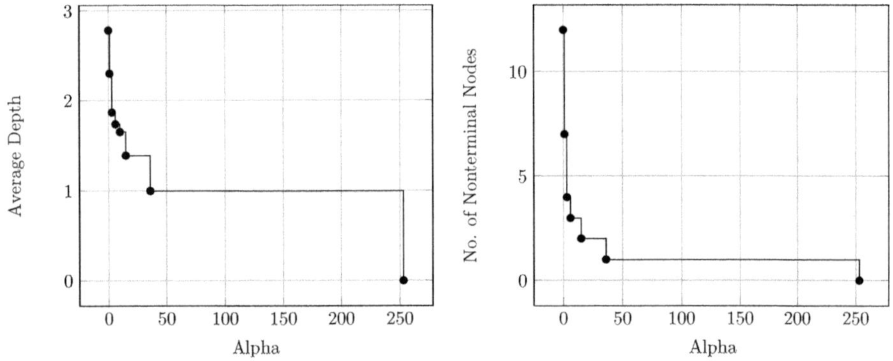

Fig. 6. TEETH dataset: (8 attributes and 33 rows)

5 Conclusion

This paper is devoted to the consideration of a new tool for decision tree study. We present and explain in detail the algorithm to compute the relationships between the uncertainty measure and one of the cost functions. That is for a given uncertainty α of a decision tree, given cost function ψ, and decision table T, the presented algorithm finds the minimum cost of an α-decision tree for T relative to ψ.

Future studies will be connected with the consideration of new types of uncertainty measures.

References

1. Frank, A., Asuncion, A.: UCI Machine Learning Repository (2010)
2. Alkhalid, A., Chikalov, I., Moshkov, M.: On Algorithm for Building of Optimal α-Decision Trees. In: Szczuka, M., Kryszkiewicz, M., Ramanna, S., Jensen, R., Hu, Q. (eds.) RSCTC 2010. LNCS, vol. 6086, pp. 438–445. Springer, Heidelberg (2010)

3. Alkhalid, A., Chikalov, I., Moshkov, M.: A Tool for Study of Optimal Decision Trees. In: Yu, J., Greco, S., Lingras, P., Wang, G., Skowron, A. (eds.) RSKT 2010. LNCS, vol. 6401, pp. 353–360. Springer, Heidelberg (2010)
4. Alkhalid, A., Chikalov, I., Hussain, S., Moshkov, M.: Extensions of Dynamic Programming as a New Tool for Decision Tree Optimization. In: Ramanna, S., Jain, L.C., Howlett, R.J. (eds.) Emerging Paradigms in Machine Learning and Applications. SIST, vol. 13, pp. 11–29. Springer, Heidelberg (2012)
5. Alkhalid, A., Amin, T., Chikalov, I., Hussain, S., Moshkov, M., Zielosko, B.: Dagger: a tool for analysis and optimization of decision trees and rules. In: Computational Informatics, Social Factors and New Information Technologies: Hypermedia Perspectives and Avant-Garde Experiencies in the Era of Communicability Expansion, pp. 29–39. Blue Herons (2011)
6. Chikalov, I., Hussain, S., Moshkov, M.: Relationships between Depth and Number of Misclassifications for Decision Trees. In: Kuznetsov, S.O., Ślęzak, D., Hepting, D.H., Mirkin, B.G. (eds.) RSFDGrC 2011. LNCS, vol. 6743, pp. 286–292. Springer, Heidelberg (2011)
7. Pawlak, Z.: Theoretical Aspects of Reasoning about Data. Kluwer Academic Publishers, Dordrecht (1991)
8. Skowron, A., Rauszer, C.: The discernibility matrices and functions in information systems. In: Slowinski, R. (ed.) Intelligent Decision Support. Handbook of Applications and Advances of the Rough Set Theory, pp. 331–362. Kluwer Academic Publishers, Dordrecht (1992)
9. Nguyen, H.S.: From optimal hyperplanes to optimal decision trees. Fundamenta Informaticae 34(1-2), 145–174 (1998)

Optimization of Quadtree Representation and Compression

Xiang Yin and Ryszard Janicki

Department of Computing and Software, McMaster University,
Hamilton, Canada L8S 4K1
{yinx5,janicki}@mcmaster.ca

Abstract. The quadtrees are a popular representation method for spatial data. In 2009, a heuristic algorithm, called CORN (Choosing an Optimal Root Node), for finding a root node of a region quadtree, has been proposed. It substantially reduces the number of leaf nodes when compared with the standard quadtree decomposition. In this paper, some approximation ideas are applied to improve the CORN algorithm. The empirical results indicate that the new proposed algorithm improves the quadtree representation and data compression.

1 Introduction

Due to an ever-increasing requirement for information storage, spatial data representation and compression have become one of the most popular and significant issues in computer graphics and image processing applications.

Quadtrees, introduced in the early 1970s [1], allow recursive decomposition of space, and have become a major representation method for spatial data. The scenario for a region quadtree is as follows: Given a window consisting of black cells situated in a fixed image area A of dimension $2^m \times 2^m$. Area A is recursively partitioned into equal sized quadrants until each quadrant consists entirely of unicolored cells. The process can be represented by a tree each non-leaf node of which has four children, corresponding to the four quadrants NW, NE, SW, and SE. Each descendant of a node g represents a quadrant in the plane whose origin is g and which is bounded by the quadrant boundaries of the previous step.

The results of [8] have indicated that a good choice of the quadtree root node improves substantially both the quadtree representation and the final image compression, especially for large size images. To find such a 'good' root node, a heuristic algorithm, called CORN (Choosing an Optimal Root Node), has been proposed and analyzed. It reduces space greatly if compared with standard method based on quadtree concept [8].

In this paper, we will enhance CORN algorithm by applying quadtrees approximations techniques [2,6,7]. The result is a new algorithm, ACORN, and tests have shown that ACORN can reduce space by $30-40\%$ when compared with CORN.

2 CORN Algorithm

It appears that the space cost of representing quadtrees is mainly measured by the number of their leaf nodes, as the number of nodes in a quadtree is directly proportional to the number of its leaves (c.f. [3]).

J.T. Yao et al. (Eds.): RSCTC 2012, LNAI 7413, pp. 198–205, 2012.

Essentially, the total number of leaf nodes of a quadtree mainly depends on two factors: the size of the coordinate grid and the position of the image on the grid. Both these two crucial factors are heavily dependent on the root node choice of the quadtree.

Standardly, the root node is always placed in the center of the chosen image area [6,7] (We call it 'Center' algorithm). It provides a decent compression ratio [9], but very often still uses more space than it should intuitively be required [8].

In [8], a heuristic algorithm (CORN) for selecting an optimal root node A has been proposed. It is assumed that the cells of the bounding box are scanned from left to right, row by row, starting at the upper left hand corner of the bounding box of I. The pseudocode of the CORN algorithm looks as follows:

1. **repeat**
   Progress to the next unvisited black cell b_i and record its position.
   Find a block s_i of maximal size (i.e., $2^n \times 2^n, n \geq 0$) that contains b_i, and maximizes the number of unvisited cells.
   Change black color of the cells of s_i into, say, red.
   until all black cells have been visited.
   $\{$At this stage, we get sets $B = \{b_1, \ldots, b_n\}$ and $S = \{s_1, \ldots, s_n\}$.$\}$
2. Find the maximal blocks, say, $s_1, \ldots, s_n$ corresponding to each cell b_i ($i = 0, 1, 2, \ldots, n$).
3. Find all maximal contact sets[1] $S_1, \ldots, S_m$ from S containing at least one of $s_1, \ldots, s_k$ and for each S_j the set C_i of points in the intersection.
4. Order the points in $\bigcup\{C_i : 1 \leq i \leq m\}$ by their neighbor numbers in non-increasing order, say $A_1, A_2, \ldots, A_k$ such that $n_{A_1} \geq n_{A_2} \geq \ldots \geq n_{A_k}$.
5. Let t be the smallest number such that $n, m \leq 2^t$; in other words, a block with side length t is a smallest block that can contain I.
6. Let $i = 0, j = t$.
   repeat
      repeat Increase i by 1.
      Decompose the image with A_i as a root node. If the resulting block has side length 2^j then choose A_i as root node and stop.
      until $i = k$.
      $\{$At this stage, none of the A_i will allow using 2^j as a side length, so we try the next smallest block.$\}$
   Increase j by 1.
   until FALSE

Its heuristics are based on the following three prioritizing criteria: ('one' is the highest priority)

1. A block[2] of maximal size will be one leaf node in the finale quadtree representation.
2. The size of neighbors[3] of A which are blocks is maximized.

---

[1] There are three types of contact: two blocks contain a common cell, or a common edge of a cell, or a common corner point [8].

[2] We call a (closed) square of size $2^k \times 2^k$ containing only black cells a *block*.

[3] If A is a point (i.e., the intersection of two orthogonal edges), the *neighbor number* n_A *of* A is the sum of the size of all blocks of which A is a corner point.

3. The choice of A minimizes the size of the image space $2^t \times 2^t$ containing I.

It has been shown that minimizing the number of black leaf nodes[4] results in a better (lossless) compression of the original image. The CORN algorithm is in principle an implementation of these three criteria.

3 Quadtree Approximation Methods

The CORN algorithm provides a kind of *lossless representation* [8]. *No information of original images is lost during the compression and decompression operations.* But simultaneously, more storage space is required than it might really be needed. One of possible solutions is to use *approximation* instead of exact representation.

Approximations are often the only solutions when incompleteness or noise prevent from getting exact representations, or when exact solutions are intractable. Even when the exact solution or representation is available, an approximation may yield very close results in much smaller time and using much less space. Moreover, in the image processing field, we usually only attention to the global perception rather than each detail.

The left two images in Figure 1 displays two representations of the same image called "Cloud" (from [9], p. 231). The first one uses the CORN algorithm, the second one uses the new algorithm, ACORN, proposed in the next section. There is almost no difference for us to understand them, while by using the new algorithm we get almost 40% reduction of memory space. Therefore, if we can tolerate losing some information, an approximation may provide more efficient representation and compression.

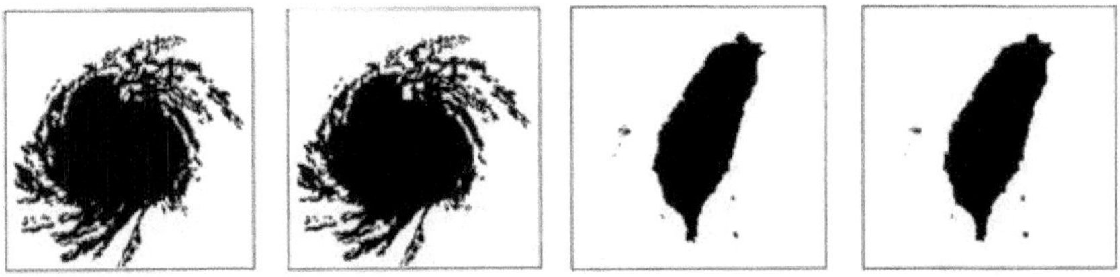

Fig. 1. Binary images – Cloud & Taiwan. Left with CORN (or Center), right with ACORN.

The quadtree approach provides two different approximation techniques, one based on *hierarchy* [6,7], and another based on *forest decomposition* [2]. We will embed both techniques into the CORN algorithm of [8].

[4] The images containing most black leaf nodes as samples were always used [8]. If an image contains more white leaf nodes than black leaf nodes, we can switch to use white leaf nodes as measurement criteria.

- **Hierarchical Approximation Method**

In principle, the hierarchical approximation is a sequence of inner and outer approximations[5] [6], where inner approximations consider grey nodes[6] as white nodes, whereas outer approximations treat them as black nodes. More precisely (Samet [7]):

> "Given an image I, the inner approximation, $IB(k)$ is a binary image defined by the black nodes at levels $\geq k$; the outer approximation, $OB(k)$ is a binary image defined by black nodes at levels $\geq k$ and the grey nodes at level k."

For example, the left side image of Figure 2 shows the quadtree decomposition of image I. Two left images in Figure 3 show $IB(2)$ and $OB(2)$, and the other two show $IB(1)$ and $OB(1)$ for I of Figure 2.

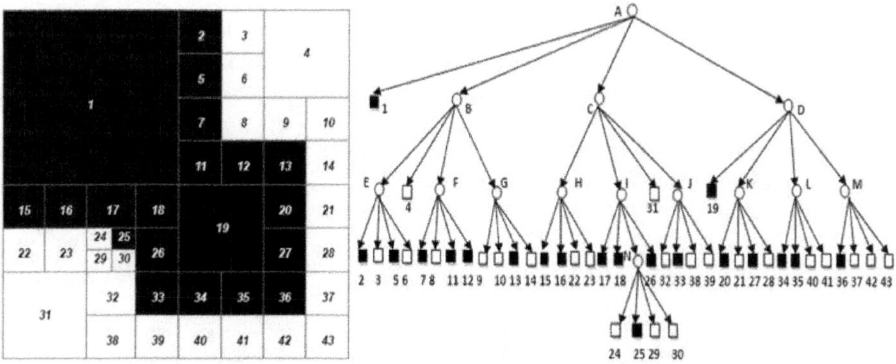

Fig. 2. Quadtree decomposition and representation of image I

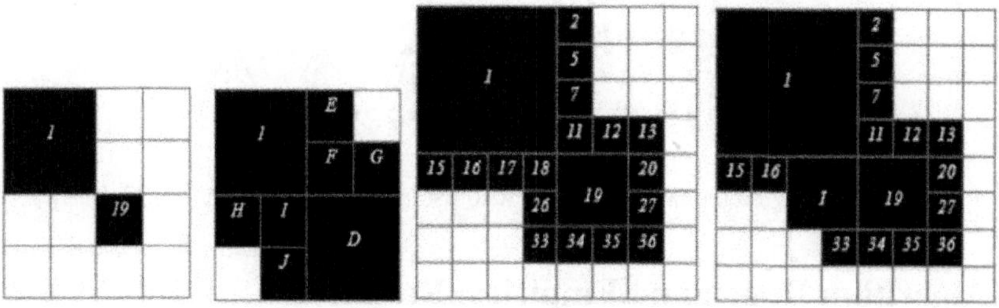

Fig. 3. IB(2), OB(2), IB(1) and OB(1) for image I

- **Forest-Based Approximation Method**

A forest of quadtrees is a decomposition of a quadtree into a collection of subquadtrees. Each of subquadtree corresponds to a maximal square of an image I, and maximal squares could be identified by refining classification of internal nodes.

---

[5] Closely related to a sequence of α-lower and α-upper rough approximations proposed in [4], special versions of Pawlak's rough approximations [5].

[6] In the quadtree representation, we regard all the non-leaf (internal) nodes as grey nodes.

The forest of quadtrees could be seen as a refinement of quadtree data structure that can help us to develop a proper sequence of binary image approximation and, at the end provide space savings [2].

There are four important concepts in this approximation method: GB node, GW node, black forest and white forest. A grey internal node is said to be of type GB node if at least two of its sons are black nodes or of type GB nodes. Otherwise, the node is said to be of type GW node.

For example, considering the right image of Figure 2, the nodes E, F, H, I, K, and L are of type GB node and nodes G, J, M, and N are of type GW node. One can show that each black node or an internal node with a label GB can be regarded as a maximal square (c.f. [2]).

A tree of black forest (similarly for white forest) is defined as follows:

- It contains the minimal set of maximal squares.
- All black or GB nodes in each maximal square are not included in any other square.
- The squares in this minimal set would cover the whole black area of the image.
- The tree is identified by its root.

For example, the black forest of Figure 2 is the singleton set of nodes $\{A\}$[7]. An example of a non-singleton black forest is in Figure 4.

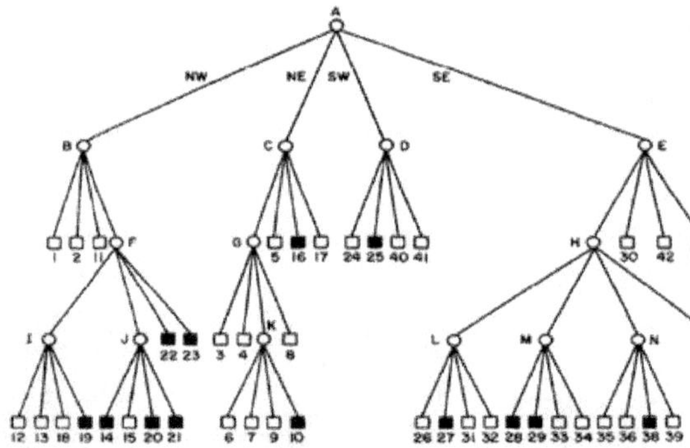

Fig. 4. Non-singleton example: the back forest is the set $\{F, 10, 16, 25, 27, M, 38\}$

4 Algorithm Design and Empirical Results

In this section we will present the main result of this paper, an algorithm called *ACORN* (*Approximations and CORN*), followed by some supporting empirical data.

In principle we added the concepts of GB and GW nodes from the forest-based approximation to the ideas of hierarchical approximation, and mixed the outcome with the CORN algorithm of [8].

[7] The grey node A is a GB node: node 1 is black, node B is GB node (since its sons E and F are GB nodes). Similarly, nodes C and D are GB nodes too. Also any square with size less than the size of A cannot cover the whole black area of the original image, i.e., $\{A\}$ is the minimal set of maximal squares.

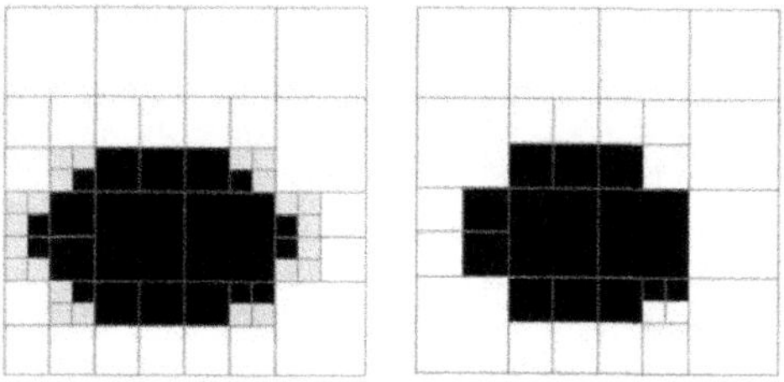

Fig. 5. Quadtree representation optimization

Following the definitions of GB and GW nodes, we say that a grey node is of type GBO (GWO) node if *three* of its sons are either black (white) nodes or of type GBO (GWO) nodes.

We now define the ACORN algorithm as follows:

1. Find the outer approximation $OB(1)$ of the quadtree representation yielded by using the CORN algorithm. For each grey node in the $OB(1)$, it contains two kind of nodes: the black nodes of its sons and *yellow* nodes which correspond to the "*Maybe*" region[8] of the model used for CORN [8] (i.e., the white nodes of their sons in the original quadtree representation). For example, the left image in Figure 5 ('yellow' represented by 'light grey') shows the outer approximation $OB(1)$.
2. Check each grey node in the $OB(1)$: If it is a type of GBO (GWO) node, then it will turn to a black(white) node in the optimization of the final quadtree representation. Otherwise, it will maintain the same format.
3. Merge the leaf nodes to simplify the quadtree structure and update the quadtree representation. For the same example, the right image in Figure 5 shows the new quadtree decomposition after optimization.

Our tests have shown that the ACORN algorithm can substantially reduce the space needed for an image representation. One simple example is shown in Figure 5 where the final number of black, white and all nodes is reduced to 12 from 18, 22 from 45, and 34 from 63, respectively.

Clearly, the loss or alteration of spatial data is an inevitable problem. However, when the size of image is large enough, the difference is negligible, which makes it useful in such areas of applications as computer vision or graphics engine design. Among others we applied the ACORN algorithm to two images analyzed in [9], "Cloud" and "Taiwan" (Figure 1). In both cases the ACORN resulted in a substantial space reduction, with almost no real quality loss. The comparison results are shown in Table 1.

Note that ACORN outperforms other algorithms more for more complex "Cloud" than for simpler "TaiWan" (38.9% versus 31.7% when compared with CORN and 42.3% versus 33.1% when compared with Center). Our other experiments have shown the similar results. In other words, the more sophisticated an image is, the more space savings ACORN can provide.

---

[8] It corresponds to the idea of "*maybe*" from Rough Sets [5].

Table 1. Data compression rate comparison for binary images TaiWan and Cloud. Both Center and CORN are *lossless*, ACORN is a an *approximation*. C_r is a compression ratio.

Binary image	Method	No. of bits needed for original image (*NBO*)	No. of bits needed for compressed image (*NBC*)	$C_r = \frac{NBO}{NBC}$	Space reduction when using ACORN instead of Center or CORN
TaiWan	Center	65536	3058	21.43	33.1%
	CORN	65536	2997	21.88	31.7%
	ACORN	65536	2047	32.02	N/A
Cloud	Center	65536	19024	3.44	42.3%
	CORN	65536	17774	3.69	38.9%
	ACORN	65536	10999	5.96	N/A

5 Conclusion and Future Research

ACORN is a mixture of quite old ([1,2,6,7], 1974-85) and rather new approximation and representation ideas ([8], 2009). Even though it does not explicitly rely on the concept of Rough Sets [5], the approximation parts of ACORN resemble sequences of α-lower and α-upper rough approximations proposed in [4] for relations, but implicitly used for quadtrees in ACORN.

The future research could go in two directions, the further refinement and improvement of ACORN algorithm, and modifying ACORN to deal with color images.

The current version of ACORN considers only the data of $OB(1)$. It really reduces a great amount of space, but unfortunately, it also leads to the borders of main objects too sharp compared with the original image. To improve this, we may try to use the data of $OB(2)$, or even $OB(3)$. Another possible improvement is to refine the definitions of GBO and GWO nodes. In the current version of ACORN, if a square of size 64 has at least 48 black cells (or 48 white cells), then the whole square could be changed to black (or white). We may try to refine the original definitions, so the number 48 might be replaced by for example 56. Both changes will improve image quality but they might slightly reduce space savings.

To extend ACORN for dealing with color images we first have to transform color image into grey-scale image (many softwares can implement this process, such as GIMP, MATLAB), and meanwhile record the coordinate and color information of each pixel which might be needed in the image recovery phase. We may then transform grey-scale image into black-white image by replacing all nodes in grey-scale $0-127$ with white nodes and $128-255$ with black nodes. After this we may apply the existing ACORN algorithm. The last step will be recovering color information for all leaves of the quadtree produced by ACORN. Some changes of colors might occur in this step.

References

1. Finkel, R., Bentley, J.: Quad trees: A data structure for retrieval on composite keys. Acta Informatica 4, 1–9 (1974)
2. Gautier, N.K., Iyengar, S.S., Lakhani, N.B., Manohar, M.: Space and time efficiency of the forest-of-quadtrees representation. Image and Vision Computing 3(2), 63–70 (1985)

3. Horowitz, E., Sahni, S.: Fundamentals of Data Structures. Computer Science Press, Potomac (1976)
4. Janicki, R.: Approximations of Arbitrary Binary Relations by Partial Orders: Classical and Rough Set Models. In: Peters, J.F., Skowron, A., Chan, C.-C., Grzymala-Busse, J.W., Ziarko, W.P. (eds.) Transactions on Rough Sets XIII. LNCS, vol. 6499, pp. 17–38. Springer, Heidelberg (2011)
5. Pawlak, Z.: Rough Sets. Kluwer, Dordrecht (1991)
6. Ranade, S., Rosenfeld, A., Samet, H.: Shape approximation using quadtrees. Pattern Recognition 15(1), 31–40 (1982)
7. Samet, H.: Data Structures for Quadtree Approximation and Compression. Image Processing and Computer Vision 28(9), 973–993 (1985)
8. Yin, X., Düntsch, I., Gediga, G.: Choosing the root node of a quadtree. In: Proc. of GrC 2009 (4th IEEE Intern. Conf. on Granular Computing), Nanchang, China, pp. 721–726 (2009)
9. Yang, Y.H., Chuang, K.L., Tsai, Y.H.: A compact improved quadtree representation with image manipulations. Image and Vision Computing 19(3), 223–231 (1999)

Outlier Mining in Rule-Based Knowledge Bases

Agnieszka Nowak-Brzezińska

University of Silesia, Institute of Computer Science,
ul. Będzińska 39, 41-200 Sosnowiec, Poland
agnieszka.nowak@us.edu.pl

Abstract. The paper presents the problem of outlier detection in rule-based knowledge bases. Unusual (rare) rules, regarded here as deviation, should be the subject of experts' and knowledge engineers' analysis because they allow influencing on the efficiency of inference in decision support systems. A different approaches to find outliers and the results of the experiments are presented.

Keywords: outliers, cluster analysis, rules knowledge bases, inference process efficiency.

1 Introduction

Data mining, in general, deals with the discovery of non-trivial, hidden and interesting knowledge from different type of data. The results of such analysis are then used for making a decision by a human or program. One of the basic problems of data mining is the outlier detection. It can be seen that it found practical application i.e. detection unusual signs of disease, unauthorised tampering to servers, defective series production or even finding innovations in the texts (e.g. new articles on the subject). Finding outliers too late can distort the proper analysis of the domain and affect the errors in decision-making. An outlier is an observation of the data that deviates so much from other observations that it arouses suspicions of being generated by a different mechanism [9].

1.1 Outliers in Rule Knowledge Bases

Outliers in rule knowledge bases (KB) are rules which don't match to the rest of the knowledge (rules) in a given KB. Such rules are rare and should be the subject of deeper penetration and searching by experts, who should then work harder to extend such KB. Only thanks to such work the efficiency of a given decision support system may icreases. Lets take as an example the simple KB consisting of 23 rules with a different number of premises:

r1:h # 4 if a # 1 & b # 1
r2:h # 4 if a # 1 & b # 1 & c # 2
r3:h # 4 if a # 1 & b # 2
r4:h # 4 if a # 1 & b # 3
r5:h # 3 if b # 3 & c # 2
r6:h # 3 if a # 1 & b # 4

J.T. Yao et al. (Eds.): RSCTC 2012, LNAI 7413, pp. 206–211, 2012.

r7:h # 3 if a # 1 & c # 2
r8:h # 3 if a # 2 & c # 2
r9:h # 2 if a # 2 & b # 3
r10:h # 2 if a # 2 & b # 1
r11:h # 2 if b # 2 & c # 2
r12:h # 2 if a # 2 & b # 2
r13:h # 1 if a # 1 & c # 1
r14:h # 1 if a # 1 & c # 3
r15:h # 1 if a # 1 & c # 4
r16:h # 1 if a # 1 & c # 5
r17:h # 1 if a # 1 & c # 6
r18:h # 1 if a # 1 & c # 7
r19:h # 5 if a # 2 & c # 1
r20:h # 5 if a # 2 & c # 2
r21:h # 6 if d # 7
r22:h # 6 if d # 4 & e # 3
r23:h # 7 if f # 7 & g # 7

At first glance three last rules differ greatly from others. They are potential outliers.

1.2 Why to Mine the Outliers in Rules ?

There are different reasons for existing the outliers in rules. Such rare (unusual) rules may represent exceptional and specific cases or may be just the result of a modification of KB which was thoughtless. Outliers affect the quality of the decisions and therefore the efficiency of a decision support system. Finding rules or groups in rules should arouse interest of knowledge engineer. Thanks to contact with domain expert it will be possible to complement the areas that are not discovered enough. The problem is much difficult if we try to find outliers in complex KB[3–5]. In order to optimize the efficiency of inference processes it is possible to cluster similar rules (with the same or at least similar enough premises). It will help to make the inference process faster because instead of searching KB we only need to find a proper rules cluster. Cluster analysis is a data mining method which is very helpful when we want to find groups of similar objects (rules) [1, 2]. It should be emphasized that finding outliers in rules before clustering may increase the quality of representants of created rules clusters. Thus, the overall efficiency will grow because the created clusters are better separated what makes the process of searching such a structure easier and faster.

2 Methods for Mining Outliers

In domain literature following methods of finding outliers are well known: distribution-based, distance-based, density-based and clustering-based [6–8]. In this section most representative methods from each group are presented with their advantages and disadvantages.

2.1 Distribution-Based Method

This statistical approach to detect outliers in data works well for spaces with small number of dimensions. Two techniques are most popular: based on *mean and standard deviation* and based on *interquartile range*. Having the value of mean ($\overline{x}$) and standard deviation of a given data (*std_dev*), each object which falls outside the range $\langle \overline{x} \pm k * std\_dev \rangle$[1] we may call an outlier. The second method is based on values of first and third quartiles ($Q1$,$Q3$) for which we calculate so called *interquartile range* ($IQR = Q3 - Q1$). Then, the outlier is each data that falls outside the range $\langle Q1 - k * IQR; Q3 + k * IQR \rangle$[2]. Of course the higher the value of the k ratio is, the wider the range of so called normal data is what makes difficult to find outliers. A serious problem appears when we start to consider multi-dimensional data with various types. Very well known *Pawlak's* method of *generalized components* presented in [10] lets to convert the multivalued representant of a given rule to one number (called *rule value*) without loosing real relation in original data. If $A = \{a_1, a_2, \dots a_n\}$ is a non-empty finite set of conditional and decisional attributes, and for every $a \in A$ the set V_a is called the set of values of attribute a: $V_a = \{v_1^a, v_2^a, \dots, v_m^a\}$, we enumerate values of each attribute a_i to natural numbers: 1 to m, where m is the number of values of attribute a (0-attribute has not occurred in the rule). Thus each rule $(a_1, v_1) * (a_2, v_2)*, \dots, *(a_n, v_n)$ is converted to the vector $b_1, b_2, \dots, b_n$. Then for each rule we assign the unique value called rule's value using the formula: $b_1, b_2, ..., b_n \Rightarrow \sum_{i=1}^{n} b_i * U_i$. The values U_i are called attributes' weights and should be as follows: $U_1 = m_2 * m_3 * \dots * m_n$, $U_2 = m_3 * \dots * m_n$, ..., $U_{n-1} = m_n$, $U_n = 1$, where $m_i = cardV_{a_i}$. Thanks to this each rule in knowledge base has its individual value. Taking into account the first rule: $h\#4ifa\#1\&b\#1$ and having values of each conditional attribute ($a : 823543$, $b : 117649$), we can calculate the value of such rule as *r1*: 941192. In such case, rule $r23$ is noted as outlier because it has a value (392) much less than the treshold calculated as i.e. $\langle \overline{x} \pm k * std\_dev \rangle$ with $k = 2$.

2.2 Distance Based Methods

The method distinguishes potential outliers from data by the analysis of distance d of a given object and objects from its neighborhood. Two techniques are popular: *Ramaswamy* method[3] and *Angiulli & Pizzutti*[4].

2.3 Density Based Methods

Local Outlier Factor (LOF) method proposed by *Breunig* [11] finds local outliers based on the local density of an object's neighborhood. Small density provides objects less similar what suggest that data don't create a natural cluster so it is rather possible that they are outlier like. In a multidimensional dataset it is

[1] k the most often is equal to 2 or 3.

[2] k the most often is equal to 1.5 or 3.

[3] Based on distance d calculated for a given object and objects in its k-nearest neighbourhood outliers are M % of objects with maximum distance.

[4] Based on distance d calculated for a given object and objects in its k-nearest neighbourhood we calculate sum of distances of a given object to the rest of objects and then choose the first M maximal sums of distances as outliers [6–8].

more meaningful to assign for each object a degree of being an outlier instead of classifying object as a outlier or normal data. If the q is an object for which we calculate the value of LOF, while p is each another object with the neighbourhood of q ($N_{k-distance}$), the algorithm of LOF goes according to the following steps. First, for each object q define distance d to its k-nearest neighbours (k-distance) and determine neighbourhood of such object($N_{k-distance(q)}(q)$). Then, so-called *reachability distance* ($reach-dist$) of each object according to its neighborhood is defined. After that, so-called *local reachability density* (lrd) of each object as inversion of average value of reachability distance is also defined. Finally, for each object q we calculate $LOF(q)$: $LOF_k(q) = \frac{\sum_{p \in N_{k-distance(q)}} \frac{lrd_k(p)}{lrd_k(q)}}{|N_{k-distance(q)}|}$ As long as the LOF for a given object os is less than 1 it is not noted as outlier.

2.4 Clustering Based Methods

It is possible to find outliers in results of clustering data as a single rule or small cluster. Of course the type of clustering algorithm (hierarchical or k-optimal) brings different techniques to find outliers[5]. It would be much better to find outliers before we start clustering data, because the clustering will be faster (less number of iterations) and will built well separated clusters.

3 The Experiments

The experiments presented in this section are based on the example of rule knowledge base presented in section 1. It should be emphasized that were done on different (bigger) knowledge bases as well (the restrictions of the size of this paper don't let to include everything). The aim of the experiments were to find outliers in such a knowledge base using methods presented in section (23 different cases)2. Table 1 presents the results. The meanings of abbreviations used in this table are as follows: R-rule, A_i-methods based on values of *mean* and *standard deviation*[6], B_i- method based on *interquartile range*[7], C_i- *Rammaswamy* method[8], D_i- *Angiulli & Pizzuti*[9], E_i for LOF method[10]. As the abbreviations

[5] If we used hierarchical clustering algorithm (AHC) as outlier we treat each object or group of objects which are clustered in the last few steps. They are clustered because the algorithm says that, not because they are similar enough. If we used nonhierarchical algoritms like k-means outliers in data may distort the knowledge of the dataset a lot. It means, that if we have at least one outlier in a given dataset, and we choose this object as an initial cluster centre it is very possible to get unproper results of such clustering.

[6] Where in A_1 we take a parameter equal to 1, in A_2: 1.5, in A_3: 2 and in A_4: 3.

[7] Where $B1$ means that we take a proper parameter equal to 1, B_2: 1.5, B_3: 2 and for B_4: 3.

[8] Where C_1 is for $k = 4$, C_2 for $k = 3$ and C_3 for $k = 2$.

[9] Where D_1 is for $k = 2$, D_2 for $k = 3$, D_3 for $k = 4$.

[10] Where E_1 is for $k = \sqrt{rules\_number}$, E_2 for $k = 2$, E_3 for $k = 3$, E_4 for $k = 4$ and E_5 for $k = 5$.

Table 1. The results of the experiments

rule	A_1	B_1	A_2	B_2	A_3	B_3	A_4	B_4	C_1	C_2	C_3	D_1	D_2	D_3	E_1	E_2	E_3	E_4	E_5	F_1	F_2	F_3	F_4
r0																+	+					+	+
r1																+					+	+	+
r2															+	+		+	+				+
r3															+		+	+	+				+
r4	+	+													+	+	+	+	+			+	+
r5															+			+	+			+	+
r6																			+				
r7	+	+													+		+	+					
r8	+	+	+	+		+									+	+	+	+	+				+
r9	+	+		+											+	+		+	+				+
r10	+	+		+												+	+		+			+	+
r11	+	+		+												+	+						
r12															+		+	+					+
r13															+	+	+	+					+
r14																			+				+
r15																			+				+
r16																			+				+
r17									+					+	+	+		+	+			+	+
r18	+	+													+			+	+				
r19	+	+																					
r20	+	+	+	+		+			+	+	+	+	+	+	+	+	+	+	+		+	+	+
r21	+	+	+	+		+			+	+		+	+	+	+	+	+	+	+		+	+	+
r22	+	+	+	+		+			+	+	+	+	+	+	+	+	+	+	+		+	+	+

of F_i we mean the method based on percentages of objects with higher distance value to all other objects in dataset[11] In the table a symbol of + means that the rule was considered as an outlier. Methods based on clustering were not analyzed during the experiments. It is planed to present such results in future research. Taking into account the results presented in the table it can be easily counted how often a given rule is considered as outlier. For example, the number of cases where the first rule (labeled as $r0$) is noted as outlier is equal to 4 which means that the frequency of being outlier for this rules equals 17.39%. Respectively for the other rules frequency shall be: "$r1$" : (4, 17.39%), "$r2$" : (5, 21.74%), "$r3$" : (5, 21.74%), "$r4$" : (9, 39.13%), "$r5$" : (5, 21.74%), "$r6$" : (1), 4.35%), "$r7$" : (5, 21.74%), "$r8$" : (11, 47.83%), "$r9$" : (8, 34.78%), "$r10$" : (8, 34.78%), "$r11$" : (5, 21.74%), "$r12$" : (4), 17.39%), "$r13$" : (5, 21.74%), "$r14$" : (2, 8.70%), "$r15$" : (2, 8.70%), "$r16$" : (2, 8.70%), "$r17$" : (8, 34.78%), "$r18$" : (5, 21.74%), "$r19$" : (2, 8.70%), "$r20$" : (19, 82.61%), "$r21$" : (18, 78.26%), "$r22$" : (19, 82.61%). If we treat as outlier each rule, which frequency of being outlier is higher or equal to 80% in such a knowledge base, we will finally find three rules $r20$, $r21$ and $r22$. If we decrease the threshold to 40% the number of rules noted as outlier will increase to 5: $r20$, $r21$, $r22$ and also $r4$ as well as $r8$.

[11] Here by F_1 we mean parameter $PCT = 5\%$, F_2: $PCT = 10\%$, F_3: $PCT = 15\%$ and F_4: $PCT = 20\%$.

4 Summary

In large *KB* identification of rare cases or exceptions is increasingly needed. Nowadays, so called *outlier mining*, becomes to be a separate method of data mining. It is very crucial to find outliers in data. Maybe the data are real mistakes but maybe they are rare cases (very important kind of knowledge) which should be discovered deeper. The problem is extremely difficult especially for multidimensional data collections, because it requires taking into account many aspects such as: a different type of data, missing values in the data, the fact that data was wrong, the unique data and so on. Diversity problems related to this problem is already reflected in a huge number of researches in this field. The aim of this paper was to present a practical approach to discover outliers in rule-based knowledge bases. It is very important to know which rules are rare (and outlies from other rules) because it let the knowledge engineer to extend knowledge by working with domain experts. Finding outliers in such datasets is rather new. The very begining steps were done in this matter: introduce the concept of the finding outliers in multidimensional data, describing some methods of outlier detection, and a very first experiments on this subject.

References

1. Kaufman, L., Rousseeuw, P.J.: Finding Groups in Data: An Introduction to Cluster Analysis. John Wiley Sons, New York (1990)
2. Koronacki, J., Cwik, J.: Statistical learning systems. WNT, Warszawa (2005) (in Polish)
3. Nowak, A.: Complex knowledge bases: the structure and the inference processes, PhD thesis, Silesian University, Katowice, Poland (2009) (in Polish)
4. Nowak-Brzeziska, A., Wakulicz-Deja, A.: The choice of similarity measure and the efficiency of clustering rules in complex knowledge bases. Studia Informatica 31(2A(89)), 189–202 (2010) (in Polish)
5. Nowak-Brzeziska, A.: Mining knowledge and the effectiveness of decision support systems. Studia Informatica 32(2A(96)), 403–416 (2011) (in Polish)
6. Pearson Ronald, K.: Mining imperfect data - dealing with contamination and incomplete records, pp. I–X, 1–305. SIAM (2005)
7. Seo, S.: A Review and Comparison of Methods for Detecting Outliers in Univariate Data Sets. University of Pittsburgh (2006)
8. Cherednichenko, S.: Outlier Detection in Clustering, University of Joensuu, Department of Computer Science, Master's Thesis (2005)
9. Hawkins, D.: Identification of Outliers. Chapman and Hall (1980)
10. Pawlak, Z., Wiktor, M.: Information storage and retrieval system - mathematical foundations. Computation Center Polish Academy of Sciences (CC PAS), Warsaw (1974)
11. Breunig, et al: LOF: Identifying Density-Based Local Outliers. In: KDD (2000)

Relationships between Number of Nodes and Number of Misclassifications for Decision Trees

Igor Chikalov, Shahid Hussain, and Mikhail Moshkov

Mathematical and Computer Sciences & Engineering Division,
King Abdullah University of Science and Technology,
Thuwal 23955-6900, Saudi Arabia
{igor.chikalov,shahid.hussain,mikhail.moshkov}@kaust.edu.sa

Abstract. This paper describes a new tool for the study of relationships between number of nodes and number of misclassifications for decision trees. In addition to the algorithm the paper also presents the results of experiments with datasets from UCI ML Repository [1].

Keywords: decision trees, no. of nodes, no. of misclassifications.

1 Introduction

We have created a software system for decision trees (as well as decision rules) called DAGGER— a tool based on dynamic programming which allows us to optimize decision trees (and decision rules) relative to various cost functions such as depth (length), average depth (average length), total number of nodes, and number of misclassifications sequentially [2]. In this paper, we consider a new tool (an extension to our software) which allows us to study relationships between the number of nodes and the number of misclassifications of a decision tree. We consider the work of this tool on decision tables from UCI ML Repository [1].

The presented algorithm and its implementation in the software tool DAGGER together with similar algorithms devised by the authors (see for example [3]) can be useful for investigations in Rough Sets [4,5] where decision trees are used as classifiers [6].

2 Basic Notions

A *decision table* T is a rectangular table with m columns labeled with conditional attributes $f_1, \ldots, f_m$. The entries of the table T are nonnegative integers as the value of attributes $f_1, \ldots, f_m$. Rows of the table are pairwise different and each row is labeled with a nonnegative integer describing the value of the decision attribute d. We denote by $E(T)$ the set of attributes (columns of the table T), each of which contains different values. For $f_i \in E(T)$, let $E(T, f_i)$ be the set of values from the column f_i.

J.T. Yao et al. (Eds.): RSCTC 2012, LNAI 7413, pp. 212–218, 2012.

Let $f_{i_1}, \ldots, f_{i_t} \in \{f_1, \ldots, f_m\}$ and $a_1, \ldots, a_t$ be nonnegative integers. We denote by $T(f_{i_1}, a_1) \ldots (f_{i_t}, a_t)$ the subtable of the table T, which consists of such and only such rows of T that at the intersection with columns $f_{i_1}, \ldots, f_{i_t}$ have numbers $a_1, \ldots, a_t$ respectively. Such nonempty tables (including the table T) will be called *separable subtables* of the table T. For a subtable Θ of the table T, we will denote by $R(\Theta)$ the number of unordered pairs of rows that are labeled with different decisions. A minimum decision value which is attached to the maximum number of rows in a nonempty subtable Θ will be called the *most common decision* for Θ.

A *decision tree Γ over* the table T is a finite directed tree with root in which each terminal node is labeled with a decision. Each nonterminal node is labeled with a conditional attribute, and for each nonterminal node the outgoing edges are labeled with pairwise different nonnegative integers. Let v be an arbitrary node of Γ. We now define a subtable $T(v)$ of the table T. If v is the root then $T(v) = T$. Let v be a node of Γ that is not the root, nodes in the path from the root to v be labeled with attributes $f_{i_1}, \ldots, f_{i_t}$, and edges in this path be labeled with values $a_1, \ldots, a_t$ respectively. Then $T(v) = T(f_{i_1}, a_1), \ldots, (f_{i_t}, a_t)$.

Let Γ be a decision tree over T. We will say that Γ is a *decision tree for T* if any node v of Γ satisfies the following conditions:

- If $R(T(v)) = 0$ then v is a terminal node labeled with the most common decision for $T(v)$;
- Otherwise, either v is a terminal node labeled with the most common decision for $T(v)$, or v is labeled with an attribute $f_i \in E(T(v))$ and if $E(T(v), f_i) = \{a_1, \ldots, a_t\}$, then t edges leave node v, and these edges are labeled with $a_1, \ldots, a_t$ respectively.

Let Γ be a decision tree for T. For any row r of T, there exists exactly one terminal node v of Γ such that r belongs to the table $T(v)$. Let v be labeled with the decision b. We will say about b as about the *result of the work of decision tree Γ on r*.

3 Representation of Sets of Decision Trees

Consider an algorithm for construction of a graph $\Delta(T)$, which represents the set of all decision trees for the table T. Nodes of this graph are some separable subtables of the table T. During each step we process one node and mark it with the symbol *. We start with the graph that consists of one node T and finish when all nodes of the graph are processed.

Let the algorithm have already performed p steps. We now describe the step number $(p+1)$. If all nodes are processed then the work of the algorithm is finished, and the resulting graph is $\Delta(T)$. Otherwise, choose a node (table) Θ that has not been processed yet. Let b be the most common decision for Θ. If $R(\Theta) = 0$, label the considered node with b, mark it with symbol * and proceed to the step number $(p+2)$. If $R(\Theta) > 0$, then for each $f_i \in E(\Theta)$ draw a bundle of edges from the node Θ (this bundle of edges will be called

Fig. 1. Trivial decision tree

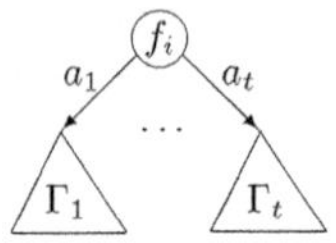

Fig. 2. Aggregated decision tree

f_i*-bundle*). Let $E(\Theta, f_i) = \{a_1, \ldots, a_t\}$. Then draw t edges from Θ and label these edges with pairs $(f_i, a_1), \ldots, (f_i, a_t)$ respectively. These edges enter into nodes $\Theta(f_i, a_1), \ldots, \Theta(f_i, a_t)$. If some of the nodes $\Theta(f_i, a_1), \ldots, \Theta(f_i, a_t)$ are not present in the graph then add these nodes to the graph. Mark the node Θ with the symbol * and proceed to the step number $(p + 2)$.

Now for each node Θ of the graph $\Delta(T)$, we describe the set of decision trees corresponding to the node Θ. We will move from terminal nodes, which are labeled with numbers, to the node T. Let Θ be a node, which is labeled with a number b. Then the only trivial decision tree depicted in Fig. 1 corresponds to the node Θ.

Let Θ be a nonterminal node (table) then there is a number of bundles of edges starting in Θ. We consider an arbitrary bundle and describe the set of decision trees corresponding to this bundle. Let the considered bundle be an f_i-bundle where $f_i \in (\Theta)$ and $E(\Theta, f_i) = \{a_1, \ldots, a_t\}$. Let $\Gamma_1, \ldots, \Gamma_t$ be decision trees from sets corresponding to the nodes $\Theta(f_i, a_1), \ldots, \Theta(f_i, a_t)$. Then the decision tree depicted in Fig. 2 belongs to the set of decision trees, which correspond to this bundle. All such decision trees belong to the considered set, and this set does not contain any other decision trees. Then the set of decision trees corresponding to the node Θ coincides with the union of sets of decision trees corresponding to the bundles starting in Θ and the set containing one decision tree depicted in Fig. 1, where b is the most common decision for Θ. We denote by $D(\Theta)$ the set of decision trees corresponding to the node Θ.

The following proposition shows that the graph $\Delta(T)$ can represent all decision trees for the table T.

Proposition 1. *Let T be a decision table and Θ a node in the graph $\Delta(T)$. Then the set $D(\Theta)$ coincides with the set of all decision trees for the table Θ.*

4 Relationships

Let T be a decision table with N rows and m columns labeled with $f_1, \ldots, f_m$ and $D(T)$ be the set of all decision trees for T. For a decision tree $\Gamma \in D(T)$ we denote by $L(\Gamma)$ *the total number of nodes of* Γ and the *number of misclassifications for decision tree* Γ *for the table* T, denoted as $\mu(\Gamma)$ is the number of rows r in T for which the result of the work of decision tree Γ on r does not equal to the decision attached to the row r. It is clear that the minimum value of L and μ on $D(T)$ are equal to one and zero, respectively, whereas $2N - 1$ and N are the respective upper bound on L and μ.

We denote $B_L = \{1, 2, \ldots, 2N-1\}$ and $B_\mu = \{0, 1, \ldots, N\}$. We now define two functions $\mathcal{G}_T : B_L \to B_\mu$ and $\mathcal{F}_T : B_\mu \to B_L$ as following:

$$\mathcal{G}_T(n) = \min\{\mu(\Gamma) : \Gamma \in D(T), L(\Gamma) \leq n\}, \quad \forall n \in B_L, \quad \text{and}$$
$$\mathcal{F}_T(n) = \min\{L(\Gamma) : \Gamma \in D(T), \mu(\Gamma) \leq n\}, \quad \forall n \in B_\mu.$$

We now describe the algorithm to construct the function $\mathcal{G}_\Theta$ for every node (subtable) Θ from the graph $\Delta(T)$. For simplicity, we assume that $\mathcal{G}_\Theta$ is defined on the set B_L. We begin from the terminal nodes of $\Delta(T)$ and move upward to the root node T.

Let Θ be a terminal node. It means that all nodes of Θ are labeled with the same decision b and the decision tree Γ_b as depicted in Fig. 1 belongs of $D(\Theta)$. It is clear that $L(\Gamma_b) = 1$ and $\mu(\Gamma_b) = 0$ for the table Θ. Therefore, $\mathcal{G}_\Theta(n) = 0$ for any $n \in B_L$.

Let us consider a nonterminal node Θ and a bundle of edges labeled with pairs $(f_i, a_1), \ldots, (f_i, a_t)$, which start from this node. Let these edges enter into nodes $\Theta(f_i, a_1), \ldots, \Theta(f_i, a_t)$, respectively, to which the functions $\mathcal{G}_{\Theta(f_i,a_1)}, \ldots, \mathcal{G}_{\Theta(f_i,a_t)}$ are already attached.

We correspond to this bundle (f_i-bundle) the function $\mathcal{G}_\Theta^{f_i}$, for any $n \in B_L$, $n > t$, $\mathcal{G}_\Theta^{f_i}(n) = \min \sum_{j=1}^{t} \mathcal{G}_{\Theta(f_i,aj)}(n_j)$ where the minimum is taken over all $n_1, \ldots, n_t$ such that $1 \leq n_j \leq 2N-1$ for $j = 1, \ldots, t$ and $n_1 + \cdots + n_t + 1 \leq n$. [Computing $\mathcal{G}_\Theta^{f_i}$ is a nontrivial task. We describe the method in detail in the following Sect. 4.1.] The minimum number of nodes for decision tree such that f_i is attached to the root is $t+1$ therefore for $n \in B_L$, such that $1 \leq n \leq t$, $\mathcal{G}_\Theta^{f_i}(n) = N(\Theta) - N_{\mathrm{mcd}}(\Theta)$, where $N_{\mathrm{mcd}}(\Theta)$ is the number of nodes in Θ labeled with *the most common decision* for Θ.

It is not difficult to prove that for all $n \in B_L, n > t$,

$$\mathcal{G}_\Theta(n) = \min\left\{\mathcal{G}_\Theta^{f_i}(n) : f_i \in E(\Theta)\right\}.$$

We can use the following proposition to construct the function $\mathcal{F}_T$ (we omit the proof).

Proposition 2. *For any $n \in B_\mu$, $\mathcal{F}_T(n) = \min\{p \in B_L : \mathcal{G}_T(p) \leq n\}$.*

Note that to find the value $\mathcal{F}_T(n)$ for some $n \in B_\mu$ it is enough to make $O(\log|B_L|) = O(\log(2N-1))$ operations of comparisons.

4.1 Computing $\mathcal{G}_\Theta^{f_i}$

Let Θ be a nonterminal node in $\Delta(T)$, $f_i \in E(\Theta)$, and $E(\Theta, f_i) = \{a_1, \ldots, a_t\}$. Furthermore, we assume that the functions $\mathcal{G}_{\Theta(f_i,a_j)}$, $j = 1, \ldots, t$, have already been computed. Let the values for $\mathcal{G}_{\Theta(f_i,a_j)}$ be given by the tuple of pairs $\left((1, \mu_1^j), (2, \mu_2^j), \ldots, (2N-1, \mu_{2N-1}^j)\right)$. We need to compute $\mathcal{G}_\Theta^{f_i}$ for all $n \in B_L$:

$$\mathcal{G}_\Theta^{f_i}(n) = \min \sum_{j=1}^{t} \mathcal{G}_{\Theta(f_i,a_j)}(n_j) \ \text{ for } 1 \leq n_j \leq 2N-1, \quad \text{s.t.,} \quad \sum_{i=1}^{t} n_i + 1 \leq n.$$

We construct a layered directed acyclic graph (DAG) $\delta(\Theta, f_i)$ to compute $\mathcal{G}_\Theta^{f_i}$ as following.

The DAG $\delta(\Theta, f_i)$ contains nodes arranged in $t+1$ layers $(l_0, l_1, \ldots, l_t)$. Each node has a pair of labels and each layer l_j $(1 \leq j \leq t)$ contains at most $j(2N-1)$ nodes. The first entry of labels for nodes in a layer l_j is an integer from $\{1, 2, \ldots, j(2N-1)\}$. The layer l_0 contains only one node labeled with $(0, 0)$.

Each node in a layer l_j $(0 \leq j < t)$ has exactly $2N-1$ outgoing edges to nodes in layer l_{j+1}. These edges are labeled with the corresponding pairs in $\mathcal{G}_{\Theta(f_i, a_{j+1})}$. A node with label x as a first entry in its label-pair in a layer l_j connects to nodes with labels $x+1$ to $x+2N-1$ (as a first entry in their label-pairs) in layer l_{j+1}, with edges labeled as $(1, \mu_1^{j+1}), (2, \mu_2^{j+1}), \ldots, (2N-1, \mu_{2N-1}^{j+1})$, respectively.

The function $\mathcal{G}_\Theta^{f_i}(n)$ for $n \in B_L$ can be easily computed using the DAG $\delta(\Theta, f_i)$ for $\Theta \in \Delta(T)$ and for the considered bundle of edges for the attribute $f_i \in E(\Theta)$ as following:

Each node in layer l_1 gets its second value copied from the corresponding second value in incoming edge label to the node (since there is only one incoming edge for each in layer l_1). Let (k, μ) be a node in layer l_j, $2 \leq j \leq t$. Let $E = \{(v_1, \mu_1), (v_2, \mu_2), \ldots, (v_r, \mu_r)\}$ be the set of incoming nodes to (k, μ) such that $(\alpha_1, \beta_1), (\alpha_2, \beta_2), \ldots, (\alpha_r, \beta_r)$ are the labels of these edges between the nodes in E and $(k, \cdot)$, respectively. It is clear that $k = v_i + \alpha_i$, $1 \leq i \leq r$. Then $\mu = \min_{1 \leq i \leq r}\{\mu_i + \beta_i\}$. We do this for every node layer-by-layer till all nodes in $\delta(\Theta, f_i)$ have received their second label.

Once we finish computing the second value of label pairs of layer l_t, we can use these labels to compute $\mathcal{G}_\Theta^{f_i}(n)$. It is clear that the nodes in layer l_t have labels as $(t, \mu(t)), \ldots, (t(2N-1), \mu(t(2N-1)))$, respectively. For $n \in B_L$ such that $1 \leq n \leq t$, $\mathcal{G}_\Theta^{f_i}(n) = N(\Theta) - N_{\mathrm{mcd}}(\Theta)$. For $n \in B_L$, such that $t < n \leq 2N-1$, $\mathcal{G}_\Theta^{f_i}(n) = \min_{t < k \leq n} \mu(k-1)$.

Let us consider the time complexity of the considered algorithm. The DAG $\delta = \delta(\Theta, f_i)$ has $t+1$ layers and each layer l_j has at most $j(2N-1)$ nodes. Therefore total number of nodes in δ is $O(t^2 N)$. Since every node has $2N-1$ outgoing edges (except the nodes in layer l_t), the number of edges in δ is $O(t^2 N^2)$. So, to build the graph δ, we need $O(t^2 N^2)$ time (proportional to the number of nodes and edges in δ). To find the second labels we need a number of additions and comparisons bounded from above by the number of edges – $O(t^2 N^2)$. Similarly, to find values of $\mathcal{G}_\Theta^{f_i}$ we need $O(N^2)$ comparisons. Therefore, the total time is $O(t^2 N^2)$.

5 Experimental Results

We performed several experiments on datasets (decision tables) acquired from UCI ML Repository [1]. The resulting plots are depicted in Figs. 3 and 4.

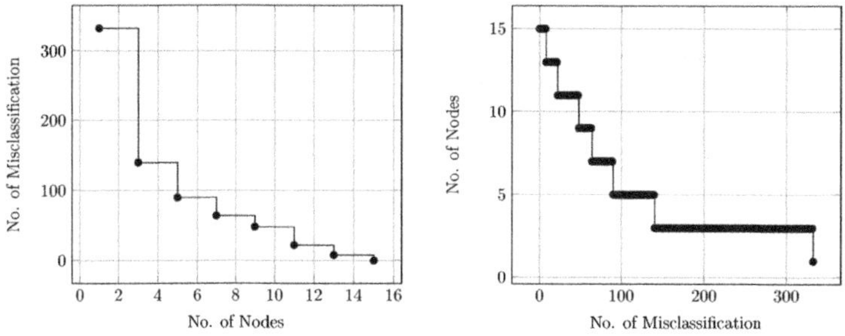

Fig. 3. Relationship plots for TIC-TAC-TOE dataset (9 attributes and 959 rows)

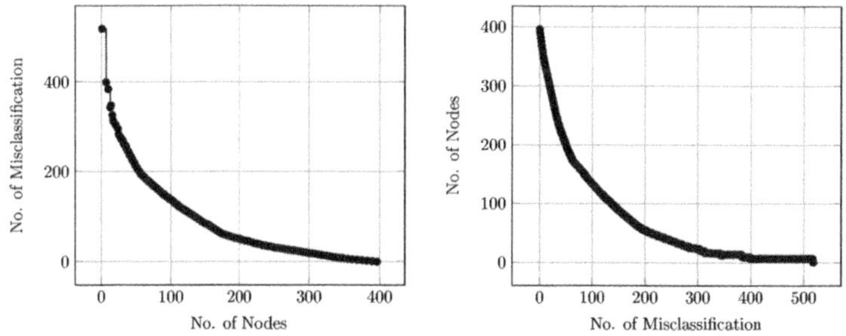

Fig. 4. Relationship plots for CARS dataset (6 attributes and 1729 rows)

6 Conclusions

This paper presents a tool for studying the relationships between the number of nodes and number of misclassifications for decision trees. Further studies will be connected with the extensions of this tool to other cost functions such as average depth and number of terminal nodes.

References

1. Frank, A., Asuncion, A.: UCI Machine Learning Repository (2010)
2. Alkhalid, A., Amin, T., Chikalov, I., Hussain, S., Moshkov, M., Zielosko, B.: Dagger: a tool for analysis and optimization of decision trees and rules. In: Computational Informatics, Social Factors and New Information Technologies: Hypermedia Perspectives and Avant-Garde Experiencies in the Era of Communicability Expansion, pp. 29–39. Blue Herons (2011)
3. Chikalov, I., Hussain, S., Moshkov, M.: Relationships between Depth and Number of Misclassifications for Decision Trees. In: Kuznetsov, S.O., Ślęzak, D., Hepting, D.H., Mirkin, B.G. (eds.) RSFDGrC 2011. LNCS, vol. 6743, pp. 286–292. Springer, Heidelberg (2011)
4. Pawlak, Z.: Theoretical Aspects of Reasoning about Data. Kluwer Academic Publishers, Dordrecht (1991)

5. Skowron, A., Rauszer, C.: The discernibility matrices and functions in information systems. In: Slowinski, R. (ed.) Intelligent Decision Support. Handbook of Applications and Advances of the Rough Set Theory, pp. 331–362. Kluwer Academic Publishers, Dordrecht (1992)
6. Nguyen, H.S.: From optimal hyperplanes to optimal decision trees. Fundamenta Informaticae 34(1-2), 145–174 (1998)

Study on Feature Trajectory for Web Video Event Mining

Xiao Wu, Chengde Zhang, and Qiang Peng

School of Information Science and Technology,
Southwest Jiaotong University, Chengdu, 610031, P.R. China
{wuxiaohk,qpeng}@home.swjtu.edu.cn, dogromantic@sina.com

Abstract. The explosive growth of web videos prompts an urgent demand on efficient grasping the major events. Unfortunately, the unique characteristics of web video scenario, such as the limited number of features, the unavoidable errors of near-duplicate keyframe detection, the noisy text information, make web video event mining a challenging task. In this paper, we first explore the properties of textual feature trajectory from title/tags and visual feature trajectory induced from near-duplicate keyframes. Based on the study, we propose web video event mining solution by fusion of textual and visual feature trajectories, which takes into account peak time difference, time span overlap, and trajectory distance. Experiments on a large number of web videos from YouTube demonstrate the proposed method achieves good performance for web video event mining.

Keywords: Feature trajectory, near-duplicate keyframes, event mining, web videos.

1 Introduction

Recently, feature trajectory has been demonstrated promising performance for topic detection and tracking [4],[6], which is modeled as a burst of activities by incorporating temporal information. The evolution of text is modeled as a bunch of feature trajectories in a 2D space of time and frequency for event clustering. Nevertheless, different from traditional text documents, the text information (title and tags) of web videos is much less than documents. In addition, they are usually noisy, ambiguous, incomplete and even misleading. Hot words also lead to poor representation of web videos. Therefore, the textual feature trajectory derived from limited title and tags of web videos may have different properties from traditional documents.

On the contrary, important visual shots are frequently inserted into related videos as a reminder or support of viewpoints, acting as hot terms in text field. These near-duplicate shots/keyframes carry useful video content, and can be used to group videos of similar theme into events. Although a trial of visual feature trajectory based on near-duplicate keyframes has been explored in [12] to mine the event structure of web videos, the feasibility and robustness of visual

J.T. Yao et al. (Eds.): RSCTC 2012, LNAI 7413, pp. 219–228, 2012.

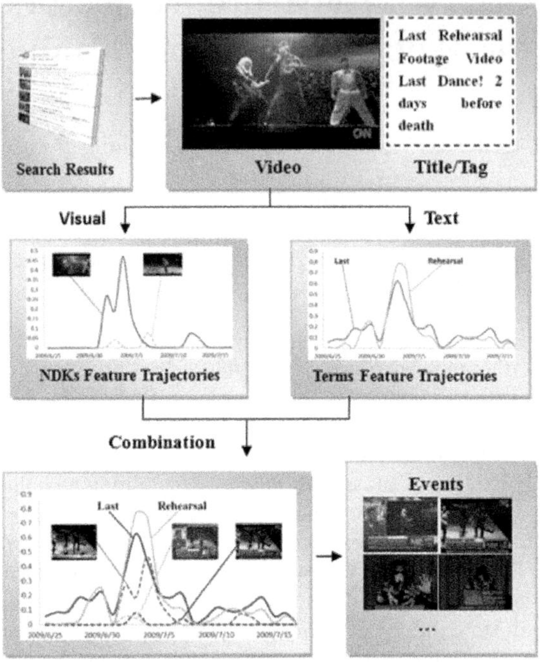

Fig. 1. Framework of web video event mining by textual and visual feature trajectories

feature trajectory hasn't been fully analyzed. Moreover, the relationship between textual and visual feature trajectories has not yet been researched.

In this paper, we analyze the characteristics of textual and visual feature trajectories, explore their feasibilities for web video event mining, and finally combine them to improve the performance. The framework of the proposed work is shown in Fig. 1.

2 Related Work

2.1 Topic Detection and Tracking

Event mining belongs to the task of topic detection and tracking (TDT) in information retrieval. The goal of TDT is to detect new topics and track known events in text news streams. Many works have been done in text areas [2],[4],[5],[6],[11]. Recently, TDT research has been extended to multimedia area. Topics were tracked with visual duplicates and semantic concepts [7]. With textual correlation and keyframe matching, topic clusters were grouped in [3] and news stories from different TV channels were linked in [6]. Topic discovery was deployed by constructing the duality between stories and textual-visual concepts through bipartite graph [8],[14]. With the assistance of near-duplicate keyframe constraints, news stories were clustered into topics by constraint based co-clustering [13].

2.2 Feature Trajectory

Feature trajectory is an important feature in the text field. In [6], the characteristics of word trajectory were analyzed to identify important and less-report, periodic and aperiodic words, from the perspective of time-series word signal. The idea of mining hot terms by timeline analysis was presented in [2]. Hot topics were further extracted using multidimensional sentence modeling grounded on hot terms. In a similar spirit, Google Trends was used to predict the milestone events of a topic in [10]. A parameter free probabilistic model was proposed to analyze time-varying features and detect bursty events from text streams [4]. However, these works are based on full-text analysis especially on news articles, in which the text articles are more informative and with less noise. For the web video field, events were discovered by text co-occurrence and visual near-duplicate feature trajectory in [12]. A trajectory-based approach was presented in [1] to discover, track, monitor and visualize web video topics. These works focused on textual or visual feature trajectory separately. There is no a comprehensive study on their properties and without an exploration about the relationship between them.

3 Textual vs. Visual Feature Trajectory

Feature trajectory is modeled as the feature distribution along the timeline in a two dimensional space with one dimension as time, and the other as feature weight. The feature trajectory y_f can be written as the sequence:

$$y_f =< y_f(t_i), y_f(t_{i+1}), \ldots, y_f(t_{i+n}) > \tag{1}$$

where $y_f(t_i)$ is a measure of feature f at time unit t_i and $n \geq 0$, a time unit is set to a time span, such as one day, $y_f(t_i)$ is calculated according to *df-idf*:

$$y_f(t_i) = \frac{df_f(t_i)}{N(t_i)} \times \log \frac{N}{df_f} \tag{2}$$

where $df_f(t_i)$ is the number of videos containing feature f at day t_i, df_f is the total number of videos including feature f over all times, $N(t_i)$ is the number of videos on day t_i, and N is the total number of videos over all time.

Under the scenario of web videos, we will study the properties of textual feature trajectory derived from text words and visual feature trajectory from near-duplicate keyframes (NDKs) in the following sections.

3.1 Studies on Textual Feature Trajectory

Title and tags are the most direct and representative feature to briefly summarize the content of a web video. An event can be concisely described by a few representative words. For example, “last”, “rehearsal”, “London” and “concert” could be the key terms of “Michael Jackson’s last rehearsal for London

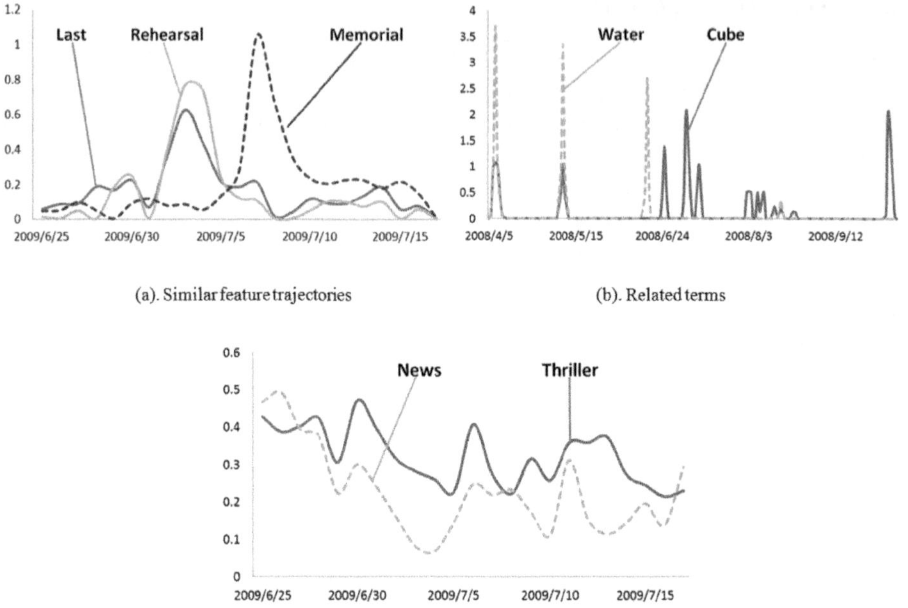

(a). Similar feature trajectories

(b). Related terms

(c). Similar trends express different themes

Fig. 2. Some examples of textual feature trajectories

concert" event. However, the text information of web videos is much less than documents and they are usually noisy, ambiguous, and incomplete. The textual feature trajectory for web videos may demonstrate different characteristics. Some observations are listed as follows:

• Closely related terms have similar feature trajectories

Some closely related terms are appeared frequently. For example, terms "Last" and "Rehearsal" are commonly accompanied for videos related to the event "Michael Jackson's last rehearsal video was released". As shown in Fig. 2(a), we can see that their trajectories show a couple of bursts at several time units, and are basically consistent. On the contrary, the trajectory of another theme on "Memorial" has different patterns with "Last" and "Rehearsal".

• The feature trajectories of related terms may demonstrate inconsistent trends

For the landmark of Beijing Olympics Games, "Water Cube", the terms "water" and "cube" have close relationship. However, their feature trajectories are totally different, as shown in Fig. 2(b). This building can be expressed in multiple forms, such as water stadium, cube center, cube stadium, and so on, causing disparate trajectories.

• Unrelated terms may have similar feature trajectories

There is the possibility that terms are completely irrelevant even though their feature trajectories look similar, such as terms "news" and "thriller" shown in Fig. 2(c).

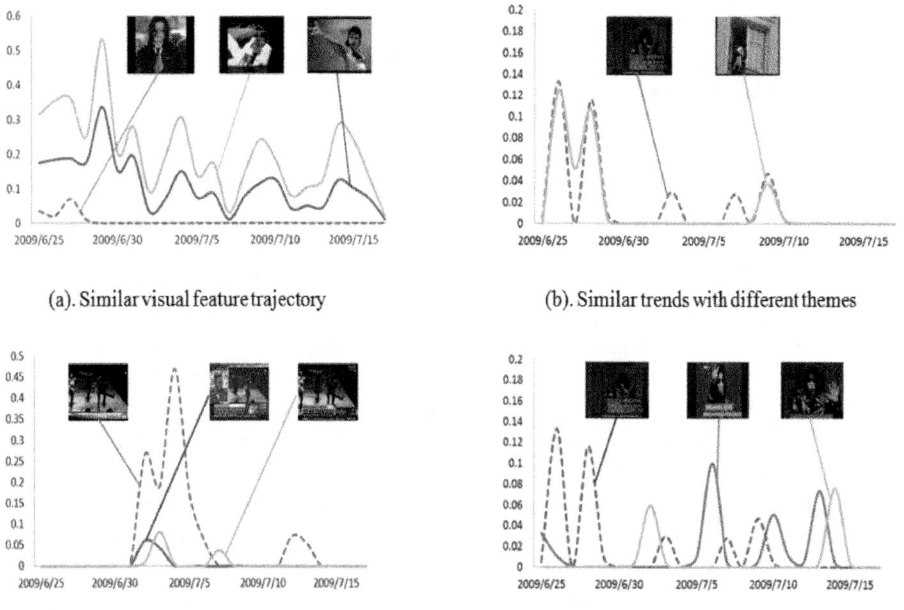

(a). Similar visual feature trajectory

(b). Similar trends with different themes

(c). Similar scenes have multiple trajectories

(d). Similar scenes have multiple trajectories

Fig. 3. Some examples of visual feature trajectory

3.2 Studies on Visual Feature Trajectory

In addition to title and tags, the essential part of web videos is the visual content itself. However, there are three major differences between keyframes and traditional words: (1) the number of NDKs is significantly smaller than the number of words in documents; (2) there exist a very large number of Non-NDK that only appear once in the corpus; (3) due to the error caused by NDK detection, some NDKs may be either missed or separated into several clusters, while others may be falsely detected. The impact of these differences has not been studied previously, which could cause the visual feature trajectory demonstrated certain distinct properties. In this section, we will explore the characteristics of visual feature trajectory.

• Relevant NDKs have similar visual feature trajectories

Different NDKs on a same event could have similar visual feature trajectories. The keyframes of "Jackson is dancing in a MTV" and "Cover of MTV" share similar visual feature trajectory distributions over time, as shown in Fig. 3(a). They belong to the event "A tribute of Michael Jackson death". In contrast, the keyframe of "Jackson is praying" shows different feature distributions from them, which belongs to the event "Sadness of Michael Jackson death".

- Similar trends represent different themes

Similar trends do not guarantee the closeness between two NDKs. The features shown in Fig. 3(b) demonstrate somewhat consistent trends. However, they belong to two themes.

- Similar scenes have multiple trajectories

Since NDK detection is still a challenging task, the detected NDK groups are not perfect. In one hand, some NDKs belonging to one group may be falsely detected to form several separated clusters. On the other hand, a few NDKs are missed and falsely treated as Non-NDK. Fig. 3(c) shows an example of three NDKs on "Jackson is dancing in the last rehearsal". Unfortunately, they are falsely treated as three NDK clusters after NDK detection, leading to three separated trajectories. It makes the current visual feature trajectories deviated from the ideal one. Similarly, another example on "News press for London Concert" is shown in Fig. 3(d).

We can see that either textual or visual information might not be sufficient for web video event mining, because both of them display inconsistent characteristics. These features make web video event mining a challenging task. A good news is that the textual and visual feature trajectories could have good overlap in terms of timeline and trajectory curves. For example, in Fig. 4, textual feature trajectories "Last" and "Rehearsal" show similar trends with visual feature trajectory (the scene of "Michael Jackson is dancing"). At the same time, textual and visual information could complement each other. Robust and reasonable approaches should combine both textual and visual features to mine the related web videos into events.

4 Web Video Event Mining by Fusion of Feature Trajectories

In this section, we explore the feasibility of mining events by combining both textual and visual feature trajectories. A burst event is usually accompanied with a burst of related web videos. Generally speaking, a burst feature will exhibit peak-like trajectory, indicating a large number of hot terms or critical video scenes appear. It is apparent that a feature becomes meaningful at special time points, and grouping feature trajectories with similar distributions of feature peaks gives clue to arising of events. Textual and visual feature trajectories provide constructive hints, either overlap or complement, for event mining. In this way, meaningful events could be identified by grouping textual and visual feature trajectories.

To measure the similarity of two feature trajectories, three factors are taken into consideration: peak time difference (*ptd*), time span overlap (*tso*), and trajectory distance (*td*). The peak time of a feature trajectory potentially indicates the time point that an event was broken out. The peak time difference should be minor if two features are closely related, denoting that they are happened at almost the same time. The time span overlap measures the time coverage of feature trajectories, implying the degree of time co-occurrence for two features.

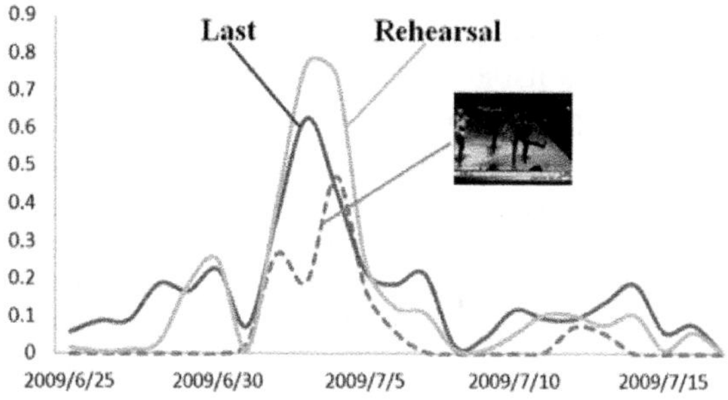

Fig. 4. Textual and visual feature trajectories have good overlap

Representative terms and video shots of an event should have a good overlap in timeline, since they appeared together. The trajectory distance evaluates the overall difference of feature trajectories.

In this paper, textual and visual feature trajectories are treated equally. They are combined to mine the web video events. The similarity of two feature trajectories (either textual or visual) y_{fi} and y_{fj} is calculated as follows, which is the product of the three factors:

$$S(y_{fi}, y_{fj}) = ptd(y_{fi}, y_{fj}) \times tso(y_{fi}, y_{fj}) \times td(y_{fi}, y_{fj}) \quad (3)$$

$$S(y_{fi}, y_{fj}) = |P_{fi} - P_{fj}| \times \frac{|y_{fi} \bigcap y_{fj}|}{max(|y_{fi}|, |y_{fj}|)} \times td(y_{fi}, y_{fj}) \quad (4)$$

$$td(y_{fi}, y_{fj}) = \sqrt{\sum_{k=1}^{m} (y_{fi}(t_k) - y_{fj}(t_k))^2} \quad (5)$$

where P_{fi} and P_{fj} represent the peak time of feature trajectory y_{fi} and y_{fj} respectively, that is the time point where the value of feature trajectory reaches the maximum along the timeline. $|y_{fi}|$ is the number of days where feature $y_{fi} > 0$. $|y_{fi} \bigcap y_{fj}|$ is the number of overlapped days containing both features. $td(y_{fi}, y_{fj})$ is the trajectory distance of trajectories of y_{fi} and y_{fj}, which is measured by the *Euclidean* distance. The value of $y_{fi}(t_k)$ is a *df-idf* score.

According to the feature trajectory similarity calculated by the above formula, *K-means* clustering algorithm is adopted to group feature trajectories into clusters. The related videos corresponding to these clustered feature trajectories are finally grouped into the events.

5 Experiment

5.1 Dataset

The dataset used in [12] is the dataset for evaluation, which consists of 19,972 web videos. We selected 10 topics from this dataset to evaluate the proposed approach. The videos were collected from YouTube by issuing multiple text queries and removing repeated videos. Most videos were collected in May, 2009. The dataset information is listed in Table 1.

To guarantee the good performance of near-duplicate keyframe detection, local keypoint based detection solution was deployed. Local points were detected by Harris-Laplace and described by SIFT [9]. Clustering was performed to quantize the keypoints into a visual dictionary (20,000 clusters). Each keyframe was encoded as a bag of words. The public available tool proposed in [15] was adopted to detect the NDKs. The detected NDKs were further grouped to form clusters by transitive closure, which represent the same visual scene. Due to noisy user-supplied tag information, text words were undergone a serial of data preprocessing steps (such as word stemming, special character removal, and so on). The words and NDKs with low frequency were removed since most of them are noises or trivial ones.

We use *Precision* and *Recall* to evaluate the performance of event mining, which is defined as:

$$Precision = \frac{G_i^+}{C_i}, Recall = \frac{G_i^+}{G^i} \tag{6}$$

where G_i^+ is the number of correctly grouped positive videos for cluster C_i, and G_i is the number of positive samples in ground truth.

5.2 Performance Evaluation

To evaluate the performance of web video event mining, a simplified version of the proposed method in [6] is treated as the baseline. Word feature trajectories are first extracted based on a normalized *df-idf* score. Highly correlated word features are grouped to construct events by mapping word sets to video sets. We also compare the performance of the approach in [12]. We test the performance of event mining using only textual feature trajectory (TFT), visual feature trajectory (VFT) and the combination (TFT+VFT). The clustering is performed by *K-means* algorithm, where k is designed according to the number of events in ground truth.

The performance comparison is listed in Table 1. We can see that the performance of the baseline is poor, which indicates that text information alone is not sufficient to mine events. Generally, the results based on visual feature trajectory have a higher precision. A set of representative visual shots are often accompanied with an important event. Therefore, the NDKs are good cues for grouping related videos into events. On the contrary, approaches with text feature trajectory have a higher recall. Text words are relatively general, broad, and noisy. Clustering with texts could bring more related videos while inducing noises at the same time. It has better recall. The baseline and TFT are both on textual feature trajectory. TFT has better performance, since it takes into account peak time difference, time span overlap and trajectory distance, which improves the performance, while baseline only considers the trajectory distance. Overall, the combination of textual and visual feature trajectory outperforms the baseline and technique in [12]. The recall

Table 1. Performance Evaluation

Topic	Video	[6]		[12]		TFT		VFT		TFT+VFT	
		P	R	P	R	P	R	P	R	P	R
Beijing Olympics	1,098	0.54	0.11	0.64	0.18	0.37	0.41	0.79	0.49	0.63	0.50
Mumbai terror attack	423	0.31	0.14	0.49	0.19	0.31	0.55	0.49	0.23	0.37	0.53
Russia Georgia war	749	0.58	0.11	0.72	0.15	0.53	0.32	0.65	0.14	0.60	0.35
Virginia tech massacre	683	0.76	0.05	0.73	0.33	0.76	0.32	0.82	0.23	0.75	0.53
Beijing Olympic torch relay	652	0.52	0.41	0.52	0.20	0.32	0.66	0.58	0.21	0.47	0.51
Sichuan earthquake	1,458	0.52	0.05	0.76	0.47	0.51	0.20	0.77	0.34	0.70	0.47
California wildfires	426	0.46	0.12	0.68	0.18	0.46	0.33	0.69	0.15	0.60	0.36
Kosovo independence	524	0.66	0.07	0.78	0.09	0.81	0.19	0.91	0.07	0.83	0.23
Iran nuclear program	1,056	0.60	0.07	0.83	0.10	0.62	0.41	0.66	0.07	0.63	0.47
Michael Jackson death	2,850	0.64	0.08	0.83	0.11	0.54	0.31	0.61	0.12	0.56	0.37
Average		0.56	0.12	0.70	0.20	0.52	0.37	0.70	0.21	0.61	0.43

has been apparently improved, though precision slightly drops. It proves that textual and visual information complement each other. The combination can group more highly related videos together.

6 Conclusion

Due to the unique characteristics of web video scenario, such as the limited number of features, the unavoidable error of near-duplicate keyframe detection, the noisy text information, makes web video event mining a challenging task. At the same time, these properties make the textual and visual feature trajectories demonstrated different appearance with traditional documents. In this paper, we study the textual and visual feature trajectories separately, and then explore their combination for web video event mining. Experiments on a large scale web video dataset from YouTube demonstrate the good potential of feature trajectory for event mining.

While encouraging, we can see that the recall is still unsatisfactory, partially because we ignore the features with relatively low frequency. In addition to the limited and noisy text and visual resources from web videos themselves, we can resort to external corpus, such as news websites for more useful information. There are a couple of issues not being addressed in our current work and worth for future consideration.

Acknowledgement. The work described in this paper was supported by the National Natural Science Foundation of China (No. 61071184, 60972111, 6103600), Research Funds for the Doctoral Program of Higher Education of China (No. 20100184120009, 20120184110001), Program for Sichuan Provincial Science Fund for Distinguished Young Scholars (No. 2012JQ0029), the Fundamental Research Funds for the Central Universities (Project no. SWJTU09CX032, SWJTU10CX08, SWJTU11ZT08) and Open Project Program of the National Laboratory of Pattern Recognition (NLPR).

References

1. Cao, J., Ngo, C.-W., Zhang, Y.-D., Li, J.-T.: Tracking web video topics: discovery, visualization and monitoring. J. IEEE TCSVT (2011)
2. Chen, K.Y., Luesukprasert, L., Chou, S.T.: Hot topic extraction based on timeline analysis and multi-dimensional sentence modeling. J. IEEE TKDE 19(8), 1016–1025 (2007)
3. Duygulu, P., Pan, J.-Y., Forsyth, D.A.: Towards auto-documentary: tracking the evolution of news Stories. In: ACM MM 2004, pp. 820–827 (2004)
4. Fung, G.P.C., Yu, J.X., Yu, P.S., Lu, H.: Parameter free bursty events detection in text streams. In: VLDB, pp. 181–192 (2005)
5. Fung, G.P.-C., Yu, J.X., Liu, H., Yu, P.S.: Time-dependent event hierarchy construction. In: KDD 2007, pp. 300–309 (2007)
6. He, Q., Chang, K., Lim, E.-P.: Analyzing feature trajectories for event detection. In: SIGIR 2007, pp. 207–214 (2007)
7. Hsu, W.H., Chang, S.-F.: Topic tracking across broadcast news videos with visual duplicates and semantic concepts. In: ICIP 2006, pp. 141–144 (2006)
8. Liu, L., Sun, L.-F., Rui, Y., et al.: "Web video topic discovery and tracking via bipartite graph reinforcement model. J. WWW 2008, 1009–1018 (2008)
9. Lowe, D.: Istinctive image features from scale-invariant key points. J. IJCV 60, 91–110 (2004)
10. Tan, S., Tan, H.-K., Ngo, C.-W.: Topical summarization of web videos by visual-text time-dependent alignment. In: ACM MM 2010, pp. 1095–1098 (2010)
11. Wang, X.-H., Zhai, C.-X., Hu, X., Sproat, R.: Mining correlated bursty topic patterns from coordinated text streams. In: KDD 2007, pp. 784–793 (2007)
12. Wu, X., Lu, Y.-J., Ngo, C.-W., Peng, Q.: Mining Event Structures from Web Vide-os. J. IEEE Multimedia 18(1), 38–51 (2011)
13. Wu, X., Ngo, C.-W., Hauptmann, A.G.: Multimodal news story clustering with pairwise wisual near-duplicate constraint. J. IEEE TMM 10(2), 188–199 (2008)
14. Wu, X., Ngo, C.-W., Li, Q.: Threading and autodocumenting news topics. J. IEEE Signal Processing Magazine, 59–68 (2006)
15. Zhao, W.-L., Wu, X., Ngo, C.-W.: On the annotation of web videos by efficient near-duplicate search. J. IEEE TMM 12(5), 448–461 (2010)
16. Zhai, Y., Shah, M.: Tracking news stories across different sources. In: ACM MM 2005, pp. 2–10 (2005)

The Triangle Inequality versus Projection onto a Dimension in Determining Cosine Similarity Neighborhoods of Non-negative Vectors*

Marzena Kryszkiewicz

Institute of Computer Science, Warsaw University of Technology
Nowowiejska 15/19, 00-665 Warsaw, Poland
mkr@ii.pw.edu.pl

Abstract. In many applications, objects are represented by non-negative vectors and cosine similarity is used to measure their similarity. It was shown recently that the determination of the cosine similarity of two vectors can be transformed to the problem of determining the Euclidean distance of normalized forms of these vectors. This equivalence allows applying the triangle inequality to determine cosine similarity neighborhoods efficiently. Alternatively, one may apply the projection onto a dimension to this end. In this paper, we prove that the triangle inequality is guaranteed to be a pruning tool, which is not less efficient than the projection in determining neighborhoods of non-negative vectors.

1 Introduction

In many applications, especially in text mining, biomedical engineering and chemistry, cosine similarity is often used to find objects (nearest neighbors) most similar to a given one. Objects themselves are frequently represented by non-negative vectors. The determination of nearest neighbors is challenging if analyzed vectors are high dimensional. In the case of distance metrics, one may apply the triangle inequality to quickly prune large numbers of objects that certainly are not nearest neighbors of a given vector [1,3,4,5,6]. While the cosine similarity does not preserve the triangle inequality, it was shown recently that the problem of determining the cosine similarity of two vectors can be transformed to the problem of determining the Euclidean distance of normalized forms of these vectors [2]. This equivalence allows applying the triangle inequality to determine cosine similarity neighborhoods efficiently. Alternatively, one may apply the projection onto a dimension to this end. In this paper, we prove that the triangle inequality is guaranteed to be a pruning tool, which is not less efficient than the projection in determining neighborhoods of non-negative vectors.

* This work was supported by the National Centre for Research and Development (NCBiR) under Grant No. SP/I/1/77065/10 devoted to the Strategic scientific research and experimental development program: 'Interdisciplinary System for Interactive Scientific and Scientific-Technical Information'.

J.T. Yao et al. (Eds.): RSCTC 2012, LNAI 7413, pp. 229–236, 2012.

Our paper has the following layout. In Section 2, we recall basic notions and the equation relating the Euclidean distance and the cosine similarity [2]. In Section 3, we recall how the problem of determining a cosine similarity neighborhood can be transformed to the problem of determining a neighborhood w.r.t the Euclidean distance [2]. The usage of the triangle inequality for efficient pruning of non-nearest neighbors is recalled in Section 4 [3,4]. In Section 5, we present an analogous approach to pruning such vectors by using the projection of vectors onto a dimension. Section 6 contains the main contribution of this paper, which consists in showing that for any dimension one may apply pruning of non-nearest neighbors in a set of non-negative normalized vectors by means of the triangle inequality, which is not less efficient than the projection onto this dimension. Section 7 summarizes our work.

2 The Euclidean Distance and the Cosine Similarity

In the paper, we consider vectors of the same dimensionality, say n. A vector u will be also denoted as $[u_1, \ldots, u_n]$, where u_i is the value of the i-th dimension of u, $i = 1..n$. A vector is called *non-negative* if all its dimensions are not negative.

The *Euclidean distance* between vectors u and v is denoted by $Euclidean(u,v)$ and is defined as $\sqrt{\sum_{i=1..n}(u_i - v_i)^2}$. The Euclidean distance preserves *the triangle inequality*; that is, for any vectors u, v, and r, $Euclidean(u,r) \leq Euclidean(u, v) + Euclidean(v,r)$ (or, alternatively $Euclidean(u,v) \geq Euclidean(u, r) - Euclidean(v, r)$).

The *cosine similarity* between vectors u and v is denoted by $cosSim(u,v)$ and is defined as the cosine of the angle between them; that is,

$$cosSim(u,v) = \frac{u \cdot v}{|u||v|}, \text{ where :}$$

- $u \cdot v$ is the *standard vector dot product of vectors u and v* and equals $\sum_{i=1..n} u_i v_i$;
- $|u|$ is *the length of vector* u and equals $\sqrt{u \cdot u}$.

In fact, the cosine similarity and Euclidean distance are related by an equation:

Lemma 1 [2]. Let u, v be non-zero vectors. Then:

$$cosSim(u,v) = \frac{|u|^2 + |v|^2 - Euclidean^2(u,v)}{2|u||v|}.$$

Clearly, the cosine similarity between any vectors u and v depends solely on the angle between the vectors and does not depend on their lengths, hence the calculation of the $cosSim(u,v)$ may be carried out on their *normalized forms*:

A *normalized form of a vector* u is denoted by $NF(u)$ and is defined as the ratio of u to its length $|u|$. A vector u is defined as a *normalized vector* if $u = NF(u)$. Obviously, the length of a normalized vector equals 1.

Theorem 1 [2]. Let u, v be non-zero vectors. Then:

$$cosSim(u,v) = cosSim(NF(u), NF(v)) = \frac{2 - Euclidean^2(NF(u), NF(v))}{2}.$$

Theorem 1 allows deducing that checking whether the cosine similarity between any two vectors exceeds a threshold ε, where $\varepsilon \in [-1, 1]$, can be carried out as checking if the Euclidean distance between the normalized forms of the vectors is less than the associated threshold $\varepsilon' = \sqrt{2 - 2\varepsilon}$:

Corollary 1 [2]. Let u, v be vectors, $\varepsilon \in [-1, 1]$ and $\varepsilon' = \sqrt{2 - 2\varepsilon}$. Then:

$$cosSim(u,v) \geq \varepsilon \text{ iff } Euclidean(NF(v),\ NF(u)) \leq \varepsilon'.$$

3 Euclidean Distance Neighborhood and Cosine Similarity Neighborhood

ε-Euclidean neighborhood of a vector p in D is denoted by $\varepsilon\text{-}NB^D_{Euclidean}(p)$ and is defined as the set of all vectors in dataset $D \backslash \{p\}$ that are distant in the Euclidean sense from p by no more than ε. *ε-cosine similarity neighborhood of a vector p in D* is denoted by $\varepsilon\text{-}SNB^D_{cosSim}(p)$ and is defined as the set of all vectors in dataset $D \backslash \{p\}$ that are cosine similar to p by no less than ε.

Corollary 1 allows transforming the problem of determining a cosine similarity neighborhood of a given vector u within a set of vectors D to the problem of determining an Euclidean neighborhood of $NF(u)$ within the vector set D' consisting of the normalized forms of the vectors from D.

Theorem 2 [2]. Let D be a set of m vectors $\{p_{(1)}, \ldots, p_{(m)}\}$, D' be the set of m vectors $\{u_{(1)}, \ldots, u_{(m)}\}$ such that $u_{(i)} = NF(p_{(i)})$, $i = 1..m$, $\varepsilon \in [-1, 1]$ and $\varepsilon' = \sqrt{2 - 2\varepsilon}$. Then, $\varepsilon\text{-}SNB^D_{cosSim}(p_{(i)}) = \{p_{(j)} \in D | u_{(j)} \in \varepsilon'\text{-}NB^{D'}_{Euclidean}(u_{(i)})\}$.

Example 1. Let us consider the determination of ε-cosine similarity neighborhood of any vector $p_{(i)}$ in dataset $D = \{p_{(1)}, \ldots, p_{(8)}\}$ from Figure 1 for $\varepsilon = 0.9856$

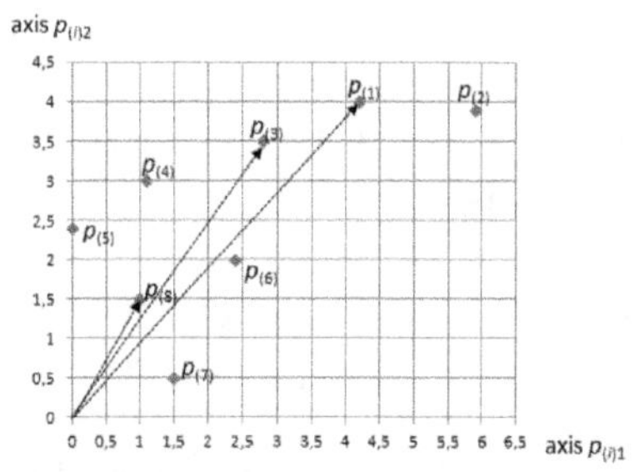

Fig. 1. Sample set D of vectors

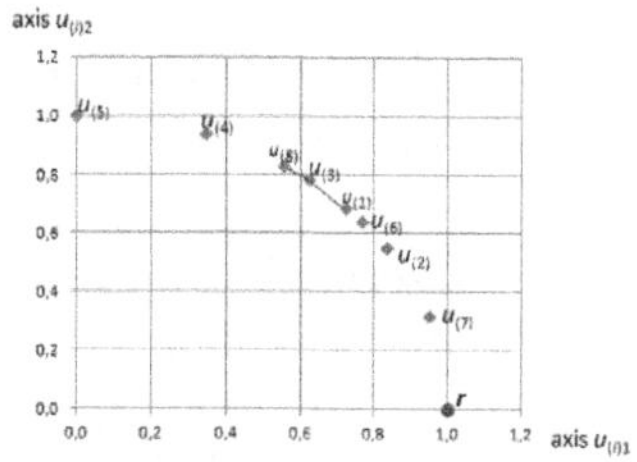

Fig. 2. Set D' containing normalized forms of vectors from D

(which roughly corresponds to the angle of 9.74^o). This task can be transformed to the task of determining ε'-Euclidean neighborhood of $u_{(i)} = NF(p_{(i)})$ in the set $D' = \{u_{(1)}, \dots, u_{(8)}\}$ containing normalized forms of the vectors from D, provided $\varepsilon' = \sqrt{2 - 2\varepsilon} \approx 0.17$. Set D' is presented in Figure 2. □

4 The Triangle Inequality in Determining Euclidean Distance Neighborhoods

We will recall now the method of determining ε-Euclidean neighborhoods as proposed in [3]. We start with Lemma 2, which follows from the triangle inequality.

Lemma 2 [3]. Let D be a set of vectors. For any vectors u, $v \in D$ and any vector r: $Euclidean(u,r) - Euclidean(v,r) > \varepsilon \Rightarrow Euclidean(u,v) > \varepsilon \Rightarrow v \notin \varepsilon\text{-}NB^D_{Euclidean}(u) \wedge u \notin \varepsilon\text{-}NB^D_{Euclidean}(v)$.

Let us consider vector q such that $Euclidean(q,r) > Euclidean(u,r)$. If $Euclidean(u,r) - Euclidean(v,r) > \varepsilon$, then $Euclidean(q,r) - Euclidean(v, r) > \varepsilon$, and thus one may conclude that $v \notin \varepsilon\text{-}NB^D_{Euclidean}(q)$ and $q \notin \varepsilon\text{-}NB^D_{Euclidean}(v)$ without calculating the real distance between q and v. This observation provides the intuition behind Theorem 3.

Theorem 3 [3]. Let r be any vector and D be a set of vectors ordered in a non-decreasing way w.r.t their distances to r. Let $u \in D$, f be a vector following vector u in D such that $Euclidean(f,r) - Euclidean(u,r) > \varepsilon$, and p be a vector preceding vector u in D such that $Euclidean(u,r) - Euclidean(p,r) > \varepsilon$. Then:

a) f and all vectors following f in D do not belong to $\varepsilon\text{-}NB^D_{Euclidean}(u)$;

b) p and all vectors preceding p in D do not belong to $\varepsilon\text{-}NB^D_{Euclidean}(u)$.

The experiments reported in [3] showed that the determination of ε-Euclidean neighborhoods by means of Theorem 3 was always faster than their determination by means of the R-Tree index, and in almost all cases speeded up the clustering process by at least an order of magnitude, also for high dimensional large vector sets consisting of hundreds of dimensions and tens of thousands of vectors.

5 Vector Projection onto a Dimension in Determining Euclidean Distance Neighborhoods

It is easy to observe that for any dimension $l, l \in [1, \dots, n]$, and any two vectors u and v, the following holds: $|u_l - v_l| = \sqrt{(u_l - v_l)^2} \leq \sqrt{\sum_{i=1..n}(u_l - v_l)^2} = Euclidean(u,v)$. Hence, if $|u_l - v_l| > \varepsilon$, then $Euclidean(u,v) > \varepsilon$; that is, $u \notin \varepsilon\text{-}NB^D_{Euclidean}(v) \wedge v \notin \varepsilon\text{-}NB^D_{Euclidean}(u)$. This observation implies Proposition 1.

Proposition 1. Let l be an index of a dimension l, where $l \in [1, \dots, n]$, and D be a set of vectors ordered in a non-decreasing way w.r.t the values of their l-th

dimension. Let $u \in D$, f be a vector following vector u in D such that $f_l - u_l > \varepsilon$, and p be a vector preceding vector u in D such that $u_l - p_l > \varepsilon$. Then:
a) f and all vectors following f in D do not belong to $\varepsilon\text{-}NB^D_{Euclidean}(u)$;
b) p and all vectors preceding p in D do not belong to $\varepsilon\text{-}NB^D_{Euclidean}(u)$.

6 The Triangle Inequality versus Projection onto a Dimension for Non-negative Normalized Vectors

In this section, we will denote $\mid u_l - v_l \mid$ by $\Delta_{dim\_l}(u,v)$ and $\mid Euclidean(u,r) - Euclidean(v,r) \mid$ by $\Delta_{ref\_r}(u,v)$. $\Delta_{dim\_l}(u,v)$ can be perceived as a pessimistic estimation of the Euclidean distance between u and v obtained by applying the projection onto l-th dimension, whereas $\Delta_{ref\_r}(u,v)$ can be perceived as a pessimistic estimation of $Euclidean(u,v)$ by means of the triangle inequality applied to reference vector r. In the following, we will focus on reference vectors of a special type. A vector will be called an *$l(a)$-vector* if its l-th coordinate equals a and all remaining coordinates equal 0. If r is an $l(a)$-vector, then $\Delta_{ref\_r}(u,v)$ will be also denoted as $\Delta_{ref\_l(a)}(u,v)$.

Proposition 2. Let $a \in R$, r be an $l(a)$-vector, u and v be non-negative normalized vectors. Then:

a) $Euclidean(u,r) = \sqrt{1 + a^2 - 2au_l}$.

b) If $u_l = v_l$, then $\Delta_{ref\_l(a)}(u,v) = \Delta_{dim\_l}(u,v) = 0$.

c) $Euclidean(u,r) = 0 \Leftrightarrow \sqrt{1 + a^2 - 2au_l} = 0 \Leftrightarrow a = u_l = 1$.

Proof. Ad a) $Euclidean(u,r) = \sqrt{\sum_{i=1..n, i \neq l}(u_i - 0)^2 + (u_l - a)^2} = \sqrt{\sum_{i=1..n} u_i^2 + a^2 - 2au_l} = \sqrt{1 + a^2 - 2au_l}$. Ad b, c) Follow from Proposition 2a. □

Lemma 3. Let $a = 1$, r be an $l(a)$-vector, u and v be non-negative normalized vectors such that $0 \leq u_l \leq v_l = 1$. Then, $\Delta_{ref\_l(a)}(u,v) \geq \Delta_{dim\_l}(u,v)$.

Proof. $\Delta_{ref\_l(a)}(u,v) \geq \Delta_{dim\_l}(u,v) \Leftrightarrow \sqrt{1 + a^2 - 2au_l} - \sqrt{1 + a^2 - 2av_l} \geq v_l - u_l \Leftrightarrow \sqrt{2 - 2u_l} \geq 1 - u_l \Leftrightarrow \sqrt{2(1 - u_l)} \geq 1 - u_l$, which is fulfilled as $1 - u_l \in [0,1]$ and in such a case $\sqrt{1 - u_l} \geq 1 - u_l$. □

Lemma 4. Let $a > 0$, r be an $l(a)$-vector, u and v be non-negative normalized vectors such that $0 < u_l \leq v_l$ and either $v_l \neq 1$ or $a \neq 1$. Then, $\Delta_{ref\_l(a)}(u,v) \geq \Delta_{dim\_l}(u,v)$ for $a \geq 1/(2u_l)$.

Proof. $\Delta_{ref\_l(a)}(u,v) \geq \Delta_{dim\_l}(u,v) \Leftrightarrow \sqrt{1 + a^2 - 2au_l} - \sqrt{1 + a^2 - 2av_l} \geq v_l - u_l \Leftrightarrow \sqrt{1 + a^2 - 2au_l} + u_l \geq \sqrt{1 + a^2 - 2av_l} + v_l$. Let $x \in (0,1]$ and $f(x) = \sqrt{1 + a^2 - 2ax} + x$. Then, $\Delta_{ref\_l(a)}(u,v) \geq \Delta_{dim\_l}(u,v) \Leftrightarrow f(u_l) \geq f(v_l)$. We will prove now that $f(u_l) \geq f(v_l)$ if $a > 1/(2u_l)$, by investigating the monotonicity of $f(x)$. $f'(x) = \frac{-a}{\sqrt{1+a^2-2ax}} + 1$ (by Proposition 2c, $\sqrt{1 + a^2 - 2ax} \neq 0$). Hence,

$f'(x) \le 0$ (and so, $f(x)$ is non-increasing) for $x \ge 1/(2a)$. Thus, $f(u_l) \ge f(v_l)$ if $\min\{u_l, v_l\} \ge 1/(2a)$. Therefore, $\Delta_{ref\_l(a)}(u,v) \ge \Delta_{dim\_l}(u,v)$ if $a \ge 1/(2u_l)$. □

Lemma 5. Let $a > 0$, r be an $l(a)$-vector, u, v be non-negative normalized vectors and $0 < u_l \le v_l$. Then, $\Delta_{ref\_l(a)}(u,v) \ge \Delta_{dim\_l}(u,v)$ for $a \ge 1/(2u_l)$.

Proof. Follows from Lemma 3 and Lemma 4. □

Theorem 4. Let $a > 0$, r be an $l(a)$-vector, D be a set of non-negative normalized vectors none of which has dimension l equal to 0. Then for any $u, v \in D$:

$$\Delta_{ref\_l(a)}(u,v) \ge \Delta_{dim\_l}(u,v) \text{ for } a \ge \frac{1}{2\mu}, \text{ where } \mu = \min\{u_l | u \in D\}.$$

Proof. Follows from Lemma 5 and the fact that $\Delta_{ref\_l(a)}(v,u) = \Delta_{ref\_l(a)}(u,v)$ and $\Delta_{dim\_l}(v,u) = \Delta_{dim\_l}(u,v)$. □

Theorem 4 tells us that for $a \ge 1/(2\mu)$, where $\mu = \min\{u_l | u \in D\}$, the pessimistic estimation of the Euclidean distance between two vectors by means of the triangle inequality applied to $l(a)$-reference vector is not less accurate than the pessimistic estimation of their Euclidean distance by means of the projection onto dimension l, provided the distance is calculated only among non-negative normalized vectors none of which has l-th dimension equal to 0.

Lemma 6. Let $a > 0$, r be an $l(a)$-vector, u, v be non-negative normalized vectors and $0 = u_l < v_l$. Then, $\Delta_{ref\_l(a)}(u,v) \ge \Delta_{dim\_l}(u,v)$ for $a \ge 1/v_l - v_l/4$.

Proof. $\Delta_{ref\_l(a)}(u,v) \ge \Delta_{dim\_l}(u,v) \Leftrightarrow \sqrt{1 + a^2 - 2au_l} - \sqrt{1 + a^2 - 2av_l} \ge v_l - u_l \Leftrightarrow \sqrt{1+a^2} - v_l \ge \sqrt{1 + a^2 - 2av_l} \Leftrightarrow 1 + a^2 + v_l^2 - 2v_l\sqrt{1+a^2} \ge 1 + a^2 - 2av_l \Leftrightarrow \sqrt{1+a^2} \le v_l/2 + a \Leftrightarrow 1 + a^2 \le v_l^2/4 + a^2 + av_l \Leftrightarrow a \ge 1/v_l - v_l/4$. □

Lemma 7. Let $\mu \in (0,1)$. Then $1/\mu - \mu/4 \ge 1/(2\mu)$.

Proof. $4 \ge 2\mu^2$. Hence, $1/(2\mu) \ge \mu/4$. Thus, $1/\mu - \mu/4 \ge 1/(2\mu)$. □

Theorem 5. Let $a > 0$, r be an $l(a)$-vector, D be a set of non-negative normalized vectors. Then for any vectors u, v in D:

$$\Delta_{ref\_l(a)}(u,v) \ge \Delta_{dim\_l}(u,v) \text{ for } a \ge \frac{1}{\mu} - \frac{\mu}{4}, \text{ where } \mu = \min\{u_l | u \in D \wedge u_l \ne 0\}.$$

Proof. Follows from Theorem 4, Lemma 6, Lemma 7, Proposition 2b and the fact that $\Delta_{ref\_l(a)}(v,u) = \Delta_{ref\_l(a)}(u,v)$ and $\Delta_{dim\_l}(v,u) = \Delta_{dim\_l}(u,v)$. □

Theorem 5 tells us that for $a \ge 1/\mu - \mu/4$, where $\mu = \min\{u_l \mid u \in D \wedge u_l \ne 0\}$, the pessimistic estimation of the Euclidean distance between two vectors by means of the triangle inequality applied to $l(a)$-reference vector is not less accurate than the pessimistic estimation of this distance by means of the projection onto dimension l, provided the distance is calculated only among non-negative normalized vectors.

Example 2. Let us consider the set D of vectors from Example 1 and the set $D' = \{u_{(1)}, \ldots, u_{(8)}\}$ of their normalized forms. In Table 1, we present the values of the vectors in D'. Let us compare the results of using the projection of vectors onto dimension 1 and of using reference vector $r_0 = [1, 0]$ (i.e., 1(1.00)-reference vector). We can see that $\Delta_{ref\_r0}(u_{(i)}, u_{(3)})$ is greater than $\Delta_{dim\_1}(u_{(i)}, u_{(3)})$ in the case of vectors $u_{(i)} = u_{(7)}$, $u_{(2)}$, $u_{(6)}$, $u_{(1)}$, identical for vectors $u_{(i)} = u_{(3)}$, $u_{(8)}$, $u_{(4)}$, and less for vector $u_{(i)} = u_{(5)}$ (please see Table 1).

Let $\mu = \min\{u_{(i)1} | u_{(i)} \in D \wedge u_{(i)1} \neq 0\} = 0.34$. Then $1/(2\mu) \approx 1.470588 < 1.48$. Let r_1 be 1(1.48)-reference vector. By Theorem 4, $\Delta_{ref\_r1}(u_{(i)}, u_{(3)}) \geq \Delta_{dim\_1}(u_{(i)}, u_{(3)})$ for all vectors in D' that have non-zero dimension 1.

Now, $1/\mu - \mu/4 \approx 2.856176 < 2.86$. Let r_2 be 1(2.86)-reference vector. By Theorem 5, $\Delta_{ref\_r2}(u_{(i)}, u_{(3)}) \geq \Delta_{dim\_1}(u_{(i)}, u_{(3)})$ for all vectors $u_{(i)}$ in D'. □

Table 1. Normalized vectors in set $D' = \{u_{(1)}, \ldots, u_{(8)}\}$ and pessimistic estimations of distances between $u_{(3)}$ and vectors in D' by means of the projection onto dimension 1 and the triangle inequality w.r.t reference vectors: $r_0 = [1, 0]$, $r_1 = [1.48, 0]$, $r_2 = [2.86, 0]$

Vector			*Euclidean*	Δ_{ref_r0}	Δ_{ref_r1}	Δ_{ref_r2}	Δ_{dim_1}
$u_{(i)}$	$u_{(i)1}$	$u_{(i)2}$	$(u_{(i)}, r_0)$	$(u_{(i)}, u_{(3)})$	$(u_{(i)}, u_{(3)})$	$(u_{(i)}, u_{(3)})$	$(u_{(i)}, u_{(3)})$
$u_{(7)}$	0.95	0.32	0.32	**0.55**	**0.54**	**0.44**	0.33
$u_{(2)}$	0.83	0.55	0.58	**0.29**	**0.31**	**0.27**	0.21
$u_{(6)}$	0.77	0.64	0.68	**0.19**	**0.21**	**0.19**	0.15
$u_{(1)}$	0.72	0.69	0.74	**0.13**	**0.13**	**0.12**	0.01
$u_{(3)}$	**0.62**	**0.78**	0.87	0.00	0.00	0.00	0.00
$u_{(8)}$	0.55	0.83	0.94	**0.07**	**0.09**	**0.08**	0.07
$u_{(4)}$	0.34	0.94	1.15	0.28	**0.32**	**0.32**	0.28
$u_{(5)}$	0.00	1.00	1.41	0.54	0.63	**0.66**	0.62

7 Summary

The problem of determining a cosine similarity neighborhood of a given vector u within a set of vectors D can be transformed to an equivalent problem of determining an Euclidean neighborhood of the normalized form of u within the vector set D' consisting of vectors of length 1 being the normalized forms of the vectors from D [2]. The triangle inequality applied to an arbitrary (reference) vector can be used for pruning when looking for an Euclidean neighborhood [3,4]. The projection onto an arbitrary dimension can be also used to this end. We proved that for any dimension l and for any set of non-negative normalized vectors, one may always determine a reference vector that guarantees not worse pruning efficiency when looking for an Euclidean neighborhood by means of the triangle inequality than the efficiency achievable by using the projection onto l.

References

1. Elkan, C.: Using the Triangle Inequality to Accelerate k-Means. In: Proc. of ICML 2003, Washington, pp. 147–153 (2003)
2. Kryszkiewicz, M.: Efficient Determination of Neighborhoods Defined in Terms of Cosine Similarity Measure. ICS Research Report 4, Institute of Computer Science. Warsaw University of Technology, Warsaw (2011)
3. Kryszkiewicz, M., Lasek, P.: TI-DBSCAN: Clustering with DBSCAN by Means of the Triangle Inequality. In: Szczuka, M., Kryszkiewicz, M., Ramanna, S., Jensen, R., Hu, Q. (eds.) RSCTC 2010. LNCS, vol. 6086, pp. 60–69. Springer, Heidelberg (2010)
4. Kryszkiewicz, M., Lasek, P.: A Neighborhood-Based Clustering by Means of the Triangle Inequality. In: Fyfe, C., Tino, P., Charles, D., Garcia-Osorio, C., Yin, H. (eds.) IDEAL 2010. LNCS, vol. 6283, pp. 284–291. Springer, Heidelberg (2010)
5. Moore, A.W.: The Anchors Hierarchy: Using the Triangle Inequality to Survive High Dimensional Data. In: Proc. of UAI, Stanford, pp. 397–405 (2000)
6. Patra, B.K., Hubballi, N., Biswas, S., Nandi, S.: Distance Based Fast Hierarchical Clustering Method for Large Datasets. In: Szczuka, M., Kryszkiewicz, M., Ramanna, S., Jensen, R., Hu, Q. (eds.) RSCTC 2010. LNCS, vol. 6086, pp. 50–59. Springer, Heidelberg (2010)

The Way of Improving PSO Performance: Medical Imaging Watermarking Case Study

Mona M. Soliman, Aboul Ella Hassanien, and Hoda M. Onsi

Cairo University, Faculty of Computers and Information, Cairo, Egypt
Scientific Research Group in Egypt (SRGE)
http://www.egyptscience.net

Abstract. Particle Swarm Optimization (PSO) and Genetic Algorithms (GA) are population based heuristic search techniques which can be used to solve the optimization problems modeled on the concept of evolutionary approach. In this paper we incorporate PSO with GA in hybrid technique called GPSO. This paper proposes the use of GPSO in designing an adaptive medical watermarking algorithm. Such algorithm aim to enhance the security, confidentiality , and integrity of medical images transmitted through the Internet. The experimental results show that the proposed algorithm yields a watermark which is invisible to human eyes and is robust against a wide variety of common attacks.

1 Introduction

Genetic algorithms (GAs) developed by Holland [1] are a family of computational models who is inspired by evolution. These algorithms encode a potential solution to a specific problem on a simple chromosome like data structure and apply recombination operators to these structures so as to preserve critical information.

Particle swarm optimization (PSO) is also an evolutionary technique introduced by Kennedy and Eberhart [2]. In PSO, each potential solution is assigned a randomized velocity, and the potential solutions, called particles, fly through the problem space by following the current best particles. As a relatively new evolutionary algorithm, PSO has been successfully applied to unconstrained and constrained optimization, artificial neural network training, parameter optimization, and feature selection [3].

Inspired by PSO and GA, we introduce a hybrid Genetic Particle Swarm Optimization (GPSO) technique by combining the advantages of both PSO and GA. The algorithm starts by applying PSO procedure in the search space and allow particles to adjust their velocity and position according to PSO equations, then in the next step we select a certain number of particles according to GA selection methods. The particles are matched into couples. Each couple reproduces two children by crossover. Then some children are adjusted by applying mutation process. These children are used to replace their parents of the previous particles to keep the number of particles unchanged. By combination of PSO and GA, evolution process is accelerated by flying behaviour and the population diversity is enhanced by genetic mechanism.

J.T. Yao et al. (Eds.): RSCTC 2012, LNAI 7413, pp. 237–242, 2012.

The basic idea behind digital watermarking is to embed a watermark signal into the host data with the purpose of copyright protection, access control, broadcast monitoring etc. Improvements in performance of watermarking schemes can be obtained by several methods. One way is to make use of artificial intelligence techniques by considering image watermarking problem as an optimization problem [4]. The proposed GPSO is modeled to solve such optimization problem of medical image watermarking. This work is compared to our previous work in [5], Which developed an adaptive watermarking procedure for medical images based on swarm intelligence.

The remainder of this paper is organized as follows. Section (2) reviews the genetic algorithm and swarm intelligence. Section (3) illustrate the hyprid (GPSO). Section (4) discusses the proposed watermarking scheme using GPSO in details. Section (5) shows the experimental results. Conclusions are discussed in Section (6).

2 Preliminaries

2.1 Genetic Algorithm

In GA, a candidate solution for a specific problem is called an individual or a chromosome and consists of a linear list of genes. Each individual represents a point in the search space, and hence a possible solution to the problem. A population consists of a finite number of individuals. Each individual is decided by an evaluating mechanism to obtain its fitness value. Based on this fitness value and undergoing genetic operators, a new population is generated iteratively with each successive population referred to as a generation. The GAs use three basic operators (reproduction, crossover, and mutation) to manipulate the genetic composition of a population [6].

2.2 Swarm Intelligence

Kennedy and Eberhart [2], considering the behavior of swarms in the nature, such as birds, fish, etc. developed the PSO algorithm. The PSO has particles driven from natural swarms with communications based on evolutionary computations. PSO combines self-experiences with social experiences [7]. The position of the particle i is represented with a position vector pi = $(p_{i1}, p_{i2}, \ldots, p_{iD})$ and a velocity vector vi = $(v_{i1}, v_{i2}, \ldots, v_{iD})$. In every time step t, particle i changes its velocity and position according to the following equations [8]:

$$v_i(t+1) = wv_i(t) + c_1r_1(pbest_i - p_i(t)) + c_2r_2(gbest - p_i(t)) \quad (1)$$

$$p_i(t+1) = p_i(t) + v_i(t+1). \quad (2)$$

where w is the inertial weight, and c_1 and c_2 are positive acceleration coefficients used to scale the contribution of cognitive and social components, respectively. r_1 and r_2 are uniform random variables in range [0,1]. $pbest_i$ is the best personal position of particle i which has been visited during the lifetime of the particle. $gbest$ is the global best position that is the best position of all particles in the swarm.

3 Hybrid Genetic Particle Swarm Optimization

To overcome the limitations of PSO, hybrid algorithms with GA are proposed. The basis behind this is that such a hybrid approach is expected to have merits of PSO with those of GA.The idea behind GA is using genetic operators: crossover and mutation. By applying crossover operation, information can be swapped between two particles to have the ability to fly to the new search area. The purpose of applying mutation to PSO is to increase the diversity of the population and the ability to have the PSO avoid the local maxima [9]. In this work we propose a hybrid algorithm based on work mentioned in [3], where evolution process is divided into two stages: PSO-Algorithm followed by GA algorithm. This hybrid algorithm is used to solve medical image watermarking problem as a case-study. Fig.1 shows the GPSO algorithm based on combination of first and second stages mentioned before.

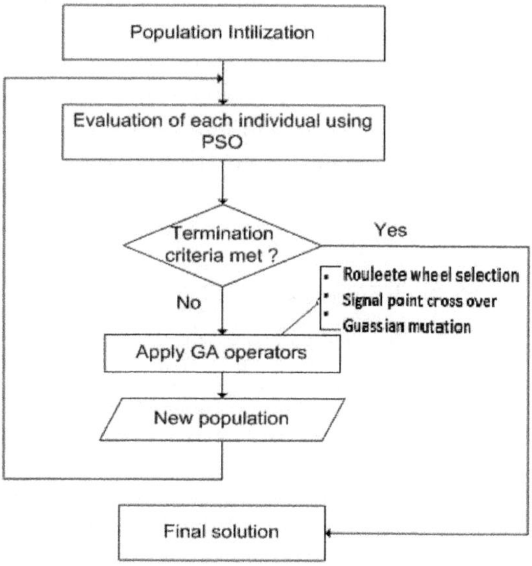

Fig. 1. GPSO algorithm

4 An Adaptive Watermarking Approach for Medical Imaging Using GPSO

In past years, a singular value decomposition SVD-based watermarking technique and its variations have been proposed [4]. Based on our work [5], The watermark can be embedded into the host image by three steps. First, DWT is performed on the host image. Second, the low performed on DCT. Then a set of final quantization steps of each block is determined to ensure a high perceptual quality of watermarked image and a low bit error rate of the detected watermark. Finally, watermark is embedded into the singular values vector of each block by adaptive and optimized quantization steps. Optimization algorithm helps search proper basic step of each block in order to optimize watermark embedding process. An efficient and optimal algorithm is required for achieving both invisibility and robustness. Here we use GPSO to automatically determine these values without making any assumption.

5 Experimental Results and Discussion

In our simulation study, we assume a group of five medical professionals,which are x-ray images of size $512x512$. The length of the watermark is $32x32$ binary bits set. The proposed algorithm is developed in $MATLAB7.6$ environment. In order to resist the normal signal processing and other different attacks, we wish the quantization step to be as high as possible. However, because the watermark directly affects the host image, it is obvious that the higher the quantization step, the lower the quality of the watermarked image will be. In other words, the robustness and the imperceptibility of the watermark are contradictory to each other. The results of our proposed watermarked method using hybrid GA and PSO (GPSO) is compared to our pervious work that depend on using basic PSO with time varying inertia weight (TPSO)[5].

The PSNR values used for quality comparison between the original and the watermark images are utilized in Table 1.

Table 1. PSNR values

Image	TPSO	GPSO
Chest-1	51.606	51.667
Chest-2	52.268	52.298
Kidney	51.554	51.677
Skull	51.535	51.752
Liver	52.241	52.351

To investigate the robustness of watermark schemes, each watermarked image is attacked using JPEG compression, Gaussian noise, Salt and Pepper noises, Gaussian filter, median filter,and geometrical attacks like image cropping and scaling. Normalized Correlation (NC) is adopted for the evaluating the robustness of the watermarking scheme. Table 2,3 shows the detailed values of NC for different types of attack that is performed on tested images, followed by Fig.2 which shows extracted watermarks after different types of attacks.

Table 2. Robustness against JPEG compression

Method	QF	Chest-1	Chest-2	Kidney	Skull	Liver
TPSO	50	0.999	0.939	0.999	0.999	0.884
GPSO		0.999	0.928	1.00	1.00	0.873
TPSO	40	0.999	0.929	0.999	0.999	0.868
GPSO		0.999	0.924	0.999	1.00	0.866
TPSO	30	0.994	0.912	0.9996	0.997	0.864
GPSO		0.996	0.921	0.998	0.0.998	0.876

Table 3. Robustness for different noise attacks

Kind of attacks	Method	Chest-1	Chest-2	Kidney	Skull	Liver
Guassian **Var=0.001**	TPSO	0.995	0.844	0.998	0.997	0.815
Var=0.005		0.812	0.641	0.825	0.821	0.617
Var=0.001	GPSO	0.996	0.842	0.997	0.996	0.805
Var=0.005		0.802	0.648	0.817	0.800	0.621
Salt and pepper **den=0.001**	TPSO	0.997	0.930	0.998	0.998	0.898
den=0.01		0.955	0.843	0.971	0.971	0.818
den=0.001	GPSO	0.997	0.919	0.998	0.998	0.889
den=0.01		0.951	0.842	0.968	0.967	0.823
Guass Filter 5×5	TPSO	0.999	0.951	0.999	0.999	0.913
7×7		0.999	0.997	0.999	0.999	0.913
5×5	GPSO	1.00	0.941	1.00	1.006	0.905
7×7		1.00	0.940	1.00	1.00	0.906
Median Filter 3×3	TPSO	0.999	0.914	0.999	0.981	0.886
5×5		0.941	0.763	0.993	0.927	0.748
3×3	GPSO	0.999	0.899	1.00	0.975	0.876
5×5		0.938	0.963	0.992	0.922	0.742
Scaling 25%	TPSO	0.999	0.834	0.999	0.971	0.692
50%		0.999	0.898	0.999	0.999	0.877
25%	GPSO	0.999	0.826	0.999	0.966	0.681
50%		1.00	0.897	1.00	0.99	0.874
Cropping 25%	TPSO	0.975	0.934	0.975	0.975	0.905
35%		0.914	0.868	0.905	0.913	0.837
25%	GPSO	0.975	0.925	0.975	0.975	0.898
35%		0.908	0.858	0.905	0.908	0.930

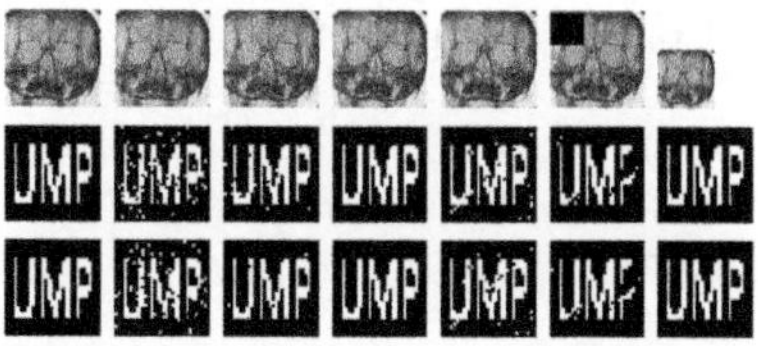

Fig. 2. From left to right different types of attack (JPEG, Gnoise, SAPnoise, Gfilter, Mfilter, Cropping, Scaling); from top to bottom attacked image ,extracted TPSO-WM, extracted GPSO-WM

6 Conclusions

This paper introduces a hybrid GPSO algorithm that is developed to design a robust watermarking for protecting medical images . GPSO approach is used to get basic quantization steps which are optimally varied to achieve the most suitable locations for various images with different frequency characteristics. The experimental results reveal that our method can improve the quality of the watermarked image and increase the robustness of the embedded watermark against various attacks.

References

1. Holland, J.H.: Adaptation in Natural and Artificial Systems. University of Michigan Press (1975)
2. Kennedy, J., Eberhart, R.C.: Particle Swarm Optimization. In: Proceedings of the IEEE International Conference on Neural Networks, vol. VI, pp. 1942–1948 (1995)
3. Yang, B., Chen, Y., Zhao, Z.: A hybrid Evolutionary Algorithm by Combination of PSO and GA for Unconstrained and Constrained Optimization Problems. In: IEEE International Conference on Control and Automation, Guangzhou, China, pp. 166–170 (2007)
4. Fakhari, P., Vahedi, E., Lucas, C.: Protecting Patient Privacy From Unauthorized Release of Medical Images Using a Bio-nspired Wavelet-based Watermarking Approach. Digital Signal Processing 21, 433–446 (2011)
5. Soliman, M.M., Ghali, N.I., Hassanien, A.E., Onsi, H.M.: An Adaptive Watermarking Approach for Medical Imaging Using Swarm Intelligent. International Journal of Smart Home 6(1), 37–51 (2012)
6. Juang, C.F.: A Hybrid of Genetic Algorithm and Particle Swarm Optimization for Recurrent Network Design. IEEE Transactions on Systems, Man, and Cypernetics Part B: Cybernetics 34(2), 997–1006 (2004)
7. Sedighizadeh, D., Masehian, E.: Particle Swarm Optimization Methods, Taxonomy and Applications. International Journal of Computer Theory and Engineering 1(5), 486–502 (2009)
8. Pant, M., Thangaraj, R., Abraham, A.: Particle Swarm Optimization: Performance Tuning and Empirical Analysis. In: Abraham, A., Hassanien, A.-E., Siarry, P., Engelbrecht, A. (eds.) Foundations of Computational Intelligence Volume 3. SCI, vol. 203, pp. 101–128. Springer, Heidelberg (2009)
9. Premalatha, K., Natarajan, A.M.: Hybrid PSO and GA for Global Maximization. Int. J. Open Problems Compt. Math. 2(4) (2009)

Toward Information Systems over Ontological Graphs

Krzysztof Pancerz

Institute of Biomedical Informatics,
University of Information Technology and Management in Rzeszów, Poland
kpancerz@wsiz.rzeszow.pl

Abstract. In the paper, we try to incorporate ontologies into information systems. In the classic information systems, there is a lack of semantics explaining meaning of data. Semantics enables us to extract some new and valuable knowledge which can be used in data analysis, rule generation, reasoning, etc. In order to cover the meaning of data, information systems over ontological graphs are defined. Next, some important relations between attribute values in such systems are investigated and an exemplary benefit of the proposed approach is briefly described.

Keywords: information systems, relation, ontological graph, semantics.

1 Introduction

For the whole decade, an increasing attention has been focused on ontologies and ontological engineering [1]. Ontologies are widely used in knowledge engineering. One of the first definitions of an ontology was given by Neches et al. [5]. They defined the ontology as the basic terms and relations comprising the vocabulary of a topic area, as well as the rules for combining terms and relations to define extensions to the vocabulary. One can find a number of other definitions in [1].

Recent research in the area of data mining shows that, in many situations, data alone are not sufficient. There is a need to add some expert knowledge about relationships within data expressing the meaning of data. Such knowledge is included in ontologies. Therefore, in this paper, we try to incorporate ontologies into information systems. In Section 2, we guide the readers from a classic definition of an information (decision) system and some relations defined over their sets of objects to new definitions of information systems over ontological graphs enabling us to take into consideration the knowledge about the meaning of data. In the new information systems, we can consider different interesting types of relations between values of attributes. Such relations can support data analysis, rule generation, reasoning, etc. In this case, attribute values are not treated individually, but they are considered in terms of semantic spaces associated with attributes. A small example showing the simple benefit of the proposed approach is presented in Section 3.

J.T. Yao et al. (Eds.): RSCTC 2012, LNAI 7413, pp. 243–248, 2012.

2 Definitions

A series of definitions (from the known to the new ones) given in this section explains our new approach to information systems.

Definition 1 (Information system). *An information system IS is a quadruple $IS = (U, A, V, f)$, where U is a nonempty, finite set of objects, A is a nonempty, finite set of attributes, $V = \bigcup_{a \in A} V_a$, where V_a is a set of values of the attribute a, and $f : U \times A \to V$ is an information function such that $f(u, a) \in V_a$ for each $u \in U$ and $a \in A$.*

In the classic approach, attribute values in information systems can be both categorical (e.g., nominal, binary, ordinal) and continuous (e.g., integer, interval-scaled, ratio-scaled), representing measured, observed or specified properties (features) of objects.

Definition 2 (Decision system). *A decision system DS is a tuple $DS = (U, C, D, V_c, V_d, c, d)$, where U is a nonempty, finite set of objects, C is a nonempty, finite set of condition attributes, D is a nonempty, finite set of decision attributes, $V_c = \bigcup_{a \in C} V_a$, where V_a is a set of values of the condition attribute a, $V_d = \bigcup_{a \in D} V_a$, where V_a is a set of values of the decision attribute a, $c : U \times C \to V_c$ is an information function such that $f(u, a) \in V_a$ for each $u \in U$ and $a \in C$, and $d : U \times D \to V_d$ is a decision function such that $f(u, a) \in V_a$ for each $u \in U$ and $a \in D$.*

A set of attributes in an information system IS determines an equivalence relation on U, called an indiscernibility relation. An indiscernibility relation on $U \times U$ is defined as $IR(A) = \{(u, v) \in U \times U : \forall_{a \in A} f(u, a) = f(v, a)\}$. For numerical attribute values, we may define a distance (dissimilarity) measure $d(u, v) \geq 0$ for two objects $u, v \in U$. In typical situations, we use several distance measures, e.g., city, Euclidean, Tchebyschev, Minkowski. In many real-life problems, the ordering properties of the considered attributes play an important role. In [2] Greco, Matarazzo and Słowiński proposed a generalization of the rough set approach to problems where ordering properties should be taken into account. In that approach a dominance relation instead of an indiscernibility relation is used. The approach presented in [2] is closely related to our approach.

In this paper, we propose to consider attribute values in the ontological (semantic) space. Our approach is based on the definitions of ontology given by Neches et al. [5] and Köhler [3]. That is, ontology is constructed on the basis of a controlled vocabulary and the relationships of the concepts in the controlled vocabulary. Formally, the ontology can be represented by means of graph structures. Graphs are a powerful tool used in information processing. In our approach, the graph representing the ontology $\mathcal{O}$ is called the ontological graph. In such a graph, each node represents one concept from $\mathcal{O}$, whereas each edge represents a relation between two concepts from $\mathcal{O}$.

Definition 3 (Ontological graph). *Let $\mathcal{O}$ be a given ontology. An ontological graph is a quadruple $OG = (\mathcal{C}, E, \mathcal{R}, \rho)$, where $\mathcal{C}$ is a nonempty, finite set of nodes representing concepts in the ontology $\mathcal{O}$, $E \subseteq \mathcal{C} \times \mathcal{C}$ is a finite set of edges representing relations between concepts from $\mathcal{C}$, $\mathcal{R}$ is a family of semantic descriptions (in natural language) of types of relations (represented by edges) between concepts, $\rho : E \to \mathcal{R}$ is a function assigning a semantic description of the relation to each edge.*

Relations are very important components in ontology modeling as they describe the relationships that can be established between concepts. In the proposed approach, we take into consideration the following family of semantic descriptions of relations between concepts (cf. [6]):

$$\mathcal{R} = \{\text{“is synonymous with”}, \text{“is generalized by”}, \text{“is specified by”}, \\ \text{“is generalized with devaluation by”}, \text{“is specified with revaluation by”}\}.$$

A special meaning of relations “is generalized with devaluation by” and “is specified with revaluation by” will be noted in Section 3. We will use the following notation: $R_{\sim}$ - “is synonymous with”, $R_{\lhd}$ - “is generalized by”, $R_{\rhd}$ - “is specified by”, $R_{\lhd\downarrow}$ - “is generalized with devaluation by”, $R_{\rhd\uparrow}$ - “is specified with revaluation by”.

We assume that the ontological graph $OG = (\mathcal{C}, E, \mathcal{R}, \rho)$ represents the whole domain $\mathcal{D}$ of a given attribute. The local ontological subgraph of OG represents a segment of the domain $\mathcal{D}$ (a small piece of reality) connected with a given attribute.

Definition 4 (Local ontological graph). *Let $OG = (\mathcal{C}, E, \mathcal{R}, \rho)$ be an ontological graph. A local ontological (sub)graph LOG of OG is a graph $LOG = (\mathcal{C}_L, E_L, \mathcal{R}_L, \rho_L)$, where $\mathcal{C}_L \subseteq \mathcal{C}$, $E_L \subseteq E$, $\mathcal{R}_L \subseteq \mathcal{R}$, and ρ_L is a function ρ restricted to E_L.*

We can create information systems over the ontological graphs. It can be done in different ways. In our investigation, we will focus on two approaches: (1) attribute values of a given information system are concepts from ontologies assigned to attributes - a simple information system over ontological graphs, (2) attribute values of a given information system are local ontological graphs of ontologies assigned to attributes - a complex information system over ontological graphs.

Definition 5 (Simple information system over ontological graphs). *A simple information system SIS^{OG} over ontological graphs is a quadruple $SIS^{OG} = (U, A, \{OG_a\}_{a \in A}, f)$, where U is a nonempty, finite set of objects, A is a nonempty, finite set of attributes, $\{OG_a\}_{a \in A}$ is a family of ontological graphs associated with attributes from A, $f : U \times A \to \mathcal{C}$, where $\mathcal{C} = \bigcup_{a \in A} \mathcal{C}_a$, is an information function such that $f(u, a) \in \mathcal{C}_a$ for each $u \in U$ and $a \in A$, where $\mathcal{C}_a$ is a set of concepts from the graph OG_a.*

Definition 6 (Complex information system over ontological graphs) *A complex information system CIS^{OG} over ontological graphs is a quadruple*

$CIS^{OG} = (U, A, \{OG_a\}_{a\in A}, f)$, where U is a nonempty, finite set of objects, A is a nonempty, finite set of attributes, $\{OG_a\}_{a\in A}$ is a family of ontological graphs associated with attributes from A, $f : U \times A \to \mathbb{LOG}$, where $\mathbb{LOG} = \bigcup_{a\in A} \mathbb{LOG}_a$, is an information function such that $f(u, a) \in \mathbb{LOG}_a$ for each $u \in U$ and $a \in A$, where $\mathbb{LOG}_a$ is a family of all local ontological graphs of the graph OG_a.

We can extend definitions of information systems over ontological graphs to decision systems over ontological graphs. In a simple case, decision attribute values can be treated individually. Therefore, a decision function can be defined as in Definition 2. In the remaining part of the paper, we will consider simple information systems over ontological graphs. Complex information systems over ontological graphs will be investigated in the future research. Many methods are proposed to measure semantic similarity between concepts (e.g., [4], [8], [9]). In our approach, we propose to consider some other relations defined over sets of attribute values in simple information systems over ontological graphs. In defined relations, we use some additional knowledge about relationships between attribute values which is included in ontological graphs.

Let $OG = (\mathcal{C}, E, \mathcal{R}, \rho)$ be an ontological graph. We will use the following notation: $[c_i, c_j]$ is a simple path in OG between $c_i, c_j \in \mathcal{C}$, $\mathcal{E}([c_i, c_j])$ is a set of edges from E belonging to the simple path $[c_i, c_j]$, $\mathcal{P}(OG)$ is a set of all simple paths in OG. In the literature, there are different definitions for a simple path in the graph. In this paper, we follow the definition in which a path is simple if no node or edge is repeated, with the possible exception that the first node is the same as the last. Therefore, the path $[c_i, c_j]$, where $c_i, c_j \in \mathcal{C}$ and $c_i = c_j$ can be also a simple path in OG.

Definition 7 (Relations over attribute value sets). *Let an ontological graph $OG_a = (\mathcal{C}_a, E_a, \mathcal{R}, \rho_a)$ be associated with the attribute a in a simple information system, where $\mathcal{R} = \{R_\sim, R_\triangleleft, R_\triangleright, R_{\triangleleft\downarrow}, R_{\triangleright\uparrow}\}$.*

- *An exact meaning relation between $c_1, c_2 \in \mathcal{C}_a$ is defined as $EMR(a) = \{(c_1, c_2) \in \mathcal{C}_a \times \mathcal{C}_a : c_1 = c_2\}$.*
- *A synonym meaning relation between $c_1, c_2 \in \mathcal{C}_a$ is defined as $SMR(a) = \{(c_1, c_2) \in \mathcal{C}_a \times \mathcal{C}_a : (c_1, c_2) \in E_a \wedge \rho((c_1, c_2)) = R_\sim\}$.*
- *A general meaning relation $GMR^k(a)$ of at most k-th order is a set of all pairs $(c_1, c_2) \in \mathcal{C}_a \times \mathcal{C}_a$ satisfying the following condition. There exists $c_3 \in \mathcal{C}_a$ for which the following holds: there exists $[c_1, c_3] \in \mathcal{P}(OG_a)$ such that $\forall_{e\in\mathcal{E}([c_1,c_3])} \rho(e) \in \{R_\sim, R_\triangleleft\}$ and $card(\{e' \in \mathcal{E}([c_1, c_3]) : \rho(e') = R_\triangleleft\}) \leq k$ as well as there exists $[c_2, c_3] \in \mathcal{P}(OG_a)$ such that $\forall_{e\in\mathcal{E}([c_2,c_3])} \rho(e) \in \{R_\sim, R_\triangleleft\}$ and $card(\{e' \in \mathcal{E}([c_2, c_3]) : \rho(e') = R_\triangleleft\}) \leq k$.*
- *A proper generalization relation $PGR^k(a)$ of at most k-th order is a set of all pairs $(c_1, c_2) \in \mathcal{C}_a \times \mathcal{C}_a$ satisfying the following condition: there exists $[c_1, c_2] \in \mathcal{P}(OG_a)$ such that $\forall_{e\in\mathcal{E}([c_1,c_2])} \rho_a(e) = R_\triangleleft$ and $card(\mathcal{E}([c_1, c_2])) \leq k$.*
- *A generalization relation $GR^k(a)$ of at most k-th order is a set of all pairs $(c_1, c_2) \in \mathcal{C}_a \times \mathcal{C}_a$ satisfying the following condition: there exists $[c_1, c_2] \in$*

$\mathcal{P}(OG_a)$ *such that* $\forall_{e \in \mathcal{E}([c_1,c_2])} \rho_a(e) \in \{R_{\sim}, R_{\lhd}\}$ *and* $card(\{e \in \mathcal{E}([c_1,c_2]) : \rho_a(e) = R_{\lhd}\}) \leq k$.

- *A proper specification relation* $PSR^k(a)$ *of at most k-th order is a set of all pairs* $(c_1, c_2) \in \mathcal{C}_a \times \mathcal{C}_a$ *satisfying the following condition: there exists* $[c_1, c_2] \in \mathcal{P}(OG_a)$ *such that* $\forall_{e \in \mathcal{E}([c_1,c_2])} \rho_a(e) = R_{\rhd}$ *and* $card(\mathcal{E}([c_1,c_2])) \leq k$.
- *A specification relation* $SR^k(a)$ *of at most k-th order is a set of all pairs* $(c_1, c_2) \in \mathcal{C}_a \times \mathcal{C}_a$ *satisfying the following condition: there exists* $[c_1, c_2] \in \mathcal{P}(OG_a)$ *such that* $\forall_{e \in \mathcal{E}([c_1,c_2])} \rho_a(e) \in \{R_{\sim}, R_{\rhd}\}$ *and* $card(\{e \in \mathcal{E}([c_1,c_2]) : \rho_a(e) = R_{\rhd}\}) \leq k$.

EMR, SMR, and GMR^k are equivalence relations whereas PGR^k, GR^k, PSR^k, and SR^k are reflexive and transitive relations.

3 An Exemplary Benefit

Due to restricted space, let us consider a simple example to show some benefit of the proposed approach. Further papers will be devoted to discussion of the problem in details. A decision system DS is given in Table 1a. Let us treat DS as a classic decision system. In this system the following rules are valid: (1) IF *Belongings* = *SUV* and *Properties* = *House*, THEN *Credit* = *Issued*, (2) IF *Belongings* = *Minivan*, THEN *Credit* = *Not issued*. Now, let us treat DS as a decision system over ontological graphs. Let us have in the ontological graph associated with attribute *Belongings* that concept *SUV* is generalized directly by concept *Car*. Analogously, concept *Minivan* is generalized directly by concept *Car*. In this case $(SUV, Minivan) \in GMR^1(Belongings)$ (see Definition 7). Therefore, let us replace both attribute values *SUV* and *Minivan* in DS by *Car* (see Table 1b). In this case rules presented earlier are transformed into: (1) IF *Belongings* = *Car* and *Properties* = *House*, THEN *Credit* = *Issued*, (2) IF *Belongings* = *Car*, THEN *Credit* = *Not issued*, and both rules are not valid. In this example, an ontological graph delivers us some new knowledge about the meaning of data and this knowledge can be used to verify our data base. In a similar way, a synonym meaning relation can change a look at knowledge included in a data base. At the same time, generalization made by us is sometimes not eligible, especially, in case of creditworthiness. Both concepts *SUV* and *Minivan* are generalized with devaluation by concept *Car*. For example, one can possess a cheap car and there is a difference (in terms of values) between this car and *SUV*

Table 1. (a) Original decision system, (b) Transformed decision system

a)

U/A	*Belongings*	*Properties*	*Credit*
u_1	*SUV*	*House*	*Issued*
u_2	*Minivan*	*House*	*Not issued*
u_3	*SUV*	*Flat*	*Not issued*
u_4	*Minivan*	*Flat*	*Not issued*

b)

U/A	*Belongings*	*Properties*	*Credit*
u_1	*Car*	*House*	*Issued*
u_2	*Car*	*House*	*Not issued*
u_3	*Car*	*Flat*	*Not issued*
u_4	*Car*	*Flat*	*Not issued*

or *Minivan*. Hence, *Car* is specified with revaluation by *SUV* or *Minivan*, and such more specific information is desired, for example, by the bank.

4 Conclusions

The approach presented in this paper constitutes the first attempt to use ontologies within information systems. We can distinguish the following main features of this approach: five basic types of semantic relations between concepts are determined, only relations between attribute values are considered, and only simple information systems over ontological graphs are investigated. These features set directions for further work. We will try to extend the approach presented here to: relations between values of attributes in complex information systems over ontological graphs, relations between objects (attribute value vectors) in simple and complex information systems over ontological graphs, extraction of rules of different types and reasoning in simple and complex information systems over ontological graphs.

References

1. Gomez-Perez, A., Fernandez-Lopez, M., Corcho, O.: Ontological Engineering. Springer, London (2004)
2. Greco, S., Matarazzo, B., Słowiński, R.: Rough sets theory for multicriteria decision analysis. European Journal of Operational Research 129(1), 1–47 (2001)
3. Köhler, J., Philippi, S., Specht, M., Rüegg, A.: Ontology based text indexing and querying for the semantic web. Knowledge-Based Systems 19(8), 744–754 (2006)
4. Li, Y., Bandar, Z., Mclean, D.: An approach for measuring semantic similarity between words using multiple information sources. IEEE Transactions on Knowledge and Data Engineering 15(4), 871–882 (2003)
5. Neches, R., Fikes, R., Finin, T., Gruber, T., Patil, R., Senator, T., Swartout, W.: Enabling technology for knowledge sharing. AI Magazine 12(3), 36–56 (1991)
6. Pancerz, K., Grochowalski, P.: Matching ontological subgraphs to concepts: a preliminary rough set approach. In: Hassanien, A., et al. (eds.) Proceedings of the ISDA 2010, Cairo, Egypt, pp. 1394–1399 (2010)
7. Pawlak, Z.: Rough Sets. Theoretical Aspects of Reasoning about Data. Kluwer Academic Publishers, Dordrecht (1991)
8. Rada, R., Mili, H., Bicknell, E., Blettner, M.: Development and application of a metric on semantic nets. IEEE Transactions on Systems, Man, and Cybernetics 19(1), 17–30 (1989)
9. Richardson, R., Smeaton, A., Murphy, J.: Using wordnet as a knowledge base for measuring semantic similarity between words. Techinical report working paper ca-1294, School of Computer Applications, Dublin City University (1994)

Two-Step Gene Feature Selection Algorithm Based on Permutation Test*

Chuanjiang Luo, Guoyin Wang, and Feng Hu

Institute of Computer Science & Technology,
Chongqing University of Posts and Telecommunications,
Chongqing, 400065, P.R. China
luochuanjiang@yahoo.com.cn, wanggy@ieee.org, hufeng@cqupt.edu.cn

Abstract. In order to filter noisy and redundant genes, this paper presents a two-step gene feature selection algorithm based on permutation Test. The proposed algorithm can select genes efficiently and process large dataset quickly due to the permutation test technique. Twelve datasets of RSCTC 2010 Discovery Challenge and two famous classifiers SVM and PAM are adopted to evaluate the performance of the proposed algorithm. The experiment results show that the small gene subset with high discriminant and low redundancy can be selected efficiently by the proposed algorithm.

Keywords: DNA Microarray, Feature Selection, Permutation Test.

1 Introduction

In recent years, the development of DNA microarray technology has made possible to analyze tens of thousands of genes simultaneously. However, when analyzing DNA microarray data, researchers have to face the few-objects-many-attributes problem. Many standard algorithms have difficulties in handling such highly dimensional data, because there are not only noisy genes, but also redundant genes. Moreover, usually only a small subset of genes is relevant in the context of a given task, for example, there is study [2] suggesting that only a few genes are usually sufficient. For these reasons, gene feature selection is a crucial part in analyzing DNA microarray data. When a small number of genes are selected, computation is reduced while prediction accuracy is increased. Their biological relationship with the target diseases is more easily identified. These marker genes thus provide additional scientific understanding of the problem.

There are two general approaches to gene feature selection: filter and wrapper. Filter method is essentially data pre-processing. Features are selected based on the intrinsic characteristics, which determine their relevance with regard to the target classes. Simple methods based on statistical tests (T-test, F-test, Wilcoxon

* Part of this work is supported by National Natural Science Foundation of China (No. 61073146), Cooperation Project between China and Poland in Science and Technology ([2010]179)

J.T. Yao et al. (Eds.): RSCTC 2012, LNAI 7413, pp. 249–258, 2012.

and Kruskal-Wallis) have been shown to be effective. In addition, many modified methods are proposed, such as Significance Analysis of Microarray (SAM) [3], Prediction Analysis for Microarrays (PAM) [4]. In wrapper methods, feature selection is wrapped around a learning method and the feature set is directly judged by the estimated accuracy of the learning method. One can often get a feature set with a small number but non redundant features, which gives high prediction accuracy. Wrapper methods typically require extensive computation to search the best features.

However, one common practice of current methods is to simply select the top-ranked genes. One deficiency of this approach is that the number of features, K, retained in the feature set is set by human intuition with trial-and error. Another one is that the features could be correlated among themselves leading redundancy. For these reasons, minimum redundancy maximum relevance (MRMR) [5] framework have been proposed, which take into account removing noisy and redundant genes at the same time. Moreover, Jaeger et al. [6] grouped genes with fuzzy clustering, and subsequently redundant genes in each group can be removed. Mitra et al [7] partitioned the original gene set into a number of clusters with k-NN, and selected a representative one from each such cluster.

In this paper, we presented a two-steps gene feature selection algorithm based on permutation test. The proposed algorithms consist of two steps, removing noisy and redundancy, which is proved more efficient according to the experiment results. The rest of this paper is organized as follows. In section 2, related works of gene feature selection are reviewed. In section 3, a novel gene feature selection algorithm is proposed. Experiments and analysis are introduced in section 4. Finally, conclusion is drawn in section 5.

2 Related Works

In this section, related work named mRMR framework is reviewed in order to compare to our work. Peng [5] proposed a minimal redundancy maximal relevance feature selection framework, which two simply combination criteria are considered as $\max(V_S - W_S)$ and $\max(V_S/W_S)$, where

$$V_S = \frac{1}{|S|}\sum_{i\in S}\phi(g_i,h) \text{ and } W_S = \frac{1}{|S|^2}\sum_{i,j\in S}\varphi(g_i,g_j)$$

$\phi(g_i,h)$ denote the relevance between gene g_i and class h , and $\varphi(g_i,g_j)$ denote the redundancy between gene g_i and g_j . Incremental search methods are used to find the near-optimal features. Let G denote all gene set, S denote already-selected gene subset. Then, for gene $g_i \in \{G\} - \{S\}$, its weight can be measured by the following two equations,

$$w_i = \phi(g_i,h) - \frac{1}{|S|}\sum_{j\in S}\varphi(g_i,g_j) \tag{1}$$

$$w_i = \phi(g_i,h)/\frac{1}{|S|}\sum_{j\in S}\varphi(g_i,g_j) \tag{2}$$

This framework demonstrates that good feature subsets contain features both highly correlated within the target class, yet uncorrelated with each other [8-9]. However, mRMR require large time complexity by searching the candidate gene subset. Moreover, it cannot remove redundancy exactly, see our experiment. Then, in this paper, we separate the framework into two steps, and adopt permutation test to determine the number of gene to select.

3 Gene Feature Selection Algorithm

3.1 Removing Noisy Gene

It is obvious that a gene is informative if it has smaller within-group variation as well as larger between-groups diffidence. In other words, a good gene is strongly different from different classes, while a noisy gene has expressions randomly or uniformly distributed in different classes. Hence, a gene should be selected as a feature as long as significant differences exist between different classes. The significant differences can be measured by *F-statistic*, which has the following form,

$$F(g_i) = \left[\sum_k n_k(\bar{g}_k - \bar{g})/(K-1) \right] /\sigma^2 \tag{3}$$

where $\bar{g}$ is the mean value of g_i in all samples, $\bar{g}_k$ is the mean value of g_i within the $k-th$ class, K is the number of classes, and σ^2 is the pooled variance, which is defined as following,

$$\sigma^2 = \left[\sum_k (n_k - 1)\sigma_k^2 \right] /(n-K) \tag{4}$$

where n_k and σ_k are the size and the variance of the k class. It is obvious that if a gene is better, it should have a larger $F(g_i)$.

3.2 Removing Redundant Gene

One important goal of gene feature selection is to select gene subset with the least number and the most information. Moreover, the selected gene subset should represent the characteristics of the whole relevant space. It is crucially important since narrow regions of the relevant space will mean the information loosing of special class type. For example, in order to illustrate the problem, 100 top ranked genes of DCOG dataset (from EBI, E-GEOD-13351) are selected. They are hierarchically clustered using Cluster and TreeView software. The results are shown in Fig 1.

Fig. 1 indicates that a large number of similar genes are clustered together. They have the same information and can identify only one class type. Obviously, they are redundant since they are either positively related or negatively related. In this context, only 4 genes are truly independent and representative, the 96 highly correlated genes can be deleted with keeping effectively the performance of

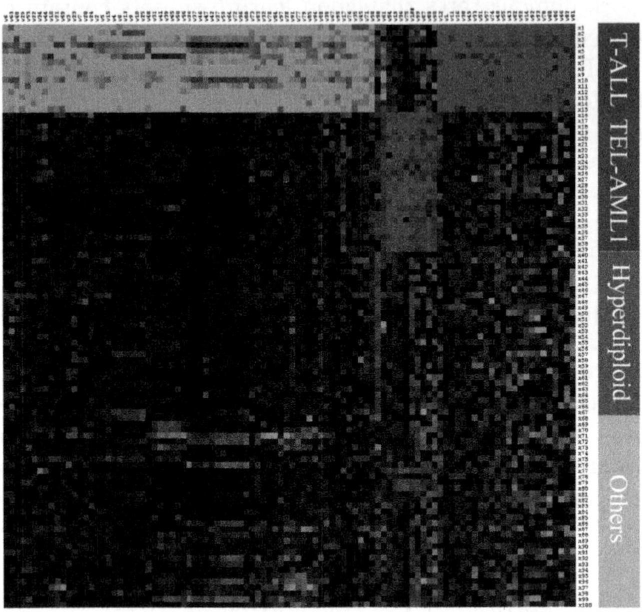

Fig. 1. Hierarchical cluster result of DCOG dataset

the prediction. Moreover, Gene 228057_at, whose ranked number is 90, is the only one that can identify Hyperdiploid. Thus, if the number of the selected genes is less than 90, gene 228057_at will be ignored, which causes the information losing. However, if the number of the chose genes is over than 100, the 96 redundant ones will be remained, which causes classifiers over fitting. One possible solution is to remove the 96 highly correlated genes and remain the 4 uncorrelated ones.

There are several criteria for measuring similarity between two genes [7], for example, *Correlation Coefficient*, *Least Square Regression Error*, *Maximal Information Compression Index*. It is known that if the data is linearly separable in the original representation, the data is still linearly separable if all but one of the linearly dependent features is removed. Then, in this paper, *Correlation Coefficient* is adopted to remove noisy genes, which is defined as,

$$C(g_i, g_j) = \frac{cov(g_i, g_j)}{\sqrt{var(g_i) * var(g_j)}} \tag{5}$$

where $var(g_i)$ denotes the variance of g_i and $\text{cov}(g_i, g_j)$ the covariance between g_i and g_j.

3.3 Permutation Test

F-Statistic and *Correlation Coefficient* require all samples following the normal distribution. When using Lilliefors [10] test, we find that a large number of them disobey normal distribution, see Table 1. Hence, *F-Statistic* and *Correlation Coefficient* cannot be used to determine the number of genes directly. However, if a gene is noisy, it has expressions randomly or uniformly distributed, and the values are exchangeable. In the same way, if two genes are uncorrelated, the values of them are exchangeable too. Based on this idea, we adopt permutation test [11] to determine the number of gene for every step.

Table 1. The rate of gene disobeying normal distribution

data1	data2	data3	data4	data5	data6	data7	data8	data9	data10	data11	data12
0.22	0.25	0.37	0.31	0.28	0.16	0.24	0.29	0.39	0.46	0.43	0.67

Def. 1. (p-value) Let F_{obs} and C_{obs} denote the observed value of F and C statistic respectively. Let F_{perm} and C_{perm} denote the permuted value of F and C statistic respectively. Let $no.\ of\ perms$ denote the permuted times, which should be larger than 1000, and 5000 is enough, see [11]. The *p-value* is defined as,

$$p{-}value(F) = \frac{no.\ of\ F_{perm} \geq F_{obs}}{total\ no.\ of\ F_{perm}} \tag{6}$$

$$p{-}value(C) = \frac{no.\ of\ C_{perm} \geq C_{obs}}{total\ no.\ of\ C_{perm}} \tag{7}$$

Def. 2. (critical value) Let F_{perms} and F_{perms} denote the arrays of F_{perm} and C_{perm} in descending order. Give two significance level α and β, the *critical value* is defined as,

$$F_{crit} = F_{perms}[no.\ of\ perms * \alpha] \tag{8}$$

$$C_{crit} = C_{perms}[no.\ of\ perms * \beta] \tag{9}$$

It is obvious that, if a gene is noise, it follow random distribution. Then, in the first step, we performs a permutation test of the default null hypothesis that gene comes from a distribution in the random family, against the alternative that it does not come from a random distribution. If $p-value$ is less than 1%, we reject the null hypothesis, and regard it as an informative gene. In the same way, we also perform a permutation test to test whether two genes are corrected in the second step. The algorithm is showing as following, which the time complexity is $O(knm)$. (n is the number of original genes; m is the original number of samples; and k is the final number of gene to select).

4 Experiment Design and Analysis

4.1 Prediction Methods

Two classifiers are adopted to evaluate the performance of our algorithm, including: SVM (Support Vector Machine) and PAM (Prediction Analysis for Microarrays) [4]. For SVM, we use LIBSVM (http://www.csie.ntu.edu.tw/ cjlin/libsvm/),

Algorithm 1. Two-Step Gene Feature Selection.

Initialization:

Set $FeatureSet = \{g_1, g_2, \cdots, g_n\}$; $\alpha = 1\%$; $\beta = 0.1\%$

Begin:

Computing $F_{crit}(\alpha)$ by definition 2;
Computing $F_{obs}(g_i)$ by equation 3;
if $F_{obs}(g_i) < F_{crit}(\alpha)$ **then**
  $FeatureSet = FeatureSet - \{g_i\}$
end if
Sorting $FeatureSet$ in descending order, and get $FeatureSet = g_1^{'}, g_2^{'}, \cdots, g_k^{'}$
Computing $C_{crit}(\beta)$ by definition 2;
Computing $C_{obs}(g_i, g_j)$ by equation 5;
if $C_{obs}(g_i^{'}, g_j^{'}) \geq C_{crit}(\beta) \&\& rank(g_i^{'}) > rank(g_j^{'})$ **then**
  $FeatureSet = FeatureSet - \{g_j^{'}\}$
end if
return $FeatureSet$

which all parameters are default, except linear kernel and $C = 100$. For PAM, we sampled it as following,

$$f(g) = \arg\min_k \left(\sum_{i=1}^{n} \frac{(g_i - \overline{g}_{ik})^2}{s_i^2}\right. \tag{10}$$

where s_i^2 is the pooled variance $s_i^2 = \sum_k \sum_{j \in C_k} (g_{ij} - \overline{g}_{ik})^2$

4.2 DNA Microarray Datasets

We explore the performance of our algorithm on twelve DNA microarray datasets, which are also used for RSCTC'2010 Discovery Challenge [1]. The twelve datasets are described in Table 2.

4.3 Experiment Analysis

First, we focus on three statuses in our algorithm. Let $Crr(S)$ denote the correct recognition rate of gene subset S. Let S_{wfs}, S_{orn}, and S_{rnr} denote gene subset of three statuses in our algorithm: without feature selection(wfs), only removed noisy(orn), and removed noise and redundancy(rnr) respectively. From Figure 2, it can be easily found that the correct recognition rate shows a ladder-upward trend, with $Crr(S_{wfs}) \leq Crr(S_{orn}) \leq Crr(S_{rnr})$. That is because, when adding informative genes, redundant genes are also added, leading classifiers over fitting. Moreover, let $Num(S)$ denote the number of gene subset S. From Table 3, our algorithm removed large number of noise and redundant genes for each step, and $Num(S_{wfs}) \geq Num(S_{orn}) \geq Num(S_{rnr})$. Li et al [2]. believe that "due to the small sample size and the presence of strong predictors, the number of genes used in a discriminant analysis in gene data sets can be much smaller than 50".

Table 2. Gene Datasets

Data	EBI	Gene	Sample	Class
data1	E-GEOD-10334	54674	123	2
data2	E-GEOD-5406	22282	105	3
data3	E-GEOD-13425	22276	95	5
data4	E-GEOD-13904	54674	113	5
data5	E-GEOD-4290	54612	89	4
data6	E-GEOD-9635	59004	92	5
data7	E-GEOD-6861	61358	160	2
data8	E-GEOD-4475	22282	221	3
data9	E-GEOD-14323	22276	124	4
data10	E-TABM-310	45100	216	7
data11	E-GEOD-9891	54620	284	3
data12	E-MTAB-37	54674	773	9

Therefore, no matter how many genes the original gene date sets have, even five thousands or fifty thousands, the number of remained genes by our algorithm is eventually about 50-200 in the experiments.

Nikulin et al. [1] constructed an ensemble criterion using WXN and FDC. The gene feature selection problem was conducted according to the rule $ENS(WXN+FDC) \geq \Delta$. Then, PAM classifier and LOOCV test method are adopted to evaluate the performance of the proposed algorithm on the first six datasets. The results are showed in table 4, which all data come from [1]. Moreover, Artiemjew et al. [12] proposed an algorithm based on experimental A Statistics, called SAM5. They applied their classifier 8_v1.4, as well as LOOCV test method, on the last six dataset. The results are showed in table 4, which all data come from [10]. Tables 4 indicate that our algorithm can achieve better performances in most cases. That is because Nikulin et al. and Artiemjew et al. only removed noisy genes but redundant ones.

We also compared our algorithm with mRMR framework. Let $\phi(g_i) = F(g_i)$ and $\varphi(g_i, g_j) = C(g_i, g_j)$. First, we select 200 top ranked genes by mRMR using

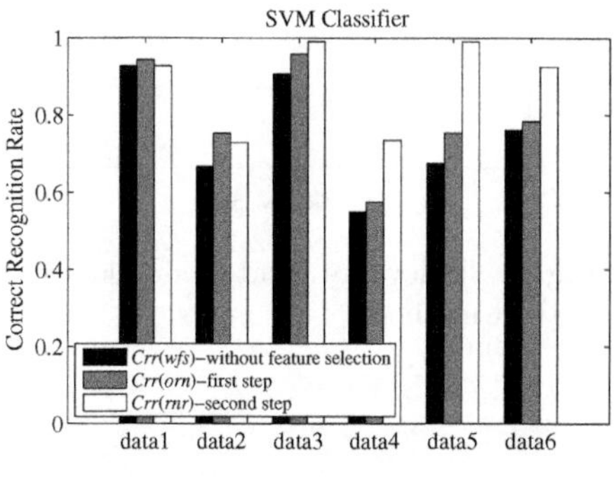

Fig. 2. The $Crr(S)$ on SVM classifier

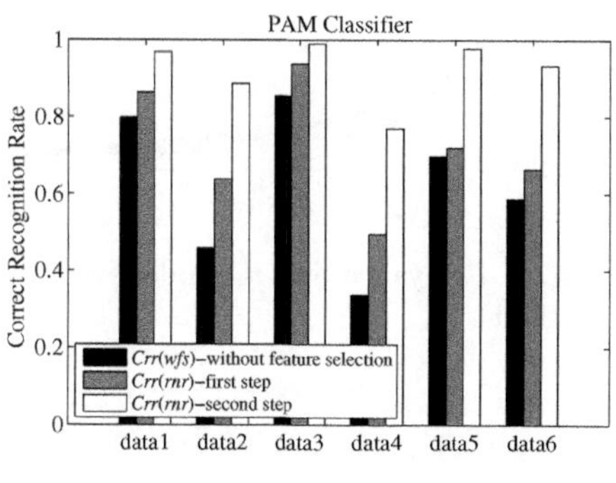

Fig. 3. The $Crr(S)$ on PAM classifier

Table 3. The number of genes of three statues

Methods	data1	data2	data3	data4	data5	data6
All Gene	54675	22283	22277	54675	54613	59004
First Step	19737	1698	5226	5028	20396	19017
Second Step	76	57	276	57	222	102

Table 4. The Correct recognition rate on data1-data12

Method		data1	data2	data3	data4	data5	data6	data7	data8	data9	data10	data11	data12
Others	P	90.88	82.73	97.75	54.33	74.32	71.27	86.6	93.2	91.0	58.1	91.3	87.6
	N	171	44	1542	1031	123	679	500	500	500	500	500	500
Ours	P	96.75	88.57	98.95	76.99	97.75	90.22	88.68	94.09	93.50	81.31	97.88	89.03
	N	76	57	276	57	222	102	47	70	70	97	125	130
Original		79.67	45.71	85.26	33.63	69.66	58.70	50.94	78.18	82.93	44.86	82.69	74.93

equation 1 on data1 (equation 2 got the same result). At the same time, 76 top ranked genes are selected by our algorithm. Then, Relevance V_S and redundancy W_S are computed, see fig 4. From fig 4, we find that the relevance and redundancy decrease quickly; when 50 genes are selected, they do not change so much. However, in fig 5, though mRMR selects relevance gene every time, it also selects some redundancy. Relevance and redundancy decrease slowly. Moreover, we also evaluate the performance of the proposed algorithm on the twelve datasets with PAM classifier, using LOOCV test method. Their correct recognition rate and computing time are shown in table 5. From table 5, we can find that our algorithm not only get higher accuracy, but also lower time.

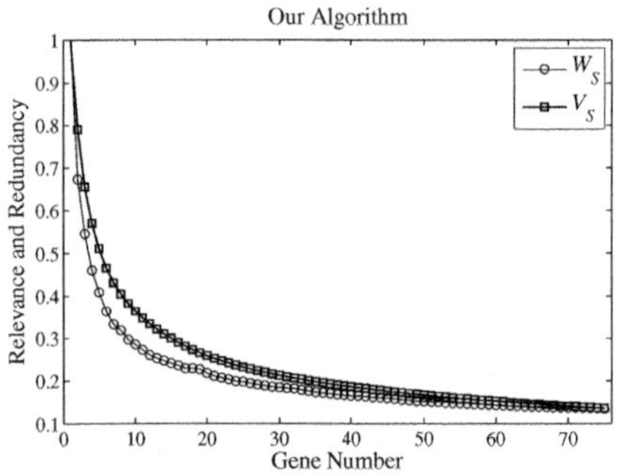

Fig. 4. Relevance and redundancy of our method

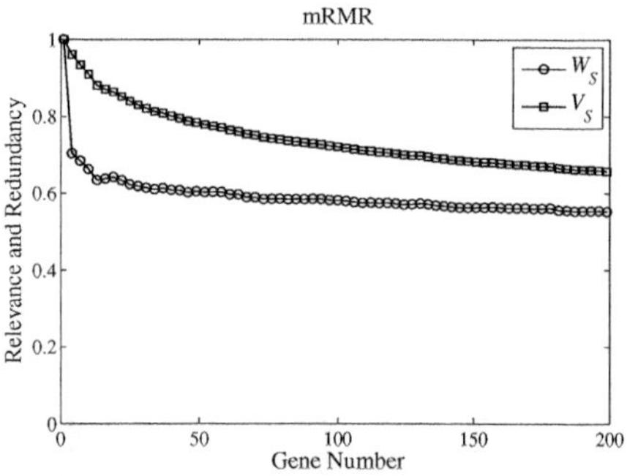

Fig. 5. Relevance and redundancy of mRMR method

Table 5. The Correct recognition rate on data1-data12

Method		data1	data2	data3	data4	data5	data6	data7	data8	data9	data10	data11	data12
mRMR	P	93.50	91.43	97.89	69.03	80.90	83.70	71.70	94.09	89.43	65.89	91.17	85.73
	T	1758	515	467	1467	1110	1236	19508	9745	4296	19943	42178	49666
Ours	P	96.75	88.57	98.95	76.99	97.75	93.22	88.68	94.09	93.50	81.31	97.88	89.03
	T	8	3	3	6	6	5	9	5	4	11	17	34

5 Conclusion

In this paper, two-step gene feature selection and permutation test are studied. An algorithm combining removing noisy gene and redundant gene is proposed. An experiment analyzes the similarity between genes and verifies the correctness of the proposed algorithm. Compared experiment results show that the effectiveness and high efficiency of the proposed algorithm for large gene dataset. In the future, gene feature selection based on statistic theory will be studied continually.

References

1. Wojnarski, M., Janusz, A., Nguyen, H.S., Bazan, J., Luo, C., Chen, Z., Hu, F., Wang, G., Guan, L., Luo, H., Gao, J., Shen, Y., Nikulin, V., Huang, T.-H., McLachlan, G.J., Bošnjak, M., Gamberger, D.: RSCTC'2010 Discovery Challenge: Mining DNA Microarray Data for Medical Diagnosis and Treatment. In: Szczuka, M., Kryszkiewicz, M., Ramanna, S., Jensen, R., Hu, Q. (eds.) RSCTC 2010. LNCS (LNAI), vol. 6086, pp. 4–19. Springer, Heidelberg (2010)
2. Li, W.T., Yang, Y.: How many genes are needed for a discriminant microarray data analysis? In: Critical Assessment of Techniques for Microarray Data Mining Workshop, pp. 137–150 (2002)
3. Tusher, V.G., Tibshirani, R., Chu, G.: Significance analysis of microarrays applied to the ionizing radiation response. PNAS 98(9), 5116–5121 (2001)
4. Tibshirani, R., Hastie, T., Narasimhan, B., et al.: Diagnosis of multiple cancer types by shrunken centroids of gene expression. PNAS 99(10), 6567–6572 (2002)
5. Peng, H.C., Long, F.H., Ding, C.H.: Feature selection based on mutual information: criteria of max-dependency, max-relevance, and min-redundancy. IEEE Transactions on Pattern Analysis and Machine Intelligence 27(8), 1226–1238 (2005)
6. Jaeger, J., Sengupta, R., Ruzzo, W.L.: Improved Gene Selection for Classification of Microarrays. In: Pacific Symposium on Biocomputing, vol. 8, pp. 53–64 (2003)
7. Mitra, P., Murthy, C.A., Pal, S.K.: Unsupervised Feature Selection Using Feature Similarity. IEEE Transactions on Pattern Analysis and Machine Intelligence 24(3), 301–311 (2002)
8. Hall, M.A., Smith, L.A.: Feature Selection for Machine Learning: Comparing a Correlation-Based Filter Approach to the Wrapper. In: Proceedings of the Twelfth International Florida Artificial Intelligence Research Society Conference, pp. 235–239 (1999)
9. Hall, M.A.: Correlation-based Feature Selection for Discrete and Numeric Class Machine Learning. In: Proceeding ICML 2000 Proceedings of the Seventeenth International Conference on Machine Learning, pp. 359–366 (2000)

10. Wikipedia, http://en.wikipedia.org/wiki/Lilliefors_test
11. Anderson, M.J.: Permutation tests for univariate or multivariate analysis of variance and regression. Canadian Journal of Fisheries and Aquatic Sciences 58(3), 626–639 (2001)
12. Artiemjew, P.: The Extraction Method of DNA Microarray Features Based on Experimental *A* Statistics. In: Yao, J., Ramanna, S., Wang, G., Suraj, Z. (eds.) RSKT 2011. LNCS, vol. 6954, pp. 642–648. Springer, Heidelberg (2011)

Minimal Test Cost Feature Selection with Positive Region Constraint

Jiabin Liu[1,2], Fan Min[2,⋆], Shujiao Liao[2], and William Zhu[2]

[1] Department of Computer Science, Sichuan University for Nationalities, Kangding 626001, China.
liujiabin418@163.com

[2] Lab of Granular Computing, Zhangzhou Normal University, Zhangzhou 363000, China
minfanphd@163.com

Abstract. Test cost is often required to obtain feature values of an object. When this issue is involved, people are often interested in schemes minimizing it. In many data mining applications, due to economic, technological and legal reasons, it is neither possible nor necessary to obtain a classifier with 100% accuracy. There may be an industrial standard to indicate the accuracy of the classification. In this paper, we consider such a situation and propose a new constraint satisfaction problem to address it. The constraint is expressed by the positive region; whereas the objective is to minimize the total test cost. The new problem is essentially a dual of the test cost constraint attribute reduction problem, which has been addressed recently. We propose a heuristic algorithm based on the information gain, the test cost, and a user specified parameter λ to deal with the new problem. Experimental results indicate the rational setting of λ is different among datasets, and the algorithm is especially stable when the test cost is subject to the Pareto distribution.

Keywords: cost-sensitive learning, positive region, test cost, constraint, heuristic algorithm.

1 Introduction

When industrial products are manufactured, they must be inspected strictly before delivery. Testing equipments are needed to classify the product as qualified, unqualified, etc. Each equipment costs money, which will be averaged on each product. Generally, we should pay more to obtain better classification accuracy. However, in real world applications, due to economic, technological and legal reasons, it is neither possible nor necessary to obtain a classifier with 100% accuracy. There may be an industrial standard to indicate the accuracy of the classification, such as 95%. Consequently, we are interested in a set of equipments with minimal cost meeting the standard. In this scenario, there are two issues: one is the equipment cost, and the other is the product classification accuracy. They are called test cost and classification accuracy, respectively. Since the classification accuracy only needs to meet the industrial standard, we can choose some testing equipments to make the total cost minimal.

⋆ Corresponding author.

J.T. Yao et al. (Eds.): RSCTC 2012, LNAI 7413, pp. 259–266, 2012.

Feature selection plays an important role in machine learning and pattern recognition application [1]. It has been defined by many authors by booking at it from various angles [11]. Minimal reducts have the best generalization ability, hence there are many existing feature selection reduction algorithms based on rough set to deal with it, such as [15,16,18,20,23] are devoted to find one of them.

The positive region is a widely used concept in rough set [12]. We use this concept instead of the classification accuracy to specify the industrial standard. We formally define this problem and call it the minimal test cost feature selection with positive region constraint (MTPC) problem. The new problem is essentially a dual of the optimal sub-reduct with test cost constraint (OSRT) problem, which has been defined in [7] and studied in [7,8,9]. The OSRT problem considers the test cost constraint, while the new problem considers the positive region constraint.

As will be discussed in the following text, the classical reduct problem can be viewed as a special case of the MTPC problem. Since the classical reduct problem is NP-hard, the new problem is at least NP-hard. Consequently, we propose a heuristic algorithm to deal with it. The heuristic information function is based on both the information gain and the test cost. This algorithm is tested on four UCI datasets with various test cost settings. Experimental results indicate the rational setting of λ is different among datasets, and the algorithm is especially stable when the test cost is subject to the Pareto distribution.

The rest of this paper is structured as follows. Section 2 describes related concepts in the rough set theory and defines the MTPC problem formally. In Section 3, a heuristic algorithm based on λ-weighted information gain is presented. Section 4 illustrates some results on four UCI datasets with detailed analysis. Finally, Section 5 concludes.

2 Preliminaries

In this section, we define the MTPC problem. First, we revisit the data model on which the problem is defined. Then we review the concept of positive region. Finally we propose the minimal test cost feature selection with positive region constraint problem.

2.1 Test-Cost-Independent Decision Systems

Decision systems are fundamental in machine learning and data mining. A decision system is often denoted as $S = (U, C, D, \{V_a|a \in C \cup D\}, \{I_a|a \in C \cup D\})$, where U is a finite set of objects called the universe, C is the set of conditional attributes, D is the set of decision attributes, V_a is the set of values for each $a \in C \cup D$, and $I_a : U \to V_a$ is an information function for each $a \in C \cup D$. We often denote $\{V_a|a \in C \cup D\}$ and $\{I_a|a \in C \cup D\}$ by V and I, respectively. A decision system is often stored in a relational database or a text file.

A *test-cost-independent decision system* (TCI-DS) [6] is a decision system with test cost information represented by a vector. It is the most simple form of the test-cost-sensitive decision system and defined as follows.

Definition 1. *[6] A test-cost-independent decision system (TCI-DS) S is the 6-tuple:*

$$S = (U, C, D, V, I, c), \tag{1}$$

where U, C, D, V and I have the same meanings as in a decision system, and $c : C \to R^+ \cup \{0\}$ is the test cost function. It can easily be represented by a vector **c**$=[c(a_1), c(a_2), \cdots, c(a_{|C|})]$. Test costs are independent of one another, that is, $c(B) = \sum_{a \in B} c(a)$ for any $B \subset C$.

2.2 Positive region

Let $S = (U, C, D, V, I)$ be a decision system. Any $\emptyset \neq B \subseteq C \cup D$ determines an indiscernibility relation on U. A partition determined by B is denoted by U/B. Let $\underline{B}(X)$ denote the B-*lower approximation* of X.

Definition 2. *[13] Let $S = (U, C, D, V, I)$ be a decision system, $\forall B \subset C$, the positive region of D with respect to B is defined as*

$$POS_B(D) = \bigcup_{X \in U/D} \underline{B}(X), \tag{2}$$

where U, C, D, V and I have the same meanings as in a decision system.

In other words, D is totally (partially) dependent on B, if all (some) elements of the universe U can be uniquely classified to blocks of the partition U/D, employing B [12].

2.3 Problem definition

Attribute reduction is the process of choosing an appropriate subset of attributes from the original dataset [17]. There are numerous reduct problems which have been defined on the classical [14], the covering-based [21,22,23], the decision-theoretical [19], and the dominance-based [2] rough set models. Respective definitions of relative reducts also have been studied in [3,15].

Definition 3. *[13] Let $S = (U, C, D, V, I)$ be a decision system. Any $B \subseteq C$ is called a decision relative reduct (or a relative reduct for brevity) of S iff:*

(1) $POS_B(D) = POS_C(D)$, and
(2) $\forall a \in B, POS_{B-\{a\}}(D) \neq POS_B(D)$.

Definition 3 implies two issues. One is that the reduct is jointly sufficient, the other is that the reduct is individually necessary for preserving a particular property (positive region in this context) of the decision systems [4]. The set of all relative reducts of S is denoted by $Red(S)$. The core of S is the intersection of these reducts, namely, $core(S) = \cap Red(S)$. Core attributes are of great importance to the decision system and should never be removed, except when information loss is allowed [19].

In this paper, due to the positive region constraint, it is not necessary to construct a reduct. On the other side, we never want to select any redundant test. Therefore we propose the following concept.

Definition 4. *Let $S = (U, C, D, V, I)$ be a decision system. Any $B \subseteq C$ is a positive region sub-reduct of S iff $\forall a \in B$, $POS_{B-\{a\}}(D) \neq POS_B(D)$.*

According to the Definition 4, we observe the following:

(1) A reduct is also a sub-reduct, and
(2) A core attribute may not be included in a sub-reduct.

Here we are interested those feature subsets satisfying the positive region constraint, and at the same time, with minimal possible test cost. We adopt the style of [5] and propose the following problem.

Problem 1. The minimal test cost feature selection with positive region constraint (MTPC) problem.
Input: $S = (U, C, d, V, I, c)$, the positive region lower bound pl;
Output: $B \subseteq C$;
Constraint: $|POS_B(D)|/|POS_C(D)| \geq pl$;
Optimization objective: $\min c(B)$.

In fact, the MTPC problem is more general than the minimal test cost reduct problem, which is defined in [4]. In case where $pl = 1$, it coincides with the later. The minimal test cost reduct problem is in turn more general than the classical reduct problem, which is NP-hard. Therefore the MTPC problem is at least NP-hard, and heuristic algorithms are needed to deal with it. Note that the MTPC is different with the variable precision rough set model. The variable precision rough set model changes the lower approximation by varying the accuracy, but in our problem definition, it is unchanged.

3 The Algorithm

Similar to the heuristic algorithm to the OSRT problem [7], we also design a heuristic algorithm to deal with the new problem. We firstly analyze the heuristic function which is the key issue in the algorithm. Let $B \subset C$ and $a_i \in C - B$, the information gain of a_i with respect to B is

$$f_e(B, a_i) = H(\{d\}|B) - H(\{d\}|B \cup \{a_i\}), \tag{3}$$

where $d \in D$ is a decision attribute. At the same time, the λ-weighted function is defined as

$$f(B, a_i, c, \lambda) = f_e(B, a_i)c_i^{\lambda}. \tag{4}$$

where λ is a non-positive number.

Our algorithm is listed in Algorithm 1. It contains two main steps. The first step contains lines 3 through 8. Attributes are added to B one by one according to the heuristic function indicated in Equation (4). This step stops while the positive region reaches the lower bound. The second step contains lines 9 through 15. Redundant attributes are removed from B one by one until all redundant have been removed. As discussed in Section 2.3, our algorithm has not a stage of core computing.

Algorithm 1. A heuristic algorithm to the MTPC problem

Input: $S = (U, C, D, V, I, c), p_{con}, \lambda$
Output: A sub-reduct of S
Method: MTPC

$B = \varnothing$; //the sub-reduct
$CA = C$; //the unprocessed attributes
while $(|POS_B(D)| < p_{con})$ **do**
  For any $a \in CA$ compute $f(B, a, c, \lambda)$
  //Addition
  Select a' with maximal $f(B, a, c, \lambda)$;
  $B = B \cup \{a'\}$;
  $CA = CA - \{a'\}$;
end while
//Deletion, B must be a sub-reduct
$CD = B$; //sort attribute in CD according to respective test cost in a descending order
while $CD \neq \varnothing$ **do**
  $CD = CD - \{a'\}$; //where a' is the first element in CD
  if $(POS_{B-\{a'\}}(D) = POS_B(D))$ **then**
    $B = B - \{a'\}$;
  end if
end while
return B

4 Experiments

To study the effectiveness of the algorithm, we have undertaken experiments using our open source software Coser [10] on 4 different datasets from the UCI library. To evaluate the performance of the algorithm, we need to study the quality of each sub-reduct which it computes. This experiment should be undertaken by comparing each sub-reduct to an optimal sub-reduct with the positive region constraint. Unfortunately, the computation of an optimal sub-reduct with test positive region constraint is more complex than that of a minimal reduct, or that of a minimal test cost reduct. In this paper, we only study the influence of λ to the quality of the result.

Because of lacking the predefined test costs in the four artificial datasets, we specify them as the same setting as that of [4] to produce test costs within [1, 100]. Three distributions, namely, Uniform, Normal, and bounded Pareto, are employed. In order to control the shape of the Normal distribution and the bounded Pareto distribution respectively, we must set the parameter α. In our experiment, for the Normal distribution, $\alpha = 8$, and test costs as high as 70 and as low as 30 are often generated. For the bounded Pareto distribution, $\alpha = 2$, and test costs higher than 50 are often generated.

We intentionally set $pl = 0.8$. This setting shows that we need a sub-reduct rather than a reduct.

The experimental results of the 4 datasets are illustrated in Fig 1. By running our program in different λ values, the 3 different test cost distributions are compared. We can observe the following.

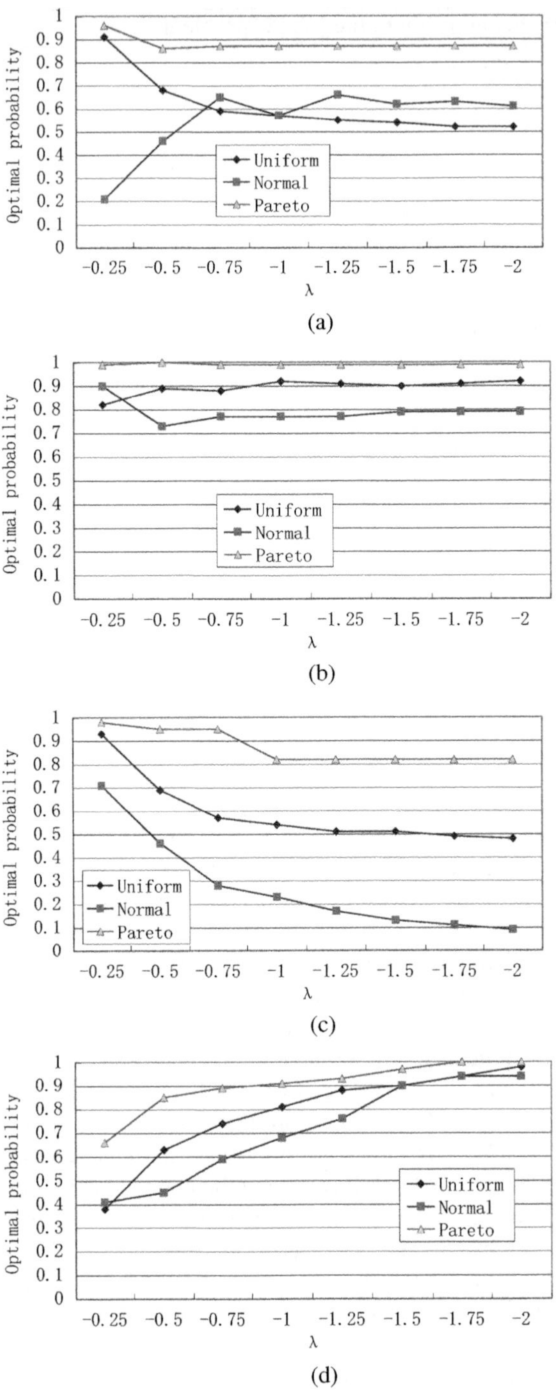

Fig. 1. Optimal probability: (a) zoo; (b) iris; (c) voting; (d) tic-tac-toe

(1) The result is influenced by the user-specified λ. The probability of obtained the best results is different with different λ values, where the "best" means the best one over the solutions we obtained, not the optimal one.
(2) The algorithm's performance is related with the test cost distribution. It is best on datasets with bounded Pareto distribution. At the same time, it is worst on datasets with Normal distribution. Consequently, if the real data has test cost subject to the Normal distribution, one may develop other heuristic algorithms to this problem.
(3) There is not a setting of λ such that the algorithm always can obtain the best result.

5 Conclusions

In this paper, we firstly proposed the MTPC problem. Then we designed a heuristic algorithm to deal with it. Experimental results indicates that the optimal solution is not easy to obtain. In the future, we will develop an exhaustive algorithm to evaluate the performance of a heuristic algorithm. We will also develop more advanced heuristic algorithms to obtain better performance.

Acknowledgements. This work is partially supported by the Natural Science Youth Foundation of Department of Education of Sichuan Province under Grant No. 2006C059, the National Science Foundation of China under Grant No. 61170128, the Natural Science Foundation of Fujian Province under Grant No. 2011J01374, and the Education Department of Fujian Province under Grant No. JA11176.

References

1. Chen, X.: An Improved Branch and Bound Algorithm for Feature Selection. Pattern Recognition Letters 24(12), 1925–1933
2. Greco, S., Matarazzo, B., Słowiński, R., Stefanowski, J.: Variable Consistency Model of Dominance-Based Rough Sets Approach. In: Ziarko, W.P., Yao, Y. (eds.) RSCTC 2000. LNCS (LNAI), vol. 2005, pp. 170–181. Springer, Heidelberg (2001)
3. Hu, Q., Yu, D., Liu, J., Wu, C.: Neighborhood rough set based heterogeneous feature subset selection. Information Sciences 178(18), 3577–3594 (2008)
4. Min, F., He, H., Qian, Y., Zhu, W.: Test-cost-sensitive attribute reduction. Information Sciences 181, 4928–4942 (2011)
5. Min, F., Hu, Q., Zhu, W.: Feature selection with test cost constraint. Submitted to Knowledge-Based Systems (2012)
6. Min, F., Liu, Q.: A hierarchical model for test-cost-sensitive decision systems. Information Sciences 179, 2442–2452 (2009)
7. Min, F., Zhu, W.: Attribute reduction with test cost constraint. Journal of Electronic Science and Technology of China 9(2), 97–102 (2011)
8. Min, F., Zhu, W.: Optimal sub-reducts in the dynamic environment. In: IEEE GrC, pp. 457–462 (2011)
9. Min, F., Zhu, W.: Optimal Sub-Reducts with Test Cost Constraint. In: Yao, J., Ramanna, S., Wang, G., Suraj, Z. (eds.) RSKT 2011. LNCS, vol. 6954, pp. 57–62. Springer, Heidelberg (2011)

10. Min, F., Zhu, W., Zhao, H., Pan, G., Liu, J.: Coser: Cost-senstive rough sets (2011), `http://grc.fjzs.edu.cn/~fmin/coser/`
11. Dash, M., Liu, H.: Feature Selection for Classification. Intelligent Data Analysis 1, 131–156 (1997)
12. Pawlak, Z.: Rough set approach to knowledge-based decision support. European Joumal of Operational Research 99, 48–57 (1997)
13. Pawlak, Z.: Rough sets. International Journal of Computer and Information Sciences 11, 341–356 (1982)
14. Pawlak, Z.: Rough sets and intelligent data analysis. Information Sciences 147(12), 1–12 (2002)
15. Qian, Y., Liang, J., Pedrycz, W., Dang, C.: Positive approximation: An accelerator for attribute reduction in rough set theory. Artificial Intelligence 174(9-10), 597–618 (2010)
16. Swiniarski, W., Skowron, A.: Rough set methods in feature selection and recognition. Pattern Recognition Letters 24(6), 833–849 (2003)
17. Wang, X., Yang, J., Teng, X., Xia, W., Jensen, R.: Feature selection based on rough sets and particle swarm optimization. Pattern Recognition Letters 28(4), 459–471 (2007)
18. Yao, J., Zhang, M.: Feature Selection with Adjustable Criteria. In: Ślęzak, D., Wang, G., Szczuka, M.S., Düntsch, I., Yao, Y. (eds.) RSFDGrC 2005. LNCS (LNAI), vol. 3641, pp. 204–213. Springer, Heidelberg (2005)
19. Yao, Y., Zhao, Y.: Attribute reduction in decision-theoretic rough set models. Information Sciences 178(17), 3356–3373 (2008)
20. Zhange, W., Mi, J., Wu, W.: Knowledge reductions in inconsistent information systems. Chinese Journal of Computers 26(1), 12–18 (2003)
21. Zhu, W.: Topological approaches to covering rough sets. Information Sciences 177(6), 1499–1508 (2007)
22. Zhu, W.: Relationship between generalized rough sets based on binary relation and covering. Information Sciences 179(3), 210–225 (2009)
23. Zhu, W., Wang, F.: Reduction and axiomization of covering generalized rough sets. Information Sciences 152(1), 217–230 (2003)

Multiple-Category Attribute Reduct Using Decision-Theoretic Rough Set Model

Xi'ao Ma[1], Guoyin Wang[1,2], and Hong Yu[2]

[1] School of Information Science and Technology,
Southwest Jiaotong University, Chengdu, 610031, China
maxiao73559@163.com, wanggy@ieee.org
[2] Chongqing Key Lab of Computational Intelligence,
Chongqing University of Posts and Telecommunications, Chongqing, 400065, China
yuhong@cqupt.edu.cn

Abstract. The main objective of this paper is to propose an approach to solve the multiple-category attribute reduct problem. The (α,β) lower approximate and (α,β) upper approximate distribution reduct are introduced into decision-theoretic rough set model. On the basis of this, the judgement theorems and discernibility matrices associated with the above two types of distribution reduct are examined as well, from which we can obtain attribute reducts. Finally, an example is used to illustrate the main ideas of the proposed approaches.

Keywords: Decision-theoretic rough set model, three-way decision, attribute reduct, probabilistic rough set model, multiple-category.

1 Introduction

Rough set theory, introduced by Pawlak[8], has become a useful mathematical tool for dealing with uncertain and inexact knowledge. One of the major limitations of the classical rough set model is that it simply considers rough set approximations as qualitative approximations of a set without considering the extent of overlap between a set and an equivalence class. To resolve this problem, many probabilistic rough set models have been proposed and studied such as the decision-theoretic rough set model(DTRS)[12] and the Bayesian rough set model(BRS)[9].

An attribute reduct is a minimum of attributes that are jointly sufficient and individually necessary for preserving a particular property of the given information table[13]. In rough set models, many reduct construction methods have been discussed. For example, Miao et al.[6] investigated three different classification properties and discusses the definitions of relative reducts in both consistent and inconsistent decision tables. Wang et al.[10] studied the relationship of the definitions of rough reduction in algebra view and information view, and developed two novel heuristic knowledge reduction algorithms which are based on conditional information entropy. Min et al.[7] posited a new research theme in regard

J.T. Yao et al. (Eds.): RSCTC 2012, LNAI 7413, pp. 267–276, 2012.

to attribute reduct and formally defined the minimal test cost reduct problem, which is to select a set of tests satisfying a minimal test cost criterion.

In practice, it seems much reasonable that admitting some level of uncertainty in the reduction process may lead to a better utilization of properties of the original data. For example, Li et al.[3] investigated the monotonicity of positive region in DTRS, and presented a new definition of attribute reduct in DTRS. Yao and Zhao[13] addressed attribute reduct in DTRS regarding different classification properties such as decision-monotocity, confidence, coverage, generality and cost. Zhao et al.[14] examined the definitions of attribute reduct and pointed out three problems of the existing definition for attribute reduct. Jia et al.[2] presented an optimization viewpoint on decision-theoretic rough set model and defined an attribute reduct based on the optimization problem.

As we have discussed, however, all decision classes are treated as the same in the interpretation and applications of approximations and three regions in the existing literatures. In other words, the same threshold or the same pair thresholds are used to define the positive, negative and boundary regions. As a natural extension to these original studies, rough set approximations for multiple-category reduct problems using different pairs of thresholds are discussed in this paper.

To combat the problem, a multiple-category attribute reduct approach using DTRS is proposed, where each class has a different pair of threshold parameters. The judgment theorems and discernibility matrices associated with these reducts are also established, from which we can obtain the approaches to attribute reduct in multiple-category classification model with DTRS.

The rest of this paper is structured as follows. In section 2, we briefly reviewed multiple-category classification model based on the three-way decision approach of DTRS. In section 3, we provide practical approaches to multiple-category attribute reduct based on DTRS and an illustration is used to show the effectiveness of the presented approaches.

2 Multi-category Classification with DTRS

DTRS provides a systematic way to calculate the two probabilistic threshold values based on the well established Bayesian decision theory[1], with the aid of more practically operable notions such as cost, risk, benefit etc.[12]. Multiple-category[4,15] classification model based on the three-way decision approach of DTRS are briefly reviewed in this section.

In general, different categories have different losses, and different threshold values should be used for different categories in a model [4]. For each decision class, the set of states is given by $\Omega = \{D_1, D_2, \cdots, D_m\}$ indicating that an object is in $D_j (j = 1, 2, \cdots, m)$. The set of action is given by $A = \{a_{P_j}, a_{B_j}, a_{N_j}\}$, where $a_{P_j}, a_{B_j}, a_{N_j}$ represent the three actions to classify an object into $POS(D_j)$, $BND(D_j)$and $NEG(D_j)$, respectively. The loss function regarding the cost of actions in different states is given by the 3×2 matrix:

$$\begin{array}{c} \\ a_{P_j} \\ a_{B_j} \\ a_{N_j} \end{array} \begin{array}{c} \begin{array}{cc} D_j & {D_j}^c \end{array} \\ \begin{bmatrix} \lambda_{P_j D_j} & \lambda_{P_j \neg D_j} \\ \lambda_{B_j D_j} & \lambda_{B_j \neg D_j} \\ \lambda_{N_j D_j} & \lambda_{N_j \neg D_j} \end{bmatrix} \end{array}$$

In the matrix, $\lambda_{P_j D_j}$ denotes the losses incurred for classifying an object in D_j into the positive region, $\lambda_{B_j D_j}$ denotes the losses incurred for classifying an object in D_j into the boundary region, $\lambda_{N_j D_j}$ denotes the losses incurred for classifying an object in D_j into the negative region. Similarly, $\lambda_{P_j \neg D_j}$, $\lambda_{B_j \neg D_j}$ and $\lambda_{N_j \neg D_j}$ denote the losses incurred for taking the same actions when the object does not belong to D_j. The expected cost $R(a_\bullet \| [x_i])$ associated with taking the individual actions $a_\bullet (\bullet = P_j, B_j, N_j)$ can be expressed as:

$$\begin{aligned} R(a_{P_j} \| [x_i]) &= \lambda_{P_j D_j} P(D_j \| [x_i]) + \lambda_{P_j \neg D_j} P(\neg D_j \| [x_i]), \\ R(a_{B_j} \| [x_i]) &= \lambda_{B_j D_j} P(D_j \| [x_i]) + \lambda_{B_j \neg D_j} P(\neg D_j \| [x_i]), \\ R(a_{N_j} \| [x_i]) &= \lambda_{N_j D_j} P(D_j \| [x_i]) + \lambda_{N_j \neg D_j} P(\neg D_j \| [x_i]). \end{aligned} \tag{1}$$

Since $P(D_j \| [x_i]]) + P(\neg D_j \| [x_i]) = 1$, we can simplify the rules based only on the probabilities $P(D_j \| [x_i])$ and the loss functions $\lambda_{\bullet\bullet}$. Then, we can easily induce the three-way decision rules when $\lambda_{P_j D_j} \leq \lambda_{B_j D_j} < \lambda_{N_j D_j}$ and $\lambda_{N_j \neg D_j} \leq \lambda_{B_j \neg D_j} < \lambda_{P_j \neg D_j}$ by using the minimum overall risk criterion:

(P) If $P(D_j \| [x_i]) \geq \alpha_j$ and $P(D_j \| [x_i]) \geq \gamma_j$, decide $x_i \in POS(D_j)$;
(B) If $P(D_j \| [x_i]) \leq \alpha_j$ and $P(D_j \| [x_i]) \geq \beta_j$, decide $x_i \in BND(D_j)$;
(N) If $P(D_j \| [x_i]) \leq \beta_j$ and $P(D_j \| [x_i]) \leq \gamma_j$, decide $x_i \in NEG(D_j)$.

where the parameter α_j, β_j and γ_j are defined as:

$$\begin{aligned} \alpha_j &= \frac{(\lambda_{P_j \neg D_j} - \lambda_{B_j \neg D_j})}{(\lambda_{P_j \neg D_j} - \lambda_{B_j \neg D_j}) + (\lambda_{B_j D_j} - \lambda_{P_j D_j})}; \\ \beta_j &= \frac{(\lambda_{B_j \neg D_j} - \lambda_{N_j \neg D_j})}{(\lambda_{N_j D_j} - \lambda_{B_j D_j}) + (\lambda_{B_j \neg D_j} - \lambda_{N_j \neg D_j})}; \\ \gamma_j &= \frac{(\lambda_{P_j \neg D_j} - \lambda_{N_j \neg D_j})}{(\lambda_{N_j D_j} - \lambda_{P_j D_j}) + (\lambda_{P_j \neg D_j} - \lambda_{N_j \neg D_j})}. \end{aligned} \tag{2}$$

In other words, from a loss function one can systematically determine the required threshold values.

3 Approach of Multiple-Category Attribute Reduct Based on DTRS

The concept of the lower and upper approximate distribution reduct was firstly presented by Mi et al. in [5] based on VPRS. In this section, we will introduce the concept of the lower and upper approximate distribution reduct into DTRS. An illustration is analyzed to indicate the validity of the proposed approaches.

3.1 Multiple-Category Attribute Reduct Method Using DTRS

Definition 1. *Given an information table* $S = (U, At = C \cup \{D\}, \{V_a | a \in At\}, \{I_a | a \in At\})$, $B \subseteq C$, $U = \{x_1, x_2, \cdots, x_n\}$, $U/IND(B) = \{[x_i]_B : x_i \in U, i = 1, 2, \cdots, n\}$, *the partition generated by* $IND(D)$ *is denoted by* $U/IND(D) = \{D_1, D_2, \cdots, D_m\}$, *the membership matrix* $M_B = (r_{ij})$, $0 \leq r_{ij} \leq 1$, *is a* $|U| \times |U/IND(D)|$ *matrix, in which the element* r_{ij} *is defined by* $P(D_j|[x_i]_B)$, *where* $i = 1, 2, \cdots, n$ *and* $j = 1, 2, \cdots, m$, *that can be denoted by:*

$$M_B = \begin{bmatrix} P(D_1|[x_1]_B) & P(D_2|[x_1]_B) & \cdots & P(D_m|[x_1]_B) \\ P(D_1|[x_2]_B) & P(D_2|[x_2]_B) & \cdots & P(D_m|[x_2]_B) \\ \vdots & \vdots & \cdots & \vdots \\ P(D_1|[x_n]_B) & P(D_2|[x_n]_B) & \cdots & P(D_m|[x_n]_B) \end{bmatrix}$$

The conditional probability[11] also be seen as rough membership value of an element belonging to X, so the rough membership function is given by the conditional probability as $\mu_B(X) = P(X|[x]_B)$. With the probabilistic interpretation of rough membership function, we indiscriminately use the rough membership function and the conditional probability in the following discussions. The membership matrix is referred to as a fuzzy matrix. In the theory of fuzzy sets, cut set and strong cut set are important notions. So, we have the following definition.

Definition 2. *Given an information table* $S = (U, At = C \cup \{D\}, \{V_a | a \in At\}, \{I_a | a \in At\})$, $B \subseteq C$, $U = \{x_1, x_2, \cdots, x_n\}$, *the partition generated by* $IND(D)$ *is denoted by* $U/IND(D) = \{D_1, D_2, \cdots, D_m\}$, *let* $\delta = (\delta_1, \delta_2, \cdots, \delta_m)$, $0 \leq \delta_i \leq 1$ *the* δ*-cutting matrix of the membership matrix* $M_B = (r_{ij})_{n\times m}$ *is defined by* $(M_B)_\delta = (r_{ij}(\delta_j))_{n\times m}$ *, where*

$$r_{ij}(\delta_j) = \begin{cases} 1 & r_{ij} \geq \delta_j \\ 0 & r_{ij} < \delta_j \end{cases} \tag{3}$$

the strong δ*-cutting matrix of the membership matrix* $M_B = (r_{ij})_{n\times m}$ *is defined by* $(M_B)_{\delta^+} = (r_{ij}(\delta_j))_{n\times m}$, *where*

$$r_{ij}(\delta_j) = \begin{cases} 1 & r_{ij} > \delta_j \\ 0 & r_{ij} \leq \delta_j \end{cases} \tag{4}$$

The δ-cutting matrix of the membership matrix is Boolean matrix, it determines a general relation from U to $U/IND(D)$. we can denote it $r_{ij}(\delta_j) = (M_B)_\delta(x_i, D_j)$ for $i = 1, 2, \cdots, n$ and $j = 1, 2, \cdots, m$. Namely, $(M_B)_\delta(x_i, D_j) = 1 \Leftrightarrow M_B(x_i, D_j) \geq \delta_j$.

Definition 3. *Given an information table* $S = (U, At = C \cup \{D\}, \{V_a | a \in At\}, \{I_a | a \in At\})$, $B \subseteq C$, $U = \{x_1, x_2, \cdots, x_n\}$, *the partition generated by* $IND(D)$ *is denoted by* $U/IND(D) = \{D_1, D_2, \cdots, D_m\}$, *let* $\alpha = (\alpha_1, \alpha_2, \cdots, \alpha_m)$ *and* $\beta = (\beta_1, \beta_2, \cdots, \beta_m)$, $M_B = (r_{ij})_{n\times m}$, *we define:*

(1) The (α, β) *lower approximate distribution matrix* $M_B^{\underline{apr}_{(\alpha,\beta)}} = (M_B)_\alpha = (r_{ij}(\alpha_j))_{n\times m}$ *denoted by* $M_B^{\underline{apr}_{(\alpha,\beta)}} = (r_{ij}^{\underline{a}})_{n\times m}$,

(2) The (α, β) *upper approximate distribution matrix* $M_B^{\overline{apr}_{(\alpha,\beta)}} = (M_B)_{\beta+} = (r_{ij}(\beta_j))_{n\times m}$ *denoted by* $M_B^{\overline{apr}_{(\alpha,\beta)}} = (r_{ij}^{\overline{a}})_{n\times m}$.

Definition 4. *Given an information table* $S = (U, At = C \cup \{D\}, \{V_a | a \in At\}, \{I_a | a \in At\})$, $B \subseteq C$, $U = \{x_1, x_2, \cdots, x_n\}$, *let* $\alpha = (\alpha_1, \alpha_2, \cdots, \alpha_m)$ *and* $\beta = (\beta_1, \beta_2, \cdots, \beta_m)$. *Then:*

(1) B is a (α, β) *lower approximate distribution consistent set iff* $M_B^{\underline{apr}_{(\alpha,\beta)}} = M_C^{\underline{apr}_{(\alpha,\beta)}}$, *B is a* (α, β) *lower approximate distribution reduct iff* $M_B^{\underline{apr}_{(\alpha,\beta)}} = M_C^{\underline{apr}_{(\alpha,\beta)}}$ *and* $M_A^{\underline{apr}_{(\alpha,\beta)}} \neq M_C^{\underline{apr}_{(\alpha,\beta)}}$ *for* $\forall A \subseteq B$.

(2) B is a (α, β) *upper approximate distribution consistent set iff* $M_B^{\overline{apr}_{(\alpha,\beta)}} = M_C^{\overline{apr}_{(\alpha,\beta)}}$ *is a* (α, β) *upper approximate distribution reduct iff* $M_B^{\overline{apr}_{(\alpha,\beta)}} = M_C^{\overline{apr}_{(\alpha,\beta)}}$ *and* $M_A^{\overline{apr}_{(\alpha,\beta)}} \neq M_C^{\overline{apr}_{(\alpha,\beta)}}$ *for* $\forall A \subseteq B$.

A (α, β) lower (upper) approximate distribution consistent set is a subset of attribute set that preserves the (α, β) lower(upper) approximate of all decision classes.

Definition 5. *Given an information table* $S = (U, At = C \cup \{D\}, \{V_a | a \in At\}, \{I_a | a \in At\})$, $U = \{x_1, x_2, \cdots, x_n\}$, $B \subseteq C$, *the partition generated by* $IND(D)$ *is denoted by* $U/IND(D) = \{D_1, D_2, \cdots, D_m\}$, *let* $\alpha = (\alpha_1, \alpha_2, \cdots, \alpha_m)$ *and* $\beta = (\beta_1, \beta_2, \cdots, \beta_m)$, *let us denote:*

$$\underline{\mathrm{A}}_B^{(\alpha,\beta)}(x_i) = \{D_j | M_B^{\underline{apr}_{(\alpha,\beta)}}(x_i, D_j) = 1\}, i = 1, 2, \cdots, n,$$
$$\overline{\mathrm{A}}_B^{(\alpha,\beta)}(x_i) = \{D_j | M_B^{\overline{apr}_{(\alpha,\beta)}}(x_i, D_j) = 1\}, i = 1, 2, \cdots, n.$$

Define:

$$D_{\underline{apr}_{(\alpha,\beta)}}([x],[y]) = \begin{cases} \{a \in C : I_a([x]) \neq I_a([y])\}, & \underline{\mathrm{A}}_B^{(\alpha,\beta)}([x]) \neq \underline{\mathrm{A}}_B^{(\alpha,\beta)}([y]) \\ C & \underline{\mathrm{A}}_B^{(\alpha,\beta)}([x]) = \underline{\mathrm{A}}_B^{(\alpha,\beta)}([y]) \end{cases},$$
$$D_{\overline{apr}_{(\alpha,\beta)}}([x],[y]) = \begin{cases} \{a \in C : I_a([x]) \neq I_a([y])\}, & \overline{\mathrm{A}}_B^{(\alpha,\beta)}([x]) \neq \overline{\mathrm{A}}_B^{(\alpha,\beta)}([y]) \\ C & \overline{\mathrm{A}}_B^{(\alpha,\beta)}([x]) = \overline{\mathrm{A}}_B^{(\alpha,\beta)}([y]) \end{cases} \quad (5)$$

where $x, y \in U$.

Theorem 1. *(Judgement Theorem of knowledge reduct) Given an information table* $S = (U, At = C \cup \{D\}, \{V_a | a \in At\}, \{I_a | a \in At\})$, $B \subseteq C$, *let* $\alpha = (\alpha_1, \alpha_2, \cdots, \alpha_m)$ *and* $\beta = (\beta_1, \beta_2, \cdots, \beta_m)$. *Then:*

(1) B is a (α, β) *lower approximate distribution consistent set iff* $B \cap D_{\underline{apr}_{(\alpha,\beta)}}([x],[y]) \neq \emptyset$, *for all* $\underline{\mathrm{A}}_B^{(\alpha,\beta)}([x]) \neq \underline{\mathrm{A}}_B^{(\alpha,\beta)}([y])$.

(2) B is a (α, β) *upper approximate distribution consistent set iff* $B \cap D_{\overline{apr}_{(\alpha,\beta)}}([x],[y]) \neq \emptyset$, *for all* $\underline{\mathrm{A}}_B^{(\alpha,\beta)}([x]) \neq \underline{\mathrm{A}}_B^{(\alpha,\beta)}([y])$.

Definition 6. *Given an information table $S = (U, At = C \cup \{D\}, \{V_a | a \in At\}, \{I_a | a \in At\})$, $B \subseteq C$, let $\alpha = (\alpha_1, \alpha_2, \cdots, \alpha_m)$ and $\beta = (\beta_1, \beta_2, \cdots, \beta_m)$, the (α, β) lower approximate(upper approximate) distribution discernibility functions are defined respectively:*

$$\wedge\{\vee\{a : a \in D_{\underline{apr}_{(\alpha,\beta)}}([x],[y])\} : \underline{A}_B^{(\alpha,\beta)}([x]) \neq \underline{A}_B^{(\alpha,\beta)}([y])\},$$

$$\wedge\{\vee\{a : a \in D_{\overline{apr}_{(\alpha,\beta)}}([x],[y])\} : \overline{A}_B^{(\alpha,\beta)}([x]) \neq \overline{A}_B^{(\alpha,\beta)}([y])\}. \quad (6)$$

From the (α, β) lower approximate(upper approximate) distribution discernibility function, the (α, β) lower approximate(upper approximate) distribution reduct, which is a prime implicant of the reduced disjunctive form of the discernibility function, can be obtained.

3.2 An Example

Consider an information table $S = (U, At = C \cup \{D\}, \{V_a | a \in At\}, \{I_a | a \in At\})$ showed in Table 1, where $U = \{x_1, x_2, \cdots, x_{12}\}$, $C = \{a_1, a_2, a_3, a_4, a_5\}$, $D = \{d\}$.

It can be easily calculated that the equivalence classes of condition attribute and decision attribute are as follows:

$U/IND(C) = \{C_1, C_2, C_3, C_4, C_5, C_6\}$,

where $C_1 = \{x_1, x_2\}$, $C_2 = \{x_3, x_8, x_{10}\}$, $C_3 = \{x_5\}$, $C_4 = \{x_4, x_6, x_7, x_{12}\}$, $C_5 = \{x_9\}$, $C_6 = \{x_{11}\}$.

$U/IND(d) = \{D_1, D_2, D_3\}$,

where $D_1 = \{x_1, x_2, x_8, x_{10}\}$, $D_2 = \{x_3, x_5, x_{12}\}$, $D_3 = \{x_4, x_6, x_7, x_9, x_{11}\}$.

According to Definition 1, we have the membership matrix in Figure 1.

For simplicity, we suppose $\alpha = \{0.6, 0.7, 0.7\}$ and $\beta = \{0.3, 0.3, 0.5\}$ for the different categories, respectively. These coefficients can be computed from the loss functions.

Table 1. An information table

U	a_1	a_2	a_3	a_4	a_5	d
x_1	0	0	1	0	1	1
x_2	0	0	1	0	1	1
x_3	0	0	0	1	1	2
x_4	0	1	1	0	1	3
x_5	0	0	0	0	0	2
x_6	0	1	1	0	1	3
x_7	0	1	1	0	1	3
x_8	0	0	0	1	1	1
x_9	1	1	1	1	1	3
x_{10}	0	0	0	1	1	1
x_{11}	1	1	1	0	1	3
x_{12}	0	1	1	0	1	2

According to the definitions of the (α, β) lower approximate and (α, β) upper approximate distribution matrix, we can obtain them in Figure 2.

In terms of the (α, β) lower approximate distribution matrix, we can obtain $\underline{\mathbf{A}}_B^{(\alpha,\beta)}(x_1) = \underline{\mathbf{A}}_B^{(\alpha,\beta)}(x_2) = \underline{\mathbf{A}}_B^{(\alpha,\beta)}(x_3) = \underline{\mathbf{A}}_B^{(\alpha,\beta)}(x_8) = \underline{\mathbf{A}}_B^{(\alpha,\beta)}(x_{10}) = \{D_1\}$, $\underline{\mathbf{A}}_B^{(\alpha,\beta)}(x_5) = \{D_2\}$,

$\underline{\mathbf{A}}_B^{(\alpha,\beta)}(x_4) = \underline{\mathbf{A}}_B^{(\alpha,\beta)}(x_6) = \underline{\mathbf{A}}_B^{(\alpha,\beta)}(x_7) = \underline{\mathbf{A}}_B^{(\alpha,\beta)}(x_9) = \underline{\mathbf{A}}_B^{(\alpha,\beta)}(x_{11}) = \underline{\mathbf{A}}_B^{(\alpha,\beta)}$ $(x_{12}) = \{D_3\}$.

$$M_C = \begin{bmatrix} 1 & 0 & 0 \\ 1 & 0 & 0 \\ 0.67 & 0.33 & 0 \\ 0 & 0.25 & 0.75 \\ 0 & 1 & 0 \\ 0 & 0.25 & 0.75 \\ 0 & 0.25 & 0.75 \\ 0.67 & 0.33 & 0 \\ 0 & 0 & 1 \\ 0.67 & 0.33 & 0 \\ 0 & 0 & 1 \\ 0 & 0.25 & 0.75 \end{bmatrix}.$$

Fig. 1. The membership matrix

$$M_B^{\underline{apr}_{(\alpha,\beta)}} = \begin{bmatrix} 1 & 0 & 0 \\ 1 & 0 & 0 \\ 1 & 0 & 0 \\ 0 & 0 & 1 \\ 0 & 1 & 0 \\ 0 & 0 & 1 \\ 0 & 0 & 1 \\ 1 & 0 & 0 \\ 0 & 0 & 1 \\ 1 & 0 & 0 \\ 0 & 0 & 1 \\ 0 & 0 & 1 \end{bmatrix}, M_B^{\overline{apr}_{(\alpha,\beta)}} = \begin{bmatrix} 1 & 0 & 0 \\ 1 & 0 & 0 \\ 1 & 1 & 0 \\ 0 & 0 & 1 \\ 0 & 1 & 0 \\ 0 & 0 & 1 \\ 0 & 0 & 1 \\ 1 & 1 & 0 \\ 0 & 0 & 1 \\ 1 & 1 & 0 \\ 0 & 0 & 1 \\ 0 & 0 & 1 \end{bmatrix}.$$

Fig. 2. The (α, β) lower approximate and (α, β) upper approximate distribution matrix

By computation we have the following

$D_{\underline{apr}_{(\alpha,\beta)}}(C_1, C_3) = \{a_3, a_5\}$, $D_{\underline{apr}_{(\alpha,\beta)}}(C_1, C_4) = \{a_2\}$, $D_{\underline{apr}_{(\alpha,\beta)}}(C_1, C_5) = \{a_1, a_2, a_4\}$,

$D_{\underline{apr}_{(\alpha,\beta)}}(C_1, C_6) = \{a_1, a_2\}$, $D_{\underline{apr}_{(\alpha,\beta)}}(C_2, C_3) = \{a_4, a_5\}$, $D_{\underline{apr}_{(\alpha,\beta)}}(C_2, C_4) = \{a_2, a_3, a_4\}$,

$D_{\underline{apr}_{(\alpha,\beta)}}(C_2, C_5) = \{a_1, a_2, a_3\}$, $D_{\underline{apr}_{(\alpha,\beta)}}(C_2, C_6) = \{a_1, a_2, a_3, a_4\}$,

$D_{\underline{apr}_{(\alpha,\beta)}}(C_3, C_4) = \{a_2, a_3, a_5\}$, $D_{\underline{apr}_{(\alpha,\beta)}}(C_3, C_5) = \{a_1, a_2, a_3, a_4, a_5\}$,

$D_{\underline{apr}_{(\alpha,\beta)}}(C_3, C_6) = \{a_1, a_2, a_3, a_5\}$.

From the (α, β) lower approximate distribution discernibility function, we conclude that $\{a_2, a_5\}$ and $\{a_2, a_3, a_4\}$ are two (α, β) lower approximate distribution reducts of S.

Similarly, we can also conclude that $\{a_2, a_3, a_4\}$, $\{a_2, a_3, a_5\}$ and $\{a_2, a_4, a_5\}$ are three (α, β) upper approximate distribution reducts of S.

4 Conclusion

The decision-theoretic rough set model provided a systematic way to calculate the required probabilistic threshold values based on the well established Bayesian decision theory. In this paper, the (α, β) lower approximate and (α, β) upper approximate distribution reduct are introduced into DTRS. The new approaches to attribute reduct in multiple-category classification model with DTRS are obtained as well, and the algorithm will be devised in the further work.

Acknowledgments. This work was supported in part by the China NNSFC grant(No.61073146) and the Chongqing CSTC grant (No.KJ110522).

References

1. Duda, R.O., Hart, P.E.: Pattern Classification and Scene Analysis. Wiley Press, New York (1973)
2. Jia, X.Y., Li, W.W., Shang, L., Chen, J.J.: An Optimization Viewpoint of Decision-Theoretic Rough Set Model. In: Yao, J., Ramanna, S., Wang, G., Suraj, Z. (eds.) RSKT 2011. LNCS, vol. 6954, pp. 457–465. Springer, Heidelberg (2011)
3. Li, H.X., Zhou, X.Z., Zhao, J.B., Liu, D.: Attribute Reduction in Decision-Theoretic Rough Set Model: A Further Investigation. In: Yao, J., Ramanna, S., Wang, G., Suraj, Z. (eds.) RSKT 2011. LNCS, vol. 6954, pp. 466–475. Springer, Heidelberg (2011)
4. Liu, D., Li, T.R., Hu, P., Li, H.X.: Multiple-Category Classification with Decision-Theoretic Rough Sets. In: Yu, J., Greco, S., Lingras, P., Wang, G., Skowron, A. (eds.) RSKT 2010. LNCS (LNAI), vol. 6401, pp. 703–710. Springer, Heidelberg (2010)
5. Mi, J.S., Wu, W.Z., Zhang, W.X.: Approaches to knowledge reduction based on variable precision rough set model. Information Sciences 159, 255–272 (2004)
6. Miao, D.Q., Zhao, Y., Li, H.X., Xu, F.F.: Relative reducts in consistent and inconsistent decision tables of the pawlak rough set model. Information Sciences 179, 4140–4150 (2009)
7. Min, F., He, H., Qian, Y., Zhu, W.: Test-cost-sensitive attribute reduction. Information Sciences 181, 4928–4942 (2011)
8. Pawlak, Z.: Rough sets. International Journal of Computer and Information Science 11, 341–356 (1982)
9. Slezak, D., Ziarko, W.: The investigation of the bayesian rough set model. International Journal Approximate Reasoning 40, 81–89 (2005)
10. Wang, G.Y., Yu, H., Yang, D.: Decision table reduction based on conditional information entropy. Chinese Journal of Computers 25, 759–766 (2002)
11. Wong, S.K.M., Ziarko, W.: Comparison of the probabilistic approximate classification and the fuzzy set model. Fuzzy Sets and Systems 21, 357–362 (1987)
12. Yao, Y.Y., Wong, S.K.M., Lingras, P.: A decision-theoretic rough set model. In: Ras, Z.W., Zemankova, M., Emrich, M.L. (eds.) Methodologies for Intelligent Systems 5, pp. 17–24. North-Holland, New York (1990)
13. Yao, Y.Y., Zhao, Y.: Attribute reduction in decision-theoretic rough set model. Information Sciences 178, 3356–3373 (2008)
14. Zhao, Y., Wong, S.K.M., Yao, Y.: A Note on Attribute Reduction in the Decision-Theoretic Rough Set Model. In: Peters, J.F., Skowron, A., Chan, C.-C., Grzymala-Busse, J.W., Ziarko, W.P. (eds.) Transactions on Rough Sets XIII. LNCS, vol. 6499, pp. 260–275. Springer, Heidelberg (2011)
15. Zhou, B.: A New Formulation of Multi-category Decision-Theoretic Rough Sets. In: Yao, J., Ramanna, S., Wang, G., Suraj, Z. (eds.) RSKT 2011. LNCS, vol. 6954, pp. 514–522. Springer, Heidelberg (2011)

Appendix: The Proof of Theorem 1

To prove Theorem 1 we need to prove the following two lemmas first.

Lemma 1. *Given an information table $S = (U, At = C \cup \{D\}, \{V_a | a \in At\}, \{I_a | a \in At\})$, $B \subseteq C$, $U = \{x_1, x_2, \cdots, x_n\}$, the partition generated by $IND(D)$ is denoted by $U/IND(D) = \{D_1, D_2, \cdots, D_m\}$, let $\alpha = (\alpha_1, \alpha_2, \cdots, \alpha_m)$ and $\beta = (\beta_1, \beta_2, \cdots, \beta_m)$. denoted by:*

$$\underline{\mathrm{A}}_B^{(\alpha,\beta)}(x_i) = \{D_j | M_B^{\underline{apr}(\alpha,\beta)}(x_i, D_j) = 1\}, i = 1, 2, \cdots, n,$$
$$\overline{\mathrm{A}}_B^{(\alpha,\beta)}(x_i) = \{D_j | M_B^{\overline{apr}(\alpha,\beta)}(x_i, D_j) = 1\}, i = 1, 2, \cdots, n.$$

Then:

(1) B is a (α, β) lower approximate distribution consistent set iff $\underline{\mathrm{A}}_B^{(\alpha,\beta)}(x_i) = \underline{\mathrm{A}}_C^{(\alpha,\beta)}(x_i)$, $i = 1, 2, \cdots, n$.

(2) B is a (α, β) upper approximate distribution consistent set iff $\overline{\mathrm{A}}_B^{(\alpha,\beta)}(x_i) = \overline{\mathrm{A}}_C^{(\alpha,\beta)}(x_i)$, $i = 1, 2, \cdots, n$.

Proof. (1)"$\Rightarrow$"if B is a (α, β) lower approximate distribution consistent set, by Definition 1, we have $M_B^{\underline{apr}(\alpha,\beta)} = M_C^{\underline{apr}(\alpha,\beta)}$, then $M_B^{\underline{apr}(\alpha,\beta)}(x_i, D_j) = M_C^{\underline{apr}(\alpha,\beta)}$ (x_i, D_j) for $\forall j = 1, 2, \cdots, m$ and $\forall i = 1, 2, \cdots, n$. When $M_B^{\underline{apr}(\alpha,\beta)}(x_i, D_j) = 1$, we have $M_C^{\underline{apr}}(x_i, D_j) = 1$, i.e., $D_j \in \underline{\mathrm{A}}_B^{(\alpha,\beta)}(x_i) \Leftrightarrow D_j \in \underline{\mathrm{A}}_C^{(\alpha,\beta)}(x_i)$. So $\underline{\mathrm{A}}_B^{(\alpha,\beta)}(x_i) = \underline{\mathrm{A}}_C^{(\alpha,\beta)}(x_i)$ for $i = 1, 2, \cdots, n$.

"$\Leftarrow$"since $\underline{\mathrm{A}}_B^{(\alpha,\beta)}(x_i) = \underline{\mathrm{A}}_C^{(\alpha,\beta)}(x_i)$ for $\forall i = 1, 2, \cdots, n$, then $D_j \in \underline{\mathrm{A}}_B^{(\alpha,\beta)}(x_i) \Leftrightarrow D_j \in \underline{\mathrm{A}}_C^{(\alpha,\beta)}(x_i)$ for $1 \leq j \leq m$, i.e., $M_B^{\underline{apr}(\alpha,\beta)}(x_i, D_j) = M_C^{\underline{apr}(\alpha,\beta)}(x_i, D_j) = 1$. Again, it follow that $D_j \notin \underline{\mathrm{A}}_B^{(\alpha,\beta)}{}_B^{(\alpha,\beta)}(x_i) \Leftrightarrow D_j \notin \underline{\mathrm{A}}_C^{(\alpha,\beta)}(x_i)$ for $1 \leq j \leq m$, i.e., $M_B^{\underline{apr}(\alpha,\beta)}(x_i, D_j) = M_C^{\underline{apr}(\alpha,\beta)}(x_i, D_j) = 0$, then we have $M_B^{\underline{apr}(\alpha,\beta)} = M_C^{\underline{apr}(\alpha,\beta)}$, according to Definition 4, B is a (α, β) lower approximate distribution consistent set.

Lemma 2. *Given an information table $S = (U, At = C \cup \{D\}, \{V_a | a \in At\}, \{I_a | a \in At\})$, $B \subseteq C$, $U = \{x_1, x_2, \cdots, x_n\}$, let $\alpha = (\alpha_1, \alpha_2, \cdots, \alpha_m)$ and $\beta = (\beta_1, \beta_2, \cdots, \beta_m)$. Then:*

(1) B is a (α, β) lower approximate distribution consistent set iff x_i and x_k satisfy $\underline{\mathrm{A}}_C^{(\alpha,\beta)}(x_i) = \underline{\mathrm{A}}_C^{(\alpha,\beta)}(x_k)$, then $[x_i]_B \cap [x_k]_B = \emptyset$, where $i = 1, 2, \cdots, n$ and $k = 1, 2, \cdots, n$.

(2) B is a (α, β) upper approximate distribution consistent set iff x_i and x_k satisfy $\overline{\mathrm{A}}_C^{(\alpha,\beta)}(x_i) = \overline{\mathrm{A}}_C^{(\alpha,\beta)}(x_k)$, then $[x_i]_B \cap [x_k]_B = \emptyset$, where $i = 1, 2, \cdots, n$ and $k = 1, 2, \cdots, n$.

Proof. (1)"$\Rightarrow$" suppose that $[x_i]_B \cap [x_k]_B \neq \emptyset$ for $i = 1, 2, \cdots, n$ and $k = 1, 2, \cdots, n$, then there must be $[x_i]_B = [x_k]_B$, thus $M_B^{\overline{apr}_{(\alpha,\beta)}}(x_i, D_j) = M_B^{\overline{apr}_{(\alpha,\beta)}}(x_k, D_j)$ hold, that is to say $\underline{A}_B^{(\alpha,\beta)}(x_i) = \underline{A}_B^{(\alpha,\beta)}(x_k)$. Since B is a (α, β) lower approximate distribution consistent set, by Lemma 1, we have the following $\underline{A}_B^{(\alpha,\beta)}(x_i) = \underline{A}_C^{(\alpha,\beta)}(x_i)$, $\underline{A}_B^{(\alpha,\beta)}(x_k) = \underline{A}_C^{(\alpha,\beta)}(x_k)$, so $\underline{A}_C^{(\alpha,\beta)}(x_i) = \underline{A}_C^{(\alpha,\beta)}(x_k)$, this is contrary to the assumption.

"$\Leftarrow$" for $i = 1, 2, \cdots, n$ and $k = 1, 2, \cdots, n$, if $[x_k]_C \subseteq [x_i]_B$, then $[x_i]_B \cap [x_k]_B \neq \emptyset$, namely $[x_i]_B = [x_k]_B$, in terms of the assumption, it follow that $\underline{A}_C^{(\alpha,\beta)}(x_i) = \underline{A}_C^{(\alpha,\beta)}(x_k)$. For $\forall 1 \le j \le m$, if $M_B^{\overline{apr}_{(\alpha,\beta)}}(x_i, D_j) = 1$, then for any $y \in [x_i]_B$, we have $M_B^{\overline{apr}_{(\alpha,\beta)}}(y, D_j) = 1$, so for $\forall 1 \le k \le n$, if $[x_k]_C \subseteq [x_i]_B$, then $M_B^{\overline{apr}_{(\alpha,\beta)}}(x_k, D_j) = 1$, by Lemma 1, $D_j \in \underline{A}_C^{(\alpha,\beta)}(x_k)$, i.e., $M_C^{\overline{apr}_{(\alpha,\beta)}}(x_i, D_j) = 1$.

On the other hand, if $M_C^{\overline{apr}_{(\alpha,\beta)}}(x_i, D_j) = 1$, then we have $D_j \in \underline{A}_C^{(\alpha,\beta)}(x_i)$, so for $\forall 1 \le k \le n$, if $[x_k]_C \subseteq [x_i]_B$, then there must be $M_C^{\overline{apr}_{(\alpha,\beta)}}(x_k, D_j) = 1$, namely, $P(D_j|[x_k]_C) \ge \alpha_j$. Thus, we obtain that

$$\begin{aligned} P(D_j|[x_i]_B) &= \sum |[x_k]_C \cap D_j| / |[x_i]_B| \\ &= \sum P(D_j|[x_k]_C) \frac{|[x_k]_C|}{|[x_i]_B|} \\ &\ge \alpha_j \sum \frac{|[x_k]_C|}{|[x_i]_B|} = \alpha_j \end{aligned}$$

Where $[x_k]_C \subseteq [x_i]_B$.

Finally, we have $M_B^{\overline{apr}_{(\alpha,\beta)}}(x_i, D_j) = 1$, therefore we can conclude that $M_B^{\overline{apr}_{(\alpha,\beta)}} = M_C^{\overline{apr}_{(\alpha,\beta)}}$, according to Definition 4, B is a (α, β) lower approximate distribution consistent set.

The proof of Theorem 1 is as follows.

Proof. (1)"$\Rightarrow$" suppose B is a (α, β) lower approximate distribution consistent set. For any $\underline{A}_C^{(\alpha,\beta)}(x) \neq \underline{A}_C^{(\alpha,\beta)}(y)$. According to Lemma 2, we obtain that $[x]_B \cap [y]_B \neq \emptyset$, that is to say, there exists $a \in B$ such that $I_a([x]) \neq I_a([y])$, so, it follows that $a \in D_{\underline{apr}_{(\alpha,\beta)}}([x], [y])$, i.e., $B \cap D_{\underline{apr}_{(\alpha,\beta)}}([x], [y]) \neq \emptyset$.

"$\Leftarrow$" Suppose $\underline{A}_C^{(\alpha,\beta)}([x]) = \underline{A}_C^{(\alpha,\beta)}([y])$. It should be noticed that $\underline{A}_C^{(\alpha,\beta)}(x) \neq \underline{A}_C^{(\alpha,\beta)}(y)$. For any $a \in B$, when $a \notin D_{\underline{apr}_{(\alpha,\beta)}}([x], [y])$, we know $I_a([x]) = I_a([y])$. Accordingly, $I_a(x) = I_a(y)$ which implies $[x]_B = [y]_B$. In terms of Lemma 2, we can conclude that B is not a (α, β) lower approximate distribution consistent set. Thus, if $B \cap D_{\underline{apr}_{(\alpha,\beta)}}([x], [y])) \neq \emptyset$ for all $\underline{A}_C^{(\alpha,\beta)}([x]) = \underline{A}_C^{(\alpha,\beta)}([y])$, then B is a (α, β) lower approximate distribution consistent set. This completes the proof.

Three-Way Decisions Method for Overlapping Clustering

Hong Yu and Ying Wang

Chongqing Key Lab of Computational Intelligence,
Chongqing University of Posts and Telecommunications,
Chongqing, 400065, P.R. China
yuhong@cqupt.edu.cn

Abstract. Most of clustering methods assume that each object must be assigned to exactly one cluster, however, overlapping clustering is more appropriate than crisp clustering in a variety of important applications such as the network structure analysis and biological information. This paper provides a three-way decision strategy for overlapping clustering based on the decision-theoretic rough set model. Here, each cluster is described by an interval set that is defined by a pair of sets called the lower and upper bounds. Besides, a density-based clustering algorithm is proposed using the new strategy, and the results of the experiments show the strategy is effective to overlapping clustering.

Keywords: overlapping clustering, three-way decision, decision-theoretic rough set theory, data mining.

1 Introduction

In recent years, clustering has been widely used as a powerful tool to reveal underlying patterns in many areas such as data mining, web mining, geographical data processing, medicine and so on. Most of clustering methods assume that each object must be assigned to exactly one cluster. However, in a variety of important applications such as network structure analysis, wireless sensor networks and biological information, overlapping clustering is more appropriate[3].

Many researchers have proposed some overlapping clustering methods for different application background. For example, Takaki and Tamura et al. [10] propose a method of overlapping clustering for network structure analysis, Aydin and Naït-Abdesselam et al. [1] propose an overlapping clusters algorithm used in the mobile Ad hoc networks. Lingras and Bhalchandra et al. [5] compare crisp and fuzzy clustering in the mobile phone call dataset. Obadi and Drážďilová et al. [8] propose an overlapping clustering method for DBLP datasets based on rough set theory.

The rough set theory [9] approximates a concept by three regions, namely, the positive, boundary and negative regions, which immediately leads to the notion of three-way decision clustering approach. Three-way decisions constructed from

J.T. Yao et al. (Eds.): RSCTC 2012, LNAI 7413, pp. 277–286, 2012.

the three regions are associated with different actions and decisions. In fact, the three-way decision approach has been achieved in some areas as the email spam filtering [15], three-way investment decisions [6], and so on [4] [12].

To combat the overlapping clustering, this paper proposes a new three-way decision clustering strategy based on the decision-theoretic rough set model [13]. Yao and Lingras et al. [14] had represented each cluster by an interval set instead of a single set as the representation of a cluster. Chen and Miao [2] study the clustering method represented as interval sets, wherein the rough k-means clustering method is combined. Inspired by the representation, the cluster in our strategy is also represented by an interval set, which is defined by a pair of sets called the lower and upper bounds. Objects in the lower bound are typical elements of the cluster and objects between the upper and lower bounds are fringe elements of the cluster.

Furthermore, the solutions to obtain the lower and upper bounds are formulated based on the three-way decisions in this paper. Then, a density-based clustering algorithm is proposed, and we demonstrate the effectiveness of the algorithm through experiments.

2 Formulation of Clustering

2.1 Decision-Theoretic Rough Set Model

The decision-theoretic rough set model [13], DTRS shorted, applies the Bayesian decision procedure for the construction of probabilistic approximations.

Let $\Omega = \{A, A^c\}$ denote the set of states indicating that an object is in A and not in A, respectively. Let $Action = \{a_P, a_N, a_B\}$ be the set of actions, where a_P, a_N, and a_B represent the three actions in classifying an object, deciding $POS(A)$, deciding $NEG(A)$ and deciding $BND(A)$, respectively. Let $i = P, N, B$, and $\lambda_{iP}(a_i|A)$ and $\lambda_{iN}(a_i|A^c)$ denote the loss (cost) for taking the action a_i when the state is A, A^c, respectively. For an object with description $[x]$, suppose an action a_i is taken. The expected loss $R(a_i|[x])$ associated with taking the individual actions can be expressed as:

$$
\begin{aligned}
R(a_P|[x]) &= \lambda_{PP}P(A|[x]) + \lambda_{PN}P(A^c|[x]), \\
R(a_N|[x]) &= \lambda_{NP}P(A|[x]) + \lambda_{NN}P(A^c|[x]), \\
R(a_B|[x]) &= \lambda_{BP}P(A|[x]) + \lambda_{BN}P(A^c|[x]).
\end{aligned}
$$

where the probabilities $P(A|[x])$ and $P(A^c|[x])$ are the probabilities that an object in the equivalence class $[x]$ belongs to A and A^c, respectively.

2.2 Extend DTRS for Clustering

To define our framework, we will assume $\mathbf{C} = \{C_1, \cdots, C_k, \cdots, C_K\}$, where $C_k \subseteq U$, is a family of clusters of a universe $U = \{x_1, \cdots, x_n\}$.

In order to interpret clustering, let's extend the DTRS model firstly. The set of states is given by $\Omega = \{C, \neg C\}$, the two complement states indicate that an object is in a cluster C and not in a cluster C, respectively. The set of action is given by $A = \{a_P, a_B, a_N\}$, where a_P, a_B and a_N represent the three actions in classifying an object, a_P represents that we will take the description of an object x into the domain of the cluster C; a_B represents that we will take the description of an object x into the boundary domain of the cluster C; a_N represents that we will take the description of an object x into the negative domain of the C.

Let λ_{PP} ,λ_{BP}, λ_{NP}, λ_{PN}, λ_{BN}, λ_{NN} denote the loss (cost) for taking the action a_P, a_B and a_N when the state is C, $\neg C$, respectively. For an object x with description $[x]$, suppose an action a_i is taken. According to Subsection 2.1, the expected loss associated with taking the actions can be expressed as:

$$\begin{array}{l} Risk(a_P|[x]) = \lambda_{PP}Pr(C|[x]) + \lambda_{PN}Pr(\neg C|[x]); \\ Risk(a_B|[x]) = \lambda_{BP}Pr(C|[x]) + \lambda_{BN}Pr(\neg C|[x]); \\ Risk(a_N|[x]) = \lambda_{NP}Pr(C|[x]) + \lambda_{NN}Pr(\neg C|[x]). \end{array} \tag{1}$$

Where $Pr(C|[x])$ represents the probability that an object x in the description $[x]$ belongs to the cluster C, and $Pr(C|[x]) + Pr(\neg C|[x]) = 1$. The Bayesian decision procedure leads to the following minimum-risk decision:

$$\begin{array}{l} (P) If\ Risk(a_P|[x]) \le Risk(a_N|[x])\ and\ Risk(a_P|[x]) \le Risk(a_B|[x]), \\ decide\ POS(C); \\ (B) If\ Risk(a_B|[x]) < Risk(a_P|[x])\ and\ Risk(a_B|[x]) < Risk(a_N|[x]), \\ decide\ BND(C); \\ (N) If\ Risk(a_N|[x]) \le Risk(a_P|[x])\ and\ Risk(a_N|[x]) \le Risk(a_B|[x], \\ decide\ NEG(C); \end{array} \tag{2}$$

Consider a special kind of loss functions with $\lambda_{PP} \le \lambda_{BP} < \lambda_{NP}$ and $\lambda_{NN} \le \lambda_{BN} < \lambda_{PN}$. That is, the loss of classifying an object x belonging to C into the positive region $POS(C)$ is less than or equal to the loss of classifying x into the boundary region $BND(C)$, and both of these losses are strictly less than the loss of classifying x into the negative region $NEG(C)$. The reverse order of losses is used for classifying an object x that does not belong to C, namely the object x is a negative instance of C. For this type of loss function, the above minimum-risk decision rules can be written as:

$$\begin{array}{l} (P) If\ Pr(C|[x]) \ge \alpha\ and\ Pr(C|[x]) \ge \gamma, decide\ POS(C); \\ (B) If\ Pr(C|[x]) < \alpha\ and\ Pr(C|[x]) > \beta, decide\ BND(C); \\ (N) If\ Pr(C|[x]) \le \beta\ and\ Pr(C|[x]) \le \gamma, decide\ NEG(C); \end{array} \tag{3}$$

Where:

$$\begin{array}{l} \alpha = \frac{(\lambda_{PN}-\lambda_{BN})}{(\lambda_{PN}-\lambda_{BN})+(\lambda_{BP}-\lambda_{PP})} = (1 + \frac{(\lambda_{BP}-\lambda_{PP})}{(\lambda_{PN}-\lambda_{BN})})^{-1} \\ \gamma = \frac{(\lambda_{PN}-\lambda_{NN})}{(\lambda_{PN}-\lambda_{NN})+(\lambda_{NP}-\lambda_{PP})} = (1 + \frac{(\lambda_{NP}-\lambda_{PP})}{(\lambda_{PN}-\lambda_{NN})})^{-1} \\ \beta = \frac{(\lambda_{BN}-\lambda_{NN})}{(\lambda_{BN}-\lambda_{NN})+(\lambda_{NP}-\lambda_{BP})} = (1 + \frac{(\lambda_{NP}-\lambda_{BP})}{(\lambda_{BN}-\lambda_{NN})})^{-1} \end{array} \tag{4}$$

In this paper, we consider that the cluster have the boundary, so we just discuss the relationship between thresholds α and β as $\alpha > \beta$. According to Eq.(4), it follows that $\alpha > \gamma > \beta$. After tie-breaking, the following simplified rules (P)-(N) are obtained:

$$\begin{array}{l}(P)\ If\ Pr(C|[x]) \geq \alpha, decide\ POS(C);\\(B)\ If\ \beta < Pr(C|[x])) < \alpha, decide\ BND(C);\\(N)\ If\ Pr(C|[x]) \leq \beta, decide\ NEG(C).\end{array} \quad (5)$$

Obviously, rules (P)-(N) give a three-way decision method for clustering. That is, an object belongs to a cluster definitely if it is in $POS(C)$ based on the available information; an object may be a fringe member if it is in $BND(C)$, we can decide whether it is in a cluster through further information. Clustering algorithms can be devised according to the rules (P)-(N).

On the other hand, according to the rough set theory [9] and the rules (P)-(N), for a subset $C \subseteq U$, we can define its lower and upper approximations as follows.

$$\begin{array}{l}\underline{apr}(C) = POS(C) = \{x|Pr(C|[x]) \geq \alpha\};\\\overline{apr}(C) = POS(C) \cup BND(C) = \{x|Pr(C|[x]) > \beta\}.\end{array} \quad (6)$$

2.3 Re-formulation of Clustering Using Interval Set

Yao and Lingras et al.[14] had formulated the clustering using the form of interval sets. It is naturally that the region between the lower and upper bound of an interval set means the overlapping region.

Assume $\mathbf{C} = \{C_1, \cdots, C_k, \cdots, C_K\}$ is a family of clusters of a universe $U = \{x_1, \cdots, x_n\}$. Formally, we can define a clustering by the properties:

$$(i)\ C_k \neq \emptyset, 0 \leq k \leq K; \quad (ii)\ \bigcup_{C_k \in \mathbf{C}} C_k = U.$$

Property (i) requires that each cluster cannot be empty. Property (ii) states that every $x \in U$ belongs to at least one cluster. Furthermore, if $C_i \cap C_j = \emptyset, i \neq j$, it is a crisp clustering, otherwise it is an overlapping clustering.

As we have discussed, we may use an interval set to represent the cluster in $\mathbf{C}$, namely, C_k is represented by an interval set $[C_k^l, C_k^u]$. Combine the conclusion in the above subsection, we can represent the lower and upper bound of the interval set as the lower and upper approximate, that is, C_k is represented by an interval set $[\underline{apr}(C_k), \overline{apr}(C_k)]$.

Any set in the family $[\underline{apr}(C_k), \overline{apr}(C_k)] = \{X|\underline{apr}(C_k) \subseteq X \subseteq \overline{apr}(C_k)\}$ may be the actual cluster C_k. The objects in $\underline{apr}(C_k)$ may represent typical objects of the cluster C_k, objects in $\overline{apr}(C_k) - \underline{apr}(C_k)$ may represent fringe objects, and objects in $U - \overline{apr}(C_k)$ may represent the negative objects. With respect to the family of clusters $\mathbf{C} = \{C_1, \cdots, C_k, \cdots, C_K\}$, we have the following family of interval set clusters:

$$\mathbf{C} = [\underline{apr}(C_1), \overline{apr}(C_1)], \ldots, [\underline{apr}(C_k), \overline{apr}(C_k)], \ldots, [\underline{apr}(C_K), \overline{apr}(C_K)].$$

Corresponding to Property (i) and (ii), we adopt the following properties for a clustering in the form of interval set:

$$(i)\ \underline{apr}(C_k) \neq \emptyset, 0 \leq k \leq K; \quad (ii)\ \bigcup \overline{apr}(C_k) = U.$$

Property (i) requires that the lower approximate must not be empty. It implies that the upper approximate is not empty. It is reasonable to assume that each cluster must contain at least one typical object and hence its lower bound is not empty. In order to make sure that a clustering is physically meaningful, Property (ii) states that any object of U belongs to the upper approximate of a cluster, which ensures that every object is properly clustered.

According to Eq.(6), the family of clusters $\mathbf{C}$ give a three-way decision clustering. Namely, objects in $\underline{apr}(C_k)$ are decided definitely to belong to the cluster C_k, objects in $U - \overline{apr}(C_k)$ can be decided not to belong to the cluster C_k. Set $BND(C_k) = \overline{apr}(C_k) - \underline{apr}(C_k)$. Objects in the region $BND(C_k)$ may be belong to the cluster or not.

There exists $k \neq t$, it is possible that $\underline{apr}(C_k) \cap \underline{apr}(C_t) \neq \emptyset$, or $BND(C_k) \cap BND(C_t) \neq \emptyset$. In other words, it is possible that an object belongs to more than one cluster.

3 Clustering Algorithm Using Three-Way Decision

Density-based clustering analysis is one kind of clustering analysis methods that can discover clusters with arbitrary shape and is insensitive to noise data. Therefore, according to the three-way decision rules (P)-(N) in Subsection 2.2, a density-based clustering algorithm will be proposed in this section to combat the overlapping clustering.

Considering the discovery area, set the center is p and Rth is the radius, the number of points in the area is called the density of p relative to Rth, denoted by $Density(p, Rth)$. The concepts are defined as follows [7].

Reference points: For any node p, distance Rth and threshold mth in the space, if $Density(p, Rth) \leq mth$, then p is a reference point and mth is the density threshold value.

The reference points are fictional points, not the points in the dataset. Threshold value mth represents a reference number. When the density of p greater than mth, p is an intensive point, otherwise it is a sparse point.

Representing Region: Every reference point p is the representative of a circular area where the point is the center of the area and the radius is Rth, and the region is the representing region of the reference point p.

All points(objects) in the representing region of a reference point p are seen as an equivalence class. In order to cluster objects(points) in the space, we need to give the method to calculate the probability in Eq.(6).

Probability: $[x]$ is a description of an object x, the $Pr(C|[x])$ is:

$$Pr(C|[x]) = \frac{|C \cap [x]|}{|[x]|}. \tag{7}$$

General speaking, the equivalence class $[x]$ of an object x can be used as a description of the object. That is, Eq.(7) gives a computing method for the

Algorithm 1. Density-based Clustering Algorithm Using Three-way Decision

Input : a universe $U = \{x_1, \cdots, x_n\}$.
Output: the clustering result $\mathbf{C}$.
begin
Step 1. Initial: Set $UN = \emptyset$, $RF = \emptyset$, the possibility $Pr(C_k|RF_t) = 0$.
Step 2. Find all candidate reference points:
$RF_1 \leftarrow x_1$; $\mathbf{RF} = \mathbf{RF} \cup \{RF_1\}$;
for *every* x_i **do**
$temp = \min_{RF_t} |RF_t - x_i|$; $k = arg(\min_{RF_t} |RF_t - x_i|)$;
If $temp > Rth$ then $\{ RF_{T+1} \leftarrow x_i$; $\mathbf{RF} = \mathbf{RF} \cup RF_{T+1}; \}$
Else alter RF_k based on object x_i;
end
Step 3. Choice the reference points and the noise points from the candidates:
for *every* $RF_t \in \boldsymbol{RF}$ **do**
For (every x_i) do $\{$ If $|RF_t - x_i| < Rth$ then $RF_t = RF_t \cup x_i$; $\}$
If $|RF_t| < mth$ then $\{ UN = UN \cup RF_t$; $RF = RF - RF_t$; $\}$
end
Step 4. Clustering the reference points according to three-way rules (P)-(N):
for *every* $\underline{apr}(C_k)$ **do**
$\underline{apr}(C_k) = RF_k$; $\overline{apr}(C_k) = RF_k$;
for *every* RF_t **do**
$Pr(C_k|RF_t) = \frac{|C_k \cap RF_t|}{|RF_t|}$ // Eq.(7)
If $Pr(C_k|RF_t) \geq \alpha$ then
$\underline{apr}(C_k) = \underline{apr}(C_k) \cup RF_t$; $\overline{apr}(C_k) = \overline{apr}(C_k) \cup RF_t$;
If $\beta < Pr(C_k|RF_t) < \alpha$ then $\overline{apr}(C_k) = \overline{apr}(C_k) \cup RF_t$,
end
end
$\mathbf{C} = [\underline{apr}(C_1), \overline{apr}(C_1)], \ldots, [\underline{apr}(C_k), \overline{apr}(C_k)], \ldots, [\underline{apr}(C_K), \overline{apr}(C_K)]$.
for *every* $\underline{apr}(C_k)$ **do**
If $\underline{apr}(C_k) \supseteq \underline{apr}(C_j)$ then $\overline{apr}(C_k) = \overline{apr}(C_k) \cup \overline{apr}(C_j)$; $\mathbf{C} = \mathbf{C} - C_j$;
end
Step 5. Clustering the noise points.
for *every* $\underline{apr}(C_k)$ **do**
for *every* UN_s **do**
If $UN_s \subseteq \underline{apr}(C_k)$ then
$\underline{apr}(C_k) = \underline{apr}(C_k) \cup UN_s$; $\overline{apr}(C_k) = \overline{apr}(C_k) \cup UN_s$;
Else $\{\mathbf{C} = \mathbf{C} \cup UN_S;\}$
end
end
end

probability, then we can devise the algorithm based on three-way decision. In other words, the different algorithms can be developed based on the different approaches of computing probability.

In this paper, a density-based clustering algorithm using three-way decision is proposed as follows. Here, UN and $\mathbf{RF} = \{RF_1, \cdots, RF_t, \cdots, RF_T\}$ means the noise data set and the family of reference points sets, respectively.

In the above algorithm, Step 1 to Step 3 obtain an initial clustering result by choosing the reference points and representing regions according to the relative concepts. Step 4 modifies the clusters according to three-way decision rules (P) to (N).

4 Experiments

The new algorithm is performed by Visual C++. Firstly, some UCI datasets [11] are used to test the different thresholds such as the distance threshold value Rth, density threshold mth, α and β. Obviously, Rth and mth are decided by the characteristics of the dataset. However, there is an interesting result that the clustering result seems good when $\alpha = 0.8$ and $\beta = 0.4$ in most cases. Thus, the result is accepted in the later experiments. On the other hand, it enlightens us we should think a formal way to define the α and β in the further work.

4.1 Synthetic Data Set

The synthetic data set is tested to illustrate the ideas presented in the previous section. The two dimensions data set is depicted in Fig.1, which have 374 points, and Fig.2 gives the clustering result. Here, the thresholds are $Rth = 1.75$, $mth = 10$, $\alpha = 0.8$ and $\beta = 0.4$.

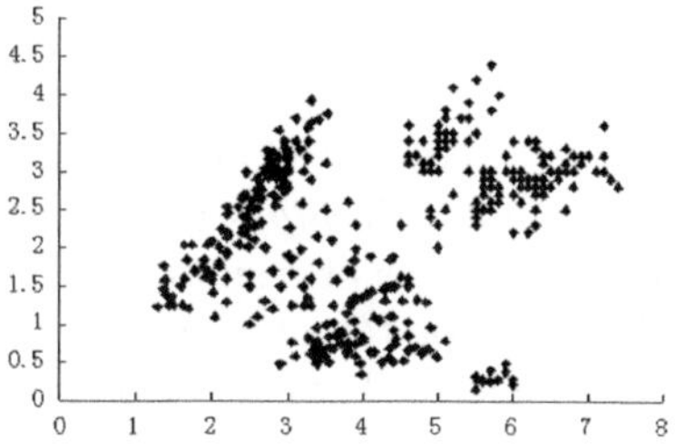

Fig. 1. A synthetic data set

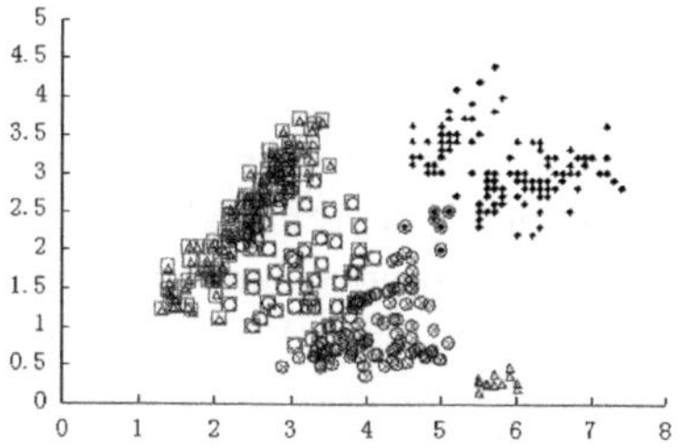

Fig. 2. The clustering result of the data set

From Fig.2, we can see that these points are clustered into three clusters. That is, the cross points means the lower and bounder regions of C_1, respectively; the circular points and tangle points means the lower and bounder regions of C_2, respectively; the dots means the lower regions of C_3. Here, the boundary of C_3 is empty.

Observe Fig.2, the lower approximations of cluster C_2 and C_3 are overlapping, and the number of the overlapping objects(points) is 6, which can be denoted by $\underline{apr}(C_2) \cap \underline{apr}(C_3) \neq \emptyset$, and $|\underline{apr}(C_2) \cap \underline{apr}(C_3)| = 6$. In addition, $\underline{apr}(C_1) \cap \underline{apr}(C_3) = \emptyset$, $|\underline{apr}(C_1) \cap \underline{apr}(C_2)| = 66$, and $\overline{apr}(C_1) \neq \overline{apr}(C_2)$.

The conclusions from Fig.2 are positive to clustering. For example, when the dataset represents the network structure, where the C_1 and C_2 have so many overlapping users. Obviously, it looks reasonable to build a new cluster composed by the uniting $\underline{apr}(C_1) \cup \underline{apr}(C_2)$. Otherwise, since the number of the overlapping objects between C_2 and C_3 is 6, we needn't to unite the two clusters. How to formal the idea is our further work.

4.2 UCI Data Set

More experiments on some standard data sets from UCI repository [11] are tested in this subsection, and results are shown in Table 1. In order to measure the test's accuracy, both the precision and the recall of the test are considered, and the F-measure is extended as follows.

Assume there is a data set $U = \{x_1, \ldots, x_i, \ldots, x_n\}$, and the objects in U are clustered into $T = \{T_1, \ldots, T_m, \ldots, T_M\}$. On the other hand, the result of clustering by the clustering algorithm based on three-way decision is: $\mathbf{C} = \{[\underline{apr}(C_1), \overline{apr}(C_1)], \ldots, [\underline{apr}(C_k), \overline{apr}(C_k)], \ldots, [\underline{apr}(C_K), \overline{apr}(C_K)]\}$.

Table 1. The CPU time and Results of the Algorithm

Database	$\|\mathbf{U}\|$	$\|A\|$	*distance*	Thresholds			Results	
				α, β	Rth	mth	$F-measure$	CPU(S)
iris	150	4	2.53	0.8,0.4	1.53	30	0.778	0.015
Letter1	1655	16	11.3	0.8,0.4	10	200	0.609	0.5
Poker1	199	10	11.2	0.8,0.4	11	80	0.595	0.078
Poker2	1188	10	12	0.8,0.4	11	900	0.503	1.296
White	4535	11	53	0.8,0.4	52	500	0.601	2.734

Precision is the number of correct upper approximate results divided by the number of all returned upper approximate results. *Recall* is the number of correct lower approximate results divided by the number of results that should have been returned. The $F-measure$ can be interpreted as a weighted as a weighted as a weighted average of the precision and recall, where an F-measure reaches its best value at 1 and worst value at 0. That is, the $F-measure$ can be denoted as the following equation.

$$F-measure = 2 \times \frac{Precision \times Recall}{Precision + Recall} \tag{8}$$

In Table 1, the $|\mathbf{U}|$ and $|A|$ are the number of objects and the number of attributes in the data set, respectively. *distance* means the average distance among objects in the data set.

Here, we choose some clusters from Letter and Porker datasets in UCI repository to generate some datasets used in Tabel 1.

Letter1 Dataset, which is composed of 1665 objects from Letter data set. There are 567 objects belong to decision attribute 'A', 570 objects belong to decision attribute 'E', and 528 objects belong to decision attribute 'O'.

Poker1 Dataset, which has 199 objects from Poker-hand-training-true data set. There are 93 objects belong to decision attribute '4', 54 objects belong to decision attribute ‘5’, 36 objects belong to decision attribute ‘6’, 6 objects belong to ‘7’, 5 objects belong to ‘8’, and 4 objects belong to ‘9’.

Poker2 Dataset, which concludes 1188 objects from Poker-hand-training-true. There are 403 objects belong to decision attribute ‘0’, 568 objects belong to ‘1’, 165 objects belong to ‘3’, 36 objects belong to ‘6’, 6 objects belong to ‘7’, 5 objects belong to ‘8’, and 4 objects belong to ‘9’.

From Table 1, we can see that the CPU runtime and the F-measure are accredited. Actually, the results of clustering are changed with the change of the parameters such as α, β, Rth and mth. Through the experiments, we find out that the result would be better when the value of Rth close to the average distance. However, the accuracy of the algorithm need to improve.

5 Conclusion

In many applications such as network structure analysis, wireless sensor networks and biological information, an object should belong to more than one cluster, and as a result, cluster boundaries necessarily overlap. Three-way decisions rules constructed from the decision-theoretic rough set model are associated with different regions. This paper provides a three-way decision strategy for overlapping clustering. Here, each cluster is described by an interval set that is defined by a pair of sets called the lower and upper bounds. In addition, a density-based clustering algorithm is proposed and tested by using the new strategy. The analysis of the example indicates the strategy is effective to overlapping clustering. How to use less parameters and improve the accuracy of the algorithm is the further work.

Acknowledgments. This work was supported in part by the China NSFC grant(No.61073146) and the Chongqing CSTC grant (No.2009BB2082).

References

1. Aydin, N., Naït-Abdesselam, F., Pryyma, V., Turgut, D.: Overlapping Clusters Algorithm in Ad hoc Networks. In: 2010 IEEE Global Telecommunications Conference (2010)
2. Chen, M., Miao, D.Q.: Interval set clustering. Expert Systems with Application 38, 2923–2932 (2011)
3. Fu, Q., Banerjee, A.: Multiplicative Mixture Models for Overlapping Clustering. In: IEEE International Conference on Data Mining, pp. 791–797 (2003)
4. Herbert, J.P., Yao, J.T.: Learning Optimal Parameters in Decision-Theoretic Rough Sets. In: Wen, P., Li, Y., Polkowski, L., Yao, Y., Tsumoto, S., Wang, G. (eds.) RSKT 2009. LNCS, vol. 5589, pp. 610–617. Springer, Heidelberg (2009)
5. Lingras, P., Bhalchandra, P., Khamitkar, S., Mekewad, S., Rathod, R.: Crisp and Soft Clustering of Mobile Calls. In: Sombattheera, C., Agarwal, A., Udgata, S.K., Lavangnananda, K. (eds.) MIWAI 2011. LNCS, vol. 7080, pp. 147–158. Springer, Heidelberg (2011)

6. Liu, D., Yao, Y.Y., Li, T.R.: Three-way investment decisions with decision-theoretic rough sets. International Journal of Computational Intelligence Systems 4(1), 66–74 (2011)
7. Ma, S., Wang, T.J., Tang, S.W., Yang, D.Q., Gao, J.: A Fast Clustering Algorithm Based on Reference and Density. Journal of Softwar. 14(6), 1089–1095 (2003) (in Chinese)
8. Obadi, G., Dráždilová, P., Hlaváček, L., Martinovič, J., Snášel, V.: A Tolerance Rough Set Based Overlapping Clustering for the DBLP Data. In: IEEE/WIC/ACM International Conference on Web Intelligence and Intelligent Agent Technology, pp. 57–60 (2010)
9. Pawlak, Z.: Rough sets. International Journal of Computer and Information Science 11(5), 341–356 (1982)
10. Takaki, M.: A Extraction Method of Overlapping Cluster based on Network Structure Analysis. In: IEEE/WIC/ACM International Conferences on Web Intelligence and Intelligent Agent Technology, pp. 212–217 (2007)
11. UCIrvine Machine Learning Repository, `http://archive.ics.uci.edu/ml/`
12. Yao, Y.Y.: The-Superiority of Three-way Decisions in Probablistic Rough Set Models. Information Sciences 181, 1080–1096 (2011)
13. Yao, Y.Y., Wong, S.K.M.: A decision theoretic framework for approximating concepts. International Journal of Man-machine Studies 37(6), 793–809 (1992)
14. Yao, Y.Y., Lingras, P., Wang, R.Z., Miao, D.Q.: Interval Set Cluster Analysis: A Re-formulation. In: Sakai, H., Chakraborty, M.K., Hassanien, A.E., Ślęzak, D., Zhu, W. (eds.) RSFDGrC 2009. LNCS (LNAI), vol. 5908, pp. 398–405. Springer, Heidelberg (2009)
15. Zhou, B., Yao, Y.Y., Luo, J.G.: A Three-Way Decision Approach to Email Spam Filtering. In: Farzindar, A., Kešelj, V. (eds.) Canadian AI 2010. LNCS, vol. 6085, pp. 28–39. Springer, Heidelberg (2010)

Three-Way Decisions Solution to Filter Spam Email: An Empirical Study

Xiuyi Jia[1,4], Kan Zheng[2], Weiwei Li[3], Tingting Liu[2], and Lin Shang[4]

[1] School of Computer Science and Technology,
Nanjing University of Science and Technology, Nanjing, China, 210094
jiaxy@mail.njust.edu.cn
[2] School of Mechanical Engineering,
Nanjing University of Science and Technology, Nanjing, China, 210094
zhengkan@mail.njust.edu.cn, liutingtingwy@163.com
[3] College of Computer Science and Technology,
Nanjing University of Aeronautics and Astronautics, Nanjing, China, 210016
amy.vivilee@gmail.com
[4] State Key Laboratory for Novel Software Technology,
Nanjing University, Nanjing, China, 210093
shanglin@nju.edu.cn

Abstract. A three-way decisions solution based on Bayesian decision theory for filtering spam emails is examined in this paper. Compared to existed filtering systems, the spam filtering is no longer viewed as a binary classification problem. Each incoming email is accepted as a legitimate or rejected as a spam or undecided as a further-exam email by considering the misclassification cost. The three-way decisions solution for spam filtering can reduce the error rate of classifying a legitimate email to spam, and provide a more meaningful decision procedure for users. The solution is not restricted to a specific classifier. Experimental results on several corpus show that the three-way decisions solution can get a better total cost ratio value and a lower weighted error.

Keywords: Decision-theoretic rough set model, spam filtering, three-way decisions solution.

1 Introduction

Spam filtering is often considered as a binary classification problem. Many machine learning algorithms were employed in different filters to classify an incoming email as a legitimate email or a spam, such as Naive bayesian classifier [11], memory based classifier (k-nn) [1], SVM based classifier [5] and so on [2,12]. In this paper, we will treat spam filtering as an optimum decision problem under the framework of cost sensitive learning.

It is easy to understand that Spam filtering is a cost sensitive learning problem. The cost for misclassifying a legitimate email as spam far outweighs the cost of marking a spam email as legitimate [11]. Many machine learning approaches to spam filtering papers considered the different costs of two types of

J.T. Yao et al. (Eds.): RSCTC 2012, LNAI 7413, pp. 287–296, 2012.

misclassifications (*legit* $\rightarrow$ *spam* and *spam* $\rightarrow$ *legit*). In [11], 99.9% was used as the certain threshold for classifying test email as spam to reflect the asymmetric cost of errors. The threshold was set manually. In [1], three scenarios($\lambda = 1$, $\lambda = 9$, $\lambda = 999$) were discussed, while *legit* $\rightarrow$ *spam* is λ times costly than *spam* $\rightarrow$ *legit*. These parameters were applied in the cost sensitive evaluation procedure only, which can measure the efficiency of the learning algorithm, but can not help learn a better result because they did not consider them in the training procedure. In these papers, appropriate settings of parameters were not discussed. We will discuss this point in this paper.

It is intuitively to treat spam filtering as a three-way decision problem [15]. Besides the usual decisions include *accept* an email as a legitimate one and *reject* an email as a spam, the third type decision *further-exam* for those suspicious spam emails is also considered in three-way decisions solution. The idea of three-way decisions making can be found in many areas. In [3], an optimum rejection scheme was derived to safeguard against excessive misclassification in pattern recognition system. In clinical decision making for a certain disease, with options of treating the conditional directly, not treating the condition, or performing a diagnose test to decide whether or not to treat the condition [9]. Yao et al. [13,14] introduced decision theoretic rough set model(DTRS) based on three-way decisions, which considered the cost of each error and Bayesian decision theory. Based on DTRS and Naive Bayesian classifier, a three-way decision approach to email spam filtering was proposed recently [15]. Our work is continuation of them. Different with their work, our three-way decision solution is not only suit to Naive Bayesian classifier, but also k-nn, SVM and other classifiers.

The main advantage of three-way decisions solution is that it allows the possibility of refusing to make a direct decision, which means it can convert some potential misclassifications into rejections, and these emails will be further-examed by users. Several cost functions are defined to state how costly each decision is, and the final decision can make the overall cost minimal filtering. We apply the three-way decisions solution on several classical classifiers which are used in spam filtering, including Naive Bayesian classifier, k-nn classifier and SVM classifier. The tests on several benchmark corpus include Ling-Spam, PU-Corpora and Enron-Spam[1] show the efficiency of the three-way decisions solution. We can get a lower *weighted error* and a better total cost ratio (*TCR*).

2 Three-Way Decisions Solution to Filter Spam

2.1 Decision-Theoretic Rough Set Model

Decision-theoretic rough set model was proposed by Yao et al. [13], which is based on Bayesian decision theory. The basic ideas of the theory [14] are reviewed.

Let $\Omega = \{\omega_1, \ldots, \omega_s\}$ be a finite set of s states and let $\mathcal{A} = \{a_1, \ldots, a_m\}$ be a finite set of m possible actions. Let $\lambda(a_i|\omega_j)$ denote the cost, for taking

[1] All corpus are available from
`http://labs-repos.iit.demokritos.gr/skel/i-config/`

action a_i when the state is ω_j. Let $p(\omega_j|x)$ be the conditional probability of an incoming email x being in state ω_j, suppose action a_i is taken. The expected cost associated with taking action a_i is given by:

$$R(a_i|x) = \sum_{j=1}^{s} \lambda(a_i|\omega_j) \cdot p(\omega_j|x). \tag{1}$$

In rough set theory [10], a set C is approximated by three regions, namely, the positive region POS(C) includes the objects that are sure belong to C, the boundary region BND(C) includes the objects that are possible belong to C, and the negative region NEG(C) includes the objects that are not belong to C. In spam filtering, we have a set of two states $\Omega = \{C, C^c\}$ indicating that an email is in C (i.e., legitimate) or not in C (i.e., spam), respectively. The set of emails can be divided into three regions, POS(C) includes emails that are legitimate emails, BND(C) includes emails that need further-exam, and NEG(C) includes emails that are spam. With respect to these three regions, the set of actions is given by $\mathcal{A} = \{a_P, a_B, a_N\}$, where a_P, a_B and a_N represent the three actions in classifying an email x, namely, deciding $x \in$ POS(C), deciding $x \in$ BND(C), and deciding $x \in$ NEG(C). Six cost functions are imported, λ_{PP}, λ_{BP} and λ_{NP} denote the costs incurred for taking actions a_P, a_B, a_N, respectively, when an email belongs to C, and λ_{PN}, λ_{BN} and λ_{NN} denote the costs incurred for taking these actions when the email does not belong to C.

The expected costs associated with taking different actions for email x can be expressed as:

$$\begin{aligned} R(a_P|x) &= \lambda_{PP} \cdot p(C|x) + \lambda_{PN} \cdot p(C^c|x), \\ R(a_B|x) &= \lambda_{BP} \cdot p(C|x) + \lambda_{BN} \cdot p(C^c|x), \\ R(a_N|x) &= \lambda_{NP} \cdot p(C|x) + \lambda_{NN} \cdot p(C^c|x). \end{aligned} \tag{2}$$

The Bayesian decision procedure suggests the following minimum-cost decision rules:

(P) If $R(a_P|x) \leq R(a_B|x)$ and $R(a_P|x) \leq R(a_N|x)$, decide $x \in$ POS(C);
(B) If $R(a_B|x) \leq R(a_P|x)$ and $R(a_B|x) \leq R(a_N|x)$, decide $x \in$ BND(C);
(N) If $R(a_N|x) \leq R(a_P|x)$ and $R(a_N|x) \leq R(a_B|x)$, decide $x \in$ NEG(C);

Consider a special kind of cost functions with:

$$\begin{aligned} &\lambda_{PP} \leq \lambda_{BP} < \lambda_{NP}, \\ &\lambda_{NN} \leq \lambda_{BN} < \lambda_{PN}. \end{aligned} \tag{3}$$

That is, the cost of classifying an email x being in C into the positive region POS(C) is less than or equal to the cost of classifying x into the boundary region BND(C), and both of these costs are strictly less than the cost of classifying x into the negative region NEG(C). The reverse order of costs is used for classifying an email not in C. Since $p(C|x) + p(C^c|x) = 1$, under above condition, we can simplify decision rules (P)-(N) as follows:

(P) If $p(C|x) \geq \alpha$ and $p(C|x) \geq \gamma$, decide $x \in \text{POS}(C)$;
(B) If $p(C|x) \leq \alpha$ and $p(C|x) \geq \beta$, decide $x \in \text{BND}(C)$;
(N) If $p(C|x) \leq \beta$ and $p(C|x) \leq \gamma$, decide $x \in \text{NEG}(C)$.

Where

$$\alpha = \frac{(\lambda_{PN} - \lambda_{BN})}{(\lambda_{PN} - \lambda_{BN}) + (\lambda_{BP} - \lambda_{PP})},$$
$$\beta = \frac{(\lambda_{BN} - \lambda_{NN})}{(\lambda_{BN} - \lambda_{NN}) + (\lambda_{NP} - \lambda_{BP})},$$
$$\gamma = \frac{(\lambda_{PN} - \lambda_{NN})}{(\lambda_{PN} - \lambda_{NN}) + (\lambda_{NP} - \lambda_{PP})}. \tag{4}$$

Each rule is defined by two out of the three parameters. The conditions of rule (B) suggest that $\alpha > \beta$ may be a reasonable constraint; it will ensure a well-defined boundary region. If we obtain the following condition on the cost functions [14]:

$$\frac{(\lambda_{NP} - \lambda_{BP})}{(\lambda_{BN} - \lambda_{NN})} > \frac{\lambda_{BP} - \lambda_{PP}}{(\lambda_{PN} - \lambda_{BN})}, \tag{5}$$

then $0 \leq \beta < \gamma < \alpha \leq 1$. In this case, after tie-breaking, the following simplified rules are obtained:

(P1) If $p(C|x) \geq \alpha$, decide $x \in \text{POS}(C)$;
(B1) If $\beta < p(C|x) < \alpha$, decide $x \in$ BND(C);
(N1) If $p(C|x) \leq \beta$, decide $x \in \text{NEG}(C)$.

The threshold parameters can be systematically calculated from cost functions based on the Bayesian decision theory.

2.2 Three-Way Decisions Solution

Given the cost functions, we can make the proper decisions for incoming emails based on the parameters (α, β), which are computed by cost functions, and the probability of each email being a legitimate one, which is provided by the running classifier.

For an email x, if the probability of being a legitimate email is $p(C|x)$, then the three-way decisions solution is:

If $p(C|x) \geq \alpha$, then x is a legitimate email;
If $p(C|x) \leq \beta$, then x is a spam;
If $\beta < p(C|x) < \alpha$, then x needs further-exam.

3 Experiments on Several Benchmark Corpus

In this section, we will check the efficiency of the three-way decisions solution. The followings are the detail of classifiers, corpus and the evaluation methods used in our experiments.

3.1 Classifiers

Three classical classifiers (Naive Bayesian classifier, k-nn classifier and SVM classifier) are applied in the experiments.

The Naive Bayesian classifier can provide the probability of an incoming email being a legitimate email. Suppose an email x is described by a feature vector $\mathbf{x} = (\mathbf{x}_1, \mathbf{x}_2, \ldots, \mathbf{x}_n)$, where $\mathbf{x}_1, \mathbf{x}_2, \ldots, \mathbf{x}_n$ are the values of attributes of the email. Let C denote the legitimate class. Based on Bayes' theorem and the theorem of total probability, given the vector of an email, the probability of being a legitimate one is:

$$p(C|x) = \frac{p(C) \cdot p(x|C)}{p(x)}, \tag{6}$$

where $p(x) = p(x|C) \cdot p(C) + p(x|C^c) \cdot p(C^c)$. Here $p(C)$ is the prior probability of an email being in the legitimate class. $p(C)$ is commonly known as the likelihood of an email being in the legitimate class with respect to x. The likelihood $p(x|C)$ is a joint probability of $p(\mathbf{x}_1, \mathbf{x}_2, \ldots, \mathbf{x}_n|C)$.

The k-nearest neighbors algorithm(k-nn) classifier [8] predicts an email's class by a majority vote of its neighbors. Euclidean distance is usually used as the distance metric. Let $n(l)$ denote the legitimate emails' number in the k nearest neighbors, and $n(s)$ denote the spam's number. So $n(l) + n(s) = k$, then we use sigmoid function to estimate the posterior probability of an email x being in the legitimate class.

$$p(C|x) = \frac{1}{1 + \exp(-a \cdot (n(l) - n(s)))}. \tag{7}$$

In our experiments, $a = 1$ and $k = 17$.

For SVM classifier, the sigmoid function also be used to estimate the posterior probability:

$$p(C|x) = \frac{1}{1 + \exp(-a \cdot d(x))}. \tag{8}$$

While $d(x) = \frac{\mathbf{w}.x+b}{\|\mathbf{w}\|}$, where weighted vector $\mathbf{w}$ and threshold b are used to define the hyperplane. $a = 6$ in our experiments.

3.2 Benchmark Corpus

Three benchmark corpus are used in this paper, called Ling-Spam, PU-Corpora and Enron-Spam.

The corpus Ling-Spam was preprocessed as three different type of corpus, named Ling-Spam-bare, Ling-Spam-lemm and Ling-Spam-stop, respectively. Ling-Spam-bare is a kind of "words only" dataset, each attribute shows if a particular word occurs in the email. For Ling-Spam-lemm, a lemmatizer was applied to Ling-Spam, which means each word was substituted by its base form(e.g.

"earning" becomes "earn"). Ling-Spam-stop is a data set generated by considering stop-list based on Ling-Spam-lemm.

The corpus PU-Corpora contains PU1, PU2, PU3 and PUA corpora. All corpus are only in "bare" form: tokens were separated by white characters, but no lemmtizer or stop-list has been applied.

Another corpus Enron-Spam is divided to 6 datasets, each dataset contains legitimate emails from a single user of the Enron corpus, to which fresh spam emails with varying legit-spam ratios were added.

Since the corpus in Ling-Spam and PU-Corpora were divided into 10 datasets, 10-fold cross-validation is used in these corpus. Enron-Spam was divided into 6 datasets, 6-fold cross-validation is used in it's experiment. Because we just want to compare the three-way decisions solution with "two-way decisions" solution based on the same classifiers and the same corpus, then it is non-necessary to apply feature selection procedures in corpus, all attributes are used in the experiments.

3.3 Measures to Evaluate Performance

As a cost sensitive learning problem, Androutsopoulos et al. [1] suggested that using *weighted accuracy*(or *weighted error rate*) and *total cost ratio* to replace the classical *accuracy*(or *error rate*) to measure the spam filter performance is reasonable. Let N_{legit} and N_{spam} be the total numbers of legitimate and spam emails, to be classified by the filter, and $n_{legit \to spam}$ the number of emails belonging to legitimate class that the filter classified as belonging to spam, $n_{spam \to legit}$, reversely. Then the classical *accuracy* and *error rate* are defined as:

$$Acc = \frac{n_{legit \to legit} + n_{spam \to spam}}{N_{legit} + N_{spam}}, \tag{9}$$

$$Err = \frac{n_{legit \to spam} + n_{spam \to legit}}{N_{legit} + N_{spam}}. \tag{10}$$

If $legit \to spam$ is λ times more costly than $spam \to legit$, then *weighted accuracy*(*WAcc*) and *weighted error rate*(*WErr*) are defined as:

$$WAcc = \frac{\lambda \cdot n_{legit \to legit} + n_{spam \to spam}}{\lambda \cdot N_{legit} + N_{spam}}, \tag{11}$$

$$WErr = \frac{\lambda \cdot n_{legit \to spam} + n_{spam \to legit}}{\lambda \cdot N_{legit} + N_{spam}}. \tag{12}$$

As the values of *accuracy* and *error rate*(or their weighted versions) are often misleadingly high, another measure is defined to get a clear picture of a classifier's performance, the ratio of its *error rate* and that of a simplistic baseline approach. The baseline approach is the filter that never blocks legitimate emails and always passes spam emails. The *weighted error rate* of the baseline is:

$$WErr^b = \frac{N_{spam}}{\lambda \cdot N_{legit} + N_{spam}}. \tag{13}$$

The *total cost ratio*(*TCR*) is:

$$TCR = \frac{WErr^b}{WErr} = \frac{N_{spam}}{\lambda \cdot n_{legit \to spam} + n_{spam \to legit}}. \tag{14}$$

Greater *TCR* values indicate better performance. For $TCR < 1$, the baseline is better. If cost is proportional to wasted time, an intuitive meaning for *TCR* is the following: it measures how much time is wasted to delete manually all spam emails when no filter is used, compared to the time wasted to delete manually any spam emails that passed the filter plus the time needed to recover from mistakenly blocked legitimate emails.

For our three-way decisions solution, *weighted rejection rate*(*WRej*) is defined to indicate the weighted ratio of emails need further-exam.

$$WRej = \frac{\lambda \cdot n_{legit \to boundary} + n_{spam \to boundary}}{\lambda \cdot N_{legit} + N_{spam}}, \tag{15}$$

while $n_{legit \to boundary}$ and $n_{spam \to boundary}$ mean the numbers of legitimate and spam emails being classified to boundary(need further-exam), $WRej = 1 - WAcc - WErr$, actually. For classical classifiers, $WRej = 0$.

3.4 Experimental Result

In our experiments, three different λ values($\lambda = 1$,$\lambda = 3$, and $\lambda = 9$) are applied, same values were considered in [15]. For three-way decisions solution, we set $\alpha = 0.9$ and $\beta = 0.1$ to make corresponding decisions.

From the results showed in the following 8 tables, we can see that the values of *WAcc* in three-way decisions solution are lower than that in classical approaches. It is the result of moving some emails into the boundary for further-exam. We can also conclude that the three-way decisions solution decreases the values of *WErr* and increases the values of *TCR* from the results, which means the three-way decisions solution can reduce the number of "wrong classified" emails. For those "suspicious spam" emails, it is a better choice to let users decide which is a legitimate email or a spam. The increment of *TCR* shows that the three-way decisions solution gives a better performance. There exists a kind of tradeoff between the *error rate* and the *rejection rate*. Users can decrease the *error rate* by increasing the *rejection rate*.

The most important conclusion we can get from the experiments result is that, the three-way decisions solution can get a better performance than "two-way" decision solution under the same situation, it does not depend on the parameter λ or some special classifiers.

Table 1. Comparison results on corpora Ling-Spam-bare

NB		*WAcc*	*WErr*	*WRej*	*TCR*
	NB	0.8989	0.1011	0.0000	1.6808
$\lambda = 1$	Three-way	0.8646	0.0693	0.0661	2.5115
	NB	0.9512	0.0488	0.0000	1.3232
$\lambda = 3$	Three-way	0.9251	0.0320	0.0430	2.1034
	NB	0.9716	0.0284	0.0000	0.8488
$\lambda = 9$	Three-way	0.9487	0.0174	0.0339	1.6006
k-nn					
	k-nn	0.8832	0.1168	0.0000	1.4963
$\lambda = 1$	Three-way	0.7863	0.0710	0.1427	2.9550
	k-nn	0.9320	0.0680	0.0000	1.0979
$\lambda = 3$	Three-way	0.8419	0.0321	0.1260	2.2840
	k-nn	0.9511	0.0489	0.0000	0.7853
$\lambda = 9$	Three-way	0.8636	0.0169	0.1195	1.7052
SVM					
	SVM	0.9393	0.0607	0.0000	9.8972
$\lambda = 1$	Three-way	0.8886	0.0272	0.0843	23.0962
	SVM	0.9371	0.0629	0.0000	6.4686
$\lambda = 3$	Three-way	0.8932	0.0279	0.0790	13.7574
	SVM	0.9321	0.0679	0.0000	4.4352
$\lambda = 9$	Three-way	0.8920	0.0300	0.0780	8.5106

Table 2. Comparison results on corpora Ling-Spam-lemm

NB		*WAcc*	*WErr*	*WRej*	*TCR*
	NB	0.9049	0.0951	0.0000	1.7984
$\lambda = 1$	Three-way	0.8706	0.0644	0.0651	2.6775
	NB	0.9529	0.0471	0.0000	1.3821
$\lambda = 3$	Three-way	0.9289	0.0301	0.0410	2.1521
	NB	0.9717	0.0283	0.0000	0.8479
$\lambda = 9$	Three-way	0.9517	0.0167	0.0316	1.4485
k-nn					
	k-nn	0.8889	0.1111	0.0000	1.5702
$\lambda = 1$	Three-way	0.7955	0.0696	0.1349	2.9454
	k-nn	0.9375	0.0625	0.0000	1.1724
$\lambda = 3$	Three-way	0.8532	0.0308	0.1160	2.3681
	k-nn	0.9566	0.0434	0.0000	0.8375
$\lambda = 9$	Three-way	0.8758	0.0156	0.1086	1.8028
SVM					
	SVM	0.9332	0.0668	0.0000	2.6733
$\lambda = 1$	Three-way	0.9000	0.0408	0.0592	4.2062
	SVM	0.9591	0.0409	0.0000	1.7334
$\lambda = 3$	Three-way	0.9350	0.0226	0.0424	2.9567
	SVM	0.9693	0.0307	0.0000	0.8859
$\lambda = 9$	Three-way	0.9487	0.0154	0.0359	1.6749

Table 3. Comparison results on corpora Ling-Spam-stop

NB		*WAcc*	*WErr*	*WRej*	*TCR*
	NB	0.8869	0.1131	0.0000	1.5081
$\lambda = 1$	Three-way	0.8426	0.0782	0.0793	2.2700
	NB	0.9430	0.0570	0.0000	1.1375
$\lambda = 3$	Three-way	0.9124	0.0350	0.0526	1.9059
	NB	0.9650	0.0350	0.0000	0.6772
$\lambda = 9$	Three-way	0.9398	0.0181	0.0421	1.3371
k-nn					
	k-nn	0.8338	0.1662	0.0000	1.1319
$\lambda = 1$	Three-way	0.6543	0.0568	0.2890	7.8151
	k-nn	0.8374	0.1626	0.0000	0.5588
$\lambda = 3$	Three-way	0.6744	0.0537	0.2719	4.9959
	k-nn	0.8389	0.1611	0.0000	0.2683
$\lambda = 9$	Three-way	0.6823	0.0525	0.2652	3.4376
SVM					
	SVM	0.9284	0.0716	0.0000	2.4032
$\lambda = 1$	Three-way	0.8861	0.0446	0.0692	3.9029
	SVM	0.9584	0.0417	0.0000	1.6406
$\lambda = 3$	Three-way	0.9269	0.0217	0.0514	2.9681
	SVM	0.9701	0.0299	0.0000	0.8998
$\lambda = 9$	Three-way	0.9429	0.0127	0.0444	1.9990

Table 4. Comparison results on corpora PU1

NB		*WAcc*	*WErr*	*WRej*	*TCR*
	NB	0.9174	0.0826	0.0000	5.9340
$\lambda = 1$	Three-way	0.9110	0.0761	0.0128	6.4019
	NB	0.9558	0.0442	0.0000	5.3140
$\lambda = 3$	Three-way	0.9519	0.0411	0.0069	5.7487
	NB	0.9769	0.0231	0.0000	4.4430
$\lambda = 9$	Three-way	0.9744	0.0219	0.0037	4.8404
k-nn					
	k-nn	0.8807	0.1193	0.0000	4.1672
$\lambda = 1$	Three-way	0.7661	0.0560	0.1780	8.9867
	k-nn	0.8727	0.1273	0.0000	2.0298
$\lambda = 3$	Three-way	0.7468	0.0654	0.1879	3.9120
	k-nn	0.8683	0.1317	0.0000	0.8361
$\lambda = 9$	Three-way	0.7362	0.0705	0.1933	1.5428
SVM					
	SVM	0.9789	0.0211	0.0000	22.4000
$\lambda = 1$	Three-way	0.9248	0.0055	0.0697	40.0000
	SVM	0.9797	0.0203	0.0000	13.2656
$\lambda = 3$	Three-way	0.9342	0.0043	0.0615	24.0000
	SVM	0.9801	0.0199	0.0000	9.0223
$\lambda = 9$	Three-way	0.9394	0.0037	0.0570	18.6667

Table 5. Comparison results on corpora PU2

NB		*WAcc*	*WErr*	*WRej*	*TCR*
	NB	0.8423	0.1577	0.0000	1.3128
$\lambda = 1$	Three-way	0.8408	0.1577	0.0014	1.3128
	NB	0.9135	0.0865	0.0000	1.0048
$\lambda = 3$	Three-way	0.9130	0.0865	0.0005	1.0048
	NB	0.9423	0.0577	0.0000	0.6925
$\lambda = 9$	Three-way	0.9421	0.0577	0.0002	0.6925
k-nn					
	k-nn	0.8282	0.1718	0.0000	1.1586
$\lambda = 1$	Three-way	0.8239	0.1648	0.0113	1.2088
	k-nn	0.9341	0.0659	0.0000	1.1586
$\lambda = 3$	Three-way	0.9324	0.0632	0.0043	1.2088
	k-nn	0.9769	0.0231	0.0000	1.1586
$\lambda = 9$	Three-way	0.9763	0.0222	0.0015	1.2088
SVM					
	SVM	0.9521	0.0479	0.0000	6.1600
$\lambda = 1$	Three-way	0.9028	0.0169	0.0803	10.3333
	SVM	0.9697	0.0303	0.0000	4.1600
$\lambda = 3$	Three-way	0.9335	0.0108	0.0557	9.1333
	SVM	0.9768	0.0232	0.0000	3.2484
$\lambda = 9$	Three-way	0.9459	0.0083	0.0457	8.4747

Table 6. Comparison results on corpora PU3

NB		*WAcc*	*WErr*	*WRej*	*TCR*
	NB	0.9138	0.0862	0.0000	5.6911
$\lambda = 1$	Three-way	0.9107	0.0799	0.0094	6.1882
	NB	0.9465	0.0535	0.0000	4.7892
$\lambda = 3$	Three-way	0.9446	0.0494	0.0061	5.2705
	NB	0.9644	0.0356	0.0000	3.6175
$\lambda = 9$	Three-way	0.9632	0.0326	0.0042	4.0650
k-nn					
	k-nn	0.9048	0.0952	0.0000	4.8235
$\lambda = 1$	Three-way	0.8320	0.0484	0.1196	9.9110
	k-nn	0.9359	0.0641	0.0000	3.4780
$\lambda = 3$	Three-way	0.8635	0.0336	0.1029	6.8689
	k-nn	0.9529	0.0471	0.0000	2.0081
$\lambda = 9$	Three-way	0.8808	0.0255	0.0937	4.0087
SVM					
	SVM	0.9688	0.0312	0.0000	17.1363
$\lambda = 1$	Three-way	0.9334	0.0157	0.0508	37.9751
	SVM	0.9702	0.0298	0.0000	10.1460
$\lambda = 3$	Three-way	0.9359	0.0145	0.0496	26.0547
	SVM	0.9709	0.0291	0.0000	4.9085
$\lambda = 9$	Three-way	0.9372	0.0138	0.0489	18.6366

Table 7. Comparison results on corpora PUA

NB		*WAcc*	*WErr*	*WRej*	*TCR*
	NB	0.9570	0.0430	0.0000	19.8776
$\lambda = 1$	Three-way	0.9509	0.0377	0.0114	22.1418
	NB	0.9575	0.0425	0.0000	17.2113
$\lambda = 3$	Three-way	0.9500	0.0382	0.0118	19.0486
	NB	0.9577	0.0423	0.0000	16.0920
$\lambda = 9$	Three-way	0.9495	0.0384	0.0121	17.6724
k-nn					
	k-nn	0.7175	0.2825	0.0000	2.2809
$\lambda = 1$	Three-way	0.6237	0.2123	0.1632	4.0057
	k-nn	0.5877	0.4123	0.0000	0.8063
$\lambda = 3$	Three-way	0.4671	0.3197	0.2132	1.3352
	k-nn	0.5098	0.4902	0.0000	0.2745
$\lambda = 9$	Three-way	0.3732	0.3837	0.2432	0.4451
SVM					
	SVM	0.9316	0.0684	0.0000	11.7014
$\lambda = 1$	Three-way	0.8728	0.0272	0.1000	32.5111
	SVM	0.9184	0.0816	0.0000	7.8165
$\lambda = 3$	Three-way	0.8548	0.0338	0.1114	23.8442
	SVM	0.9105	0.0895	0.0000	5.9354
$\lambda = 9$	Three-way	0.8440	0.0377	0.1182	20.6893

Table 8. Comparison results on corpora Enron-Spam

NB		*WAcc*	*WErr*	*WRej*	*TCR*
	NB	0.8583	0.1417	0.0000	3.9030
$\lambda = 1$	Three-way	0.8447	0.1293	0.0260	4.5028
	NB	0.8710	0.1290	0.0000	2.6236
$\lambda = 3$	Three-way	0.8589	0.1179	0.0232	3.0856
	NB	0.8620	0.1380	0.0000	2.0322
$\lambda = 9$	Three-way	0.8503	0.1270	0.0227	2.4331
k-nn					
	k-nn	0.7726	0.2274	0.0000	37.9166
$\lambda = 1$	Three-way	0.7598	0.2143	0.0260	44.1145
	k-nn	0.7531	0.2469	0.0000	17.7413
$\lambda = 3$	Three-way	0.7359	0.2316	0.0325	18.9961
	k-nn	0.7137	0.2863	0.0000	7.0077
$\lambda = 9$	Three-way	0.6928	0.2688	0.0385	7.1975
SVM					
	SVM	0.9393	0.0607	0.0000	9.8972
$\lambda = 1$	Three-way	0.8886	0.0272	0.0843	23.0962
	SVM	0.9371	0.0629	0.0000	6.4686
$\lambda = 3$	Three-way	0.8932	0.0279	0.0790	13.7574
	SVM	0.9321	0.0679	0.0000	4.4352
$\lambda = 9$	Three-way	0.8920	0.0300	0.0780	8.5106

4 Conclusions

In this paper, email spam filtering is seen as a cost sensitive learning problem. From the point of optimum decision view, we propose a three-way decisions solution to filter spam email. For those suspicious emails, the three-way decisions solution moves them to boundary for further-exam, which can covert potential misclassification into rejections. The solution can be applied in any classifier only if the classifier can provide the probability of each mail being a legitimate. Naive Bayesian classifier, k-nn and SVM classifiers by considering three-way decisions solution are examined on several corpus, the result show the efficiency of the solution. With considering the three-way decisions solution, we can get a lower *weighted error rate* and a higher TCR.

Acknowledgments. This research is supported by the National Natural Science Foundation of China under Grant No. 60105003, 61170180, and the Fundamental Research Funds for the Central Universities under Grant No. NS2012129.

References

1. Androutsopoulos, I., Paliouras, G., Karkaletsis, V., Sakkis, G., Spyropoulos, C.D., Stamatopoulos, P.: Learning to filter spam e-mail: A comparison of a naive bayesian and a memory-based approach. In: 4th European Conference on Principles and Practice of Knowledge Discovery in Databases, pp. 1–13 (2000)
2. Carreras, X., Marquez, L.: Boosting trees for anti-spam email filtering. In: European Conference on Recent Advances in NLP (2001)
3. Chow, C.K.: On optimum recognition error and reject tradeoff. IEEE Transcations on Information Theory 16(1), 41–46 (1970)
4. Domingos, P., Pazzani, M.: Beyond independece: Conditions for the optimality of the simple Bayesian classifier. In: 13th International Conference on Machine Learning, pp. 105–112 (1996)
5. Drucker, H., Wu, D.H., Vapnik, V.N.: Support vector machines for spam categorization. IEEE Transactions on Neural Networks 10(5), 1048–1054 (1999)
6. Elkan, C.: The foundations of cost-sensitive learning. In: 17th International Joint Conference on Artificial Intelligence, pp. 973–978 (2001)
7. Metsis, V., Androutsopoulos, I., Paliouras, G.: Spam filtering with naive bayes-which naive bayes? In: 3rd Conference on Email and Anti-Spam (2006)
8. Mitchell, T.M.: Machine Learning. McGraw-Hill (1997)
9. Pauker, S.G., Kassirer, J.P.: The threshold approach to clinical decision making. New England Journal of Medicine 302, 1109–1117 (1980)
10. Pawlak, Z.: Rough sets. International Journal of Computer and Information Sciences 11, 341–356 (1982)
11. Sahami, M., Dumais, S., Heckerman, D., Horvitz, E.: A bayesian approach to filtering junk e-mail. In: Learning for Text Categorization-Papers from the AAAI Workshop, pp. 55–62 (1996)
12. Schneider, K.M.: A comparison of event models for Naive Bayes anti-spam e-mail filtering. In: 10th Conference of the European Chapter of the Association for Computational Linguistics, pp. 307–314 (2003)
13. Yao, Y., Wong, S.K.M., Lingras, P.: A decision-theoretic rough set model. Methodologies for Intelligent Systems 5, 17–24 (1992)
14. Yao, Y.: Three-way decisions with probabilistic rough sets. Information Sciences 180, 341–353 (2010)
15. Zhou, B., Yao, Y., Luo, J.: A Three-Way Decision Approach to Email Spam Filtering. In: Farzindar, A., Kešelj, V. (eds.) Canadian AI 2010. LNCS, vol. 6085, pp. 28–39. Springer, Heidelberg (2010)

A CUDA-Based Algorithm for Constructing Concept Lattices

Bo Shan[1], Jianjun Qi[1], and Wei Liu[2]

[1] School of Computer Science & Technology, Xidian University, Xi'an 710071, China
shanbo@stu.xidian.edu.cn, qijj@mail.xidian.edu.cn
[2] Institute of Software Engineering, Xidian University, Xi'an 710071, China
liuwei@xidian.edu.cn

Abstract. This paper presents a GPGPU (General-purpose computing on graphics processing units) implement of the concept lattice construction algorithm CloseByOne based on CUDA (Compute Unified Device Architecture), and compares it with the corresponding single-threaded and multi-threaded algorithms on the CPU by experiment. Experiment results show that, GPGPU can be used to generate concept lattices, but it has more restricts than CPU, the improvement of speed is not noteworthy. In the future, with the progress of graphics hardware, especially the improvement of memory capacity and branch prediction, the efficiency of GPGPU-based algorithms for generating concept lattices may be significantly increased.

Keywords: concept lattice, CloseByOne, GPGPU, CUDA.

1 Introduction

Concept lattice theory, also called formal concept analysis, was firstly proposed by German mathematician Wille in 1982 [1]. It is a kind of applied mathematics based on concepts and concept hierarchy, and provides strong support for data analysis. The core data structure of concept lattice theory is concept lattices, so a good algorithm of constructing concept lattices is necessary.

With the development of concept lattice theory, there are several algorithms for generating concepts, and the algorithms can be basically divided into batch algorithms and incremental algorithms [2, 3]. In recent years, with the development of hardware, parallel algorithms are becoming a growing concern, and have had some applications in concept lattice theory [4–6].

This paper focuses on generating formal concepts in a binary context. CloseByOne algorithm is designed respectively with CPU single-threaded programs, CPU multi-threaded programs and CUDA programs. The comparison in the running time shows that the speed of CloseByOne programs with CUDA is nearly the same as CPU single-threaded programs, and is a little bit faster.

The experiments show that memory capacity and branch prediction of CUDA need to be improved, but CUDA can be used to generate concepts. For example, for a random context in the size of 100*100, CUDA programs can get the right result and generate more than ten million concepts.

J.T. Yao et al. (Eds.): RSCTC 2012, LNAI 7413, pp. 297–302, 2012.

2 Basic Knowledge of GPU and CUDA

CUDA is a parallel computing architecture developed by Nvidia. Because CUDA is based on C language, most people who are familiar with C language can write programs executed on the GPU easily by CUDA. GPU usually has high memory bandwidth and a lot of execution units, and mostly the price is relatively cheap.

By CUDA, programs are divided into CPU part and GPU part. Firstly some programs are executed on the CPU, then the data is copied to GPU and the programs go on computing on the GPU, finally the results are copied back to CPU. GPU has much bigger ram delay, smaller cache and poorer branch prediction than CPU, besides memory of GPU is rather limited. If the program doesn't have high concurrency degree, the effect of optimization is not good. Concrete characteristics and programming principle can be found in the references [7, 8].

For generating concepts, there are a lot of circulations and judgements in programs and programs need considerable memory to save variables and store results, which are disadvantages. But GPU has a lot of cores, and its concurrency degree is high, this is also the reason why we do experiments with CUDA.

3 CPU and CUDA Programs of CloseByOne Algorithm

3.1 CPU Single-Threaded and Multi-threaded Programs

CloseByOne algorithm [5] is based on closure. For a given intent of a concept, the algorithm generates closure through adding an attribute that the concept does not contain. For example, given a concept (A, B), add an attribute $y \in M$ (M is the set containing all attributes in the formal context (G, M, I)) and $y \notin B$, $((B \bigcup \{y\})', (B \bigcup \{y\})'')$ is the closure.

Single-threaded CloseByOne algorithm starts from the biggest concept (the extent of the concept is G, G is the set containing all objects in the formal context (G, M, I)), and then adds attribute in sequence for recursion, ultimately generates all concepts. Given a concept (A, B), the algorithm adds a attribute $y \in M(y \notin B)$ to generate the concept (C, D). If (C, D) satisfies with dictionary sequence, then the algorithm adds it to the result and goes on recursion, or gives up. Dictionary sequence will not affect generation of all the concepts, but guarantees that each concept is generated only once.

Because the support for recursion in CUDA is weak and only after $2.x$ computing power and the efficiency is not high, this paper transforms the recursive algorithm into non-recursive version. The non-recursive process is consistent with the past recursion [5], *part* is used to keep status of recursion in the stack. The extra variable y in the concept is used to identify the number of next attribute which will be added.

The CPU multi-threaded programs are as same as that of reference [5]. Each L-depth concept is stored as a starting point concept, assigned into a thread in the way of Roulette. After division, each thread generates subsequent concepts of each starting point concept. Finally the algorithm ensures all the threads finished by synchronization whose function is WaitForMultipleObjects. Considering CPU memory is limited, the programs directly write the result to file. Mutual exclusion is used to ensure valid storage in the programs.

3.2 CUDA Programs of CloseByOne Algorithm

Basically the CUDA algorithm is to use the characteristic of massive cores of GPU. The core of the algorithm is to divide starting point concepts according to the CUDA thread number, then generates in parallel. Considering overtime problem of the system, it needs to record the recursive status and set maximal number of generating concepts one time so that the result can be effectively stored and the algorithm can generate all concepts by circularly solving.

```
   void CUDAcomputestart()
1.     divide starting point concepts by the count of blockx*gridx;
2.     foreach i in the set of start numbering of each block do
3.         copy the block of start numbering i to resultkc;
4.         set all values of mkk -1 to initialize recursion position;
5.         while do
6.             generateFromcc<<<gridx, blockx>>>(resultkc, resultcc,
                                      resultc, pcount-i, part, mkk);
7.             Synchronize so that all threads are finished;
8.             copy back the result to CPU;
9.             write the generated concepts to the file;
10.            judge whether the task of each thread is finished, if
each task of all threads is finished, goto line 2, or goto line 5;
11.        end
12.    end
```

The variables of the code are described below: *resultkc* stores L-depth concepts as starting point concepts on the GPU. *pcount* records the total number of starting point concepts. *resultc* stores the result generated on the GPU and ensures each thread have space to store *max_every* concepts. *resultcc* stores number of generated concepts of each thread on the GPU.

Description of the Algorithm: Line 1 divides starting point concepts into parts to run separately according to thread number. Line 6 executes on the GPU. Lines 5–11 generate all concepts of the block by circularly solving based on the kept status of recursion.

```
   __global__ void generateFromcc(resultkc, resultcc, resultc,
                                  sum, part, mkk)
      pk records the numbering of the thread;
      if pk < sum then
          set count and resultcc[pk] to 0;
          declare local variable mk;
          if mkk[pk] == -1 then
              set part[pk][0] to resultkc[pk] and mk to 0;
          else if mkk[pk] == -2 then
             return;
          else
              set mk to mkk[pk];
```

```
11.         end
12.         store part[pk][mk] in resultc;
13.         while part[pk][0].y < n do
14.             declare local variable j;
15.             for j from part[pk][mk].y upto n do
16.                 if part[pk][mk].intent[j] == false then
17.                     set cd to computeClosurec(part[pk][mk], j);
18.                     judge whether the closure satisfy
dictionary sequence, if satisfy dictionary sequence then
19.                         set part[pk][mk].y to j+1, mk to mk+1,
cd.y to j+1, and  part[pk][mk] to cd;
20.                         if count >= max_every then
21.                             set mkk[pk] to mk and  resultcc[pk] to count;
22.                             return;
23.                         else
24.                             store cd in resultc;
25.                         end
26.                         break;
27.                     end
28.                 end
29.             end
30.             if j >= n then
31.                 set mk to mk-1;
32.                 if mk < 0 then
33.                     break;
34.                 end
35.             end
36.         end
37.         set mkk[pk] to -2;
38.         set resultcc[pk] to count;
39.     end
```

The variables of generateFromcc are described below: *count* records the numbering of generated concept. *resultcc* records final number of generated concepts. *mk* records the position of recursion.

Description of the Algorithm: Lines 5–6 deal with the condition that the starting point concepts generate concepts firstly, line 7 recognizes the condition that the starting point concepts has finish the work, and lines 9–11 are used to identify the depth of recursion to continue last generating. Lines 12–38 generate the branch concepts of starting point concepts. Line 20 judges whether generated concept count of the thread reaches *max_every*, if *yes*, record the status and return. Lines 30–35 realize the function of backtracking. Line 37 marks that the task has been completed.

The above generateFromcc differs from the original one in that, it needs to mark the current position of the stack and whether or not the task of the thread has been completed, and need to update the *part* which is the stack used to record the generated way of next concept. The algorithm needs to judge whether the generated concept count has reached *max_every*, and accesses the result of each thread according to *pk*.

Here, we don't store generated concepts using atomic operation, because it is slower; on the other hand, it also needs to use *max_every*, or overtime will have a runtime error, the programs can't get the result(see the experiment).

4 Experiment Comparison of CloseByOne Algorithms

The experiment is based on VC2008, CUDA with version 3.2, GPU with GeForce GT 430, dual-core AMD Athlon II X2 250 Processor, and memory of 3.50 GB.

This paper adopts the non-recursive version of CloseByOne algorithm, and designs CPU single-threaded programs, CPU multi-threaded programs, and CUDA programs, respectively. For comparison, single-threaded programs are got by the serialization of the multi-threaded programs so that the experiment can effectively show difference of efficiency in the same structure.

Parameters are set as follows: thread count of multi-threaded programs is 3, thread count of CUDA programs is 256*64, $max\_every = 10$, the value of depth L in all algorithms is 3. In the CUDA programs, constant memory is used to store the context in order to read it quickly. If the context is bigger, pitch-linear memory should be used instead.

The data in Table 1 is based on the context that generated by random number, whose density is 0.5 (the density of the context is defined with the proportion of the value 1 in the context). The unit of time in the following table is *s*.

Table 1. Comparison of CloseByOne Algorithms

the size of context	single-threaded	multi-threaded	CUDA
50*50	10.875	12.391	8.813
70*70	126.016	115.563	100.359
100*100	2478.36	1549.766	2182.407

Table 1 shows that CUDA programs is slightly better than CPU single-threaded programs, but in the environment of dual-core processor it is not better than CPU multi-threaded programs when the context is bigger. The reason may be synchronization caused by *max_every*. If GPU can support overtime, it will be better. The GPU of the experiment does support overtime, you can check the attribute by running `deviceQuery.exe`, in my environment it shows: `Run time limit on kernels:Yes`, the endured time generally is five seconds, when the kernel is run more than the limit, you will get errors and can not get final result. Here *max_every* is set to avoid overtime, but it results in the need of synchronization which is rather slow. But to generate concepts, it is a kind lift helplessly, otherwise the storage of the results will be another problem. In sum, CloseByOne algorithm has many judgments and branches, memory is also a problem, accessing global memory on the GPU is slow, but it needs to save the stack (because the register is very limited, there is no enough memory to store all generated concepts at a time and the kernel programs need to be run many times), accessing global memory have bad influence on performance.

5 Conclusions

This paper explores the feasibility and efficiency of generating concepts with CUDA, and summarizes the limit of generating concepts with CUDA. The experiments show that CUDA can be used to generate concepts, but there are more restrictions than CPU, the improvement of the speed is not distinct while the program complexity is higher, so it needs the improvement of CUDA and further exploration.

Acknowledgments. This work was supported by grants from the National Natural Science Foundation of China (No. 60703117 and No. 11071281) and the Fundamental Research Funds for the Central Universities (No. JY10000903010 and No. K50510230005).

References

1. Wille, R.: Restructuring lattice theory: An approach based on hierarchies of concepts. In: Rival, I. (ed.) Ordered Sets, pp. 445–470. Reidel, Dordrecht (1982)
2. Carpineto, C., Romano, G.: Concept Data Analysis. Theory and Applications. Wiley, New York (2004)
3. Kuznetsov, S., Obiedkov, S.: Comparing performance of algorithms for generating concept lattices. J. Exp. Theor. Artif. Int. 14, 189–216 (2002)
4. Fu, H.G., Nguifo, E.M.: A Parallel Algorithm to Generate Formal Concepts for Large Data. In: Eklund, P. (ed.) ICFCA 2004. LNCS (LNAI), vol. 2961, pp. 394–401. Springer, Heidelberg (2004)
5. Krajca, P., Outrata, J., Vychodil, V.: Parallel Recursive Algorithm for FCA. In: Concept Lattices and Their Applications, pp. 71–82. Palacky University, Olomouc (2008)
6. Langdon, W.B., Yoo, S., Harman, M.: Formal Concept Analysis on Graphics Hardware. In: CLA 2011, Nancy, France, pp. 413–416 (2011)
7. Sanders, J., Kandrot, E.: CUDA by Example: an introduction to general purpose GPU programming. Pearson Education, Inc. (2010)
8. Zhang, S., Chu, Y.L.: The high performance computing with CUDA based on GPU. China WaterPower Press (2009) (in chinese)

A Heuristic Knowledge Reduction Algorithm for Real Decision Formal Contexts

Jinhai Li[1], Changlin Mei[1], Yuejin Lv[2], and Xiao Zhang[1]

[1] School of Mathematics and Statistics, Xi'an Jiaotong University, Xi'an 710049, P.R. China
[2] School of Mathematics and Information Sciences, Guangxi University, Nanning 530004, P.R. China

Abstract. Knowledge reduction is one of the key issues in real formal concept analysis. This study investigates the issue of developing efficient knowledge reduction methods for real decision formal contexts. A corresponding heuristic algorithm is proposed and some numerical experiments are conducted to assess its efficiency.

1 Introduction

Formal concept analysis (FCA), proposed by Wille [19], is an effective approach for data analysis and knowledge processing. FCA starts with a formal context formalized by a triple (U, A, I) where U is a set of objects, A is a set of attributes, and I is a binary relation on $U \times A$. From a formal context, some formal concepts which are basic outputs of FCA can be derived, and the set of all the formal concepts forms a complete lattice called the concept lattice of the formal context. Concept lattice has been demonstrated to be a useful tool for conceptual knowledge discovery and data analysis [6].

Knowledge reduction is one of the key issues in FCA and much attention has been paid to this topic in recent years. For example, based on the predefined arrow relations, an approach to knowledge reduction for formal contexts was introduced in [6]. In [26], a novel reduction method was proposed for formal contexts from the point of view of lattice isomorphism. Motivated by the work in [26], the authors of [13] put forward two new reduction approaches for formal contexts based on the object-oriented concept lattice [22,23] and the property-oriented concept lattice [4]. This issue was also investigated in literature [1] and [14] from the perspectives of fuzzy K-means clustering and extension equivalence, respectively. In addition, some knowledge reduction methods for decision formal contexts [25], an extension of the formal contexts, were explored in [8,9,10,17,18,20].

In the classical formal contexts, the relationship between the objects and the attributes is described by a two-valued form that can only specify whether or not an object has an attribute. In many real-world situations, however, the relationship may be needed to be fuzzy-valued or interval-valued. Therefore, some studies have recently been devoted to the generalized formal contexts or decision

J.T. Yao et al. (Eds.): RSCTC 2012, LNAI 7413, pp. 303–312, 2012.

formal contexts such as fuzzy formal contexts [2,3,5,12,16,24], real formal contexts [7], fuzzy decision formal contexts [15] and real decision formal contexts [11,21]. Knowledge reduction is still an important issue in the analysis of the generalized decision formal contexts. For instance, an approach to knowledge reduction for fuzzy decision formal contexts was proposed in [15]. By defining an implication mapping between the conditional and the decision real concept lattices, real decision formal contexts were classified in [21] into consistent and inconsistent real decision formal contexts, and a knowledge reduction approach was presented for the consistent real decision formal contexts. Since the decision rules derived directly from a real decision formal context are in general not concise or compact, we put forward a new knowledge reduction method in [11] for real decision formal contexts to make the derived decision rules more compact. Although some studies has been done on knowledge reduction in real decision formal contexts, the existing reduction approaches are all based on the discernibility matrix and Boolean function. That is to say, they are all computationally expensive and are even impossible to implement for large real decision formal contexts. Just as we pointed out in [11], efficient reduction algorithms for real decision formal contexts are still needed to be further studied. This paper develops a heuristic knowledge reduction algorithm for real decision formal contexts to speed up the implementation of the knowledge reduction.

In Section 2, we briefly recall some basic notions of real FCA and discuss several properties related to the subcontexts and the real decision formal contexts. In Section 3, attribute characteristics of a real decision formal context are investigated. In Section 4, we develop a heuristic reduction algorithm for real decision formal contexts. In Section 5, some numerical experiments are conducted to assess the efficiency of the proposed algorithm. The paper is then concluded with a brief summary.

2 Real FCA

In this section, we briefly review some basic notions of real FCA and discuss some properties related to the subcontexts and the real decision formal contexts.

2.1 Real Formal Contexts and Real Concept Lattices

Definition 1. *[7] Let $\mathfrak{R}$ be the set of real numbers. A real interval on $\mathfrak{R}$, denoted by $I = [u, v]$, $u, v \in \mathfrak{R}$, represents the set of the real numbers delimited by u and v, where u and v are called the lower and the upper bounds of I, respectively. If $u > v$, then I is said to be empty, denoted by $[,]$.*

Let $I_1 = [u_1, v_1]$ and $I_2 = [u_2, v_2]$ be two real intervals. The intersection of I_1 and I_2 is defined by $\mathrm{Inter}(I_1, I_2) = [\max(u_1, u_2), \min(v_1, v_2)]$. The closure of a real interval set $E = \{I_1, I_2, \cdots, I_n\}$ is defined by

$$\mathrm{Closure}(\{I_1, I_2, \cdots, I_n\}) = \mathrm{Closure}(\mathrm{Closure}(\{I_1, I_2\}), I_3, \cdots, I_n\}),$$

where

$$\text{Closure}(\{I_1, I_2\}) = \begin{cases} \{[\min(u_1, u_2), \max(v_1, v_2)]\}, & \text{if Inter}(I_1, I_2) \neq [,], \\ \{I_1, I_2\}, & \text{otherwise.} \end{cases}$$

Let $E = \{I_1, \cdots, I_s\}$ and $F = \{I'_1, \cdots, I'_t\}$ be two sets of real intervals. The intersection and the union of E and F are defined by

$$E \cap F = \text{Closure}(\{\text{Inter}(I_i, I'_j) \mid i = 1, \cdots, s; j = 1, \cdots, t\}),$$
$$E \cup F = \text{Closure}(\{I_1, \cdots, I_s, I'_1, \cdots, I'_t\}).$$

E is said to be largely less than F, denoted by $E \leq_L F$, if for any $[u, v] \in F$, there exists $[u', v'] \in E$ such that $u' \leq u$ and $v' \geq v$; E is said to be strictly less than F, denoted by $E \leq_S F$, if for any $[u, v] \in F$, there exists $[u', v'] \in E$ such that $u \leq u'$ and $v \geq v'$.

Definition 2. *[7] Let $U = \{x_1, x_2, \cdots, x_n\}$ be a set of objects, V be the set of all real intervals on $\Re$, and $P(V)$ be the power set of V. A real set $\widetilde{X}$ of U is defined by its characteristic function $\mu_{\widetilde{X}} : U \to P(V)$, where $\mu_{\widetilde{X}}(x)$, a set of real intervals, indicates the possible values that can be chosen for x in the real set $\widetilde{X}$. The real set $\widetilde{X}$ is denoted by*

$$\widetilde{X} = \left\{ \frac{\mu_{\widetilde{X}}(x_1)}{x_1}, \frac{\mu_{\widetilde{X}}(x_2)}{x_2}, \cdots, \frac{\mu_{\widetilde{X}}(x_n)}{x_n} \right\}$$

and the empty real set of U is denoted by $\widetilde{\emptyset}$.

For brevity, we write $\mu_{\widetilde{X}}(x)$ $(x \in U)$ as $\widetilde{X}(x)$. The restriction of $\widetilde{X}$ on a subset X of U, denoted by $\widetilde{X}|_X$, is defined by $\widetilde{X}|_X(x) = \widetilde{X}(x)$ for all $x \in X$.

For two real sets $\widetilde{X}$ and $\widetilde{Y}$ of U, two kinds of intersection, union and inclusion can be defined [7]. Concretely, if $\widetilde{X}(x) \leq_L \widetilde{Y}(x)$ for all $x \in U$, then $\widetilde{X}$ is said to be largely included in $\widetilde{Y}$, denoted by $\widetilde{X} \subseteq_L \widetilde{Y}$; the large intersection and the large union of $\widetilde{X}$ and $\widetilde{Y}$ are defined by $(\widetilde{X} \cap_L \widetilde{Y})(x) = \widetilde{X}(x) \cup \widetilde{Y}(x)$ and $(\widetilde{X} \cup_L \widetilde{Y})(x) = \widetilde{X}(x) \cap \widetilde{Y}(x)$, respectively. In addition, if $\widetilde{X}(x) \leq_S \widetilde{Y}(x)$ for all $x \in U$, then $\widetilde{X}$ is said to be strictly included in $\widetilde{Y}$, denoted by $\widetilde{X} \subseteq_S \widetilde{Y}$; the strict intersection and the strict union of $\widetilde{X}$ and $\widetilde{Y}$ are defined by $(\widetilde{X} \cap_S \widetilde{Y})(x) = \widetilde{X}(x) \cap \widetilde{Y}(x)$ and $(\widetilde{X} \cup_S \widetilde{Y})(x) = \widetilde{X}(x) \cup \widetilde{Y}(x)$, respectively.

Let U be a set of objects and A be a set of attributes. A real binary relation $\widetilde{I}$ on $U \times A$ is a mapping that takes each of its value to be a set of real intervals. That is, for each $(x, a) \in U \times A$, $\widetilde{I}(x, a)$ is a set of real intervals.

Definition 3. *[21] A real formal context is a triple $(U, A, \widetilde{I})$, where U is a set of objects, A is a set of attributes, and $\widetilde{I}$ is a real binary relation on $U \times A$.*

Definition 4. *[7] Let $\mathbb{S} = (U, A, \widetilde{I})$ be a real formal context, $\mathcal{P}(U)$ be the power set of U and $\mathcal{R}(A)$ be the set of all real sets of A. For $X \in \mathcal{P}(U)$ and $\widetilde{B} \in \mathcal{R}(A)$, four operators $\uparrow, \Box : \mathcal{P}(U) \to \mathcal{R}(A)$ and $\downarrow, \Diamond : \mathcal{R}(A) \to \mathcal{P}(U)$ are defined by*

$$X^{\uparrow} = \{\tfrac{f(a)}{a} \mid a \in A, f(a) = \bigcup_{x \in X} \widetilde{I}(x,a)\},$$
$$\widetilde{B}^{\downarrow} = \{x \in U \mid \forall a \in A, \widetilde{B}(a) \leq_L \widetilde{I}(x,a)\},$$
$$X^{\square} = \{\tfrac{g(a)}{a} \mid a \in A, g(a) = \bigcap_{x \in X} \widetilde{I}(x,a)\},$$
$$\widetilde{B}^{\diamond} = \{x \in U \mid \forall a \in A, \widetilde{B}(a) \leq_S \widetilde{I}(x,a)\}.$$

Definition 5. *[7] Let $\mathbb{S} = (U, A, \widetilde{I})$ be a real formal context, $X \in \mathcal{P}(U)$ and $\widetilde{B} \in \mathcal{R}(A)$. The pair $(X, \widetilde{B})$ is called a large real concept of $\mathbb{S}$ if $X^{\uparrow} = \widetilde{B}$ and $\widetilde{B}^{\downarrow} = X$, and it is called a strict real concept of $\mathbb{S}$ if $X^{\square} = \widetilde{B}$ and $\widetilde{B}^{\diamond} = X$. Here, X and $\widetilde{B}$ are called the extension and the intension of the large (strict) real concept $(X, \widetilde{B})$, respectively.*

According to [7], if large real concepts are ordered by $(X_1, \widetilde{B}_1) \preceq_L (X_2, \widetilde{B}_2) \Leftrightarrow X_1 \subseteq X_2 (\Leftrightarrow \widetilde{B}_2 \subseteq_L \widetilde{B}_1)$, then the set $\mathfrak{B}_L(U, A, \widetilde{I})$ of all large real concepts of $\mathbb{S} = (U, A, \widetilde{I})$ together with the order $\preceq_L$ forms a complete lattice, called the *large real concept lattice* of $\mathbb{S}$. Similarly, if strict real concepts are ordered by $(X_1, \widetilde{B}_1) \preceq_S (X_2, \widetilde{B}_2) \Leftrightarrow X_1 \subseteq X_2 (\Leftrightarrow \widetilde{B}_2 \subseteq_S \widetilde{B}_1)$, then the set $\mathfrak{B}_S(U, A, \widetilde{I})$ of all strict real concepts of $\mathbb{S}$ together with the order $\preceq_S$ also forms a complete lattice, called the *strict real concept lattice* of $\mathbb{S}$.

Let $\mathbb{S} = (U, A, \widetilde{I})$ be a real formal context and $E \subseteq A$. The restriction of $\widetilde{I}$ on $U \times E$, denoted by $\widetilde{I}_E$, is defined by $\widetilde{I}_E(x, e) = \widetilde{I}(x, e)$ for all $(x, e) \in U \times E$. The real formal context $(U, E, \widetilde{I}_E)$ is called a subcontext of $(U, A, \widetilde{I})$ [11]. Let $X^{\uparrow_E}$, $\widetilde{B}^{\downarrow_E}$, $X^{\square_E}$ and $\widetilde{B}^{\diamond_E}$ be the restriction of the four operators given in Definition 4 on the subcontext $(U, E, \widetilde{I}_E)$. Then we can, similarly, define the large and the strict real concepts in $(U, E, \widetilde{I}_E)$. Clearly, the set $\mathfrak{B}_L(U, E, \widetilde{I}_E)$ of all the large real concepts of $(U, E, \widetilde{I}_E)$ together with the order $\preceq_L$ also forms a complete lattice as well as the set $\mathfrak{B}_S(U, E, \widetilde{I}_E)$ of all the strict real concepts of $(U, E, \widetilde{I}_E)$ together with the order $\preceq_S$.

Proposition 1. *[11] Let $\mathbb{S} = (U, A, \widetilde{I})$ be a real formal context and $E \subseteq A$. Then for both the large and the strict real concepts, each extension of $(U, E, \widetilde{I}_E)$ is also an extension of $\mathbb{S}$.*

Proposition 2. *Let $\mathbb{S} = (U, A, \widetilde{I})$ be a real formal context and $E \subseteq A$. Then $\mathfrak{B}_L(U, E, \widetilde{I}_E) = \{((\widetilde{B}|_E)^{\downarrow_E}, \widetilde{B}|_E) \mid (X, \widetilde{B}) \in \mathfrak{B}_L(U, A, \widetilde{I})\}$ and $\mathfrak{B}_S(U, E, \widetilde{I}_E) = \{((\widetilde{B}|_E)^{\diamond_E}, \widetilde{B}|_E) \mid (X, \widetilde{B}) \in \mathfrak{B}_S(U, A, \widetilde{I})\}$.*

Proof. The proof is similar to that in the classical formal contexts [9].

2.2 Real Decision Formal Contexts and Their Reduction

Definition 6. *[21] A real decision formal context is a quintuple $(U, A, \widetilde{I}, D, \widetilde{J})$, where $(U, A, \widetilde{I})$ and $(U, D, \widetilde{J})$ are two real formal contexts and $A \cap D = \emptyset$. The sets A and D are called the conditional and the decision attribute sets of $(U, A, \widetilde{I}, D, \widetilde{J})$, respectively.*

Although two kinds of real concept lattices, i.e., the large real concept lattice and the strict real concept lattice, have been introduced in Section 2.1, the knowledge reduction frameworks based on them are similar to each other according to [11,21]. Therefore, to achieve the task of developing efficient reduction algorithms for real decision formal contexts, it is sufficient to consider one of them only. In this paper, we would like to propose a corresponding heuristic algorithm based on the large real concept lattice.

Let $\mathbb{K} = (U, A, \widetilde{I}, D, \widetilde{J})$ be a real decision formal context. For a subset E of A, we call $(U, E, \widetilde{I}_E, D, \widetilde{J})$ a subcontext of $\mathbb{K}$.

Definition 7. *[11] Let $\mathbb{K} = (U, A, \widetilde{I}, D, \widetilde{J})$ be a real decision formal context, $E \subseteq A$, $(X, \widetilde{B}) \in \underline{\mathfrak{B}}_L(U, E, \widetilde{I}_E)$, and $(Y, \widetilde{C}) \in \underline{\mathfrak{B}}_L(U, D, \widetilde{J})$, where $X \neq \emptyset$, $Y \neq \emptyset$, $\widetilde{B} \neq \emptyset$ and $\widetilde{C} \neq \emptyset$. If $X \subseteq Y$, then $\widetilde{B} \rightarrow \widetilde{C}$ is called an L decision rule. The real sets $\widetilde{B}$ and $\widetilde{C}$ are called the premise and the conclusion of $\widetilde{B} \rightarrow \widetilde{C}$, respectively.*

Hereinafter, we denote by $\mathfrak{R}_L(E, D)$ the set of all L decision rules derived from the subcontext $(U, E, \widetilde{I}_E, D, \widetilde{J})$. That is, $\mathfrak{R}_L(E, D)$ denotes the set of all the L decision rules generated between the large real concepts in $\underline{\mathfrak{B}}_L(U, E, \widetilde{I}_E)$ and those in $\underline{\mathfrak{B}}_L(U, D, \widetilde{J})$.

Definition 8. *[11] Let $\mathbb{K} = (U, A, \widetilde{I}, D, \widetilde{J})$ be a real decision formal context and $E \subseteq A$. For $\widetilde{B} \rightarrow \widetilde{C} \in \mathfrak{R}_L(E, D)$ and $\widetilde{B}' \rightarrow \widetilde{C}' \in \mathfrak{R}_L(A, D)$, if $\widetilde{B} \subseteq_L \widetilde{B}'|_E$ and $\widetilde{C}' \subseteq_L \widetilde{C}$, we say that $\widetilde{B}' \rightarrow \widetilde{C}'$ can be implied by $\widetilde{B} \rightarrow \widetilde{C}$. If each L decision rule of $\mathfrak{R}_L(A, D)$ can be implied by an L decision rule of $\mathfrak{R}_L(E, D)$, we say that $\mathfrak{R}_L(E, D)$ implies $\mathfrak{R}_L(A, D)$, denoted by $\mathfrak{R}_L(E, D) \Rightarrow_L \mathfrak{R}_L(A, D)$.*

Definition 9. *[11] For a real decision formal context $\mathbb{K} = (U, A, \widetilde{I}, D, \widetilde{J})$, $E \subseteq A$ is called an L consistent set of $\mathbb{K}$ if $\mathfrak{R}_L(E, D) \Rightarrow_L \mathfrak{R}_L(A, D)$; otherwise, E is called an L inconsistent set of $\mathbb{K}$. Furthermore, if E is an L consistent set of $\mathbb{K}$ and any $F \subset E$ is an L inconsistent set of $\mathbb{K}$, then E is called an L reduct of $\mathbb{K}$. The intersection of all the L reducts of $\mathbb{K}$ is called the L core of $\mathbb{K}$.*

According to [11], the knowledge reduction of a real decision formal context $\mathbb{K} = (U, A, \widetilde{I}, D, \widetilde{J})$ is to find an L reduct of $\mathbb{K}$, which can make the obtained L decision rules more compact. In preparation for developing an efficient approach to find an L reduct, we derive in the following a sufficient and necessary condition of justifying whether or not a conditional attribute set is an L consistent set.

Definition 10. *[11] Let $\mathbb{K} = (U, A, \widetilde{I}, D, \widetilde{J})$ be a real decision formal context and $E \subseteq A$. For $\widetilde{B} \rightarrow \widetilde{C} \in \mathfrak{R}_L(E, D)$, if there exists another L decision rule $\widetilde{B}_0 \rightarrow \widetilde{C}_0$ in $\mathfrak{R}_L(E, D)$ such that $\widetilde{B}_0 \rightarrow \widetilde{C}_0$ implies $\widetilde{B} \rightarrow \widetilde{C}$, then $\widetilde{B} \rightarrow \widetilde{C}$ is said to be redundant in $\mathfrak{R}_L(E, D)$; otherwise, $\widetilde{B} \rightarrow \widetilde{C}$ is said to be non-redundant in $\mathfrak{R}_L(E, D)$.*

For convenience, we denote by $\mathfrak{R}^*{}_L(E, D)$ the set of all the non-redundant L decision rules of $\mathfrak{R}_L(E, D)$.

Proposition 3. *Let $\mathbb{K} = (U, A, \widetilde{I}, D, \widetilde{J})$ be a real decision formal context. Then $E \subseteq A$ is an L consistent set of $\mathbb{K}$ if and only if $\mathfrak{R}^*{}_L(E, D) \Rightarrow_L \mathfrak{R}^*{}_L(A, D)$.*

Proof. The proof is immediate from Definitions 9 and 10.

3 Attribute Characteristics of a Real Decision Formal Context

Definition 11. *Let $\mathbb{K} = (U, A, \widetilde{I}, D, \widetilde{J})$ be a real decision formal context and $a \in A$. If $A - \{a\}$ is an L consistent set of $\mathbb{K}$, then a is called an unnecessary attribute of $\mathbb{K}$; otherwise, it is called a necessary attribute of $\mathbb{K}$.*

That is, the conditional attributes of a real decision formal context are classified into two categories: the necessary attributes and the unnecessary attributes. It can be known from Definition 11 that to remove an unnecessary attribute from a real decision formal context will not have any affect on the extraction of the L decision rules. However, each necessary attribute is essential to the rule acquisition.

In preparation for deriving in the next section a heuristic reduction algorithm for real decision formal contexts, we formulate in this section a sufficient and necessary condition of justifying whether or not an attribute is necessary with respect to a given real decision formal context. Firstly, we put forward the notion of an extension-preserved large real concept in real formal contexts.

Definition 12. *Let $\mathbb{S} = (U, A, \widetilde{I})$ be a real formal context, $a \in A$ and $E = A - \{a\}$. For $(X, \widetilde{B}) \in \mathfrak{B}_L(U, A, \widetilde{I})$, if there does not exist $(X_0, \widetilde{B}_0) \in \mathfrak{B}_L(U, A, \widetilde{I})$ such that $X \subset X_0$ and $\widetilde{B}_0|_E = \widetilde{B}|_E$, we say that $(X, \widetilde{B})$ is an extension-preserved large real concept with respect to $(U, E, \widetilde{I}_E)$.*

Proposition 4. *Let $\mathbb{S} = (U, A, \widetilde{I})$ be a real formal context, $a \in A$ and $E = A - \{a\}$. Then $(X, \widetilde{B}) \in \mathfrak{B}_L(U, A, \widetilde{I})$ is an extension-preserved large real concept with respect to $(U, E, \widetilde{I}_E)$ if and only if $(X, \widetilde{B}|_E) \in \mathfrak{B}_L(U, E, \widetilde{I}_E)$.*

Proof. ($\Rightarrow$) By Proposition 2, we have $((\widetilde{B}|_E)^{\downarrow_E}, \widetilde{B}|_E) \in \mathfrak{B}_L(U, E, \widetilde{I}_E)$. If $(\widetilde{B}|_E)^{\downarrow_E} \neq X$, then $X \subset (\widetilde{B}|_E)^{\downarrow_E}$ according to Definition 4. Based on Proposition 1, we obtain $((\widetilde{B}|_E)^{\downarrow_E}, (\widetilde{B}|_E)^{\downarrow_E\uparrow}) \in \mathfrak{B}_L(U, A, \widetilde{I})$. Therefore, $(\widetilde{B}|_E)^{\downarrow_E\uparrow}|_E = (\widetilde{B}|_E)^{\downarrow_E\uparrow_E} = \widetilde{B}|_E$. It follows from Definition 12 that $(X, \widetilde{B})$ is not an extension-preserved large real concept with respect to $(U, E, \widetilde{I}_E)$, which is in contradiction with the assumption. Thus, $(\widetilde{B}|_E)^{\downarrow_E} = X$ and consequently $(X, \widetilde{B}|_E) \in \mathfrak{B}_L(U, E, \widetilde{I}_E)$.

($\Leftarrow$) If $(X, \widetilde{B})$ is not an extension-preserved large real concept with respect to $(U, E, \widetilde{I}_E)$, then by Definition 12 there exists $(X_0, \widetilde{B}_0) \in \mathfrak{B}_L(U, A, \widetilde{I})$ such that $X \subset X_0$ and $\widetilde{B}_0|_E = \widetilde{B}|_E$. Therefore, $((\widetilde{B}_0|_E)^{\downarrow_E}, \widetilde{B}|_E) \in \mathfrak{B}_L(U, E, \widetilde{I}_E)$ and $X \subset X_0 = \widetilde{B}_0^{\downarrow} \subseteq (\widetilde{B}_0|_E)^{\downarrow_E}$ which are in contradiction with $(X, \widetilde{B}|_E) \in \mathfrak{B}_L(U, E, \widetilde{I}_E)$.

In addition, we present the notion of a key large real concept in real decision formal contexts.

Definition 13. *Let* $\mathbb{K} = (U, A, \widetilde{I}, D, \widetilde{J})$ *be a real decision formal context,* $E \subseteq A$ *and* $(X, \widetilde{B}) \in \underline{\mathfrak{B}}_L(U, E, \widetilde{I}_E)$*. If there exists* $(Y, \widetilde{C}) \in \underline{\mathfrak{B}}_L(U, D, \widetilde{J})$ *such that the L decision rule* $\widetilde{B} \to \widetilde{C}$ *is non-redundant in* $\mathfrak{R}_L(E, D)$*, then* $(X, \widetilde{B})$ *is called a key large real concept of* $\underline{\mathfrak{B}}_L(U, E, \widetilde{I}_E)$ *with respect to* $\underline{\mathfrak{B}}_L(U, D, \widetilde{J})$*.*

Proposition 5. *Let* $\mathbb{K} = (U, A, \widetilde{I}, D, \widetilde{J})$ *be a real decision formal context. Then* $E \subseteq A$ *is an L consistent set of* $\mathbb{K}$ *if and only if the extensions of all the key large real concepts of* $\underline{\mathfrak{B}}_L(U, E, \widetilde{I}_E)$ *with respect to* $\underline{\mathfrak{B}}_L(U, D, \widetilde{J})$ *are the same as those of* $\underline{\mathfrak{B}}_L(U, A, \widetilde{I})$ *with respect to* $\underline{\mathfrak{B}}_L(U, D, \widetilde{J})$*.*

Proof. The proof can easily be completed according to Definition 13 and Propositions 1, 2, and 3.

Theorem 1. *Let* $\mathbb{K} = (U, A, \widetilde{I}, D, \widetilde{J})$ *be a real decision formal context,* $a \in A$ *and* $E = A - \{a\}$*. Then a is an unnecessary attribute of* $\mathbb{K}$ *if and only if all the key large real concepts of* $\underline{\mathfrak{B}}_L(U, E, \widetilde{I}_E)$ *with respect to* $\underline{\mathfrak{B}}_L(U, D, \widetilde{J})$ *are extension-preserved large real concepts with respect to* $(U, E, \widetilde{I}_E)$*.*

Proof. The proof can easily be completed according to Definition 11 and Propositions 4 and 5.

Corollary 1. *Let* $\mathbb{K} = (U, A, \widetilde{I}, D, \widetilde{J})$ *be a real decision formal context,* $a \in A$ *and* $E = A - \{a\}$*. Then a is a necessary attribute of* $\mathbb{K}$ *if and only if there exists a key large real concept* $(X, \widetilde{B})$ *of* $\underline{\mathfrak{B}}_L(U, E, \widetilde{I}_E)$ *with respect to* $\underline{\mathfrak{B}}_L(U, D, \widetilde{J})$ *such that* $(X, \widetilde{B})$ *is not an extension-preserved large real concept with respect to* $(U, E, \widetilde{I}_E)$*.*

4 A Heuristic Reduction Algorithm for Real Decision Formal Contexts

In this section, we propose a heuristic algorithm for computing an L reduct of a real decision formal context.

Theorem 2. *Let* $\mathbb{K} = (U, A, \widetilde{I}, D, \widetilde{J})$ *be a real decision formal context and* $E \subseteq A$*. If E is an L consistent set of* $\mathbb{K}$ *and each* $e \in E$ *is a necessary attribute of* $(U, E, \widetilde{I}_E, D, \widetilde{J})$*, then E is an L reduct of* $\mathbb{K}$*.*

Proof. It is immediate from Definitions 9 and 11.

For a real decision formal context $\mathbb{K} = (U, A, \widetilde{I}, D, \widetilde{J})$, we can obtain an L reduct of $\mathbb{K}$ by the following steps: If each $a \in A$ is a necessary attribute of $\mathbb{K}$, then by Theorem 2 A is already an L reduct of $\mathbb{K}$; otherwise, choose an unnecessary attribute a_1 from $\mathbb{K}$ and consider $A - \{a_1\}$ for finding an L reduct. If each attribute of $A - \{a_1\}$ is a necessary attribute of $(U, A - \{a_1\}, \widetilde{I}_{A-\{a_1\}}, D, \widetilde{J})$, then $A - \{a_1\}$ is an L reduct of $\mathbb{K}$ since $A - \{a_1\}$ can easily be verified to be an L consistent set of $\mathbb{K}$ according to Definitions 9 and 11; otherwise, choose an unnecessary attribute a_2 from $(U, A - \{a_1\}, \widetilde{I}_{A-\{a_1\}}, D, \widetilde{J})$ and continue considering $A - \{a_1, a_2\}$ for finding an L reduct. This process is performed repeatedly and it will end in finite steps since A is a finite set of attributes. That is to say, there must exist $k < |A|$ such that $A - \{a_1, \cdots, a_k\}$ is an L reduct of $\mathbb{K}$.

Based on the above analysis, we are now ready to develop a heuristic algorithm to compute an L reduct with its cardinality being as small as possible. The reason for seeking the L reduct with small cardinality is that in general the smaller the cardinality of the reduct is, the more compact the L decision rules derived from the reduced real decision formal context are.

Algorithm 1. Computing an L reduct of a real decision formal context.

Input: A real decision formal context $\mathbb{K} = (U, A, \widetilde{I}, D, \widetilde{J})$.
Output: An L reduct of $\mathbb{K}$.

1) Initialize $E = A$.
2) If there does not exist $e \in E$ such that e is an unnecessary attribute of $(U, E, \widetilde{I}_E, D, \widetilde{J})$, then go to step 4); otherwise, go to step 3).
3) Choose such an unnecessary attribute f from E that satisfies

$$||(U, E - \{f\}, \widetilde{I}_{E-\{f\}}, D, \widetilde{J})|| = \max_{e \in E} \left\{ ||(U, E - \{e\}, \widetilde{I}_{E-\{e\}}, D, \widetilde{J})|| \right\},$$

where $|| \bullet ||$ denotes the number of all the unnecessary attributes of a real decision formal context. Then, set $E = E - \{f\}$ and go back to step 2).
4) Output E and end the algorithm.

By Theorem 2, it is easy to prove that the output set E in Algorithm 1 is an L reduct of the input real decision formal context $\mathbb{K}$. Moreover, it can easily be verified that the time complexity of Algorithm 1 is polynomial (the number of the real intervals in each value of the real binary relation $\widetilde{I}$ as well as $\widetilde{J}$ is in general supposed to be very small).

5 Numerical Experiments

In this section, we conduct some numerical experiments to assess the efficiency of Algorithm 1.

The real decision formal context $\mathbb{K} = (U, A, \widetilde{I}, D, \widetilde{J})$ in Table 1 is taken from [11], where $U = \{x_1, x_2, x_3, x_4\}$, $A = \{a_1, a_2, a_3, a_4\}$ and $D = \{d_1, d_2\}$. An auxiliary real decision formal context $\mathbb{K}_L$ is constructed by replacing the values in each row of Table 1 with $\{[5, 15]\}, \{[20, 30]\}, \{[9, 18]\}, \{[1, 7]\}, \{[2, 10]\}, \{[4, 13]\}$. Data sets 1 and 2 are then obtained by ten times of vertical concatenation of such two real decision formal contexts that are six and seven times of symmetrical mergence of $\mathbb{K}$ with respect to $\mathbb{K}_L$ (see [11] for the details of the symmetrical mergence and vertical concatenation approaches). The running time for Data sets 1 and 2 are reported in Table 2. It can be seen that Algorithm 1 is much more efficient than the Boolean reasoning-based algorithm in [11] especially for the large data set.

Table 1. A real decision formal context $\mathbb{K} = (U, A, \widetilde{I}, D, \widetilde{J})$

U	a_1	a_2	a_3	a_4	d_1	d_2
x_1	$\{[8,14]\}$	$\{[27,29]\}$	$\{[9,10],[14,17]\}$	$\{[3,6]\}$	$\{[4,9]\}$	$\{[9,12]\}$
x_2	$\{[5,9]\}$	$\{[20,23],[25,27]\}$	$\{[10,12],[13,16]\}$	$\{[1,3]\}$	$\{[2,4]\}$	$\{[4,5],[6,9]\}$
x_3	$\{[7,13]\}$	$\{[23,25]\}$	$\{[12,15]\}$	$\{[2,4]\}$	$\{[3,6]\}$	$\{[5,6]\}$
x_4	$\{[8,13]\}$	$\{[20,25]\}$	$\{[11,15]\}$	$\{[3,4]\}$	$\{[4,5]\}$	$\{[4,6]\}$

Table 2. Efficiency comparison between the algorithm in [11] and Algorithm 1

Data set	$\lvert U\rvert$	$\lvert A\rvert$	$\lvert D\rvert$	Running time (s)	
				The algorithm in [11]	Algorithm 1
Data set 1	240	24	12	1998.856	704.441
Data set 2	280	28	14	24580.341	1025.728

6 Final Remarks

Knowledge reduction is one of the key issues in real FCA. In this paper, a heuristic reduction algorithm for real decision formal contexts has been proposed to speed up the implementation of the knowledge reduction. Some numerical experiments have demonstrated that the proposed algorithm is much more efficient than the Boolean reasoning-based algorithm [11] especially for large databases.

Acknowledgments. This work was supported by the National Natural Science Foundation of China (Nos. 10971161 and 61005042).

References

1. Aswani-Kumar, C., Srinivas, S.: Concept lattice reduction using fuzzy K-Means clustering. Expert Systems with Applications 37(3), 2696–2704 (2010)
2. Bělohlávek, R.: Fuzzy Galois connections. Mathematical Logic Quarterly 45(4), 497–504 (1999)
3. Burusco, A., Fuentes-González, R.: The study of the L-fuzzy concept lattice. Mathware and Soft Computing 3, 209–218 (1994)
4. Düntsch, I., Gediga, G.: Approximation Operators in Qualitative Data Analysis. In: de Swart, H., Orłowska, E., Schmidt, G., Roubens, M. (eds.) TARSKI 2003. LNCS, vol. 2929, pp. 214–230. Springer, Heidelberg (2003)
5. Elloumi, S., Jaam, J., Hasnah, A., Jaoua, A., Nafkha, I.: A multi-level conceptual data reduction approach based on the Lukasiewicz implication. Information Sciences 163(4), 253–262 (2004)
6. Ganter, B., Wille, R.: Formal Concept Analysis, Mathematical Foundations. Springer, Berlin (1999)
7. Jaoua, A., Elloumi, S.: Galois connection, formal concepts and Galois lattice in real relations: application in a real classifier. The Journal of Systems and Software 60, 149–163 (2002)
8. Li, J., Mei, C., Lv, Y.: Knowledge reduction in decision formal contexts. Knowledge-Based Systems 24(5), 709–715 (2011)

9. Li, J., Mei, C., Lv, Y.: A heuristic knowledge-reduction method for decision formal contexts. Computers and Mathematics with Applications 61(4), 1096–1106 (2011)
10. Li, J., Mei, C., Lv, Y.: Knowledge reduction in formal decision contexts based on an order-preserving mapping. International Journal of General Systems 41(2), 143–161 (2012)
11. Li, J., Mei, C., Lv, Y.: Knowledge reduction in real decision formal contexts. Information Sciences 189, 191–207 (2012)
12. Li, L.F., Zhang, J.K.: Attribute reduction in fuzzy concept lattices based on T implication. Knowledge-Based Systems 23(6), 497–503 (2010)
13. Liu, M., Shao, M.W., Zhang, W.X., Wu, C.: Reduction method for concept lattices based on rough set theory and its application. Computers and Mathematics with Applications 53(9), 1390–1410 (2007)
14. Mi, J.S., Leung, Y., Wu, W.Z.: Approaches to attribute reduction in concept lattices induced by axialities. Knowledge-Based Systems 23(6), 504–511 (2010)
15. Pei, D., Li, M.Z., Mi, J.S.: Attribute reduction in fuzzy decision formal contexts. In: 2011 International Conference on Machine Learning and Cybernetics, pp. 204–208. IEEE Press, New York (2011)
16. Popescu, A.: A general approach to fuzzy concepts. Mathematical Logic Quarterly 50(3), 265–280 (2004)
17. Wang, H., Zhang, W.X.: Approaches to knowledge reduction in generalized consistent decision formal context. Mathematical and Computer Modelling 48(11-12), 1677–1684 (2008)
18. Wei, L., Qi, J.J., Zhang, W.X.: Attribute reduction theory of concept lattice based on decision formal contexts. Science in China: Series F—Information Sciences 51(7), 910–923 (2008)
19. Wille, R.: Restructuring lattice theory: an approach based on hierarchies of concepts. In: Rival, I. (ed.) Ordered Sets, pp. 445–470. Reidel, Dordrecht (1982)
20. Wu, W.Z., Leung, Y., Mi, J.S.: Granular computing and knowledge reduction in formal contexts. IEEE Transactions on Knowledge and Data Engineering 21(10), 1461–1474 (2009)
21. Yang, H.Z., Leung, Y., Shao, M.W.: Rule acquisition and attribute reduction in real decision formal contexts. Soft Computing 15(6), 1115–1128 (2011)
22. Yao, Y.Y.: Concept lattices in rough set theory. In: Dick, S., Kurgan, L., Pedrycz, W., Reformat, M. (eds.) Proceedings of 23rd International Meeting of the North American Fuzzy Information Processing Society, pp. 796–801. IEEE Press, New York (2004)
23. Yao, Y.Y.: A Comparative Study of Formal Concept Analysis and Rough Set Theory in Data Analysis. In: Tsumoto, S., Słowiński, R., Komorowski, J., Grzymała-Busse, J.W. (eds.) RSCTC 2004. LNCS (LNAI), vol. 3066, pp. 59–68. Springer, Heidelberg (2004)
24. Zhang, W.X., Ma, J.M., Fan, S.Q.: Variable threshold concept lattices. Information Sciences 177(22), 4883–4892 (2007)
25. Zhang, W.X., Qiu, G.F.: Uncertain decision making based on rough sets. Tsinghua University Press, Beijing (2005)
26. Zhang, W.X., Wei, L., Qi, J.J.: Attribute reduction theory and approach to concept lattice. Science in China: Series F—Information Sciences 48(6), 713–726 (2005)

A Multi-step Backward Cloud Generator Algorithm

Guoyin Wang[1,3], Changlin Xu[2,3], Qinghua Zhang[3], and Xiaorong Wang[2,3]

[1] Institute of Electronic Information Technology,
Chongqing Institute of Green & Intelligent Technology, CAS,
Chongqing 401122, China
[2] School of Information Science & Technology, Southwest Jiaotong University,
Chengdu 610031, China
[3] Institute of Computer Science and Technology,
Chongqing University of Posts & Telecommunications, Chongqing, 400065, China
xuchlin@163.com, wanggy@ieee.org

Abstract. Cloud model is an effective tool in uncertain transforming between qualitative concepts and their quantitative expressions. Backward cloud generator can transform quantitative values into qualitative concepts. In this paper, based on the theory of probability statistics, the authors make analysis of backward cloud algorithm, and construct a new algorithm of backward cloud which is more precise than the old. Finally, a simulation is given to compare the new algorithm with the old algorithms, and the results show that the new algorithm has better stability and adaptability.

Keywords: second-order normal cloud model, forward cloud generator, backward cloud generator.

1 Introduction

Knowledge representation has been a bottleneck for years in artificial intelligence. And the difficulty is uncertainty hidden in qualitative concepts, in particular, randomness and fuzziness[1]. Fuzzy sets and Rough sets use sets to depict a concept, which can be thought as the extension of a concept[2][3][4]. Considering the randomness of membership degree, Prof. Deyi Li proposed cloud model[5] as a cognitive model of uncertainty based on probability theory[6] and fuzzy sets theory in 1995. The cloud model uses three numerical characters (expectation Ex, entropy En and hyper entropy He) representing a qualitative concept to characterize the randomness and the fuzziness of uncertainty, and it has been realized the transformation between a qualitative concept and quantitative data and reveals the uncertainty of knowledge representation profoundly. It is very important meaning to understand connotation and extension of the qualitative concept. From the viewpoint of fuzzy sets, Ex is the expected sample of a concept with membership degree 1. En is used to depict the uncertainty of samples in the concept, which can be used to calculate the membership degree. He is used to depict the uncertainty of the membership degree.

J.T. Yao et al. (Eds.): RSCTC 2012, LNAI 7413, pp. 313–322, 2012.

Normal distribution exists extensively in natural and social life, and it has two parameters, expectation (Ex) and standard variance (En). Cloud model adopts the third parameter—hyper entropy (He), the uncertainty measurement of the standard variance (En), to depict the transition from normal distribution to heavy-tailed distribution. Normal cloud model based on normal distribution is an extremely important cloud model in the cloud model research. It has the universality[7]. The normal cloud model has been successfully applied in many fields, such as intelligent control, data mining, system evaluation, and image segmentation[8][9][10][11], and so on.

Cloud models can represent the randomness, fuzziness and their relations of uncertain concepts. The forward cloud generator(FCG) and the backward cloud generator(BCG) are the two most basic and critical cloud model algorithm in normal cloud model[1]. The former transforms a qualitative concept with three numerical characters (Ex, En, He) into a number of cloud drops (x_i) representing the quantitative description of the concept. In addition, it depicts the forward and direct process from thought to practice; the latter is used to transform a number of cloud drops into three numerical characters (Ex, En, He) representing a qualitative concept. It is a reversed and indirect process, and there will be errors inevitably in this process. In this paper, the existed BCG algorithms are analyzed firstly, and we find that the BCG algorithms proposed in [12][13] have defects. A new BCG algorithm is proposed in the paper, and the effectiveness of the new BCG are compared by experiments. The results show that the mean of parameter entropy En and hyper entropy He will be close to the true value, and the mean square error(MSE) of them will decrease and tend to zero with the increase of sample size.

2 The Normal Cloud Model and Cloud Generator

2.1 Cloud Model and the Second-Order Normal Cloud Model

Cloud model is a cognition model to transform between quantitative data and qualitative concepts, which can formalize a concept to three numbers. Considering the uncertainty and objectivity of the membership degree, Cloud model automatically produces the membership degrees based on probability distribution to interpret the fuzziness of concepts, thereby disclosing the relationship between randomness and fuzziness[1].

Definition 2.1. Let U be a universal set described by precise numbers, and C be the qualitative concept related to U. If there is a number $x \in U$, which randomly realizes the concept C, and the certainty degree of x for C, i.e. $z(x) \in [0, 1]$, is a random value with stabilization tendency:

$$z(x) : U \to [0, 1] \quad \forall x \in U \quad x \to z(x),$$

then the distribution of x on U is defined as a **cloud**, and each x is defined as a **cloud drop**, noted $Drop(x, z)$.

Definition 2.2. Let U be a universal set described by precise numbers, and C be the qualitative concept containing three numerical characters Ex, En, He

related to U. If there is a number $x \in U$, which is a random realization of the concept C and satisfies $x = R_N(Ex, y)$, where $y = R_N(En, He)$, and the certainty degree of x on U is

$$z(x) = e^{-\frac{(x-Ex)^2}{2y^2}}$$

then the distribution of x on U is a **second-order normal cloud**. Where $y = R_N(En, He)$ denoted a normally distributed random number with expectation En and variance He^2.

The key point in definition 2.2 is the second-order relationship, i.e. within the two normal random numbers. If $He = 0$, then the distribution of x on U will become a normal distribution. If $He = 0, En = 0$, then x will be a constant Ex and $z(x) \equiv 1$. In other words, certainty is the special case of the uncertainty. When He turns larger, the distribution of random variable X will show a heavier tail, which can be used in economic and social researches.

As shown in Fig.1, different people have different understanding about "The Young", so it is very difficult to give a crisp membership degree. However, the second-order normal cloud model can describe this uncertainty. Meanwhile, it can also demonstrate the basic certainty of uncertainty.

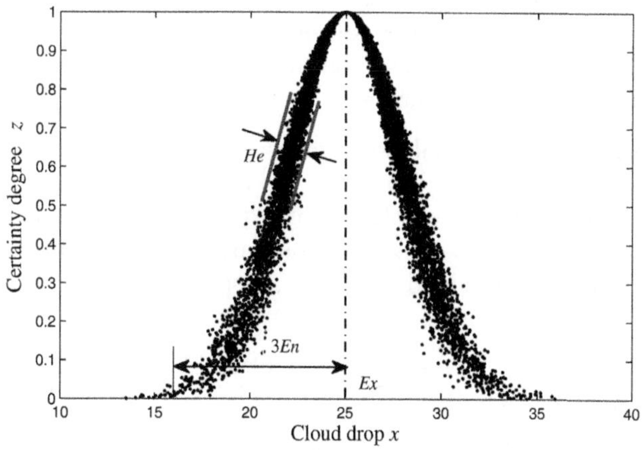

Fig. 1. Describe "The Young" by the second-order normal cloud model

In cloud model, cloud generator, which is a basic tool to realize the transformation between qualitative concepts and quantitative data, is composed of forward cloud generator and backward cloud generator.

2.2 Forward Cloud Generator

Forward Cloud Generator (FCG) algorithm[1] transforms a qualitative concept with three numerical characters (Ex, En, He) into a number of cloud drops (x_i) representing the quantitative description of the concept. It depicts the process from thought to practice. Because of the universality of normal distribution, we mainly focus on second-order normal cloud model. According to the definition 2.2, FCG algorithm is as follows.

Algorithm Forward normal cloud generator—**FCG**(Ex, En, He)
Input: (Ex, En, He), and the number of cloud drops n;
Output: n cloud drops and their certainty degrees, i.e. $Drop(x_i, z(x_i)), i = 1, 2, \cdots, n$;
Step 1: Generate a normally distributed random number y_i with expectation En and variance He^2, i.e. $y_i = R_N(En, He)$;
Step 2: Generate a normally distributed random number x_i with expectation Ex and variance y_i^2, i.e. $x_i = R_N(Ex, y_i)$;
Step 3: Calculate $z(x_i) = e^{-\frac{(x_i - Ex)^2}{2y_i^2}}$;
Step 4: x_i with certainty degree $z(x_i)$, $Drop(x_i, z(x_i))$ is a cloud drop in the domain;
Step 5: Repeat the step 1 to step 4 until n cloud drops are generated.

2.3 The Backward Cloud Generator-1

Backward Cloud Generator (BCG) is an algorithm based on probability statistics. It is used to transform a number of cloud drops (sample data) into three numerical characters representing a concept. In 2004, Liu[12] proposed a backward cloud generator algorithm based on the sample variance and the first-order absolute central moment as following.

Algorithm Backward normal cloud generator—**BCG1**
Input: Drops$(x_i), i = 1, 2, \cdots, n$
Output: (Ex, En, He) representation of a qualitative concept.
Step 1: Calculate the sample mean, sample variance and the first-order sample absolute central moment of cloud drops(x_1), respectively, i.e.

$$\hat{Ex} = \bar{X} = \frac{1}{n}\sum_{i=1}^{n} x_i, S^2 = \frac{1}{n-1}\sum_{i=1}^{n}(x_i - \bar{X})^2, E|X - \hat{Ex}| = \frac{1}{n}\sum_{i=1}^{n}\left|x_i - \bar{X}\right|,$$

Step 2: According to the character of second-order normal cloud distribution, Liu got the equations:

$$\begin{cases} S^2 = En^2 + He^2, \\ E\left|X - \hat{Ex}\right| = \sqrt{\frac{2}{\pi}}En. \end{cases} \tag{1}$$

Step 3: Calculate the estimates of En and He from (1), i.e.

$$\hat{En} = \sqrt{\frac{\pi}{2}} \times \frac{1}{n}\sum_{i=1}^{n}\left|x_i - \hat{Ex}\right|, \ \hat{He} = \sqrt{S^2 - \hat{En}^2}.$$

2.4 The Backward Cloud Generator-2

In addition to the BCG1 given by Liu, Wang[13] proposed another backward cloud generator algorithm according to the sample variance and the fourth-order sample central moment as following.

Algorithm Backward normal cloud generator—**BCG2**
Input: Drops$(x_i), i = 1, 2, \cdots, n$
Output: (Ex, En, He) representation of a qualitative concept.
Step 1: Calculate the sample mean, sample variance and the fourth-order sample central moment of cloud drops(x_i), respectively, i.e.

$$\hat{Ex} = \bar{X} = \frac{1}{n}\sum_{i=1}^{n} x_i, S^2 = \frac{1}{n-1}\sum_{i=1}^{n}(x_i - \bar{X})^2, \bar{\mu}_4 = \frac{1}{n-1}\sum_{i=1}^{n}(x_i - \bar{X})^4.$$

Step 2: According to the character of the second-order normal cloud distribution, Wang got the equations:

$$\begin{cases} S^2 = En^2 + He^2, \\ \bar{\mu}_4 = 3(3He^4 + 6He^2En^2 + En^4). \end{cases} \tag{2}$$

Step 3: Calculate the estimates of En and He from (2), i.e.

$$\hat{En} = \sqrt[4]{\frac{9\left(S^2\right)^2 - \bar{\mu}_4}{6}}, \hat{He} = \sqrt{S^2 - \hat{En}^2}.$$

The two BCG algorithms are both based on the statistical principles, and there will be errors for different cloud drops inevitably. Especially, calculate the estimate of hyper entropy He, if the sample standard deviation is less than the estimate of entropy En, that is, $S^2 - \hat{En}^2 < 0$, then $\hat{He}$ is a imaginary number. It shows that the two BCG algorithms have deficiency. In this paper, we propose a new BCG algorithm—a multi-step backward cloud generator algorithm, denoted by BCG-new.

3 A Multi-step Backward Cloud Generator Algorithm

In FCG, the cloud drops are generated by two random numbers, that is, the second-order relationship, and one is the input of the other in generation. The FCG is a transformation process from qualitative concept to quantitative number. While the BCG is a reversed process from quantitative number to qualitative concept, that is the restored process of numerical characters. So, using the sample data (cloud drops) to restore the three numerical characters Ex, En, He step by step according to the statistical properties of normal cloud as following.

Algorithm Backward normal cloud generator—**BCG-new**
Input: Drops$(x_i), i = 1, 2, \cdots, n$
Output: (Ex, En, He) representation of a qualitative concept.

Step 1: Calculate the sample mean $\hat{E}x = \bar{X} = \frac{1}{n}\sum\limits_{k=1}^{n} x_k$ from $x_1, x_2, \cdots, x_n$.

Step 2: Obtain the new sample from $x_1, x_2, \cdots, x_n$, that is, make the sample data $x_1, x_2, \cdots, x_n$ divide into m groups randomly, and each group will have r samples (i.e. $n = m \cdot r$ and n, m, r are positive integers). Calculate the sample variance $\hat{y}_i^2 = \frac{1}{r-1}\sum\limits_{j=1}^{r}(x_{ij} - \hat{E}x_i)^2 (i = 1, 2, \cdots, m)$ from each group, where, $\hat{E}x_i = \frac{1}{r}\sum\limits_{j=1}^{r} x_{ij} (i = 1, 2, \cdots, m)$. So, $y_1, y_2, \cdots, y_m$ are seen as a new random sample from a $N(En, He^2)$ distribution.

Step 3: Calculate the estimates of En^2 and He^2 from the new sample y_1^2, y_2^2, $\cdots$, y_m^2. We have

$$\begin{aligned} \hat{E}n^2 &= \tfrac{1}{2}\sqrt{4(\hat{E}Y^2)^2 - 2\hat{D}Y^2}, \\ \hat{H}e^2 &= \hat{E}Y^2 - \hat{E}n^2. \end{aligned} \tag{3}$$

Where, $\hat{E}Y^2 = \frac{1}{m}\sum\limits_{i=1}^{m}\hat{y}_i^2$, $\hat{D}Y^2 = \frac{1}{m-1}\sum\limits_{i=1}^{m}(\hat{y}_i^2 - \hat{E}Y^2)^2$.

The difference of the three BCGs is that BCG1 and BCG2 are to estimate En and He from the sample $x_1, x_2, \cdots, x_n$ directly, while the BCG-new is to estimate them indirectly through multi-step reduction according to the mutually reversed features of forward cloud generator and backward generator. At the same time, $\hat{H}e$ will not be a imaginary number in BCG-new. Correctness of the formula (3) can be proved by Theorem 3.1.

Theorem 3.1 Let $Y_1, Y_2, \cdots Y_m$ be independent and identically distributed random sample from normal distribution $N(En, He^2)$, then

$$\begin{aligned} \hat{E}n^2 &= \tfrac{1}{2}\sqrt{4(\hat{E}Y^2)^2 - 2\hat{D}Y^2}, \\ \hat{H}e^2 &= \hat{E}Y^2 - \hat{E}n^2. \end{aligned}$$

Where, $\hat{E}Y^2 = \frac{1}{m}\sum\limits_{i=1}^{m}\hat{y}_i^2$, $\hat{D}Y^2 = \frac{1}{m-1}\sum\limits_{i=1}^{m}(\hat{y}_i^2 - \hat{E}Y^2)^2$.

Proof: From the step 2 in BCG-new, if $Y_1, Y_2, \cdots, Y_m$ are a random sample from normal distribution $N(En, He)$, we have

$$EY^2 = \int_{-\infty}^{+\infty} y^2 f_Y(y)dy = He^2 + En^2, \tag{4}$$

and

$$EY^4 = \int_{-\infty}^{+\infty} y^4 f_Y(y)dy = 3He^4 + 6En^2He^2 + En^4, \tag{5}$$

where,

$$f_Y(y) = \frac{1}{\sqrt{2\pi}He} e^{-\frac{(y-En)^2}{2He^2}}.$$

Thus, from (4) and (5), we have,

$$DY^2 = EY^4 - (EY^2)^2 = 2He^4 + 4En^2He^2. \tag{6}$$

From (4) and (6), we obtain

$$2He^4 - 4EY^2He^2 + DY^2 = 0, \tag{7}$$

This is a quadratic equation about He^2. We can get the discriminant of root from (7), that is,

$$\Delta = 16(EY^2)^2 - 8DY^2 = 16En^4 \geq 0. \tag{8}$$

Hence, from (4) and (7), we have

$$He_1^2 = \frac{2EY^2 - \sqrt{4(EY^2)^2 - 2DY^2}}{2}, \; He_2^2 = \frac{2EY^2 + \sqrt{4(EY^2)^2 - 2DY^2}}{2},$$

and

$$En_1^2 = \frac{\sqrt{4(EY^2)^2 - 2DY^2}}{2}, \; En_2^2 = \frac{-\sqrt{4(EY^2)^2 - 2DY^2}}{2}.$$

However,

$$En_2^2 = \frac{-\sqrt{4(EY^2)^2 - 2DY^2}}{2} < 0.$$

Thus, the estimates $\hat{En}^2$ and $\hat{He}^2$ are as follows,

$$\hat{En}^2 = \frac{\sqrt{4(\hat{E}Y^2)^2 - 2\hat{D}Y^2}}{2},$$

$$\hat{He}^2 = \frac{2EY^2 - \sqrt{4(EY^2)^2 - 2DY^2}}{2}.$$

From (8), we know $\hat{En}^2 \geq 0$ and $\hat{He}^2 \geq 0$. So, the BCG-new algorithm can ensure that $\hat{En}$ and $\hat{He}$ will not be a imaginary number in theory. Therefore, $\hat{En}, \hat{He}$ can be obtained ($\hat{En} > 0, \hat{He} > 0$) from the formula (3). We next analyze and compare the errors of $\hat{En}$ and $\hat{He}$ in the three BCGs through experiments. Since the three BCGs make use of the same method to estimate the expectation Ex, that is, the sample mean $\hat{Ex} = \frac{1}{n}\sum_{k=1}^{n} x_k$, there will be no comparison the error of $\hat{Ex}$ in the paper.

4 Experiments and Analysis

The comparison of the three BCG algorithms illustrates the validity and advantage of the BCG-new by experiments from two aspects: (I) when the sample size is certain, execute the FCG(Ex, En, He) T times and generate n cloud drops each time, and then using the obtained n cloud drops to calculate $\hat{Ex}, \hat{En}, \hat{He}$, the mean and the MSE ($T$ times) of them by the three BCGs, respectively; (II) calculate and compare the changes of the mean and the MSE of $\hat{En}$ and $\hat{He}$ under the different sample sizes.

(I) Calculate $\hat{Ex}, \hat{En}, \hat{He}$ when the sample size n is certain.

Let FCG(Ex, En, He, n) =FCG(25,3,0.1,5000), number of executions $T = 20$, and determine the optimal values $m = 10, r = 500$. The results are shown in Fig.2. The mean and the MSE of $\hat{Ex}, \hat{En}$ and $\hat{He}$ are shown in Table 1.

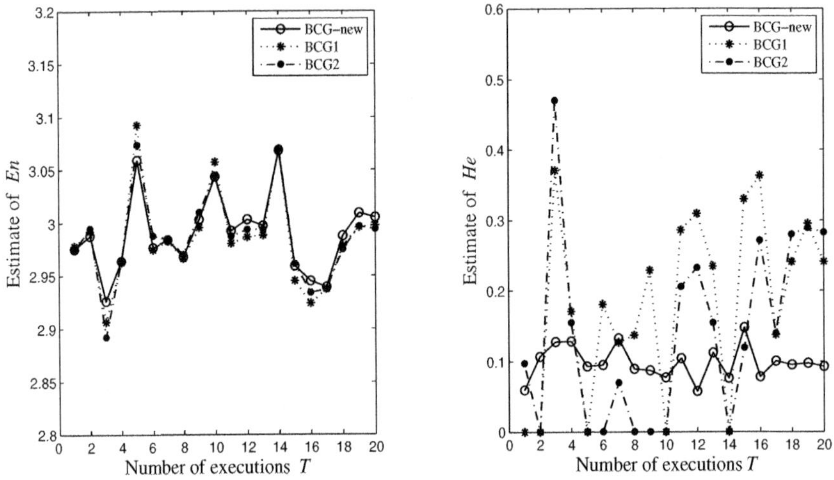

Fig. 2. The estimates of 20 times generated by the three algorithms about $\hat{En}$ and $\hat{He}$

The results indicate that $\hat{He}$ is a imaginary number sometimes in BCG1 and BCG2 when $He = 0.1$ (shown in Fig.2). These points appeared in the horizontal axis are imaginary numbers for $\hat{He}$ in Fig.2, that is, $\hat{He}^2 < 0$. We can see that the complex values of $\hat{He}$ have appeared 5 times and 7 times in 20 experiments in BCG1 and BCG2, respectively. However, this situation can be avoided by the BCG-new. In Table1, the mean and the MSE of $\hat{He}$ are obtained by excluding the complex value of $\hat{He}$. The MSE of $\hat{He}$ in BCG1 and BCG2 is larger than in BCG-new obviously. The three BCG algorithms can make a good estimate for

Table 1. The mean and MSE of $Ex = 25, En = 3, He = 0.1$

Numerical characters	Mean and MSE	BCG1	BCG2	BCG-new
Ex	Mean	25.0054	25.0054	25.0054
	MSE	0.0025	0.0025	0.0025
En	Mean	2.9976	2.9978	2.9984
	MSE	0.0011	9.0095×10^{-4}	7.5960×10^{-4}
He	Mean	0.2566	0.2037	0.0947
	MSE	0.0311	0.0157	2.7497×10^{-4}

En. The means and the MSEs of the estimates $\hat{En}$ have only samll differences in the three BCGs.

(II) Calculate the mean and the MSE(T times) of $\hat{En}, \hat{He}$ under the different sample sizes

Similar to (I), let FCG(Ex, En, He, n) =FCG(25, 3, 0.1), number of executions T = 20. Calculate the mean and the MSE of $\hat{En}, \hat{He}$ under the different sample sizes respectively. The results are shown in Fig.3.

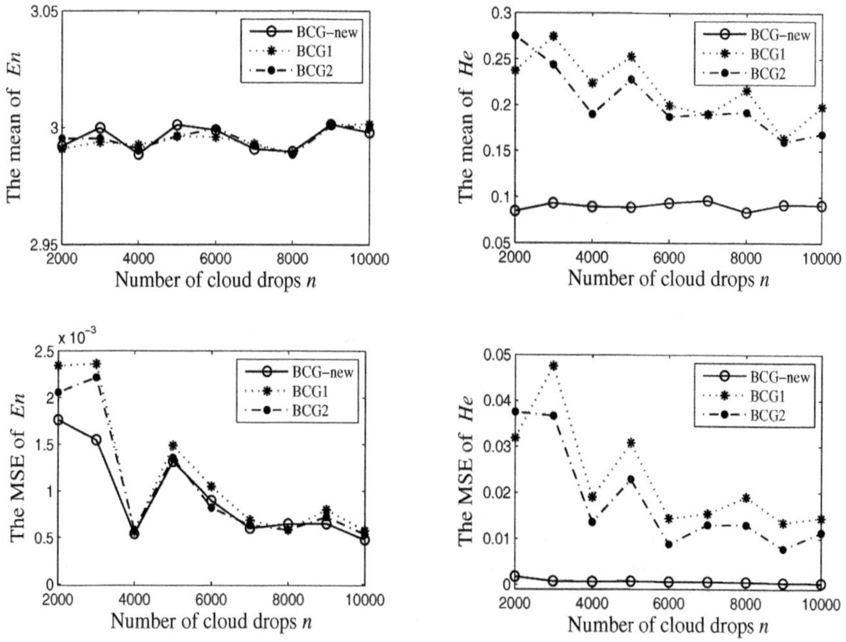

Fig. 3. The means and MSEs of $\hat{En}$ and $\hat{He}$ under the different sample sizes

From Fig.3, the mean of $\hat{He}$ is larger than 0.1 and its MSE is also larger with the increase of the sample sizes in BCG1 and BCG2. But the mean of $\hat{He}$ is very close to 0.1 and its MSE tends to zero obviously in BCG-new. There is no significant differences on the mean and the MSE of $\hat{En}$ in three BCGs. The three BCGs have the same time complexity, i.e. $O(n)$.

5 Conclusions

In this paper, the backward cloud generator algorithm was studied. The limitations of the backward cloud generators proposed in[12][13] were analyzed and a new backward cloud generator (BCG-new) was proposed. The three BCGs were compared by simulation experiments. The results show that the method was able to obtain better estimates for the parameters (i.e. entropy En and hyper entropy He).

Acknowledgments. This paper is supported by National Natural Science Foundation of P. R. China under grant 61073146, Natural Science Foundation Project of CQ CSTC under grant 2008BA2041.

References

1. Li, D.Y., Du, Y.: Artificial intelligence with uncertainty. National Defense Industry Press, Beijing (2005)
2. Zadeh, L.A.: Fuzzy sets. Information and Control 8, 338–353 (1965)
3. Wang, G.Y.: Rough set theory and knowledge acquisition. Jiao Tong University Press, Xi'an (2001)
4. Yao, Y.Y.: A comparative study of fuzzy sets and rough sets. Information Sciences 109(1-4), 227–242 (1998)
5. Li, D.Y., Meng, H.J., Shi, X.M.: Membership cloud and membership cloud generator. Journal of Computer Research and Development 32(6), 32–41 (1995)
6. Wang, Z.K.: Probability theory and its Applications. Beijing Normal University Press, Beijing (1995)
7. Li, D.Y., Liu, C.Y.: Study on the universality of the normal cloud model. Engineering Science 6(8), 28–34 (2004)
8. Li, D.Y.: The cloud control method and balancing patterns of triple link inverted pendulum systems. Engineering Sciences 1(2), 41–46 (1999)
9. Wang, S.L., Li, D.R., Shi, W.Z., et al.: Cloud model-based spatial data mining. Geographical Information Science 9(2), 67–78 (2003)
10. Lu, H.J., Wang, Y., Li, D.Y., Liu, C.Y.: The application of backward cloud in qualitative evaluation. Chinese Journal of Computers 26(8), 1009–1014 (2003)
11. Qin, K., Xu, K., Du, Y., Li, D.Y.: An image segmentation approach based on histogram analysis utilizing cloud model. In: Proceedings of the 2010 Seventh International Conference on Fuzzy Systems and Knowledge Discovery (FSKD 2010), pp. 524–528 (2010)
12. Liu, C.Y., Feng, M., Dai, X.J., Li, D.Y.: A new algorithm of backward cloud. Journal of System Simulation 16(11), 2417–2420 (2004)
13. Wang, L.X.: The basic mathematical properties of normal cloud and cloud filter. Personal Communication, May 3 (2011)

AFS-Based Formal Concept Analysis within the Logic Description of Granules

Lidong Wang*, Xiaodong Liu, and Xin Wang

Department of Mathematics,
Dalian Maritime University, Dalian 116026, People's Republic of China
wld1979@yahoo.com.cn

Abstract. AFS (Axiomatic Fuzzy Sets) -based formal concept is a generalization and development of classical concept lattice and monotone concept, which can be applied to represent the logic operations of queries in information retrieval. Granular computing is an emerging field of study that attempts to formalize and explore methods and heuristics of human problem solving with multiple levels of granularity and abstraction. The main objective of this paper is to investigate and develop AFS-based formal concept by using granule logics. Some generalized formulas of granular computing are introduced, in which AFS-based formal concept and AFS-based formal concept on multi-valued context are interpreted from the point of granular computing, respectively.

Keywords: Formal concept, concept lattice, granular computing, AFS-based formal concept.

1 Introduction

Formal concept analysis (FCA) proposed by Wille (1982) has been found useful in conceptual data analysis and knowledge processing [14]. FCA starts with the notion of a formal context consisting of a set of objects, a set of attributes, and a binary relation between the object set and attribute set. Concept lattice, or Galois lattice, forms the core of the mathematical theory of FCA, which reflects the relationship of generalization and specialization among concepts. Concept lattice is a form of concepts hierarchy, in which each node (formal concept) represents a subset of objects (extent) with their common attributes (intent). The characteristic of concept lattice theory lies in reasoning on the possible attributes of data sets [24]. As a powerful methodology for data analysis, FCA has been widely applied to machine learning, artificial intelligence, and knowledge discovery.

For requirements of real world applications, a number of different extensions of Wille's concept lattice have been proposed by combining some soft computing

* Supported by the Natural Science Foundation of China (No. 61175041), the Doctor Startup Foundation of Liaoning Province (No. 20111030), and the Fundamental Research Funds for the Central Universities (Nos. 2011QN057, 2011QN059, 2012TD032, 2011QN147).

J.T. Yao et al. (Eds.): RSCTC 2012, LNAI 7413, pp. 323–331, 2012.

methods, such as fuzzy sets, rough sets, AFS (Axiomatic Fuzzy Sets) theory, etc. In [2], Deogun and Saquer discussed some of limitations of the Wille's formal concept and proposed a monotone concept. Monotone concept is a generalization of Wille's notion of concept where disjunctions are allowed in the intent and set unions are allowed in the extent. This generalization allows an information retrieval query containing disjunctions to be understood as the intent of a monotone concept whose answer is the extent of that concept. In order to further develop monotone concept, Wang and Liu proposed AFS-based formal concept [11], which can be more conveniently applied to represent query than monotone concept in information retrieval systems, and proved the set of all AFS-based formal concepts forms a complete lattice.

Granular computing (GrC) is an emerging computing paradigm of information processing [4,9,23]. The concept of granular computing was initially called information granularity or information granulation [16,20,21,22]. The term granular computing first appeared in literature [22] as follows, "a subset of computing with words is granular computing." A central notion of granular computing is multilevel granular structures consisting of a family of interrelated and interacting granules. Granular computing focuses on problem solving by describing and representing a problem and its solution in various levels of granularity so that one can focus on things that serve a specific interest and ignore unimportant and irrelevant details. Granular computing makes use of knowledge structures and hence has a significant impact on the study of human intelligence and the design of intelligent systems [19,25]. Granular computing has been studied under various names in many different fields, such as concrete models of granular computing [1,3,5,10,13,15,17,18].

Motivated by the works of Yao and Zhou [19,25], the main objective of the paper is to make further contribution along this line by investigating AFS-based formal concept within a granular logic approach. In Section 2, some notations of AFS algebras are recalled. Some new logic formulas are introduced based on AFS logics in Section 3. In Section 4, we investigate AFS-based formal concept within the logics of granular computing, and extend AFS-based formal concept in the multi-valued context. Finally, a conclusion is drawn in Section 5.

2 AFS Algebras

AFS (Axiomatic Fuzzy Sets) theory was firstly proposed by Liu in 1998 [6,7]. In essence, AFS theory provides an effective tool to convert the information in the training examples and databases into the membership functions and their fuzzy logic operations. The following example, which employs the information table, serves as an introductory illustration of the set EM^* and EM^*/R.

Example 1. Let $X = \{x_1, x_2, ..., x_5\}$ be a set of five people with feature set $F = \{f_1, f_2, ..., f_8\}$, and which are described by real numbers (f_1: age, f_2: height, f_3: weight, f_4: salary, f_5: estate), Boolean values (f_6: male, f_7: female, f_8: MBA degree). Let $M = \{m_1^1, m_2^1, m_1^2, m_2^2, m_1^3, m_2^3, m_1^4, m_2^4, m_1^5, m_2^5, m_1^6, m_1^7, m_1^8\}$, in which m_j^i is the j-th assertion about f_i, m_1^i=large, m_2^i=small, $(i = 1, 2, 3, 4, 5)$,

Table 1. Descriptions of the information system [12]

		appearance		wealth		gender		degree
	age	height	weight	salary	estate	male	female	MBA
x_1	21	1.69	50	1	0	1	0	0
x_2	30	1.62	52	120	200	0	1	1
x_3	20	1.80	53	100	40	1	0	1
x_4	60	1.5	63	80	324	0	1	0
x_5	45	1.71	54	145	940	1	0	1

m_1^6=male, m_1^7=female, m_1^8=with MBA degree. Let F_i be the set of feature values on i-th feature f_i, and ϕ_i be the partial function $\phi_i : X \to F_i$. For each $m \in M$ is an assertion of a feature value about an object x of the form $m_i(x) =$ '$\phi_i(x)$ is some value'. For example, for object x_1, put $m_1^1(x_1) =$ '$\phi_1(x_1)$ is large', $m_1^2(x_1) =$ '$\phi_2(x_1)$ is large', and $\gamma = m_1^1 m_1^4 + m_1^1 m_1^7$ (the "+" denotes a disjunction of the assertion about features) is a complex assertion, $\gamma(x) =$ '$\phi_1(x)$ is large and $\phi_4(x)$ is large' or '$\phi_1(x)$ is large and $\phi_7(x)$ is female'. For $A_i \subseteq M, i \in I$, $\sum_{i\in I}(\prod_{m\in A_i} m)$ has a well-defined meaning such as the one we have discussed above.

Let M be a nonempty set. The set EM^* is defined by

$$EM^* = \{\sum_{i\in I}(\prod_{m\in A_i} m) \mid A_i \subseteq M, i \in I, I \text{ is any nonempty indexing set}\}. \quad (1)$$

In [6], Liu established the quotient set EM^*/R by introducing the binary relation R on EM^*. Moreover, Liu established EI algebra $(EM^*/R, \vee, \wedge)$ by introducing the algebra operations $\vee$ ("*or*") and $\wedge$ ("*and*") on the set EM^*/R.

Theorem 1. *[6] Let M be a non-empty set. Then $(EM^*/R, \vee, \wedge)$ forms a complete distributive lattice under the binary compositions $\vee$ and $\wedge$ defined as follows. For any $\sum_{i\in I}(\prod_{m\in A_i} m)$, $\sum_{j\in J}(\prod_{m\in B_j} m) \in EM^*/R$,*

$$\sum_{i\in I}(\prod_{m\in A_i} m) \vee \sum_{j\in J}(\prod_{m\in B_j} m) = \sum_{k\in I\sqcup J}(\prod_{m\in C_k} m), \quad (2)$$

$$\sum_{i\in I}(\prod_{m\in A_i} m) \wedge \sum_{j\in J}(\prod_{m\in B_j} m) = \sum_{i\in I, j\in J}(\prod_{m\in A_i\cup B_j} m), \quad (3)$$

where for any $k \in I \sqcup J$ (the disjoint union of I and J, i.e., an element in I and an element in J are always regarded as different elements in $I \sqcup J$), $C_k = A_k$ if $k \in I$, and $C_k = B_k$ if $k \in J$.

In what follows, we introduce another AFS algebra — $E^{\#}I$ algebra over X, which will play the role of the extents of AFS-based formal concepts.

The set EX^* is defined by $EX^* = \{\sum_{i\in I} a_i | a_i \in 2^X, I \text{ is any non-empty indexing set}\}$.

In [7], Liu established the quotient set EM^*/R by introducing the binary relation $R^\#$ on EX^*, and established $E^\#I$ algebra $(EX^*/R^\#, \vee, \wedge)$ by introducing the algebra operations $\vee$ and $\wedge$ as follows:

Theorem 2. *[6] For any $\sum_{i\in I} a_i$, $\sum_{j\in J} b_j \in EX^*/R^\#$, then $(EX^*/R^\#, \vee, \wedge)$ forms a complete distributive lattice under the binary compositions $\vee$, $\wedge$ defined as follows*

$$\sum_{i\in I} a_i \vee \sum_{j\in J} b_j = \sum_{k\in I\sqcup J} c_k, \quad \sum_{i\in I} a_i \wedge \sum_{j\in J} b_j = \sum_{i\in I, j\in J} (a_i \cap b_j), \tag{4}$$

where for any $k \in I \sqcup J$ (the disjoint union of I and J, i.e., an element in I and an element in J are always regarded as different elements in $I \sqcup J$), $c_k = a_k$ if $k \in I$, and $c_k = b_k$ if $k \in J$.

$(EX^*/R^\#, \vee, \wedge)$ is called $E^\#I$ algebra over X.

For $\mu = \sum_{i\in I} a_i$, $\nu = \sum_{j\in J} b_j \in EX^*/R^\#$, $\mu \le \nu \Longleftrightarrow \mu \vee \nu = \nu \Leftrightarrow \forall a_i$ $(i \in I)$, $\exists b_h$ $(h \in J)$ such that $a_i \subseteq b_h$.

Just as the Example 1, EI algebra can be represented by the fuzzy terms, the membership can be defined as follows:

Definition 1. *[6] Let X and M be sets. If for any $x \in X$, then for the complex assertion (feature) $\eta = \sum_{i\in I}(\prod_{m\in A_i} m) \in EM^*/R$, the membership function of η is defined as follows:*

$$\mu_\eta(x) = \sup_{i\in I} \frac{|A_i^{\succeq}(x)|}{|X|} \tag{5}$$

where $A_i^{\succeq}(x) = \{y \in X | x \succeq_m y, m(y) \succ 0, \forall m \in A_i\}$, $x \succeq_m y$ means the degree of x belonging to the assertion (feature) m is larger than or equal to that of y.

3 Granular Computing Based on AFS Logics

Assume that information about objects in a finite universe are given by an information table, in which objects are described by their values (or assertions) on a finite set of attributes. Formally, an information table can be expressed as:

Definition 2. *[8] Information table $S = (U, At, \{L_a | a \in At\}, \{V_l | l \in L_a\}, \{\rho_l | l \in L_a\})$, where U is a finite nonempty set of objects, At is a finite nonempty set of attributes, L_a is the set of fuzzy terms defined on attribute $a \in At$, $L = \{L_a | a \in At\}$ is called the elementary language, V_l is a nonempty set of values for $a \in At$, $\rho_l : U \to V_l$ is an information function. Each ρ_l is a total function that maps an object of U to exactly one value in V_l.*

In the elementary language L, an atomic formula is given by $l = v$, where $l \in L_a$ and $v \in V_l$ defined by Definition 2. The set of atomic formulas provides a basis on which more complex knowledge can be represented. Compound formulas can be built recursively from atomic formulas by using logic connectives [19]. If ϕ and ψ are formulas, then so are : $\neg\phi$, $\phi \vee \psi$ and $\phi \wedge \psi$ ($\phi \rightarrow \psi$, $\phi \leftrightarrow \psi$ not be considered in this paper).

In order to deal with the formal context with real number attributes, Boolean attributes and intuition order attributes, we take the generalized formula description of information table by using EI and $E^{\#}I$ algebra. In the sequel, any formula is always an element in EM^*/R. For each element in EM^*/R is a set of fuzzy terms (or assertions) with the definite semantic interpretation.

Definition 3. *Assume that the formulas ψ, $\phi \in EM^*/R$. The satisfiability of the formula ψ by an object x, written $x \models_S \psi$ or in short $x \models \psi$ if S is understood, is defined by the following conditions:*
(1) $x \models \neg\psi$ iff not $x \models \psi$;
(2) $x \models \psi \wedge \phi$ iff $x \models \psi$ and $x \models \phi$;
(3) $x \models \psi \vee \phi$ iff $x \models \psi$ or $x \models \phi$.

Definition 4. *Assume that the formulas ψ, $\phi \in EM^*/R$. The satisfiability of a formula ψ by an object x with confidence threshold δ, $\mu_\psi(x) \geq \delta$, written $x \models_\delta \psi$ or in short $x \models \psi$, is defined by the following conditions:*
(1) $x \models \psi \in L$ iff $\mu_\psi(x) \geq \delta$;
(2) $x \models \neg\psi$ iff not $x \models \psi$, i.e., $\mu_\psi(x) < \delta$;
(3) if $x \models \psi$ and $x \models \phi$, iff $x \models \psi \wedge \phi$;
(4) $x \models \psi \vee \phi$ iff $x \models \psi$ or $x \models \phi$.

Within the framework of AFS theory, the meaning of a formula ψ is therefore the logical description of sets of all objects with the semantic expressed by the formula ψ. In EI algebra, the meaning of a formula $\psi \in EM^*/R$ is a element in $EX^*/R^{\#}$, denoted as $\alpha(\psi) = \{\nu \in EX^*/R^{\#} | \nu \models \psi\}$. In other words, ψ can be viewed as the fuzzy description of the set of objects $\alpha(\psi)$. With the introduction of generalized formulas, a formal description of concepts is much richer than classical concept in semantic expressions. A concept definable in an information table is a pair $(\psi, \alpha(\psi))$, where $\psi \in EM^*/R$. More specifically, ψ is a description of $\alpha(\psi)$ in S, the intension of concept $(\psi, \alpha(\psi))$, and $\alpha(\psi)$ is the set of objects satisfying ψ. Correspondingly, the extension of concept $(\nu, \beta(\nu))$, is the logical description of sets of all attributes possed by the ν, where $\nu \in EX^*/R^{\#}$, $\beta(\nu) = \{\varphi \in EM^*/R | \nu \models \varphi\}$.

Theorem 3. *[11] For the formulas ζ, $\eta \in EM^*/R$, then the following assertions hold:*

$$(1)\ \alpha(\zeta \vee \eta) = \alpha(\zeta) \vee \alpha(\eta), \alpha(\zeta \wedge \eta) = \alpha(\zeta) \wedge \alpha(\eta),$$
$$(2)\ \zeta \leq \eta \Rightarrow \alpha(\zeta) \leq \alpha(\eta).$$

4 AFS-Based Formal Concept within the Logics of Granular Computing

4.1 AFS-Based Formal Concept Analysis

In [11], we proposed AFS-based formal concept in which the Galois connection "$'$" of context $(X, M, \mathbb{I})$ can be extended to the connection between the EI algebra $(EM^*/R, \vee, \wedge)$ and the $E^\# I$ algebra $(EX^*/R^\#, \vee, \wedge)$. Notice that

$$A_i' = \{x \in X | x \models \prod_{m \in A_i} m\}, a_j' = \{m \in M | x \models m \text{ for all } x \in a_j\},$$

we have

$$\alpha(\prod_{m \in A_i} m) = \bigcap_{m \in A_i} \alpha(m) = A_i' \in EX^*/R^\# \tag{6}$$

$$\beta(a_j) = \prod_{m \in \alpha(a_j)} m = \prod_{m \in a_j'} m \in EM^*/R \tag{7}$$

$$\alpha(\sum_{i \in I}(\prod_{m \in A_i} m)) = \sum_{i \in I} \alpha(\prod_{m \in A_i} m) = \sum_{i \in I} \bigcap_{m \in A_i} \alpha(m) = \sum_{i \in I} A_i' \in EX^*/R^\# \tag{8}$$

$$\beta(\sum_{j \in J} a_j) = \sum_{j \in J}(\prod_{m \in \alpha(a_j)} m) = \sum_{j \in J}(\prod_{m \in a_j'} m) \in EM^*/R \tag{9}$$

Definition 5. *[11] Let $(X, M, \mathbb{I})$ be a context, $\zeta = \sum_{i \in I}(\prod_{m \in A_i} m) \in EM^*/R$, $\nu \in \sum_{j \in J} a_j \in EX^*/R^\#$. (ν, ζ) is called an AFS-based formal concept of the context $(X, M, \mathbb{I})$, if $\alpha(\zeta) = \nu$, $\beta(\nu) = \zeta$. ν is called the extent of the AFS-based formal concept (ν, ζ) and ζ is called the intent of the AFS-based formal concept (ν, ζ).*

Theorem 4. *[11] Let $(X, M, \mathbb{I})$ be a context and $\mathcal{B}(EX^*/R^\#, EM^*/R, \mathbb{I})$ be the set of all AFS-based formal concepts of the context $(X, M, \mathbb{I})$. Then, for any $(\nu, \zeta) \in \mathcal{B}(EX^*/R^\#, EM^*/R, \mathbb{I})$, ν and ζ are uniquely determined by each other.*

4.2 AFS-Based Formal Concept Analysis on Multi-valued Context

In this part, we present the AFS-based formal concept analysis on multi-valued context $(X \times X, M, I_\tau)$, in which M is a set of fuzzy or crisp attributes (assertions) on X.

Definition 6. *Let X, M be two sets. A binary relation I_τ from $X \times X$ to M is defined as follows: for $(x, y) \in X \times X, m \in M$,*

$$(x, y) I_\tau m \Leftrightarrow 0 \prec m(y) \preceq m(x) \Leftrightarrow y \in m^{\succeq}(x), \tag{10}$$

we write $(x, y) \models (m, I_\tau)$.

Based on the above mentioned works, we propose AFS-based formal concept on multi-valued context, in which the Galois connection "′" of the context $(X \times X, M, I_\tau)$ can be extended to the connection between the EI algebra $(EM^*/R, \vee, \wedge)$ and the $E^{\#}I$ algebra $(E(X \times X)^*/R^{\#}, \vee, \wedge)$. Notice that

$$\alpha(\prod_{m \in A_i} m) = \{(x, y) \in X \times X | (x, y) \models (m, I_\tau) \text{ for all } m \in A_i\},$$

$$\beta(a_j) = \{m \in M | (x, y) \models (m, I_\tau) \text{ for all } (x, y) \in a_j\}.$$

So, for any $\sum_{i \in I}(\prod_{m \in A_i} m) \in EM^*/R$, $\sum_{j \in J} a_j \in E(X \times X)^*/R^{\#}$,

$$\alpha(\sum_{i \in I}(\prod_{m \in A_i} m)) = \sum_{i \in I} \alpha(\prod_{m \in A_i} m) = \sum_{i \in I} A_i' \in E(X \times X)^*/R^{\#}, \tag{11}$$

$$\beta(\sum_{j \in J} a_j) = \sum_{j \in J}(\prod_{m \in a_j'} m) \in EM^*/R, \tag{12}$$

Definition 7. *Let $\zeta = \sum_{i \in I}(\prod_{m \in A_i} m) \in EM^*/R$, $\nu \in \sum_{j \in J} a_j \in E(X \times X)^*/R^{\#}$. (ν, ζ) is called an AFS-based formal concept of the context $(X \times X, M, I_\tau)$, if $\alpha(\zeta) = \nu$, $\beta(\nu) = \zeta$. Then ν is called the extent of the AFS-based formal concept (ν, ζ), and ζ is called the intent of the AFS-based formal concept (ν, ζ).*

Theorem 5. *Let $(X \times X, M, I_\tau)$ be the context defined by Definition 6 and $\mathcal{B}(E(X \times X)^*/R^{\#}, EM^*/R, I_\tau)$ be the set of all AFS-based formal concepts of the context $(X \times X, M, I_\tau)$. Then, for any $(\nu, \zeta) \in \mathcal{B}(E(X \times X)^*/R^{\#}, EM^*/R, I_\tau)$, ν and ζ are uniquely determined by each other.*

In Example 1,

$$\gamma = m_1^1 m_1^4 + m_1^1 m_1^6, \ \xi = m_1^1 m_1^3 m_1^4 + m_1^1 m_1^6$$

are two fuzzy complex assertions in EM^*/R.

$$\begin{aligned} \alpha(m_1^1 m_1^4) &= \{(x_1, x_1), (x_2, x_1), (x_2, x_2), (x_3, x_3), (x_4, x_1), (x_4, x_4), \\ &\quad (x_5, x_1), (x_5, x_2), (x_5, x_3), (x_5, x_5)\} \\ \alpha(m_1^1 m_6^1) &= \{(x_1, x_1), (x_1, x_3), (x_3, x_3), (x_5, x_1), (x_5, x_3), (x_5, x_5)\} \\ \beta(\alpha(m_1^1 m_1^4)) &= \{m_1^1 m_1^3 m_1^4\}, \ \beta(\alpha(m_1^1 m_6^1)) = \{m_1^1 m_6^1\} \end{aligned}$$

while

$$\beta(\alpha(m_1^1 m_1^3 m_1^4)) = m_1^1 m_1^3 m_1^4.$$

So, (ν, ξ) is an AFS-based formal concept of the context $(X \times X, M, I_\tau)$, where $\nu = \{(x_1, x_1), (x_2, x_1), (x_2, x_2), (x_3, x_3), (x_4, x_1), (x_4, x_4), (x_5, x_1), (x_5, x_2), (x_5, x_3), (x_5, x_5)\} + \{(x_1, x_1), (x_1, x_3), (x_3, x_3), (x_5, x_1), (x_5, x_3), (x_5, x_5)\}$. From (5), one can get $\mu_{m_1^1 m_1^4}(x) = \mu_{m_1^1 m_1^3 m_1^4}(x)$ for any $x \in X$. So, $\mu_\zeta(x) = \mu_\xi(x)$ for any $x \in X$, the fuzzy complex assertion ζ is equivalent to ξ in Table 1.

Definition 8. *Let* $(\nu_1, \zeta_1), (\nu_2, \zeta_2) \in \mathcal{B}(E(X \times X)^*/R^\#, EM^*/R, I_\tau)$. *Define* $(\nu_1, \zeta_1) \leq (\nu_2, \zeta_2)$ *if and only if* $\nu_1 \leq \nu_2$ *in lattice* $E(X \times X)^*/R^\#$ *(or equivalently* $\zeta_1 \leq \zeta_2$ *in lattice* EM^*/R*).*

It is obvious that $\leq$ defined in Definition 8 is a partial order relation on $\mathcal{B}(E(X \times X)^*/R^\#, EM^*/R, I_\tau)$. The following theorem shows that $\mathcal{B}(E(X \times X)^*/R^\#, EM^*/R, I_\tau)$ forms a complete lattice under the relation $\leq$.

Theorem 6. *Let* $(X \times X, M, I_\tau)$ *be the context defined by Definition 6 and* $\mathcal{B}(E(X \times X)^*/R^\#, EM^*/R, I_\tau)$ *be the set of all AFS-based formal concepts of the context* $(X \times X, M, I_\tau)$. *Then* $(\mathcal{B}(E(X \times X)^*/R^\#, EM^*/R, I_\tau), \leq)$ *is a complete lattice, in which suprema and infima are given as follows: for any* $(\nu_k, \zeta_k) \in \mathcal{B}(E(X \times X)^*/R^\#, EM^*/R, I_\tau), \leq)$,

$$\vee_{k \in K}(\nu_k, \zeta_k) = (\vee_{k \in K}\alpha(\zeta_k), \beta(\vee_{k \in K}\alpha(\zeta_k))), \tag{13}$$

$$\wedge_{k \in K}(\nu_k, \zeta_k) = (\wedge_{k \in K}\alpha(\zeta_k), \beta(\wedge_{k \in K}\alpha(\zeta_k))). \tag{14}$$

where $k \in K$, K *is any non-empty indexing set.*

5 Conclusions

In this paper, a generalized logic for granular computing is introduced by using AFS theory, which can be viewed as the elementary language L in the multi-granular setting. Under the new formulas, AFS-based formal concept is investigated, and further extended to AFS-based formal concept on multi-valued context. These results are useful to study multi-granular formal concept model.

References

1. Bargiela, A., Pedrycz, W.: Granular Computing: An Introduction. Kluwer Academic Publishers, Boston (2002)
2. Deogun, J.S., Saquer, J.: Monotone Concepts for Formal Concept Analysis. Discrete Appl. Math. 144, 70–78 (2004)
3. Hirota, K., Pedrycz, W.: Fuzzy Computing for Data Mining. Proc. of the IEEE 87, 1575–1600 (1999)
4. Lin, T.Y.: Granular Computing. Announcement of the BISC Special Interest Group on Granular Computing (1997)
5. Lin, T.Y.: Granular Computing, Rough Sets, Fuzzy Sets, Data Mining, and Granular Computing. In: Wang, G., Liu, Q., Yao, Y., Skowron, A. (eds.) RSFDGrC 2003. LNCS (LNAI), vol. 2639, pp. 16–24. Springer, Heidelberg (2003)
6. Liu, X.D.: The Fuzzy Theory Based on AFS Algebras and AFS Structure. J. Math. Anal. Appl. 217, 459–478 (1998)
7. Liu, X.D., Pedrycz, W.: Axiomatic Fuzzy Set Theroy and Its Applications. Springer, Heidelberg (2009)
8. Pawlak, Z.: Rough Sets-Theoretical Aspects of Reasoning About Data. Kluwer Publishers, Boston (1991)
9. Pedrycz, W.: Granular Computing: An Emerging Paradigm. Physica-Verlag, Heidelberg (2001)

10. Qiu, G.F., Ma, J.M., Yang, H.Z., Zhang, W.X.: A Mathematical Model for Concept Granular Computing Systems. Sci China Inf. Sci. 53(7), 1397–1408 (2010)
11. Wang, L.D., Liu, X.D.: Concept Analysis via Rough Set and AFS Algebra. Inf. Sci. 178(21), 4125–4137 (2008)
12. Wang, L.D., Liu, X.D., Qiu, W.R.: Nearness Approximation Space Based on Axiomatic Fuzzy Sets. Int. J. Approx. Reason 53, 200–211 (2012)
13. Wei, L., Zhang, X.H., Qi, J.J.: Granular Reduction of Property-Oriented Concept Lattices. In: Croitoru, M., Ferré, S., Lukose, D. (eds.) ICCS 2010. LNCS, vol. 6208, pp. 154–164. Springer, Heidelberg (2010)
14. Wille, R.: Restructuring Lattice Theory: An Approach Based on Hierarchies of Concepts. In: Rival, I. (ed.) Ordered Sets, pp. 445–470. Reidel (1982)
15. Wu, W.Z., Leung, Y., Mi, J.S.: Granular Computing and Knowledge Reduction in Formal Contexts. IEEE T Knwl. Data En. 21(10), 1461–1474 (2009)
16. Yao, J.T.: A Ten-Year Review of Granular Computing. In: GrC 2007, pp. 734–739 (2007)
17. Yao, Y.Y.: Granular Computing: Basic Issues and Possible Solutions. In: Proceedings of the 5th Joint Conference on Information Sciences, pp. 186–189 (2000)
18. Yao, Y.Y.: A Partition Model of Granular Computing. In: Peters, J.F., Skowron, A., Grzymała-Busse, J.W., Kostek, B.z., Świniarski, R.W., Szczuka, M.S. (eds.) Transactions on Rough Sets I. LNCS, vol. 3100, pp. 232–253. Springer, Heidelberg (2004)
19. Yao, Y.Y., Zhou, B.: A Logic Language of Granular Computing. In: 6th IEEE International Conference on Cognitive Informatics, pp. 178–185 (2007)
20. Zadeh, L.A.: Fuzzy Sets and Information Granurity. In: Gupta, M., Ragade, R.K., Yager, R.R. (eds.) Advances in Fuzzy Set Theory and Applications, pp. 3–18. North-Holland Publishing Company (1979)
21. Zadeh, L.A.: Key Roles of Information Granulation and Fuzzy Logic in Human Reasoning, Concept Formulation and Computing with Words. In: Proceedings of IEEE 5th International Fuzzy Systems, p. 1 (1996)
22. Zadeh, L.A.: Towards a Theory of Fuzzy Information Granulation and Its Centrality in Human Reasoning and Fuzzy Logic. Fuzzy Set Syst. 90(2), 111–127 (1997)
23. Zadeh, L.A.: Some Reflections on Soft Computing, Granular Computing and their Roles in the Conception, Design And Utilization of Information/Intelligent Systems. Soft Comput. 2, 23–25 (1998)
24. Zhao, Y., Halang, W.A., Wang, X.: Rough Ontology Mapping in E-Business Integration. SCI, vol. 37, pp. 75–93. Springer, Heidelberg (2007)
25. Zhou, B., Yao, Y.Y.: A Logic Approach to Granular Computing. Int. J. Cogn. Inform. Natl. Intell. 2, 63–79 (2008)

Approximate Concepts Based on N-Scale Relation

Ling Wei* and Qing Wan

Department of Mathematics, Northwest University, Xi'an, 710069, P.R. China
wl@nwu.edu.cn, wqysbe@163.com

Abstract. To be an efficient tool for knowledge discovery, formal concept analysis has been paid more attention to and applied to many fields in recent years. Through studying the n-scale relation defined in this paper based on the formal context, we obtain some knowledge: left neighborhood sets and right neighborhood sets, which also belong to the powerset of partition on the object set, like the set of extents. Especially, when n is a special value, the corresponding left neighborhood approximate concepts and right neighborhood approximate concepts are the join-dense subsets of the property oriented concept lattice and the concept lattice of a formal context, respectively. And then, the whole lattices can be obtained.

Keywords: n-scale relation, left neighborhood set, right neighborhood set, left neighborhood approximate concept, right neighborhood approximate concept.

1 Introduction

Formal concept analysis (FCA) was proposed by German mathematician Wille R. in 1982 [1, 2]. The foundation of FCA are formal contexts, formal concepts, and the corresponding concept lattices, which depend on the binary relation between an object set and an attribute set. FCA has been widely applied to machine learning, artificial intelligence, knowledge discovery, and so on [3–7].

Rough set theory (RST) was proposed by Polish mathematician Pawlak Z. also in 1982 [8]. The basic relation of RST is the equivalence relation defined on the object set, it determines the partition of the universe, based on which, lower and upper approximation of a set are proposed. For a set, if its lower approximation and upper approximation are not equal, then it is a rough set. RST has been successfully applied to many filed [9–12]. FCA and RST are related and complementary. In recent years, many efforts have been made to compare and combine the two theories [13–16]. Combination of FCA and RST provides some new approaches for data analysis and knowledge discovery [17–22].

In a real world, people often can't obtain their perfect expectation. So, they need weaken their requirements, and select other choices, which is similar to

* Corresponding author.

J.T. Yao et al. (Eds.): RSCTC 2012, LNAI 7413, pp. 332–340, 2012.

choose a approximate value. Inspired by such real situation, we put this idea into formal concept analysis. Based on the formal context, we define the n-scale relation on the object set through considering the attributes owned by an object, and obtain some approximate concepts. At the same time, when n is supposed to be a special value, the approximate concepts have certain connections with the property oriented concept lattice and the original concept lattice. Thus, the join-dense subsets of the property oriented concept lattice and the concept lattice of a formal context can be obtained. Therefore, all the concepts can be obtained. Here, the n-scale relation weakens the equivalence relation in RST, and it connects FCA and RST efficiently.

Basic definitions of formal contexts and property oriented concept lattices are recalled in Section 2. Section 3 proposes the definition of n-scale relation, and studies the left and right neighborhood approximate concepts, based on which, the relations between left (right) neighborhood approximate concepts and the property oriented concept lattice (concept lattice) are revealed. Section 4 concludes the paper.

2 Preliminaries

This section reviews some basic definitions in formal concept analysis.

Definition 1. *[2]A formal context (G, M, I) consists of two sets G and M and a relation I between G and M. The elements of G are called the objects and the elements of M are called the attributes of the context. In order to express that an object g is in a relation I with an attribute m, we write gIm or $(g, m) \in I$ and read it as "the object g has the attribute m".*

For a formal context (G, M, I), Wille defined two operators on $A \subseteq G, B \subseteq M$ as follows:

$$A^* = \{m | m \in M, \forall g \in A, (g, m) \in I\},$$
$$B' = \{g | g \in G, \forall m \in B, (g, m) \in I\}.$$

Then we call (A, B) is a formal concept, if and only if, $A^* = B, A = B'$. All the concepts of the context (G, M, I) is a lattice, called concept lattice, and is denoted by $L(G, M, I)$. Where, the infimum and the supremum are give by:

$$(A_1, B_1) \wedge (A_2, B_2) = (A_1 \cap A_2, (B_1 \cup B_2)'^*)$$
$$(A_1, B_1) \vee (A_2, B_2) = ((A_1 \cup A_2)^{*\prime}, B_1 \cap B_2) .$$

For the convenience, $\forall g \in G$, $\forall m \in M$, g^* and m' denote $\{g\}^*$ and $\{m\}'$, respectively. In this paper, the formal context is required to be regular and finite.

The property oriented concept lattice was introduced by Duntsch and Gediga [23]. A pair $(A, B), A \subseteq G, B \subseteq M$, is called a property oriented concept if $A^{\Diamond} = B$ and $B^{\Box} = A$. Here, two approximation operators $\Diamond, \Box$ are defined as follows:

$$A^{\Diamond} = \{m \in M | m' \cap A \neq \emptyset\},$$
$$B^{\Box} = \{g \in G | g^* \subseteq B\}.$$

All property oriented concepts of the context (G, M, I) is denoted by $L_P(G, M, I)$, it is a lattice and is called property oriented concept lattice. Where, the infimum and the supremum are respectively defined as follows:

$$(A_1, B_1) \wedge (A_2, B_2) = (A_1 \cap A_2, (B_1 \cap B_2)^{\Box\Diamond})$$
$$(A_1, B_1) \vee (A_2, B_2) = ((A_1 \cup A_2)^{\Diamond\Box}, B_1 \cup B_2).$$

For a binary relation I, its complement relation is define by: $I^c = \{(g, m) | \neg(g, m)\}$. The formal context (G, M, I^c) is referred to as the complement formal context of (G, M, I)[13].

The partial orders define on $L(G, M, I)$ and $L_P(G, M, I)$ are the same, they are defined by:

$$(A_1, B_1) \leq (A_2, B_2) \Longleftrightarrow A_1 \subseteq A_2.$$

Definition 2. *[2]Let (G, M, I) be a formal context, $(g^{*\prime}, g^*)$ is called an object concept.*

Definition 3. *[24]Let L be a lattice. An element $x \in L$ is join-irreducible if*
1. $x \neq 0$(in case L has a zero);
2. $x = a \vee b$ implies $x = a$ or $x = b$ for all $a, b \in L$.

In this paper, the set of all the join-irreducible elements of a lattice L is denoted by $J(L)$.

Definition 4. *[24]Let P be an ordered set and let $Q \subseteq P$. Then Q is called join-dense in P if for every element $a \in P$ there is a subset A of Q such that $a = \bigvee_P A$.*

Lemma 1. *[24]Let L be a finite lattice. Every element is join of join-irreducible elements.*

3 Approximate Concepts Based on N-Scale Relation

This section mainly introduces n-scale relation, and left (right) neighborhood approximate concepts. Furthermore, it presents the relation between left (right) neighborhood approximate concepts and property oriented concept lattices (concept lattices) when n is a special value.

3.1 N-Scale Relation and Approximate Concepts

Definition 5. *Let (G, M, I) be a formal context, $\forall B \subset M$, we say binary relation R_B^n $(0 \leq n \leq |B|)$ is a n-scale relation on G, if*

$$R_B^n = \{(g_i, g_j) \in G \times G | g_i^* \subseteq g_j^* \subseteq B \text{ and } |g_j^*| - |g_i^*| \leq n\}$$

Obviously, R_B^n has reflexivity. We also define

$$[g]_B^{n-} = \{g_i \in G \mid (g_i, g) \in R_B^n\},$$
$$[g]_B^{n+} = \{g_j \in G \mid (g, g_j) \in R_B^n\},$$

to be the n-left neighborhood and n-right neighborhood of the object g respectively.

Remark. It is easy to see that R_M^0 is an equivalence relation on G with respect to M, which is just the equivalence relation R on G in rough set theory. Accordingly, the corresponding $[g]_R$ is an equivalence class.

Property. Let (G, M, I) be a formal context. $\forall g \in G$, $[g]_B^{n-}$ and $[g]_B^{n+}$ satisfy the following properties ($0 \leq n \leq |B|$):

(1) $[g]_B^{(n-1)-} \subseteq [g]_B^{n-}, [g]_B^{(n-1)+} \subseteq [g]_B^{n+}$ ($1 \leq n \leq |B|$) for all $g \in G$;
(2) $\forall g_i, g_j \in G$, If $g_i \in [g_j]_B^{k+}$, then $g_j \in [g_i]_B^{k-}$, where, $k = 1, 2, ..., n$;
(3) $\forall g_i, g_j \in G$, If $[g_i]_R = [g_j]_R$, then $[g_i]_B^{n-} = [g_j]_B^{n-}, [g_i]_B^{n+} = [g_j]_B^{n+}$;
(4) If $g \in [g_i]_B^{n-}$, then $[g_j]_R \subseteq [g_i]_B^{n-}$ for all $g \in [g_j]_R$;
(5) If $g \in [g_i]_B^{n+}$, then $[g_j]_R \subseteq [g_i]_B^{n+}$ for all $g \in [g_j]_R$;
(6) Suppose that (G, M, I^c) is the complement formal context of (G, M, I), then $[g]_B^{n-} = [g_c]_B^{n+}, [g]_B^{n+} = [g_c]_B^{n-}$. Where, g_c and g are the same object, using g_c is for emphasizing it is in the (G, M, I^c).

Proof. It is easy to prove (1), (2) and (3) from Definition 5.

Now, we prove (4). Let $[g_j]_R = \{g_1, g_2, \ldots, g_m\}$, then $g_1^* = g_2^* = \ldots = g_m^*$. Assume $g_1 \in [g_i]_B^{n-}$, from Definition 5, we have $g_1^* \subseteq g_i^*$. So, $g_t^* \subseteq g_i^*, t = 1, 2, \ldots, m$. Hence $g_t \in [g_i]_B^{n-}, t = 1, 2, \ldots, m$. We claim that $[g_j]_R \subseteq [g_i]_B^{n-}$.

Similarly, we can prove (5).

(6) can be proved from Definition 5 and definition of complement formal context.

Example 1. Table 1 shows a formal context (G, M, I). In which, the object set $G = \{1, 2, 3, 4, 5, 6\}$ consists of 6 different buildings, the attributes in M ($|M| = 5$) are a: price of the house, b: traffic situation, c: entertainment instrument, d: estate management, e: architectural quality. This formal context is a investigation result. $\times$ means the customer is satisfied with the item, and the space means he is not. The corresponding concept lattice and the property oriented concept lattice are shown in Fig.1 and Fig.2 respectively. In which, each set is denoted directly by the string of its elements except G, M and $\emptyset$.

Table 1. A formal context (G, M, I)

	a	b	c	d	e
1	×		×	×	×
2	×		×		
3		×			×
4		×			×
5	×				
6	×	×			×

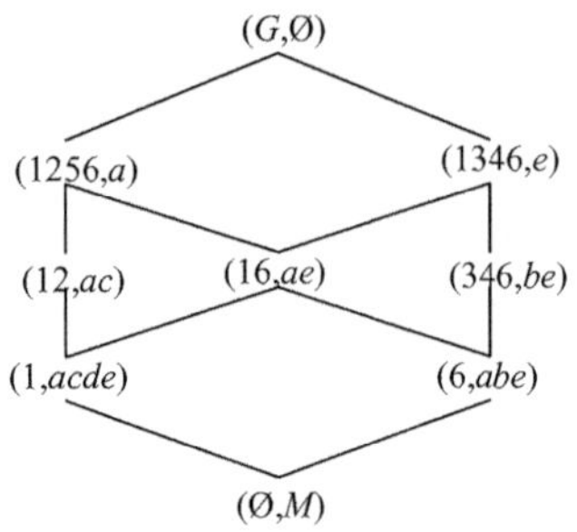

Fig. 1. $L(G, M, I)$

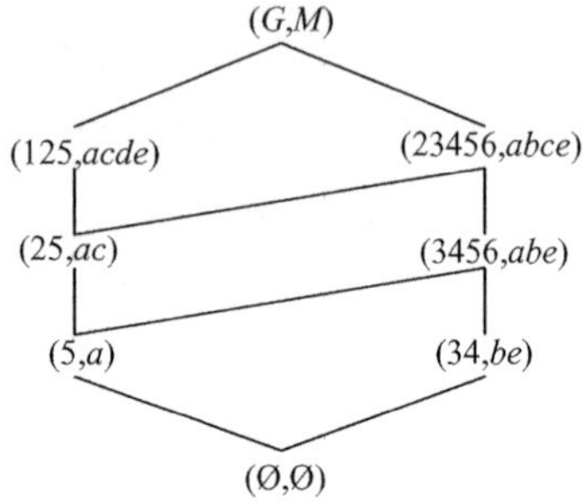

Fig. 2. $L_P(G, M, I)$

In Table 1, when $B = \{a, c, d\}, n = 1, 2, 3$, all the n-left neighborhoods of each object are as follows:

$n = 1 : [1]_B^{1-} = \{1, 2\}, [2]_B^{1-} = \{2, 5, 6\}, [3]_B^{1-} = [4]_B^{1-} = \{3, 4\}, [5]_B^{1-} = [6]_B^{1-} = \{3, 4, 5, 6\}$.

$n = 2 : [1]_B^{2-} = \{1, 2, 5, 6\}, [2]_B^{2-} = \{2, 3, 4, 5, 6\}, [3]_B^{2-} = [4]_B^{2-} = \{3, 4\}, [5]_B^{2-} = [6]_B^{2-} = \{3, 4, 5, 6\}$.

$n = 3 : [1]_B^{3-} = \{1, 2, 3, 4, 5, 6\}, [2]_B^{3-} = \{2, 3, 4, 5, 6\}, [3]_B^{3-} = [4]_B^{3-} = \{3, 4\}, [5]_B^{3-} = [6]_B^{3-} = \{3, 4, 5, 6\}$.

all the n-right neighborhoods of each object are as follows:

$n = 1 : [1]_B^{1+} = \{1\}, [2]_B^{1+} = \{1, 2\}, [3]_B^{1+} = [4]_B^{1+} = \{3, 4, 5, 6\}, [5]_B^{1+} = [6]_B^{1+} = \{2, 5, 6\}$.

$n = 2 : [1]_B^{2+} = \{1\}, [2]_B^{2+} = \{1, 2\}, [3]_B^{2+} = [4]_B^{2+} = \{2, 3, 4, 5, 6\}, [5]_B^{2+} = [6]_B^{2+} = \{1, 2, 5, 6\}$.

$n = 3 : [1]_B^{3+} = \{1\}, [2]_B^{3+} = \{1, 2\}, [3]_B^{3+} = [4]_B^{3+} = \{1, 2, 3, 4, 5, 6\}, [5]_B^{3+} = [6]_B^{3+} = \{1, 2, 5, 6\}$.

It should be noted that when $B = M$, we have $0 \le n \le |M| - 2$ since the formal context is required to be regular. Because the following content we discussed in this paper under the condition that $B = M$, for the sake of convenience, $[g]_M^{n-}$ and $[g]_M^{n+}$ are denoted by $[g]^{n-}$ and $[g]^{n+}$ respectively.

Definition 6. *Let (G, M, I) be a formal context. Denote $L_O^k = \{[g]^{k-} | g \in G\}, U_O^k = \{[g]^{k+} | g \in G\}$, where, $k = 1, 2, \ldots, n$. L_O^k and U_O^k are the coverings of G. We call L_O^k the k-left neighborhood set of G, and U_O^k the k-right neighborhood set of G. All the left neighborhood sets and right neighborhood sets are denoted by L_O and U_O respectively, that is, $L_O = \bigcup_{k=1}^{n} L_O^k, U_O = \bigcup_{k=1}^{n} U_O^k$.*

We can see that $|L_O^k| = |U_O^k| = |G/R|$.

Definition 7. *[21] Suppose L_1, L_2 are two families of sets, and $|L_1| = |L_2|$. If $\forall X \in L_1$, there exists different $Y \in L_2$ such that $X \subseteq Y$, we say L_1 is included by L_2 component-wisely, and is denoted by $L_1 \Subset L_2$.*

We can obtain the following theorem from Definition 7 and Property (1) naturally.

Theorem 1. *Let (G, M, I) be a formal context. Then,*
$$L_O^1 \Subset L_O^2 \Subset \ldots \Subset L_O^{|M|-2}, \qquad U_O^1 \Subset U_O^2 \Subset \ldots \Subset U_O^{|M|-2}.$$

Theorem 2. *Let (G, M, I) be a formal context. Then,*
$$L_O \subseteq \sigma(G/R), \qquad U_O \subseteq \sigma(G/R).$$

Proof. Because Property (4), if $g \in [g_i]_B^{n-}$, then $[g_j]_R \subseteq [g_i]_B^{n-}$ for all $g \in [g_j]_R$, so, $[g_i]_B^{n-} = \{\cup[g]_B | g \in [g_i]_B^{n-}\}$. Thus, from the definition of L_O and power set $\sigma(G/R)$, we obtain $L_O \subseteq \sigma(G/R)$. Similarly, we have $U_O \subseteq \sigma(G/R)$ since Property (5).

Definition 8. *Let (G, M, I) be a formal context, $\forall g \in G$, we call $([g]^{n-}, g^*)$ and $([g]^{n+}, g^*)$ are n-left neighborhood approximate concept and n-right neighborhood approximate concept; shortly, denoted by n-LNAC and n-RNAC.*

Example 2. For the formal context (G, M, I) in Example 1, we have the following results: $G/R = \{\{1\}, \{2\}, \{3, 4\}, \{5\}, \{6\}\}$,
$L_O^1 = \{\{1\}, \{2, 5\}, \{3, 4\}, \{5\}, \{3, 4, 6\}\}$,
$L_O^2 = \{\{1, 2\}, \{2, 5\}, \{3, 4\}, \{5\}, \{3, 4, 5, 6\}\}$,
$L_O^3 = \{\{1, 2, 5\}, \{2, 5\}, \{3, 4\}, \{5\}, \{3, 4, 5, 6\}\}$,
$U_O^1 = \{\{1\}, \{2\}, \{3, 4, 6\}, \{2, 5\}, \{6\}\}$,
$U_O^2 = \{\{1\}, \{1, 2\}, \{3, 4, 6\}, \{2, 5, 6\}, \{6\}\}$,
$U_O^3 = \{\{1\}, \{1, 2\}, \{3, 4, 6\}, \{1, 2, 5, 6\}, \{6\}\}$,
$L_O = L_O^1 \cup L_O^2 \cup L_O^3 = \{\{1\}, \{1, 2\}, \{1, 2, 5\}, \{2, 5\}, \{3, 4\}, \{5\}, \{3, 4, 6\}, \{3, 4, 5, 6\}\}$,
$U_O = U_O^1 \cup U_O^2 \cup U_O^3 = \{\{1\}, \{2\}, \{1, 2\}, \{2, 5, 6\}, \{1, 2, 5, 6\}, \{3, 4, 6\}, \{6\}, \{2, 5\}\}$.

It is easy to see that $|L_O^k| = |U_O^k| = |G/R| = 5$, $k = 1, 2, 3$; and
$L_O^1 \Subset L_O^2 \Subset L_O^3$, $U_O^1 \Subset U_O^2 \Subset U_O^3$; $L_O \subseteq \sigma(G/R)$, $U_O \subseteq \sigma(G/R)$.

Correspondingly, we can get all the neighborhood approximate concepts:

1-LNACs:$(1, acde), (25, ac), (34, be), (5, a), (346, abe)$;
2-LNACs: $(12, acde), (25, ac), (34, be), (5, a), (3456, abe)$;
3-LNACs: $(125, acde), (25, ac), (34, be), (5, a), (3456, abe)$;
1-RNACs: $(1, acde), (2, ac), (346, be), (25, a), (6, abe)$;
2-RNACs: $(1, acde), (12, ac), (346, be), (256, a), (6, abe)$;
3-RNACs: $(1, acde), (12, ac), (346, be), (1256, a), (6, abe)$.

We can explain the results detailedly.

The 1-LNAC of object No.6 is $(346, abe)$, it means: the customer is satisfied with the house price, traffic situation and architectural quality. However, under these conditions, the choice is limited, there is only No.6 meets the demand. If he does not care about the price and reduce the demands, he can select more, that is, he can choose one from No.3, No.4 and No.6.

All the right neighborhood approximate concepts of object No.6 is $(6, abe)$, it means: the customer is satisfied with the house price, traffic situation and architectural quality, and there is no other building which has more attributes. So, the customer can but choose No.6.

3.2 The Relation between Neighborhood Approximate Concepts and Two Kinds of Concept Lattices

In the above Example, we can find that 3-LNACs and 3-RNACs are included in property oriented concept lattice and concept lattice, respectively. In fact, we have the following theorem.

Theorem 3. *Let (G, M, I) be a formal context. If $n = |M| - 2$, then*

$$[g]^{n+} = g^{*\prime}, \qquad [g]^{n-} = g^{\Diamond\Box}.$$

Proof. 1)When $n = |M| - 2$, let $[g]^{n+} = \{g_1, g_2, \dots, g_s\}$. If $g = g_1$, then $g_1^* \subseteq g_l^* (l = 1, 2, \dots, s)$. $\forall g_0 \in g_1^{*\prime}$, we have $g_1^* \subseteq g_0^*$, thus, $g_0 \in [g_1]^{n+}$, so $g_1^{*\prime} \subseteq [g_1]^{n+}$, that is, $g^{*\prime} \subseteq [g]^{n+}$; $\forall g_0 \in [g]^{n+}$, we have $g^* \subseteq g_0^*$, so, $g^{*\prime} \supseteq g_0^{*\prime}$, while, $g_0 \in g_0^{*\prime}$, thus, $g_0 \in g^{*\prime}$. So, $[g]^{n+} \subseteq g^{*\prime}$.

2)Since $\forall g \in G$, we have $g^* = g^{\Diamond}$ such that $g^{\Diamond\Box} = g^{*\Box}$. So, similar to above proof, we can obtain $[g]^{n-} = g^{*\Box} = g^{\Diamond\Box}$.

This theorem shows that when $n = |M| - 2$, $([g]^{n+}, g^*)$ is not only a formal concept of the formal context (G, M, I) but also an object concept, and$([g]^{n-}, g^*)$ is a property oriented concept.

Denote $Q_1 = \{([g]^{n+}, g^*) | g \in G, n = |M| - 2\}, Q_2 = \{([g]^{n-}, g^*) | g \in G, n = |M| - 2\}$, then, Q_1 is the set of object concepts, $Q_1 \subseteq L(G, M, I), Q_2 \subseteq L_P(G, M, I)$. From Definition 6, we have $|Q_1| = |Q_2| = |G/R|$.

Now, we will show that we can obtain concept lattice $L(G, M, I)$ and property oriented concept lattice $L_P(G, M, I)$ from Q_1 and Q_2.

Lemma 2. *[24]Let (G, M, I) be a formal context, L is its concept lattice. Then $J(L) \subseteq Q_1$.*

Combining Definition 4 and Lemma 1, this lemma shows the set of object concepts Q_1 is a join-dense subset in L.

Theorem 4. *Let (G, M, I) be a formal context, L_P is its property oriented concept lattice. Then $J(L_P) \subseteq Q_2$.*

Proof. Let $P_2 = \{([g]^{n-}, \sim g^*) | g \in G, n = |M| - 2\}$. Since $L_c(G, M, I) = \{(A, \sim B) | (A, B) \in L_P(G, M, I)\}$, $Q_2 = \{([g]^{n-}, g^*) | g \in G, n = |M| - 2\}, Q_2 \subseteq L_P(G, M, I)$, we have $|P_2| = |Q_2|$, $P_2 \subseteq L_c(G, M, I)$. Since $\sim g^* = (g_c)^*$, $[g]^{n-} = [g_c]^{n+}$ by Property (6), so, we obtain that $P_2 = \{([g_c]^{n+}, (g_c)^*) | g \in G, n = |M| - 2\}$. Then, from Theorem 3 and Lemma 2, we have $J(L_c) \subseteq P_2$.

We know that $L_c(G, M, I) \cong L_P(G, M, I)$. Suppose $f : L_c(G, M, I) \to L_P(G, M, I)$ is a bijection, $f((A, \sim B)) = (A, B)$, $L_c(G, M, I)$ and $L_P(G, M, I)$ have the same partial relation: $(A_1, B_1) \leq (A_2, B_2) \Longleftrightarrow A_1 \subseteq A_2$, then, we have $|J(L_c)| = |J(L_P)|$, $f(J(L_c)) = J(L_P)$; since $|P_2| = |Q_2|$, $f(P_2) = Q_2$, f is a bijection, and $J(L_c) \subseteq P_2$, we can obtain that $J(L_P) \subseteq Q_2$.

Combining Definition 4 and Lemma 1, Theorem 4 shows that the subset Q_2 of property oriented concept lattice is also a join-dense subset in L_P. Thus, we can find all the concepts and property oriented concepts based on Q_1 and Q_2.

Example 3. We consider the formal context showed in Example 1. When $n = |M| - 2$, the join-dense subsets of $L(G, M, I)$ and $L_P(G, M, I)$ are as follows:

3-RNACs:$(1, acde)$,$(12, ac)$, $(346, be)$,$(1256, a)$,$(6, abe)$;

3-LNACs:$(125, acde)$,$(25, ac)$, $(34, be)$,$(5, a)$,$(3456, abe)$.

That is: $Q_1 = \{(1, acde), (12, ac), (346, be), (1256, a), (6, abe)\}$;

$Q_2 = \{(125, acde), (25, ac), (34, be), (5, a), (3456, abe)\}$.

From Q_1, we can obtain all the concepts of $L(G, M, I)$:

$(1, acde), (12, ac), (346, be), (1256, a), (6, abe), (1346, e), (16, ae), (G, \emptyset), (\emptyset, M)$;

From Q_2, we can obtain all the property oriented concepts of $L_P(G, M, I)$:

$(125, acde), (25, ac), (34, be), (5, a), (3456, abe), (23456, abce), (G, M), (\emptyset, \emptyset)$.

Which are the same with the previous figures.

4 Conclusion

The n-scale relation R_B^n defined in this paper just has reflexivity, it is a weakening of the equivalence relation. The knowledge discovered by which is not concrete, however, it has real significance. Similar to our social life, sometimes, too concrete means no answer. And, when $n = |M| - 2$, the corresponding n-right neighborhood approximate concepts and n-left neighborhood approximate concepts are join-dense subsets of $L(G, M, I)$ and $L_P(G, M, I)$ respectively. So, using the property of join-irreducible elements, we can find all the formal concepts and property oriented formal concepts. In fact, we can also define the n-scale relation on the attribute set since duality between object set and attribute set, and when n is supposed to be a special value, also through the method similar to that of this paper, we can obtain object oriented concept lattice.

Acknowledgments. The authors gratefully acknowledge the support of the Natural Science Foundation of China (No.11071281, No.60703117).

Bibliography

1. Wille, R.: Restructuring lattice theory: an approach based on hierarchies of concepts, vol. 83, pp. 445–470. Reidel, Dordrecht (1982)
2. Ganter, B., Wille, R.: Formal Concept Analysis. Mathematical Foundations. Springer, Berlin (1999)
3. Fan, S.Q., Zhang, W.X., Xu, W.: Fuzzy inference based on fuzzy concept lattice. Fuzzy Sets and Systems 157, 3177–3187 (2006)
4. Priss, U.: Formal concept analysis in information science. Annual Review of Information Science and Technology 40, 521–543 (2007)
5. Belohlavek, R., Vychodil, V.: Formal concept analysis with background knowledge: attribute priorities. IEEE Transactions on Systems, Man, and Cybernetics, Part C 39, 399–409 (2009)
6. Cabrera, I.P., Cordero, P., Gutirrez, G., Martnez, J., Ojeda-Aciego, M.: A coalgebraic approach to non-determinism: Applications to multilattices. Information Sciences 4323–4335 (2010)

7. Ganter, B., Stumme, G., Wille, R.: Formal Concept Analysis: Foundations and Applications. Springer, Heidelberg (2005)
8. Pawlak, Z.: Rough sets. International Journal of Computer and Information Science 11, 341–356 (1982)
9. Hu, Q.H., Liu, J.F., Yu, D.R.: Mixed feature selection based on granulation and approximation. Knowledge-Based Systems 21, 294–304 (2008)
10. Pawlak, Z., Skowron, A.: Rough sets: some extensions. Information Sciences 177, 28–40 (2007)
11. Pawlak, Z., Skowron, A.: Rough sets and Boolean reasoning. Information Sciences 177, 41–73 (2007)
12. Wang, C.Z., Wu, C.X., Chen, D.G.: A systematic study on attribute reduction with rough sets based on general binary relations. Information Sciences 178, 2237–2261 (2008)
13. Yao, Y.Y.: Concept lattices in rough set theory. In: Dick, S., Kurgan, L., Pedrycz, W., Reformat, M. (eds.) Proceedings of 2004 Annual Meeting of the North American Fuzzy Information Processing Society, June 27-30, pp. 796–801 (2004)
14. Yao, Y.: A Comparative Study of Formal Concept Analysis and Rough Set Theory in Data Analysis. In: Tsumoto, S., Słowiński, R., Komorowski, J., Grzymała-Busse, J.W. (eds.) RSCTC 2004. LNCS (LNAI), vol. 3066, pp. 59–68. Springer, Heidelberg (2004)
15. Saquer, J., Deogun, J.S.: Concept approximaions based on rough sets and similarity measures. International Joronal of Applied Mathematics and Computer Science 11, 655–674 (2001)
16. Qi, J.-J., Wei, L., Li, Z.-z.: A Partitional View of Concept Lattice. In: Ślęzak, D., Wang, G., Szczuka, M.S., Düntsch, I., Yao, Y. (eds.) RSFDGrC 2005. LNCS (LNAI), vol. 3641, pp. 74–83. Springer, Heidelberg (2005)
17. Zhang, W.X., Ma, J.M., Fan, S.Q.: Variable threshold concept lattices. Information Sciences 177, 4883–4892 (2007)
18. Saquer, J., Deogun, J.S.: Approximating monotone concepts. IOS Press, Amsterdam (2003)
19. Wang, L.D., Liu, X.D.: A new model of evaluating concept similarity. Knowledge-Based Systems 21, 842–846 (2008)
20. Zhang, W.X., Qiu, G.F.: Uncertain Decision Making Based on Rough Sets. Tsinghua University Press, BeiJing (2005) (in chinese)
21. Wei, L., Qi, J.J., Zhang, W.X.: Attribute reduction theory of concept lattice based on decision formal contexts. Science in China Series F: Information Science 51, 910–923 (2008)
22. Chen, Y.H., Yao, Y.Y.: A multiview approach for intelligent data analysis based on data operators. Information Sciences 178, 1–20 (2008)
23. Gediga, G., Duntsch, I.: Modal-style operations in qualitative data analysis. In: Proceedings of the 2002 IEEE International Conference on Data Mining, pp. 155–162 (2002)
24. Davey, B.A., Priestley, H.A.: Introduction to lattices and order. Cambridge University Press, Cambridge (2002)

Attribute Characteristics of Object (Property) Oriented Concept Lattices

Ling Wei* and Min-Qian Liu

Department of Mathematics, Northwest University, Xi'an, 710069, P.R. China
wl@nwu.edu.cn, xixi4317@163.com

Abstract. Object oriented concept lattices and property oriented concept lattices are two kinds of concept lattices and belong to formal concept analysis. In the theory of formal concept analysis, attribute reduction is one of basic problems, it can make the discovery of implicit knowledge in data easier and the representation simpler. Different attributes play different roles in reduction theory. This paper studies the attribute characteristics of object oriented concept lattices, and gives equivalent conditions for each kind of attribute characteristics. Finally, the paper shows that there are the same attribute characteristics for both object oriented concept lattices and property oriented concept lattices, so, study each one of them can obtain the other's information.

Keywords: Formal context, attribute characteristics, object oriented concept lattice, property oriented concept lattice.

1 Introduction

Formal concept analysis (FAC) was proposed by Wille R. in 1982 as an effective method for data analysis to find, order and display concepts [1]. A concept lattice shows the relationship of specialization and generalization among the formal concepts. Nowadays, it has become an efficient methodology for data analysis and knowledge discovery in various fields, such as machine learning, computer network, data mining [2–6]. The reduction of the concept lattices is one of the hot spots in recent years, which makes the discovery and expression of implied knowledge in formal contexts clearer [7–10].

Rough set theory proposed by Pawlak also provided methods for data analysis from another perspectives [11].

Duntsch and Gediga presented the property oriented concept lattice using a pair of approximation operators inspired by Rough sets and modal logics[12, 13]. Yao introduced the object oriented concept lattice and proved that object oriented concept lattice is isomorphic to the property oriented concept lattice for a same formal context [14]. Liu discussed the reduction of object oriented concept lattice and property oriented concept lattice using the lattice-preserving reduction theory proposed in reference [7], and proposed judgment approaches of consistent sets [9].

* Corresponding author.

J.T. Yao et al. (Eds.): RSCTC 2012, LNAI 7413, pp. 341–348, 2012.

The attribute set of an object (property) oriented concept lattice can be divided into three parts: core attribute set, relatively necessary attribute set, and absolutely unnecessary attribute set according to their importance to the reduction of the formal context. This paper studies the method which can distinguish the attribute characteristics.

To make the paper self-contained, basic notions are introduced in Section 2. Attribute characteristics of object oriented concept lattices and property oriented concept lattices are discussed in Section 3 and 4 respectively. Finally, the paper is concluded by Section 5.

2 Preliminaries

To facilitate our discussion, some basic notions and important propositions are introduced in this section.

Definition 1. *[15] A formal context (G, M, I) consists of two sets G and M and a relation I between G and M. The elements of G are called the objects and the elements of M are called the attributes of the context. In order to express that an object g is in a relation I with an attribute m, we write gIm or $(g, m) \in I$ and read it as "the object g has the attribute m".*

Wille defined operators $*$ and $'$ for every$X \subseteq G, Y \subseteq M$:

$$X^* = \{m \in M | (g, m) \in I \text{ for all } g \in X\} \ ,$$
$$Y' = \{g \in G | (g, m) \in I \text{ for all } m \in Y\} \ .$$

$\forall x \in G, \forall y \in M$, we denote $xI = \{y \in M | xIy\} = \{x\}^*$, $Iy = \{x \in G | xIy\} = \{y\}'$. For simplicity, we write x^* instead of $\{x\}^*$ and y' instead of $\{y\}'$ in the sequence.

In this paper, we assume that all the formal contexts are regular, that is, for every $x \in G, x^* \neq \emptyset, x^* \neq M$, and for every $y \in M, y' \neq \emptyset, y' \neq G$. And also, we assume that all the formal contexts are finite, that is, G and M are finite sets.

With respect to a formal context (G, M, I), a pair of dual approximation operators, $\Diamond : 2^G \to 2^M$ and $\Box : 2^M \to 2^G$, are defined as follows [12–14]:

Definition 2. *With respect to a formal context (G, M, I), a pair of dual approximation operators $\Box$, $\Diamond$: $2^G \to 2^M$ are defined as follows for any $X \subseteq G$:*

$$X^{\Box} = \{y \in M | \ \forall x \in G(xIy \Rightarrow x \in X)\} = \{y \in M | Iy \subseteq X\},$$
$$X^{\Diamond} = \{a \in M | \exists x \in G(xIa \wedge x \in X)\} = \{a \in M | Ia \cap X \neq \emptyset\} = \bigcup_{x \in X} xI = XI.$$

Correspondingly, the dual approximation operators $\Box$, $\Diamond$: $2^M \to 2^G$ for any $Y \subseteq M$ are:

$$Y^{\Box} = \{x \in G | \ \forall y \in M(xIy \Rightarrow y \in Y)\} = \{x \in G | xI \subseteq Y\},$$
$$Y^{\Diamond} = \{x \in G | \exists y \in M(xIy \wedge y \in Y)\} = \{x \in G | xI \cap Y \neq \emptyset\} = \bigcup_{y \in Y} Iy = IY.$$

It's easy to see that, $\forall x \in G$, $x^{\Diamond} = x^* = xI$. $\forall y \in M$, $y^{\Diamond} = y' = Iy$.

The approximation operators have the following properties: $\forall X, X_1, X_2 \subseteq G$, $\forall Y, Y_1, Y_2 \subseteq M$,

1. $X_1 \subseteq X_2 \Rightarrow X_1^{\Box} \subseteq X_2^{\Box}, X_1^{\Diamond} \subseteq X_2^{\Diamond}$; $Y_1 \subseteq Y_2 \Rightarrow Y_1^{\Box} \subseteq Y_2^{\Box}, Y_1^{\Diamond} \subseteq Y_2^{\Diamond}$.
2. $X^{\Box\Diamond} \subseteq X \subseteq X^{\Diamond\Box}$; $Y^{\Box\Diamond} \subseteq Y \subseteq Y^{\Diamond\Box}$.
3. $X^{\Box\Diamond\Box} = X^{\Box}, X^{\Diamond\Box\Diamond} = X^{\Diamond}$; $Y^{\Box\Diamond\Box} = Y^{\Box}, Y^{\Diamond\Box\Diamond} = Y^{\Diamond}$.
4. $(X_1 \cap X_2)^{\Box} = X_1^{\Box} \cap X_2^{\Box}, (X_1 \cup X_2)^{\Diamond} = X_1^{\Diamond} \cup X_2^{\Diamond}$;
   $(Y_1 \cap Y_2)^{\Box} = Y_1^{\Box} \cap Y_2^{\Box}, (Y_1 \cup Y_2)^{\Diamond} = Y_1^{\Diamond} \cup Y_2^{\Diamond}$.

Definition 3. *[14] Suppose (G, M, I) is a formal context. A pair (X, Y), $X \subseteq G, Y \subseteq M$, is called a property oriented concept, if $X = Y^{\Box}$ and $Y = X^{\Diamond}$; (X, Y) is called an object oriented concept, if $X = Y^{\Diamond}$ and $Y = X^{\Box}$. The object set X and the attribute set Y are called the extent and the intent of (X, Y) respectively.*

Denote all the object oriented concepts (property oriented concepts) of a formal context (G, M, I) as $L_O(G, M, I)$ ($L_P(G, M, I)$). Then, $L_O(G, M, I)$ is a lattice, where, $\forall (X_1, Y_1), (X_2, Y_2) \in L_O(G, M, I)$,

$$(X_1, Y_1) \wedge (X_2, Y_2) = ((X_1 \cap X_2)^{\Box\Diamond}, Y_1 \cap Y_2),$$
$$(X_1, Y_1) \vee (X_2, Y_2) = (X_1 \cup X_2, (Y_1 \cup Y_2)^{\Diamond\Box}).$$

And, $L_P(G, M, I)$ is also a lattice, where $\forall (X_1, Y_1), (X_2, Y_2) \in L_P(G, M, I)$,

$$(X_1, Y_1) \wedge (X_2, Y_2) = (X_1 \cap X_2, (Y_1 \cap Y_2)^{\Box\Diamond}),$$
$$(X_1, Y_1) \vee (X_2, Y_2) = ((X_1 \cup X_2)^{\Diamond\Box}, (Y_1 \cup Y_2)).$$

Definition 4. *[9] Let (G, M, I) be a formal context. $\forall D \subseteq M$, If $L_O(G, D, I_D) =_G L_O(G, M, I)$, then D is a consistent object oriented set; if $L_P(G, D, I_D) =_G L_P(G, M, I)$, then D is a consistent property oriented set. Moreover, $\forall d \in D$, if $L_O(G, D - \{d\}, I_{D-\{d\}}) \neq_G L_O(G, M, I)$, we say D is a reduct of $L_O(G, M, I)$; if $L_P(G, D - \{d\}, I_{D-\{d\}}) \neq_G L_P(G, M, I)$, we say D is a reduct of $L_P(G, M, I)$. Where, $I_D = I \cap (G \times D)$, the denotation $=_G$ means two lattices are equal with respect to extents.*

Example 1. Table 1 shows a formal context (G, M, I). In which, $G = \{1, 2, 3, 4, 5\}$ is the object set, $M = \{a, b, c, d, e, f, g\}$ is the attribute set.

Table 1. A formal context (G, M, I)

	a	b	c	d	e	f	g
1	×		×		×	×	
2	×	×		×			×
3			×			×	
4			×			×	
5	×	×					×

The corresponding object oriented concept lattice $L_O(G, M, I)$ and property oriented concept lattice $L_P(G, M, I)$ are shown as Fig.1 and Fig.2. Where, each set is described by the series of its elements except G, M, and $\varnothing$.

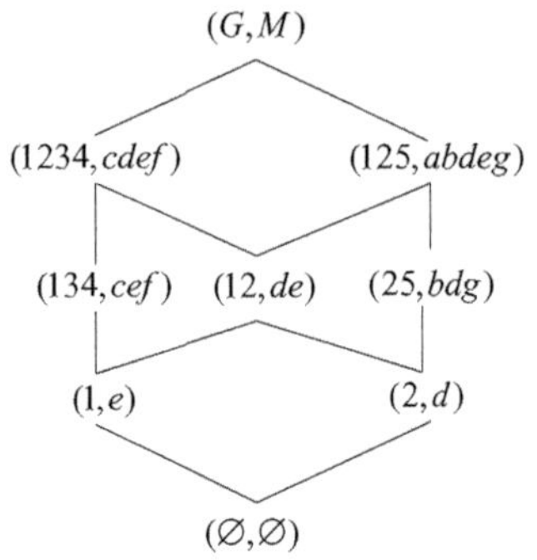

Fig. 1. $L_O(G, M, I)$

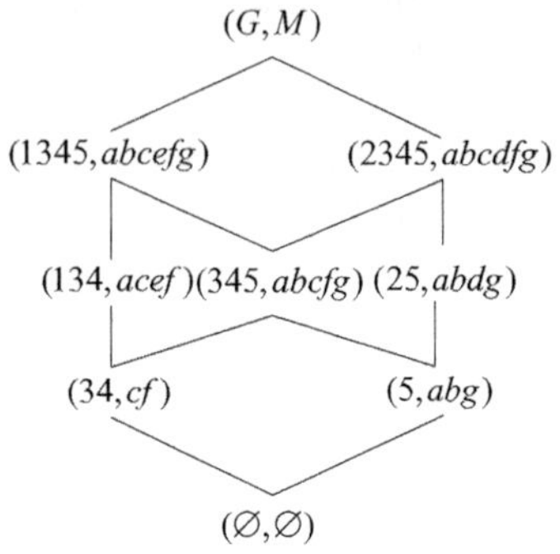

Fig. 2. $L_P(G, M, I)$

In the theory of concept lattice reduction [7], Zhang et al. has classified the attributes into three kinds, which is cited as follows.

Definition 5. *[7] Let (G, M, I) be a formal context, the set $\{D_i | D_i$ is a reduct, $i \in \tau\}$ (τ is an index set) includes all the reducts of (G, M, I). Then each element in M is classified into one of the following three kinds:*

1. *Absolutely necessary attribute (core attribute) $b : b \in \bigcap_{i \in \tau} D_i$.*
2. *Relatively necessary attribute $c : c \in \bigcup_{i \in \tau} D_i - \bigcap_{i \in \tau} D_i$.*
3. *Absolutely unnecessary attribute $d : d \in M - \bigcup_{i \in \tau} D_i$.*

Accordingly, $C = \cap_{i \in \tau} D_i$ is called the core attribute set (also called the core of (G, M, I)), $S = \cup_{i \in \tau} D_i - \cap_{i \in \tau} D_i$ is named the relatively necessary attribute set, and $K = M - \cup_{i \in \tau} D_i$ is called the absolutely unnecessary attribute set. The elements not in core are called non-core elements.

Similarly, there are the same classification in the reduction theory of object (property) oriented concept lattices. Reference [9] has given the judgment theorems about the reduction theory.

Theorem 1.[9] Let (G, M, I) be a formal context, $D \subseteq M, D \neq \emptyset, E = M - D$. Then, D is a consistent object oriented set $\Leftrightarrow \forall e \in E, (e^{\Diamond\Box} - E)^{\Diamond} = (e^{\Diamond\Box} \cap D)^{\Diamond} = e^{\Diamond}$.

Theorem 2.[9] Let (G, M, I) be a formal context, $D \subseteq M, D \neq \emptyset, E = M - D$. Then, D is a reduct of $L_O(G, M, I) \Leftrightarrow \forall e \in E$, $(e^{\Diamond\Box} - E)^{\Diamond} = (e^{\Diamond\Box} \cap D)^{\Diamond} = e^{\Diamond}$, and, $\forall d \in D$, $(d^{\Diamond\Box} - (E \cup \{d\}))^{\Diamond} = (d^{\Diamond\Box} \cap (D - \{d\}))^{\Diamond} \neq d^{\Diamond}$.

3 Attribute Characteristics of Object Oriented Concept Lattices

This section analyzes the attribute characteristics of object oriented concept lattices, and gives some judgment theorems.

Theorem 3. Let (G, M, I) be a formal context, core $K \subseteq M$, $K \neq \emptyset$, $\forall a \in M - K$, if $(a^{\Diamond\Box} \cap K)^{\Diamond} = a^{\Diamond}$, then, a is an absolutely unnecessary attribute.

Proof. Assume that a isn't an absolutely unnecessary attribute, then, there exists a reduct D such that $a \in D$.

Since $a \in M - K$, and $K \subseteq D$, we obtained that $K \subseteq D - \{a\}$. Then, $a^{\Diamond\Box} \cap K \subseteq a^{\Diamond\Box} \cap (D - \{a\})$, so, $a^{\Diamond} = (a^{\Diamond\Box} \cap K)^{\Diamond} \subseteq (a^{\Diamond\Box} \cap (D - \{a\}))^{\Diamond}$. On the other hand, $a^{\Diamond\Box} \cap (D - \{a\}) \subseteq a^{\Diamond\Box}$, so, $(a^{\Diamond\Box} \cap (D - \{a\}))^{\Diamond} \subseteq a^{\Diamond\Box\Diamond} = a^{\Diamond}$.

Then, $a^{\Diamond} = (a^{\Diamond\Box} \cap (D - \{a\}))^{\Diamond}$. So, from Theorem 2, we know that $D - \{a\}$ is a reduct of $L_O(G, M, I)$, which contradicts to "D is a reduct". Therefore, the result is proved.

Theorem 4. Let (G, M, I) be a formal context. Then, $a \in M$ is a non-core attribute $\Leftrightarrow (a^{\Diamond\Box} - \{a\})^{\Diamond} = a^{\Diamond}$.

Proof. Since a is a non-core attribute $\Leftrightarrow M - \{a\}$ is a consistent object oriented set, according to Theorem 1, it is equivalent to $\forall e \in E = M - (M - \{a\}) = \{a\}$, $(e^{\Diamond\Box} - E)^{\Diamond} = e^{\Diamond}$, that is, $(a^{\Diamond\Box} - \{a\})^{\Diamond} = a^{\Diamond}$.

Theorem 5. Let (G, M, I) be a formal context, a be a relatively necessary attribute. Then, there exists $b \in M$, $b \neq a$ such that $a^{\Diamond} = b^{\Diamond}$.

Proof. Since $a \in M$ is a relatively necessary attribute, we have $(a^{\Diamond\Box} - \{a\})^{\Diamond} = a^{\Diamond}$ from Theorem 4, and there must exist a reduct D such that $a \in D$. Assume that for any $b \in M$, $a \neq b$, there must be $a^{\Diamond} \neq b^{\Diamond}$; that is, for any $b \in a^{\Diamond\Box} \cap (M - D)$, $b^{\Diamond} \neq a^{\Diamond}$.

Combining with Theorem 1, we have $b^{\Diamond} = (b^{\Diamond\Box} \cap D)^{\Diamond} = (b^{\Diamond\Box} \cap (D - \{a\}))^{\Diamond} \subset (a^{\Diamond\Box} \cap (D - \{a\}))^{\Diamond}$, $(a^{\Diamond\Box} \cap (M - D))^{\Diamond} = \bigcup_{b \in a^{\Diamond\Box} \cap (M-D)} b^{\Diamond} \subseteq (a^{\Diamond\Box} \cap (D - \{a\}))^{\Diamond}$.
According to Theorem 2, $(a^{\Diamond\Box} \cap (D - \{a\}))^{\Diamond} \neq a^{\Diamond}$. So, $(a^{\Diamond\Box} - \{a\})^{\Diamond} = (a^{\Diamond\Box} \cap (A - D))^{\Diamond} \cup (a^{\Diamond\Box} \cap (D - \{a\}))^{\Diamond} = (a^{\Diamond\Box} \cap (D - \{a\}))^{\Diamond} \neq a^{\Diamond}$. Obviously, it's a contradiction to the fact "$(a^{\Diamond\Box} - \{a\})^{\Diamond} = a^{\Diamond}$".

Therefore,there must exists $b \in M$, $b \neq a$ such that $a^{\Diamond} = b^{\Diamond}$.

Thus, we can obtain the following theorem to check the type of each attribute in an object oriented concept lattice.

Theorem 6. Let (G, M, I) be a formal context, $T_a = \{b \in M \mid b^{\Diamond} \subset a^{\Diamond}\}$. Then, in an object oriented concept lattice, we have the following statements for any $a \in M$:

1. a is a core attribute $\Leftrightarrow (a^{\Diamond\Box} - \{a\})^{\Diamond} \neq a^{\Diamond}$.
2. a is an absolutely unnecessary attribute $\Leftrightarrow (a^{\Diamond\Box} - \{a\})^{\Diamond} = a^{\Diamond}$, and $T_a^{\Diamond} = a^{\Diamond}$.
3. a is a relatively necessary attribute $\Leftrightarrow (a^{\Diamond\Box} - \{a\})^{\Diamond} = a^{\Diamond}$, and $T_a^{\Diamond} \neq a^{\Diamond}$.

Proof

1. It is a conversion-negative proposition of Theorem 4.
2. Necessity. Firstly, from statement 1, it's easy to see that $(a^{\Diamond\Box} - \{a\})^{\Diamond} = a^{\Diamond}$. we only need to prove $T_a^{\Diamond} = a^{\Diamond}$. Firstly, definition of $T_a^{\Diamond}$ shows that $T_a^{\Diamond} \subseteq a^{\Diamond}$. Secondly, we can prove $a^{\Diamond} \subseteq T_a^{\Diamond}$. In fact, since $\forall b \in a^{\Diamond\Box} \cap D$, $b \in a^{\Diamond\Box} \Rightarrow b^{\Diamond} \subseteq a^{\Diamond}$, while, $b \in D$ and D is a reduct mean that $b \neq a$. Thus, $b^{\Diamond} \subset a^{\Diamond}$. That is, $b \in T_a$. Therefore, $a^{\Diamond\Box} \cap D \subseteq T_a$, thus, $(a^{\Diamond\Box} \cap D)^{\Diamond} = a^{\Diamond} \subseteq (T_a)^{\Diamond}$.
   Sufficiency. Assume that a isn't an absolutely unnecessary attribute, then there exists a reduct D such that $a \in D$, $(a^{\Diamond\Box} \cap (D - \{a\}))^{\Diamond} \neq a^{\Diamond}$. Since $a^{\Diamond\Box} \cap (D - \{a\}) \subseteq a^{\Diamond\Box} \Rightarrow (a^{\Diamond\Box} \cap (D - \{a\}))^{\Diamond} \subseteq a^{\Diamond}$. Thus, $(a^{\Diamond\Box} \cap (D - \{a\}))^{\Diamond} \subset a^{\Diamond}$. Since $a \notin T_a$, then T_a can be described as $T_a = (T_a \cap (D - \{a\})) \cup (T_a \cap (A - D))$. Then, $T_a^{\Diamond} = (T_a \cap (D - \{a\}))^{\Diamond} \cup (T_a \cap (A - D))^{\Diamond}$. We can show that $T_a^{\Diamond} \subset a^{\Diamond}$.
   In fact, we can show it from two cases. Firstly, $T_a = \{b \in A \mid b^{\Diamond} \subset a^{\Diamond}\} \Longrightarrow T_a^{\Diamond} \subset a^{\Diamond}$, so, $T_a \subset a^{\Diamond\Box}$. Thus, $(T_a \cap (D - \{a\}))^{\Diamond} \subset (a^{\Diamond\Box} \cap (D - \{a\}))^{\Diamond} \subset a^{\Diamond}$. Secondly, $\forall e \in (T_a \cap (A - D))$, obviously, $e \in T_a$, so $e^{\Diamond\Box} \subset a^{\Diamond\Box}$, and, $a \notin e^{\Diamond\Box}$. At the same time, $e \in (A - D)$ and D is consistent, so, $e^{\Diamond} = (e^{\Diamond\Box} \cap D)^{\Diamond} = (e^{\Diamond\Box} \cap (D - \{a\}))^{\Diamond} \subseteq (a^{\Diamond\Box} \cap (D - \{a\}))^{\Diamond}$. Moreover, $(T_a \cap (A - D))^{\Diamond} = (\bigcup_{e \in T_a \cap (A-D)} e)^{\Diamond} = \bigcup_{e \in T_a \cap (A-D)} e^{\Diamond} \subseteq (a^{\Diamond\Box} \cap (D - \{a\}))^{\Diamond}$.
   Combining the above results, we have $T_a^{\Diamond} = (T_a \cap (D - \{a\}))^{\Diamond} \cup (T_a \cap (A - D))^{\Diamond} \subset (a^{\Diamond\Box} \cap (D - \{a\}))^{\Diamond} \subseteq (a^{\Diamond\Box})^{\Diamond} = a^{\Diamond}$. That is, $T_a^{\Diamond} \subset a^{\Diamond}$.
3. The proposition is obvious according to the statements 1 and 2.

Thus, we have obtained a series of theorems about attribute characteristics of object oriented concept lattices.

Moreover, we study how to analyze the attribute characteristics of property oriented concept lattices in the next part.

Example 2. Consider the formal context shown in Example 1. According to Theorem 6, we analyze each attribute as follows:

$(a^{\Diamond\Box} - \{a\})^{\Diamond} = \{b, d, e, g\}^{\Diamond} = \{1, 2, 5\} = a^{\Diamond}$, and $T_a^{\Diamond} = \{b, d, e, g\}^{\Diamond} = \{1, 2, 5\} = a^{\Diamond}$. So, a is an absolutely unnecessary attribute.

$(b^{\Diamond\Box} - \{b\})^{\Diamond} = (\{b, d, g\} - \{b\})^{\Diamond} = \{d, g\}^{\Diamond} = \{2, 5\} = b^{\Diamond}$, and $T_b^{\Diamond} = d^{\Diamond} = \{2\} \neq b^{\Diamond}$. So, b is a relatively necessary attribute. Similarly, we can obtain that c, f, g are also relatively necessary attributes.

$(d^{\Diamond\Box} - \{d\})^{\Diamond} = \emptyset^{\Diamond} \neq d^{\Diamond}$. So, d is a core attribute. Similarly, we can obtain that e is also a core attribute.

4 Attribute Characteristics of Property Oriented Concept Lattices

Y. Y. Yao has shown that the object oriented concept lattice is isomorphic to the property oriented concept lattice since the duality of approximation operators $\Diamond$ and $\Box$, and $(X, Y) \in L_O(G, M, I) \Leftrightarrow (X^c, Y^c) \in L_P(G, M, I)$ [14].

Assume that D is a consistent object oriented set of (G, M, I).

We have:

$$\begin{aligned} & L_O(G, D, I_D) \leq L_O(G, M, I) \\ & \Leftrightarrow \forall (X, Y) \in L_O(G, M, I), \exists (X, Y_0) \in L_O(G, D, I_D) \\ & \Leftrightarrow \forall (X^c, Y^c) \in L_P(G, M, I), \exists (X^c, Y_0^c) \in L_P(G, D, I_D) \\ & \Leftrightarrow L_P(G, D, I_D) \leq L_P(G, M, I). \end{aligned}$$

It means D is also a consistent property oriented set of (G, M, I). Consequently, according to Definition 5 and 6, we know that the object oriented concept lattice and property oriented concept lattice have the same consistent sets, reduct sets and attribute classification. That is, the theorems proposed in Section 3 are also fit for the property oriented concept lattices.

5 Conclusion

Since attribute reduction theory is important in formal concept analysis, to discuss each attribute's role in reduction is also interesting. The paper has discussed the attribute characteristics of object oriented concept lattices and property oriented concept lattices, and reveal each attribute's importance in keeping lattice structure. In the real world, we can distinguish attributes' significance according to their role, and obtain the wanted meaningful attributes.

Acknowledgments. The authors gratefully acknowledge the support of the Natural Science Foundation of China (No.11071281, No.60703117, No.61005042).

References

1. Wille, R.: Restructuring lattice theory: an approach based on hierarchies of concept, ordered sets. In: Rival, I. (ed.), pp. 445–470. Reidel, Dordrecht (1982)
2. Godin, R.: Incremental concept formation algorithm based on Galois lattices. Comput. Intell. 11(2), 246–267 (1999)
3. Saquer, J., Deogun, J.S.: Formal Rough Concept Analysis. In: Zhong, N., Skowron, A., Ohsuga, S. (eds.) RSFDGrC 1999. LNCS (LNAI), vol. 1711, pp. 91–99. Springer, Heidelberg (1999)
4. Kent, R.E., Bowman, C.M.: Digital Libraries, Conceptual Knowledge Systems and the Nebula Interface. Technical Report, University of Arkansas (1995)
5. Sutton, A., Maletic, J.I.: Recovering UML class models from C++: a detailed explanation. Inf. Softw. Technol. 48(3), 212–229 (2007)
6. Ganter, B., Stumme, G., Wille, R.: Formal Concept Analysis: Foundations and Applications. Springer, Heidelberg (2005)

7. Zhang, W.X., Wei, L., Qi, J.J.: Attribute reduction theory and approach to concept lattice. Science in China Series F-Information Science 48(6), 713–726 (2005)
8. Wang, X., Ma, J.-M.: A Novel Approach to Attribute Reduction in Concept Lattices. In: Wang, G.-Y., Peters, J.F., Skowron, A., Yao, Y. (eds.) RSKT 2006. LNCS (LNAI), vol. 4062, pp. 522–529. Springer, Heidelberg (2006)
9. Liu, M.Q., Wei, L., Zhao, W.: The Reduction Theory of Object Oriented Concept Lattices and Property Oriented Concept Lattices. In: Wen, P., Li, Y., Polkowski, L., Yao, Y., Tsumoto, S., Wang, G. (eds.) RSKT 2009. LNCS, vol. 5589, pp. 587–593. Springer, Heidelberg (2009)
10. Wei, L., Zhang, X.-H., Qi, J.-J.: Granular Reduction of Property-Oriented Concept Lattices. In: Croitoru, M., Ferré, S., Lukose, D. (eds.) ICCS 2010. LNCS, vol. 6208, pp. 154–164. Springer, Heidelberg (2010)
11. Pawlak, Z.: Rough sets. International Journal of Computer and Information Science 11, 341–356 (1982)
12. Düntsch, I., Gediga, G.: Approximation Operators in Qualitative Data Analysis. In: de Swart, H., Orłowska, E., Schmidt, G., Roubens, M. (eds.) TARSKI 2003. LNCS, vol. 2929, pp. 214–230. Springer, Heidelberg (2003)
13. Gediga, G., Duntsch, I.: Modal-style operations in qualitative data analysis. In: Proceedings of the 2002 IEEE International Conference on Data Mining, pp. 155–162 (2002)
14. Yao, Y.Y.: Concept lattices in rough set theory. In: Dick, S., Kurgan, L., Pedrycz, W., Reformat, M. (eds.) Proceedings of 2004 Annual Meeting of the North American Fuzzy Information Processing Society, June 27-30, pp. 796–801 (2004)
15. Ganter, B., Wille, R.: Formal Concept Analysis, Mathematical Foundations. Springer, Berlin (1999)

Dependence Space-Based Model for Constructing Dual Concept Lattice in Sub-formal Context

Jian-Min Ma*

Department of Mathematics and Information Science
Faculty of Science, Chang'an University, Xi'an, Shaan'xi, China, 710064
cjm-zm@126.com

Abstract. This paper mainly studies a dependence space-based model for constructing dual concept lattice in sub-formal context. Based on the operators of a dual concept lattice , a $\cap-$congruence on the power set of objects is defined, and $\cap-$dependence space is then obtained. By the $\cap-$congruence, dual concept granules and an inner operator decided by any subset of attributes are introduced. It is proved that each open element of the inner operator is just the minimal element in an dual concept granule. Furthermore, the open element is also the extension of a dual concept lattice.

Keywords: Dual concept lattice, dependence space, $\cap-$congruence relation, dual concept granular.

1 Introduction

Formal concept analysis (FCA), proposed by Wille in 1982 [17], is a mathematical framework for discovery and design of concept hierarchies from a formal context. It is an embranchment of applied mathematics, which made it need mathematical thinking for applying FCA to data analysis and knowledge processing [2]. All formal concepts of a formal context with their specification and generalization form a concept lattice [3]. And the concept lattice can be depicted by a Hasse diagram, where each node expresses a formal concept. Concept lattice is the core structure of data in FCA. In essence, a formal concept represents a relationship between the extension of a set of objects and the intension of a set of attributes, and the extension and the intension are uniquely determined each other. Thus FCA is regarded as a power tool for learning problems [1,4,5,7,9].

Rough sets and formal concept analysis, two kinds of theories for knowledge representation and data analysis, offer related and complementary approaches for data analysis. Many efforts have been made to compare and combine the two theories [2,4,8,10,12, 14-16,18-24]. Novotny [15] introduced the theory of dependence space into information system, and discussed attribute reduction. Jarvinen [6] studied dependence relation and dependence function. Li [10] defined

* Corresponding author.

J.T. Yao et al. (Eds.): RSCTC 2012, LNAI 7413, pp. 349–356, 2012.

the congruence on semi-lattice, and proposed classification and reduction in a formal context. Then Ma [18] discussed object oriented concept lattice based rough set, and showed approaches to obtain extensions (or intensions) of all concepts of object-oriented concept lattice. This approach is extended to fuzzy concept lattice and concept lattice based on two universe formal context [17,19]. Then reference [20] generalized these results, and constructed dependence space model of concept lattice in a sub-formal context.

In this paper, a dependence space-based model for constructing dual concept lattice is formed in a sub-formal context. A new $\cap-$congruence based on a subset of attributes is defined. Then $\cap-$dependence space and dual concept granules are obtained. By using dual concept granules, an inner operator is proposed, and properties of the inner operator are also investigated. Last, approaches to obtain extensions of all dual concepts in a sub-formal context are given which guarantee the construction of dual concept lattice in a sub-formal context.

2 Preliminaries

In this section, basic notions and properties of a pair of dual concept lattices are given. Some connections between this pair of dual concept lattices are investigated.

A formal context is a triplet (U, A, I), where $U = \{x_1, x_2, \cdots, x_n\}$ is a nonempty finite set of objects called a universe of discourse, $A = \{a_1, a_2, \cdots, a_m\}$ is a non-empty finite set of attributes, and I is a relation between U and A. For any $x \in U$ and $a \in A$, $(x, a) \in I$, also written as xIa, means that the object x has the attribute a, or the attribute a is possessed by the object x, and $(x, a) \notin I$ means that the object x has not the attribute a. If we denote $(x, a) \in I$ by 1 and $(x, a) \notin I$ by 0, a formal context can be denoted as a table with 0 and 1 [3,17].

Let (U, A, I) be a formal context, and $\mathcal{P}(U)$ be the power set of U. For any $X \in \mathcal{P}(U)$ and $B \in \mathcal{P}(A)$, a pair data operators $^* : \mathcal{P}(U) \to \mathcal{P}(A)$ and $^* : \mathcal{P}(A) \to \mathcal{P}(U)$, called sufficiency operators [3,17], are defined by

$$X^* = \{a|\ a \in A, \forall x \in X, xIa\} \tag{1}$$

$$B^* = \{x|\ x \in U, \forall a \in B, xIa\} \tag{2}$$

where X^* is the set of attributes shared by all objects in X, and B^* is the set of objects which possess all attributes in B. For simplicity, for any $x \in U$ and $a \in A$, we use x^* and a^* to denote the sets $\{x\}^*$ and $\{a\}^*$, respectively. If $\forall x \in U$, $x^* \neq \emptyset$, $x^* \neq A$, and $\forall a \in A$, $a^* \neq \emptyset$, $a^* \neq U$, then a formal context (U, A, I) is called regular. In this paper, we suppose all formal contexts are regular.

It is obvious that for any $x \in U$ and $a \in A$, $x^* = \{a \in A :\ (x, a) \in I\}$ and $a^* = \{x \in U :\ (x, a) \in I\}$. Then $(x, a) \in I \Leftrightarrow x \in a^* \Leftrightarrow a \in x^*$.

Property 1 [3,17]. Let (U, A, I) be a formal context, $X, X_1, X_2 \in \mathcal{P}(U)$ and $B, B_1, B_2 \in \mathcal{P}(A)$. Then the following properties hold:

(1) $X_1 \subseteq X_2 \Rightarrow X_2^* \subseteq X_1^*$, $B_1 \subseteq B_2 \Rightarrow B_2^* \subseteq B_1^*$;
(2) $X \subseteq X^{**}, B \subseteq B^{**}$;
(3) $X^* = X^{***}, B^* = B^{***}$;
(4) $(X_1 \cup X_2)^* = X_1^* \cap X_2^*, (B_1 \cup B_2)^* = B_1^* \cap B_2^*$;
(5) $X \subseteq B^* \Leftrightarrow B \subseteq X^*$.

A pair (X, B) with $X \subseteq U$ and $B \subseteq A$ is called a formal concept if $X^* = B$ and $B^* = X$. X is called the extent of the concept and B is called its intent. The set of all concepts of (U, A, I), denoted by $L(U, A, I)$, forms a complete lattice called concept lattice [3,17], where the partial order $\leq$ is defined as follows: for any $(X_1, B_1), (X_2, B_2) \in L(U, A, I)$,

$$(X_1, B_1) \leq (X_2, B_2) \Leftrightarrow X_1 \subseteq X_2.$$

And the meet and join are given as follows:

$$\begin{aligned}(X_1, B_1) \wedge (X_2, B_2) &= (X_1 \cap X_2, (B_1 \cup B_2)^{**})\\ (X_1, B_1) \vee (X_2, B_2) &= ((X_1 \cup X_2)^{**}, B_1 \cap B_2).\end{aligned}$$

Consider now the dual operator $^\sharp$ of * defined by [21]:

$$\begin{aligned}X^\sharp &= X^{c*c} = [(X^c)^*]^c\\ &= \{a \in A : \exists x \in U (x \in X^c \wedge (x, a) \notin I)\}.\end{aligned} \tag{2}$$

Similarly, we can get:

$$\begin{aligned}B^\sharp &= B^{c*c} = [(B^c)^*]^c\\ &= \{x \in U : \exists a \in A (a \in B^c \wedge (x, a) \notin I)\},\end{aligned} \tag{3}$$

where, X^c denotes the complement set of X. We call the pair data operators $^\sharp : \mathcal{P}(U) \to \mathcal{P}(A)$ and $^\sharp : \mathcal{P}(A) \to \mathcal{P}(U)$ dual sufficiency operators.

Property 2 [21]. Let (U, A, I) be a formal context. Then for any $X, X_1, X_2 \subseteq U$ and $B, B_1, B_2 \subseteq A$, the following properties hold:

(1) $X_1 \subseteq X_2 \Rightarrow X_2^\sharp \subseteq X_1^\sharp$, $B_1 \subseteq B_2 \Rightarrow B_2^\sharp \subseteq B_1^\sharp$;
(2) $X \supseteq X^{\sharp\sharp}, B \supseteq B^{\sharp\sharp}$;
(3) $X^\sharp = X^{\sharp\sharp\sharp}, B^\sharp = B^{\sharp\sharp\sharp}$;
(4) $(X_1 \cap X_2)^\sharp = X_1^\sharp \cup X_2^\sharp, (B_1 \cap B_2)^\sharp = B_1^\sharp \cup B_2^\sharp$;
(5) $X \supseteq B^\sharp \Leftrightarrow B \supseteq X^\sharp$.

For any $X \subseteq U, B \subseteq A$, a pair (X, B) is called a dual formal concept if $X = B^\sharp, B = X^\sharp$. Then for any $X \subseteq U$ and $B \subseteq A$, $(X^{\sharp\sharp}, X^\sharp)$ and $(B^\sharp, B^{\sharp\sharp})$ are dual concepts. We denote by $L_d(U, A, I)$ the set of all dual concepts of (U, A, I). For any $(X_1, B_1), (X_2, B_2) \in L_d(U, A, I)$, define a binary relation $\leq_d$ as follows:

$$(X_1, B_1) \leq_d (X_2, B_2) \Leftrightarrow X_1 \subseteq X_2.$$

Then $\leq_d$ is a partial order on $L_d(U, A, I)$. And $L_d(U, A, I)$ is a complete lattice, called dual concept lattice [21], where the meet and join are given as follows:

$$\begin{aligned}(X_1, B_1) \wedge_d (X_2, B_2) &= ((X_1 \cap X_2)^{\sharp\sharp}, B_1 \cup B_2)\\ (X_1, B_1) \vee_d (X_2, B_2) &= (X_1 \cup X_2, (B_1 \cap B_2)^{\sharp\sharp}).\end{aligned}$$

Since $^\sharp$ is the dual operator of *, then we can get the relationship between concept lattice and dual concept lattice.

Then for a formal context (U, A, I), we can obtain the following:

$$(X, B) \in L(U, A, I) \Leftrightarrow (X^c, B^c) \in L_d(U, A, I).$$

That is

$$L(U, A, I) = \{(X^c, B^c) : \ (X, B) \in L_d(U, A, I)\}.$$

3 Dependence Space-Based Model of Dual Concept Lattice Lattice in Sub-formal Context

Definition 1. Let U be a finite nonempty set, and $\mathcal{R}$ is an equivalence relation on $\mathcal{P}(U)$.

(1) $\mathcal{R}$ is said to be a $\cap$−congruence relation on $\mathcal{P}(U)$, if for any $(X_1,\ Y_1) \in \mathcal{R}$ and $(X_2,\ Y_2) \in \mathcal{R}$, we have $(X_1 \cap X_2,\ Y_1 \cap Y_2) \in \mathcal{R}$.

(2) $(U, \mathcal{R})$ is said to be a $\cap$−dependence space, if U is a finite nonempty set, and $\mathcal{R}$ is a $\cap$−congruence relation on $\mathcal{P}(U)$.

Theorem 1. Let (U, A, I) be a formal context. It should be noted that

$$\mathcal{R}_B^d = \{(X, Y) : \ X^\sharp \cap B = Y^\sharp \cap B\}.$$

Then $(U, \mathcal{R}_B^d)$ is a $\cap$−dependence space.

Proof. It is easy to prove that $\mathcal{R}_B^d$ is an equivalence relation on $\mathcal{P}(U)$. Suppose $(X_i, Y_i) \in \mathcal{R}_B^d$, then $X_i^\sharp \cap B = Y_i^\sharp \cap B$ $(i = 1, 2)$. By using Property 2(4) we can obtain that, $(X_1 \cap X_2)^\sharp \cap B = (X_1^\sharp \cup X_2^\sharp) \cap B = (X_1^\sharp \cap B) \cup (X_2^\sharp \cap B) = (Y_1^\sharp \cap B) \cup (Y_2^\sharp \cap B) = (Y_1 \cap Y_2)^\sharp \cap B$. That is, $(X_1 \cap X_2, Y_1 \cap Y_2) \in \mathcal{R}_B^d$. Then $\mathcal{R}_B^d$ is a $\cap$−congruence relation on $\mathcal{P}(U)$, and $(U, \mathcal{R}_B^d)$ is a $\cap$−dependence space by Definition 1.

By Theorem 1 we know $\mathcal{R}_B^d$ is an equivalence relation on $\mathcal{P}(U)$, which generates a partition: $\mathcal{P}(U)/\mathcal{R}_B^d = \{[X]_B : \ X \in \mathcal{P}(U)\}$, where $[X]_B = \{Y \subseteq U : (X, Y) \in \mathcal{R}_B^d\}$ is called a dual concept granule decided by B, for short, a dual concept granule.

Definition 2 [6]**.** Let $\mathcal{P} = (P, \leq)$ be a poset. If the mapping $i : P \to P$ satisfies:
(1) $i(x) \leq x$ $(x \in P)$;
(2) $x \leq y \Rightarrow i(x) \leq i(y)$ $(x, y \in P)$;
(3) $i(i(x)) = i(x)$ $(x \in P)$.
we call the operator i is an inner operator on $\mathcal{P}$ or a coclosure operator.

Theorem 2. Let (U, A, I) be a formal context. It should be noted that

$$\mathcal{I}(\mathcal{R}_B^d)(X) = \cap[X]_B = \cap\{Y \subseteq U : \ (X, Y) \in \mathcal{R}_B^d\}.$$

Then the following statements hold:

(1) $\mathcal{I}(\mathcal{R}_B^d)(X) \in [X]_B$, and for any $X' \in [X]_B$, $\mathcal{I}(\mathcal{R}_d(B))(X) \subseteq X'$;
(2) if $X_1 \subseteq X' \subseteq X_2$, and $X_1, X_2 \in [X]_B$, we have $X' \in [X]_B$.

Proof. (1) Take $Y \in [X]_B$. Then $X^\sharp \cap B = Y^\sharp \cap B$. By $\mathcal{I}(\mathcal{R}_B^d)(X) = \cap[X]_B$ and Property 2 we can get that

$$\begin{aligned}\mathcal{I}(\mathcal{R}_B^d)(X)^\sharp \cap B &= (\cap[X]_B)^\sharp \cap B \\ &= (\cap\{Y \subseteq U : (X,Y) \in \mathcal{R}_B^d\})^\sharp \cap B \\ &= (\cup\{Y^\sharp : (X,Y) \in \mathcal{R}_B^d\}) \cap B \\ &= \cup\{Y^\sharp \cap B : (X,Y) \in \mathcal{R}_B^d\}) \\ &= \cup\{X^\sharp \cap B : (X,Y) \in \mathcal{R}_B^d\} \\ &= X^\sharp \cap B.\end{aligned}$$

Therefore, $\mathcal{I}(\mathcal{R}_B^d)(X) \in [X]_B$. Then for any $X' \in [X]_B$, we cam easily get that $\mathcal{I}(\mathcal{R}_B^d)(X) \subseteq X'$.

(2) Suppose $X_1 \subseteq X' \subseteq X_2$, and $X_1, X_2 \in [X]_B$. Then $X^\sharp \cap B = X_1^\sharp \cap B = X_2^\sharp \cap B$ and $(X_1, X_2) \in \mathcal{R}_B^d$. Since $\mathcal{R}_B^d$ is a $\cap-$congruence relation firstly, we have $(X', X') \in \mathcal{R}_B^d$. Thus, $(X_1 \cap X', X_2 \cap X') \in \mathcal{R}_B^d$. By $X_1 \subseteq X' \subseteq X_2$ we can obtain that $(X', X_2) \in \mathcal{R}_B^d$. Then $X' \in [X]_B$.

Theorem 3. Let (U, A, I) be a formal context. Then the operator $\mathcal{I}(\mathcal{R}_B^d) : \mathcal{P}(U) \to \mathcal{P}(U)$ is an inner operator.

Proof. Since $\mathcal{R}_B^d$ is reflexive, we have $(X, X) \in \mathcal{R}_B^d$. Then $\mathcal{I}(\mathcal{R}_B^d)(X) \subseteq X$ by Theorem 2.

For any $X_1 \subseteq X_2 \subseteq U$, by Theorem 2 we can obtain that $(X_1, \mathcal{I}(\mathcal{R}_B^d)(X_1)) \in \mathcal{R}_B^d$ and $(X_2, \mathcal{I}(\mathcal{R}_B^d)(X_2)) \in \mathcal{R}_B^d$. Since $\mathcal{R}_B^d$ is a $\cap-$congruence relation, we have $(X_1, \mathcal{I}(\mathcal{R}_B^d)(X_1) \cap \mathcal{I}(\mathcal{R}_B^d)(X_2)) \in \mathcal{R}_B^d$. Thus, $\mathcal{I}(\mathcal{R}_B^d)(X_1) \subseteq \mathcal{I}(\mathcal{R}_B^d)(X_1) \cap \mathcal{I}(\mathcal{R}_B^d)(X_2))$. And then $\mathcal{I}(\mathcal{R}_B^d)(X_1) \subseteq \mathcal{I}(\mathcal{R}_B^d)(X_2)$.

For any $X \subseteq U$, we can get that $\mathcal{I}(\mathcal{R}_B^d)(X) \subseteq U$. Then $\mathcal{I}(\mathcal{R}_B^d)(\mathcal{I}(\mathcal{R}_B^d)(X)) \subseteq \mathcal{I}(\mathcal{R}_B^d)(X)$ by Theorem 2. On the other hand, since $(X, \mathcal{I}(\mathcal{R}_B^d)(X)) \in \mathcal{R}_B^d$ and $(\mathcal{I}(\mathcal{R}_B^d)(X), \mathcal{I}(\mathcal{R}_B^d)(\mathcal{I}(\mathcal{R}_B^d)(X))) \in \mathcal{R}_B^d$, we have $(X, \mathcal{I}(\mathcal{R}_B^d)(\mathcal{I}(\mathcal{R}_B^d)(X))) \in \mathcal{R}_B^d$. Again using Theorem 2 (1) we can obtain that $\mathcal{I}(\mathcal{R}_B^d)(X) \subseteq \mathcal{I}(\mathcal{R}_B^d)(\mathcal{I}(\mathcal{R}_B^d)(X))$. Thus, $\mathcal{I}(\mathcal{R}_B^d)(X) = \mathcal{I}(\mathcal{R}_B^d)(\mathcal{I}(\mathcal{R}_B^d)(X))$.

According to Definition 2 we can get that $\mathcal{R}_B^d$ is an inner operator on $\mathcal{P}(U)$.

We denote by $F(\mathcal{I}(\mathcal{R}_B^d)) = \{X \subseteq U : \mathcal{I}(\mathcal{R}_B^d)(X) = X\}$ the set of all fixed points of the operator $\mathcal{I}(\mathcal{R}_B^d)$. Each fixed point X in $F(\mathcal{I}(\mathcal{R}_B^d))$ is referred to be a open element of $\mathcal{I}(\mathcal{R}_B^d)$.

Theorem 4. Let (U, A, I) be a formal context. Then

$$F(\mathcal{I}(\mathcal{R}_B^d)) = \{\mathcal{I}(\mathcal{R}_B^d)(X) : X \subseteq U\}.$$

Proof. It should be noted that $L = \{\mathcal{I}(\mathcal{R}_B^d)(X) : X \subseteq U\}$. Take $Y \in F(\mathcal{I}(\mathcal{R}_B^d))$. Then $\mathcal{I}(\mathcal{R}_B^d)(Y) = Y$ and then $Y \in L$. Suppose $Y \in L$. Then there exists $X \subseteq U$ such that $Y = \mathcal{I}(\mathcal{R}_B^d)(X)$. By Theorem 3 we can get that $\mathcal{I}(\mathcal{R}_B^d)(Y) = \mathcal{I}(\mathcal{R}_B^d)(\mathcal{I}(\mathcal{R}_B^d)(X)) = \mathcal{I}(\mathcal{R}_B^d)(X) = Y$. So we have $Y \in F(\mathcal{I}(\mathcal{R}_B^d))$. Therefore, we can obtain that $F(\mathcal{I}(\mathcal{R}_B^d)) = L = \{\mathcal{I}(\mathcal{R}_B^d)(X) : X \subseteq U\}$.

Theorem 4 shows that, for any $X \subseteq U$, $\mathcal{I}(\mathcal{R}_B^d)(X)$ is a open element of the operator $\mathcal{I}(\mathcal{R}_B^d)$.

Lemma 1. Let (U, A, I) be a formal context and $B \subseteq A$. Then for any $X \subseteq U$,

$$\mathcal{I}(\mathcal{R}_B^d)(X) = (X^\sharp \cap B)^\sharp.$$

Proof. Since (U, B, I_B) is a sub-formal context, for any $X \subseteq U$, we can get that $X^\sharp \cap B = ((X^\sharp \cap B)^\sharp)^\sharp \cap B$. So $(X^\sharp \cap B)^\sharp \in [X]_B$. By Theorem 2(1) we can get that $\mathcal{I}(\mathcal{R}_B^d)(X) \subseteq (X^\sharp \cap B)^\sharp$. Furthermore, take any $a \in (X^\sharp \cap B)^\sharp$. Then for any $Y \subseteq U$, $Y^\sharp \cap B = X^\sharp \cap B$, we have $a \in (X^\sharp \cap B)^\sharp = (Y^\sharp \cap B)^\sharp \subseteq Y$. By the arbitrariness of Y we can obtain that $a \in \cap\{Y \subseteq U : Y^\sharp \cap B = X^\sharp \cap B\} = \mathcal{I}(\mathcal{R}_B^d)(X)$. So we can obtain that $(X^\sharp \cap B)^\sharp \subseteq \mathcal{I}(\mathcal{R}_B^d)(X)$. That is, $\mathcal{I}(\mathcal{R}_B^d)(X) = (X^\sharp \cap B)^\sharp$.

Theorem 5. Let (U, A, I) be a formal context and $B \subseteq A$. Then

$$(X, D) \in L_d(U, B, I_B) \Leftrightarrow X \in F(\mathcal{I}(\mathcal{R}_B^d)).$$

Proof. " $\Rightarrow$ " Take $(X, D) \in L_d(U, B, I_B)$. Then $X^\sharp \cap B = D$ and $D^\sharp = X$. Thus, $(D^\sharp)^\sharp \cap B = X$. By Lemma 3.1 we can get that $\mathcal{I}(\mathcal{R}_B^d)(X) = (X^\sharp \cap B)^\sharp = X$. Therefore, we can obtain that $X \in F(\mathcal{I}(\mathcal{R}_B^d))$.

" $\Leftarrow$ " Suppose $X \in F(\mathcal{I}(\mathcal{R}_B^d))$. Then $\mathcal{I}(\mathcal{R}_B^d)(X) = X$. According to Lemma 1 we know that $\mathcal{I}(\mathcal{R}_B^d)(X) = (X^\sharp \cap B)^\sharp = X$. Take $D = X^\sharp \cap B$. We can get that $X^\sharp \cap B = D$ and $D^\sharp = X$. That is, $(X, D) \in L_d(U, B, I_B)$.

Theorem 5 shows an approach to obtain intensions of dual sub-formal concepts for a dual sub-concept lattice $L_d(U, B, I_B)$:

4 Conclusions

This paper proposed a dependence space-based model for constructing dual concept lattice in a sub-formal context. By defining a $\cap-$congruence on the power set of objects, we can get a model of dependence space. Then a partition and dual concept granules are also obtained. By the dual concept granules, an inner operator on the power set of objects is given, and properties of the operator are also examined. By the inner operator we can prove that the minimal element in a dual concept granule is just the extension of a dual concept of concept lattice in the sub-formal context, which guarantee the acquirement of the dual concept lattice for the sub-formal context.

Acknowledgement. This work is supported by the Nature Science Foundation of China (10901025) and the Special Fund for Basic Scientific Research of Central Colleges (CHD2012JC003).

References

1. Chaudron, L., Maille, N.: Generalized Formal Concept Analysis. In: Ganter, B., Mineau, G.W. (eds.) ICCS 2000. LNCS, vol. 1867, pp. 357–370. Springer, Heidelberg (2000)
2. Deogun, J.S., Saqer, J.: Monotone concepts for formal concept analysis. Discrete Applied Mathematics 144, 70–78 (2004)
3. Ganter, B., Wille, R.: Formal Concept Analysis: Mathematical Foundations. Springer, Berlin (1999)
4. Hu, K., Sui, Y., Lu, Y., Wang, J., Shi, C.: Concept Approximation in Concept Lattice. In: Cheung, D., Williams, G.J., Li, Q. (eds.) PAKDD 2001. LNCS (LNAI), vol. 2035, pp. 167–173. Springer, Heidelberg (2001)
5. Jaoua, A., Elloumi, S.: Galois connection, formal concept and Galois lattice in real binary relation. Journal of Systems and Software 60(2), 149–163 (2002)
6. Jarvinen, J.: Representations of Information Systems and Dependences Spaces, and Some Basic Algorithms. University of Turku, Turku (1997)
7. Krajci, S.: Cluster based efficient generation of fuzzy concepts. Neural Network World 5, 521–530 (2003)
8. Kent, R.E.: Rough concept analysis: a synthesis of rough sets and formal concept analysis. Fund. Inform. 27, 169–181 (1996)
9. Latiri, C.C., Elloumi, S., Chevallet, J.P., et al.: Extension of fuzzy Galois connection for information retrieval using a fuzzy quantifier. In: ACS/IEEE International Conference on Computer Systems and Applications, Tunis, Tunisia (2003)
10. Li, H.R.: Research on theory of rough set and concept lattice based on lattice topological construction, Doctor Dissertation, XI'an Jiaotong University (2006)
11. Ma, J.-M., Zhang, W.X., Cai, S.: Variable Threshold Concept Lattice and Dependence Space. In: Wang, L., Jiao, L., Shi, G., Li, X., Liu, J. (eds.) FSKD 2006. LNCS (LNAI), vol. 4223, pp. 109–118. Springer, Heidelberg (2006)
12. Ma, J.M., Zhang, W.X., Wang, X.: Dependence space of concept lattices based on rough set. In: Proceedings of the 2006 IEEE International Conference on Granular Computing, pp. 200–204 (2006)
13. Ma, J.M., Cai, S., Zhang, W.X.: An approach to construct concept lattices based on double universe formal context. In: International Conference on Machine Learning and Cybernetics, pp. 1098–1103 (2006)
14. Ma, J.M.: Dependence Space-based Model for Constructing Concept Lattice in Sub-formal Context. Computer Science 36 (8A), 113–115 (2009)
15. Novotny, M.: Dependence Spaces of Information Systems. In: Orlowska, E. (ed.) Incomplete Informations: Rough Sets Analysis, pp. 193–246. Physica-Verlag (1998)
16. Wei, L.: Reduction theory and approach to rough set and concept lattice, PhD Thesis, XI'an Jiaotong University. XI'an Jiaotong University Press, Xi'an (2005)
17. Wille, R.: Restructuring lattice theory: an approach based on hierarchies of concepts. In: Rival, I. (ed.) Ordered Sets, pp. 445–470. Reidel, Dordrecht (1982)
18. Wolff, K.E.: A Conceptual View of Knowledge Bases in Rough Set Theory. In: Ziarko, W., Yao, Y. (eds.) RSCTC 2000. LNCS (LNAI), vol. 2005, pp. 220–228. Springer, Heidelberg (2001)
19. Yang, X.B., Song, X.N., Chen, Z.H., Yang, J.Y.: On multigranulation rough sets in incomplete information system. Int. J. Mach. Learn. & Cyber. (2011), doi: 10.1007/s13042-011-0054-8
20. Yao, Y.Y.: Concept lattices in rough set theory. In: Proc. 2004 Annu. Meeting of the North American Fuzzy Information Processing Society, pp. 796-801 (2004)

21. Yao, Y.Y.: A Comparative Study of Formal Concept Analysis and Rough Set Theory in Data Analysis. In: Tsumoto, S., Słowiński, R., Komorowski, J., Grzymała-Busse, J.W. (eds.) RSCTC 2004. LNCS (LNAI), vol. 3066, pp. 59–68. Springer, Heidelberg (2004)
22. Yao, Y.Y., Chen, Y.: Rough set approximations in formal concept analysis. In: 2004 Annu. Meeting of the North American Fuzzy Information Processing Society, pp. 73-78 (2004)
23. Zhang, W.X., Qiu, G.F.: Uncertain Decision Making Based on Rough Sets. Publication of Tsinghua University, Beijing (2005)
24. Zhu, W., Wang, S.: Matroidal approaches to generalized rough sets based on relations. Int. J. Mach. Learn. & Cyber. 2(4), 273–279 (2011)

XDCKS: A RCP-Based Prototype for Formal Concept Analysis

Wei Liu[1], Jianjun Qi[2], and Bing Liang[2]

[1] Software Engineering Institute, Xidian University, Xi'an 710071, China
liuwei@xidian.edu.cn
[2] School of Computer Science & Technology, Xidian University, Xi'an 710071, China
qijj@mail.xidian.edu.cn

Abstract. Formal Concept Analysis(FCA) is an effective knowledge representation and discovery tool. Since FCA is proposed, lots of FCA prototype systems were designed and implemented. However, none of these prototype systems is relatively full-featured and strong scalability. In this paper, we propose a new FCA prototype system named XD-CKS, which is based on Eclipse Rich Client Platform(RCP) technology. The system integrates functions of file parsing, concept lattice building, concept lattice visualization, association rule mining and concept lattice applying. In order to support the expansion of system functions, all of the modules are developed based on Eclipse RCP plug-in technology and thus realize the Plug-and-play modules.

Keywords: Formal Concept Analysis, Concept Lattice, RCP, FCA System

1 Introduction

Formal Concept Analysis(FCA) is a mathematical representation theory about concept and concept hierarchy proposed by Wille in 1982 [1,2]. Since FCA was proposed, many domestic and overseas scholars make an intensive study of the basic theory and applications of FCA, they have achieved substantial results. Now FCA is crucial in knowledge representation and knowledge discovery, which has been successfully used in many fields, such as information retrieval, data mining, software engineering, semantic web, knowledge discovery and so on.

FCA prototype systems, mainly realize file parsing, concept lattices building and concept lattices visualization. And FCA also refers to association rule mining, ontology construction, data mining and information retrieval. FCA prototype systems are divided into two kinds, based on the basic theory and based on the applications.

The focal point of FCA prototype systems is concentrated on the research of basic theory, including the algorithms of concept lattices building and concept lattices visualization. The representative systems are as follows: Galicia [3] is a rich FCA integration system which contains all the key operations; FcaStone [4], a command line system, achieves the transformation of file and enhances

J.T. Yao et al. (Eds.): RSCTC 2012, LNAI 7413, pp. 357–362, 2012.

the interactional capabilities between FCA prototype system, graphic editing software and vector drawing software. Conexp-clj [5] is a general FCA system, which offers the basic operations and involves some fuzzy FCA theories. Lattice Miner [6] supports concept lattices building, concept lattices visualization and rule mining, It also integrates formal concept and association rule mining. OpenFCA [7],an integration FCA prototype system, furnishes formal context building, concept lattices visualization, attributes mining and so on.

The other kind of FCA prototype systems, based on the applications, is mainly used in image retrieval, network service, and data mining. Camelis [8] achieves the requirement of LIS (Logical Information Systems). It uses the FCA method to describe and search objects and implements information retrieval through the format of navigating. Camelis2 [9], based on Camelis, is a tool oriented towards semantic network retrieval. Camelis2 can import and scan RDFS (Resource Description Framework), which provides search, navigating, updating operations to correspond with the rich data models of RDFS.

In order to compensate for lack of functionality and poor extended performance in the existing FCA systems, this thesis proposes a new FCA prototype system, XDCKS, which integrates file parsing, concept lattices building, concept lattices visualization, association rule mining and concept lattices applications into one system. Through using the technology of Eclipse RCP plug-in, this new system can divide modules into separate plug-ins to achieve a "Plug-and-play" effect. The flexibility and extendibility is a great feature of XDCKS.

2 Basic Knowledge of Eclipse RCP

Eclipse is an integrated Development Environment developed by IBM, which works on the mechanism of plug-in. Adopting the technology of microkernel and the structure of extendible plug-in, all the plug-ins can run associated with others.

Eclipse RCP is provided by Eclipse for building rich client application. Fig.1 shows the structure of Eclipse RCP. The OSGi runtime provides the framework to run the modular application. Standard Widget Toolkit(SWT) is a library used by Eclipse and JFace and provides some convenient Application Programming Interface(API) on top of SWT. The workbench provides the framework for your application. The workbench is responsible for displaying all other User Interface(UI) components. For a headless Eclipse based applications (without UI), only the runtime is necessary.

3 Prototype System

3.1 System Functions

The purpose of XDCKS is to provide a rich functionality and good expansibility FCA prototype system, to achieve the main function of formal concept analysis with good performance in expansibility.

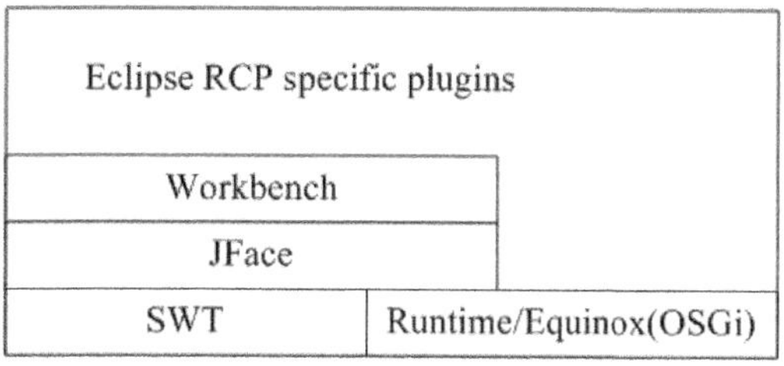

Fig. 1. The structure of RCP framework

In order to achieve this goal, the system should be designed to have the following main functions. First, file parsing function, which can resolve the formal context from different file formats; second, multi-valued background conversion function, which will transform a multi-valued background into a binary one for further research; third, concept lattice building function that can construct concept lattice through quick and effective algorithms; forth, graphic visualization function, which can show concept lattice through the graphic intuitive; fifth, some of the FCA functions expansion, such as ontology construction, mining of association rules, etc; and the last, it has good expansibility, which provides convenient to the following new functions for access to.

The system consists of file module, theoretical research module, application module and system configuration module. The file module is made up of file reading module and multi-valued background conversion module. Its main function is to convert different types of data read from local, database or interface into binary background or multi-valued background to other modules for later use. The theoretical module is composed by concept lattice building module, graphic visualization module and algorithm performance analysis module. This module's main function is to transform the binary background, in kernel, into the corresponding concept lattice, and at the same time show some of the information of algorithms, such as algorithm execution time and the concept of the algorithm to get. The application module includes two modules: association rules mining module and building ontology module. This module will realize FCA common applications. And the last one, system configuration module, mainly completes functions such as system function setting, algorithm configuration, module setting, etc.

3.2 Design of the Structure

The RCP plug-in technology is used in the development of the XDCKS, by which the system becomes a modular, dynamic management, Plug-and-play system.

There are three plug-in structures in this system– kernel level, module level and algorithm level. First, the kernel level is the foundation of the whole system, including the basic components of Eclipse RCP and run-time environment of the plug-in, also defining some elements involved in FCA, for example, object, attribute, concept and the partial order relationship between concepts, etc.

Second, the module level connects kernel level and algorithm level, provides the definition of extended point, and is the entrance of the algorithm level. Third, the algorithm level as an extension, through the implementation of extension point, realizes each function modules specifically. In this way, when new function modules or algorithms should be added, we can only extend the upper extension point without modifying the upper codes.

The Fig.2 shows the data exchange between system layers. The file reading module loads the data from local document, database or interface and analyzes it into formal context, and then, inputs the formal context into the kernel plug-in module. Theoretical research module takes formal context and lattice file as input and provide users operations such as concept lattice building, graphic visualization, algorithm performance analysis, etc. Application module has the same way as the theoretical research module to get and deal with the data, but it provides operations such as association rules mining, ontology building, etc. The system configuration module, through exchanging data with the kernel module, configures the system running state.

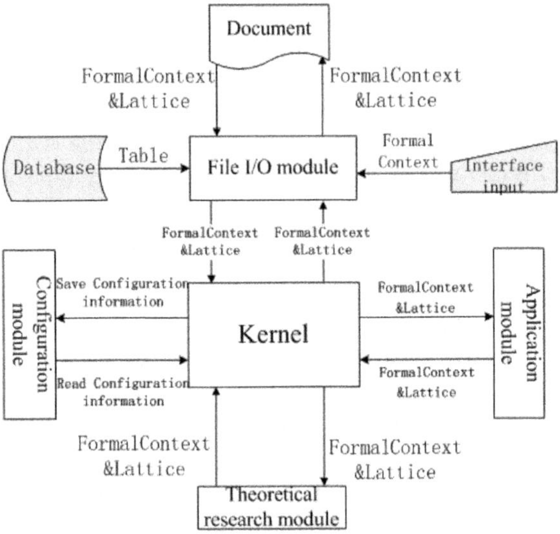

Fig. 2. Data exchange between system layers

3.3 Design of the Kernel

The kernel of XDCKS is the foundation of the entire system and the main function of the kernel includes defining the structure and detailed information of the plug-in, maintaining the plug-in extensions and extension points, loading modules dynamically according to the registered information and then generating a menu bar, a system bar, a view of the control module, etc. At the same time, some elements involved in FCA, e.g. object, attribute, concept, partial order relation between concepts and so on, are defined in the layer of kernel.

It can be seen from the Fig. 3 that classes and interfaces involved in FCA system can be divided into three parts. The basic element interface, IBasicElement, is used to output characters and judge conceptual elements. The basic collection interface, IBasicSet, is the most fundamental interface of collection, it inherits the collection interface and the copy interface for adding and removing individual element, supporting the operators of intersection, union and complement, judging the relationship between collections. The system is mainly related to six classes, they are FormalConcept, Intent, Extent, FormalContext, FormalAttribute and FormalObject. FormalConcept is used to represent concept, Intent represents the collection of attributes, Extent represents the collection of objects, FormalContext represents formal context, FormalAttribute represents the attribute of concept and FormalObject represents the object of concept.

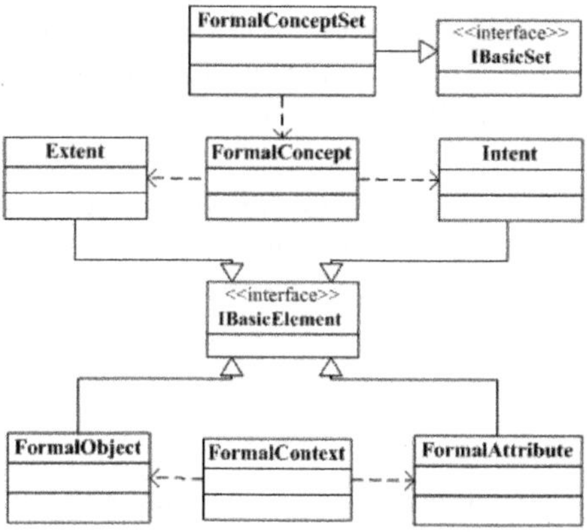

Fig. 3. Class diagram of the system

In order to represent the partial order relation between concepts, the system has defined classes and interfaces, they are IPartialOrder and Operator. IPartialOrder is used to represent partial order relation between concepts and Operator is used to capture the common attributes of an object collection or capture the common objects of an attribute collection so that we can judge whether the object collection and attribute collection form a concept.

4 Implementation of XDCKS

Based on plug-in mechanism and related classes defined in the kernel of FCA, we have designed and completed modules of FCA and implemented some common algorithms in the concept lattice building module and the association rule mining module.

XDCKS has implemented most of the main concept lattice building algorithms including FindNeighbors Algorithm, CloseByOne Algorithm, Chein Algorithm, ParallelRecursive Algorithm, Norris Algorithm, Bordat Algorithm, Lindig Algorithm, NextClosure Algorithm, Intersection Algorithm, ParallelRecurAlgoarray and Godin Algorithm.

XDCKS has implemented five main association rule mining algorithms, based on plug-in mechanism, they are Apriori Algorithm, AprioriLCS Algorithm, Mining Algorithm Based on Quantitative Concept Lattices, Apriori improved Algorithm and Apriori RuleMining Algorithm based on concept lattices.

5 Conclusions

In this paper, we have proposed a FCA prototype system XDCKS which is based on Eclipse RCP technology. The system integrates functions of file parsing, concept lattice building, concept lattice visualization, association rule mining and concept lattice applying. With the support of plug-in mechanism, we have implemented the main concept lattice building algorithms and association rule mining algorithms and analyzed the performance of the algorithms. In the future work, we will add new algorithm and make analysis.

Acknowledgments. This work was supported by grants from the National Natural Science Foundation of China (No. 60703117 and No. 11071281) and the Fundamental Research Funds for the Central Universities (No. JY10000903010 and No. K50510230005).

References

1. Ganter, B., Wille, R.: Formal Concept Analysis: Mathematical Foundations, pp. 25–36. Springer, Berlin (1999)
2. Wille, R.: Restructuring Lattice Theory: An approach based on hierarchies of concepts. Rivall. In: Ordered Sets, pp. 445–470. Reidel, Dordrecht (1982)
3. Valtchev, P., Grosser, D., Roume, C., Hacene, M.R.: Galicia: an open platform for lattices. In: de Moor, A., Ganter, B. (eds.) Using Conceptual Structures: Contributions to 11th Intl. Conference on Conceptual Structures (ICCS 2003), pp. 241–254. Shaker, Aachen (2003)
4. Priss, U.: FcaStone - FCA file format conversion and interoperability software. In: Conceptual Structures Tool Interoperability Workshop, CS-TIW 2008 (2008)
5. `http://www.math.tu-dresden.de/~borch/conexp-clj`
6. Lahcen, B., Kwuida, L.: Lattice Miner: A Tool for Concept Lattice Construction and Exploration. In: The 8th Int. Conf. on Formal Concept Analysis, Agadir, Marocco (2010)
7. Borza, P.V., Sabou, O., Sacarea, C.: OpenFCA, an open source formal concept analysis toolbox. In: IEEE International Conference on Automation Quality and Testing Robotics (AQTR), vol. 3, pp. 1–5, 28–30 (2010)
8. Ferre, S.: Camelis: a logical information system to organise and browse a collection of documents. International Journal of General Systems 38(4), 379–403 (2009)
9. `http://www.irisa.fr/LIS/ferre/camelis/camelis2.html`

An Incremental Approach for Updating Approximations Based on Set-Valued Ordered Information Systems

Chuan Luo[1], Tianrui Li[1], Hongmei Chen[1], and Dun Liu[2]

[1] School of Information Science and Technology, Southwest Jiaotong University, Chengdu, 610031, China
luochuan@my.swjtu.edu.cn, {trli,hmchen}@swjtu.edu.cn
[2] School of Economics and Management, Southwest Jiaotong University, Chengdu, 610031, China
newton83@163.com

Abstract. Incremental learning is an efficient technique for knowledge discovery in a dynamic database. Rough set theory is an important mathematical tool for data mining and knowledge discovery in information systems. The lower and upper approximations in the rough set theory may change while data in the information system evolves with time. In this paper, we focus on the incremental updating principle for computing approximations in set-valued ordered information systems. The approaches for updating approximations are proposed when the object set varies over time.

Keywords: Rough set theory, approximations, incremental learning, information systems.

1 Introduction

Granular Computing (GrC), a new concept for information processing based on Zadeh's "information granularity", is a term of theories, methodologies, techniques, and tools that make use of granules in the process of problem solving [1, 2]. With the development of artificial intelligence, the study on the theory of GrC has aroused the concern of more and more researchers. Up to now, GrC has been successfully applied to many branches of artificial intelligence. The basic notions and principles of GrC have appeared in many related fields, such as concept formation, data mining and knowledge discovery [3, 4]. Rough Set Theory (RST) is a powerful mathematical tool for dealing with inexact, uncertain or vague information [5]. It is also known as one of three primary models of GrC [6].

In real-life applications, data in information systems is generated and collected dynamically, which leads to knowledge discovered by RST needs updating [7]. The incremental technique is an effective method to update knowledge for dealing with the new added-in data set without re-implementing the original data

J.T. Yao et al. (Eds.): RSCTC 2012, LNAI 7413, pp. 363–369, 2012.

mining algorithm [8, 9]. A lot of works have been done towards the incremental learning techniques under RST. Many incremental updating approaches in RST have been developed for knowledge discovery [10–12]. For example, an efficient incremental RST based approach was presented to maintain knowledge dynamically [13]. The lower and upper approximations are the basic concepts of RST. Since the calculation of approximations is an indispensable step for knowledge representation and reduction in information systems, our focus here is to develop the incremental methods for updating approximations under the variation of the object set.

Set-valued information system is a generalization of a single-valued information system, which is always used to characterize the incomplete information, *i.e.*, the values of some attributes are unknown or multi-values [14, 15]. On the other hand, attributes in the set-valued information system sometimes are with preference-ordered domains. The ordering of attributes' values may play a crucial role. Then a so-called Set-valued Ordered Information System (SOIS) was introduced by Qian et al. to describe such situations [15]. For the problem of incremental updating approximations in SOIS, Chen et al. discussed the updating principle in the case of attribute values' coarsening and refining [16]. In this paper, we study the incremental approaches for computing approximations of SOIS while objects in the universe evolve over time.

The remainder of the paper is organized as follows. In Section 2, some basic concepts of SOIS are introduced. In Section 3, the principle of incremental updating approximations in SOIS when the object set varies with time is presented. In Section 4, we concludes the research work of this paper.

2 Preliminaries

For convenience, some basic concepts of rough sets and SOIS are reviewed in this section [11].

A set-valued information system is an ordered quadruple (U, AT, V, f), where $U = \{x_1, x_2, \ldots, x_n\}$ is a non-empty finite set of objects, called the universe. $AT = \{a_1, a_2, \ldots, a_l\}$ is a non-empty finite set of attributes. $AT = C \cup \{d\}$, where C is the set of condition attributes and d is a decision attribute with $C \cap \{d\} = \emptyset$; $V = V_C \cup V_d$, where V_C is the set of condition attributes' values and V_d is the set of decision attributes' values; f is a mapping from $U \times (C \cup \{d\})$ to V such that $f : U \times \{C\} \rightarrow 2^{V_c}$ is a set-valued mapping and $f : U \times \{d\} \rightarrow V_d$ is a single-valued mapping.

In a set-valued information system, if attribute values are ordered according to a decreasing or increasing preference, then the attribute is a criterion.

Definition 1. *A set-valued information system is called a SOIS if all the elements of condition attributes are criteria and the decision attribute is an overall preference.*

From [15], we know that the SOIS can be classified into conjunctive and disjunctive systems by the inclusion dominance relation and max-min dominance relation, respectively.

Definition 2. *Let (U, AT, V, f) be a conjunctive SOIS and $A \subseteq C$. The inclusion dominance relation in terms of A is defined as:*

$$R_A^{\wedge\geq} = \{(x, y) \in U \times U \mid f(y, a) \supseteq f(x, a), \forall a \in A\} \tag{1}$$

Definition 3. *Let (U, AT, V, f) be a disjunctive SOIS and $A \subseteq C$. The max-min dominance relation in terms of A is defined as:*

$$R_A^{\vee\geq} = \{(x, y) \in U \times U \mid max f(y, a) \geq min f(x, a), \forall a \in A\} \tag{2}$$

For convenience, we denote $R_A^{\Delta\geq}(\Delta\in \{\wedge, \vee\})$ as the dominance relation in SOIS, where $\wedge$ represents the conjunctive one and $\vee$ represents the disjunctive one. Furthermore, we denote the dominance class of an object x induced by the dominance relation $R_A^{\Delta\geq}(\Delta\in \{\wedge, \vee\})$ as: $[x]_A^{\Delta\geq} = \{y \in U \mid (x, y) \in R_A^{\Delta\geq}, \Delta\in \{\wedge, \vee\}\}$.

Let $D = \{D_1, D_2, \ldots, D_r\}$ be a finite number of classes which is a partition of U induced by the decision attribute d. We define $D_i^{\geq} = \bigcup_{i<j} D_j, 1 \leq i < j \leq r$. The statement $x \in D_i^{\geq}$ means "x belongs to at least class D_i".

The definitions of the lower and upper approximations of $D_i^{\geq}(i \leq r)$ with respect to the dominance relation $R_A^{\Delta\geq}(\Delta\in \{\wedge, \vee\})$ in SOIS are as follows:

Definition 4. *Let (U, AT, V, f) be a SOIS. $A \subseteq C$ and $D = \{D_1, D_2, \ldots, D_r\}$ is the decision classes induced by $\{d\}$. The lower and upper approximations of $D_i^{\geq}(i \leq r)$ with respect to the dominance relation $R_A^{\Delta\geq}(\Delta\in \{\wedge, \vee\})$ are defined respectively as follows.*

$$\underline{R_A^{\Delta\geq}}(D_i^{\geq}) = \{x \in U \mid [x]_A^{\Delta\geq} \subseteq D_i^{\geq}\}; \overline{R_A^{\Delta\geq}}(D_i^{\geq}) = \bigcup_{x\in D_i^{\geq}} [x]_A^{\Delta\geq}. \tag{3}$$

3 Approaches for Incremental Updating Approximations When Objects Varies with Time

In this section, we discuss the principle of incremental updating approximations based on SOIS. For convenience, we let $\underline{R_A^{\Delta\geq}}(D_i^{\geq})'$ and $\overline{R_A^{\Delta\geq}}(D_i^{\geq})'$ denote the lower and upper approximations, respectively, when the information system is updated.

3.1 Deletion of an Object

When an object x_i is deleted from the universe of SOIS, there are two cases may happen:

Case 1: $x_i \in D_i^{\geq}$.

Proposition 1. *Let (U, AT, V, f) be a SOIS. Then we have the following results for $\underline{R_A^{\Delta\geq}}(D_i^{\geq})'$:*

(1) If $x_i \in \underline{R_A^{\Delta\geq}}(D_i^{\geq})$*, then* $\underline{R_A^{\Delta\geq}}(D_i^{\geq})' = \underline{R_A^{\Delta\geq}}(D_i^{\geq}) - \{x_i\}$*;*
(2) Otherwise, $\underline{R_A^{\Delta\geq}}(D_i^{\geq})' = \underline{R_A^{\Delta\geq}}(D_i^{\geq})$.

Proposition 2. *Let* (U, AT, V, f) *be a SOIS. Then* $\overline{R_A^{\Delta\geq}}(D_i^{\geq})' = \overline{R_A^{\Delta\geq}}(D_i^{\geq}) - R \cup \{x_i\}$*, where* $R = \{x \in ((\overline{R_A^{\Delta\geq}}(D_i^{\geq}) - D_i^{\geq}) \cap [x_i]_A^{\Delta\geq}) \mid x \notin [x_j]_A^{\Delta\geq}, x_j \in D_i^{\geq} - \{x_i\}\}$.

Case2: $x_i \notin D_i^{\geq}$.

Proposition 3. *Let* (U, AT, V, f) *be a SOIS. Then we have the following results for* $\underline{R_A^{\Delta\geq}}(D_i^{\geq})'$*:*
(1) If $\overline{R_A^{\Delta\geq}}(D_i^{\geq}) - D_i^{\geq} \supseteq x_i$*, then*
(a) If $D_i^{\geq} \supseteq [x_j]_A^{\Delta\geq} - \{x_i\}$*,* $x_j \in D_i^{\geq} - \underline{R_A^{\Delta\geq}}(D_i^{\geq})$*, then* $\underline{R_A^{\Delta\geq}}(D_i^{\geq})' = \underline{R_A^{\Delta\geq}}(D_i^{\geq}) \cup \{x_j\}$*;*
(b) Otherwise, $\underline{R_A^{\Delta\geq}}(D_i^{\geq})' = \underline{R_A^{\Delta\geq}}(D_i^{\geq})$.
(2) Otherwise, $\underline{R_A^{\Delta\geq}}(D_i^{\geq})' = \underline{R_A^{\Delta\geq}}(D_i^{\geq})$.

Proposition 4. *Let* (U, AT, V, f) *be a SOIS. Then we have the following results for* $\overline{R_A^{\Delta\geq}}(D_i^{\geq})'$*:*
(1) If $x_i \in \overline{R_A^{\Delta\geq}}(D_i^{\geq})$*, then* $\overline{R_A^{\Delta\geq}}(D_i^{\geq})' = \overline{R_A^{\Delta\geq}}(D_i^{\geq}) - \{x_i\}$*;*
(2) Otherwise, $\overline{R_A^{\Delta\geq}}(D_i^{\geq})' = \overline{R_A^{\Delta\geq}}(D_i^{\geq})$.

3.2 Insertion of a New Object

When a new object x is inserted into the universe of SOIS, there are two cases may happen:

Case 1: $x \in D_i^{\geq}$, then $D_i^{\geq} = D_i^{\geq} \cup \{x\}$.

Proposition 5. *Let* (U, AT, V, f) *be a SOIS. Then we have the following results for* $\underline{R_A^{\Delta\geq}}(D_i^{\geq})'$*:*
(1) If $D_i^{\geq} \supseteq [x]_A^{\Delta\geq}$*, then* $\underline{R_A^{\Delta\geq}}(D_i^{\geq})' = \underline{R_A^{\Delta\geq}}(D_i^{\geq}) \cup \{x\}$*;*
(2) Otherwise, $\underline{R_A^{\Delta\geq}}(D_i^{\geq})' = \underline{R_A^{\Delta\geq}}(D_i^{\geq})$.

Proposition 6. *Let* (U, AT, V, f) *be a SOIS. Then* $\overline{R_A^{\Delta\geq}}(D_i^{\geq})' = \overline{R_A^{\Delta\geq}}(D_i^{\geq}) \cup [x]_A^{\Delta\geq}$.

Case 2: $x \notin D_i^{\geq}$.

Proposition 7. *Let* (U, AT, V, f) *be a SOIS. Then* $\underline{R_A^{\Delta\geq}}(D_i^{\geq})' = \underline{R_A^{\Delta\geq}}(D_i^{\geq}) - R$*, where* $R = \{y \in \underline{R_A^{\Delta\geq}}(D_i^{\geq}) \mid x \in [y]_A^{\Delta\geq'}\}$.

Proposition 8. *Let* (U, AT, V, f) *be a SOIS. Then* $\overline{R_A^{\Delta\geq}}(D_i^{\geq})' = \overline{R_A^{\Delta\geq}}(D_i^{\geq}) \cup R$*, where* $R = \{y \in D_i^{\geq} \mid x \in [y]_A^{\Delta\geq'}\}$.

4 An Illustrative Example

Let the SOIS $S = (U, AT, V, f)$ be given in Table 1, where $U = \{x_1, x_2, x_3, x_4, x_5, x_6, x_7, x_8, x_9\}$, $AT = C \cup \{d\}$, $C = \{a_1, a_2, a_3, a_4\}$, $V_C = \{e, f, g\}$, $V_d = \{G, M, P\} = \{Good, Medium, Poor\}$.

To demonstrate the validity of our method for updating approximations in SOIS, we consider the two cases as follows:

(1) The object x_9 is deleted from Table 1;
(2) The object x_{10} which shown in Table 2 is inserted into Table 1.

Table 1. A set-valued ordered decision information system

U	a_1	a_2	a_3	a_4	d
x_1	$\{e\}$	$\{e\}$	$\{f, g\}$	$\{f, g\}$	P
x_2	$\{e, f, g\}$	$\{e, f, g\}$	$\{f, g\}$	$\{e, f, g\}$	G
x_3	$\{e, g\}$	$\{e, f\}$	$\{f, g\}$	$\{f, g\}$	M
x_4	$\{e, f\}$	$\{e, g\}$	$\{f, g\}$	$\{f\}$	M
x_5	$\{f, g\}$	$\{f, g\}$	$\{f, g\}$	$\{f\}$	M
x_6	$\{f\}$	$\{f\}$	$\{e, f\}$	$\{e, f\}$	P
x_7	$\{e, f, g\}$	$\{e, f, g\}$	$\{e, g\}$	$\{e, f, g\}$	G
x_8	$\{e, f\}$	$\{f, g\}$	$\{e, f, g\}$	$\{e, g\}$	G
x_9	$\{f, g\}$	$\{g\}$	$\{f, g\}$	$\{f, g\}$	P

Table 2. The object inserted into Table 1

U	a_1	a_2	a_3	a_4	d
x_{10}	$\{e, f\}$	$\{e, g\}$	$\{f, g\}$	$\{e, f\}$	M

From Table 1, we have $D = \{D_1, D_2, D_3\}$, where $D_1 = \{x_2, x_7, x_8\}$, $D_2 = \{x_3, x_4, x_5\}$, $D_3 = \{x_1, x_6, x_9\}$. Here, we only consider the lower and upper approximations of $D_2^{\geq}$ to the validate our method. We have the following results:

(a) $D_2^{\geq} = \bigcup_{j \leq 2} D_j = D_1 \cup D_2 = \{x_2, x_3, x_4, x_5, x_7, x_8\}$;
(b) $[x_1]_C^{\Delta\geq} = \{x_1, x_2, x_3\}$, $[x_2]_C^{\Delta\geq} = \{x_2\}$, $[x_3]_C^{\Delta\geq} = \{x_2, x_3\}$, $[x_4]_C^{\Delta\geq} = \{x_2, x_4\}$, $[x_5]_C^{\Delta\geq} = \{x_2, x_5\}$, $[x_6]_C^{\Delta\geq} = \{x_6\}$, $[x_7]_C^{\Delta\geq} = \{x_7\}$, $[x_8]_C^{\Delta\geq} = \{x_8\}$, $[x_9]_C^{\Delta\geq} = \{x_2, x_9\}$.

Then, based on the definitions of the lower and upper approximations in SOIS, we have:

(a) $\underline{R_C^{\Delta\geq}}(D_2^{\geq}) = \{x_2, x_3, x_4, x_5, x_7, x_8\}$;
(b) $\overline{R_C^{\Delta\geq}}(D_2^{\geq}) = \{x_2, x_3, x_4, x_5, x_7, x_8\}$.

When the object x_9 is deleted from Table 1, the lower and upper approximations are updated as follows.

(a) From Table 1, it clear that the object x_9 satisfies: $x_9 \notin D_2^{\geq}$;
(b) Since $x_9 \notin \overline{R_C^{\Delta\geq}}(D_2^{\geq}) - D_2^{\geq} = \varnothing$, we have: $\underline{R_C^{\Delta\geq}}(D_2^{\geq})' = \underline{R_C^{\Delta\geq}}(D_2^{\geq})$;
(c) Since $x_9 \notin \overline{R_C^{\Delta\geq}}(D_2^{\geq})$, we have: $\overline{R_C^{\Delta\geq}}(D_2^{\geq})' = \overline{R_C^{\Delta\geq}}(D_2^{\geq})$.

When the object x_{10} in Table 2 is inserted into Table 1, the lower and upper approximations are updated as follows.

(a) From Table 1, it is clear that the object x_{10} satisfies: $x_{10} \in D_2$. Then we have $D_2^{\geq} = D_2^{\geq} \cup \{x_{10}\} = \{x_2, x_3, x_4, x_5, x_7, x_8, x_{10}\}$ and $[x_{10}]_C^{\Delta\geq} = \{x_2, x_{10}\}$;
(b) Since $D_2^{\geq} \supseteq [x_{10}]_C^{\Delta\geq}$, we have: $\underline{R_C^{\Delta\geq}}(D_2^{\geq})' = \underline{R_C^{\Delta\geq}}(D_2^{\geq}) \cup \{x_{10}\}$;
(c) $\overline{R_C^{\Delta\geq}}(D_2^{\geq})' = \overline{R_C^{\Delta\geq}}(D_2^{\geq}) \cup \{x_{10}\}$.

5 Conclusions

The incremental technique is an effective way to maintain knowledge in the dynamic environment. The SOIS is an important and common model of information systems. In this paper, we proposed the principle of incremental updating approximations based on SOIS when the objects in the information system vary with time. An example was given to illustrate the proposed method. Our future work will focus on the development of algorithms to validate the effectiveness of the proposed methods.

Acknowledgements. This work is supported by the National Science Foundation of China (Nos. 60873108, 61175047, 61100117), the Youth Social Science Foundation of the Chinese Education Commission (No. 11YJC630127), the Fundamental Research Funds for the Central Universities (SWJTU11ZT08, SWJTU12CX117, SWJTU12CX091) and the Scientific Research Fund of Yibin University (No. 2011Z15).

References

1. Zadeh, L.A.: Towards a Theory of Fuzzy Information Granulation and Its Centrality in Human Reasoning and Fuzzy Logic. Fuzzy Sets and Systems 90(2), 111–127 (1997)
2. Zadeh, L.A.: Fuzzy Logic=Computing with Words. IEEE Tran. On Fuzzy Systems 4(1), 103–111 (1996)
3. Yao, Y.Y., Zhong, N.: Potential Applications of Granular Computing in Knowledge Discovery and Data Mining. In: Proc. World Multiconference on Systemics Cybernetics and Informatics, pp. 573–580 (1999)
4. Yao, Y.Y.: Perspectives of Granular Computing. In: Proc. GrC, pp. 85–90 (2005)
5. Pawlak, Z.: Rough Sets. International Journal of Computer and Information Sciences 11, 341–356 (1982)
6. Yao, Y.Y.: Granular Computing: Basic Issues and Possible Solutions. In: Proc. 5th Joint Conference on Information Sciences, pp. 186–189 (2000)

7. Pedrycz, W., Weber, R.: Special Issue on Soft Computing for Dynamic Data Mining. Applied Soft Computing 8(4), 1281–1282 (2008)
8. Shan, N., Ziarko, W.: Data-based acquisition and incremental modification of classification rules. Computational Intelligence 11(2), 357–370 (1995)
9. Liu, D., Li, T., Ruan, D., Zou, W.: An incremental approach for inducing knowledge from dynamic information systems. Fundamenta Informaticae 94(2), 245–260 (2009)
10. Li, T.R., Ruan, D., Geert, W., et al.: A Rough Sets Based Characteristic Relation Approach for Dynamic Attribute Generalization in Data Mining. Knowledge-Based Systems 20, 485–494 (2007)
11. Li, T.R., Ruan, D., Song, J.: Dynamic Maintenance of Decision Rules with Rough Set under Characteristic Relation. Wireless Communications Networking and Mobile Computing, 3713–3716 (2007)
12. Chen, H.M., Li, T.R., Qiao, S.J., Ruan, D.: A Rough Set Based Dynamic Maintenance Approach for Approximations in Coarsening and Refining Attribute Values. International Journal of Intelligent Systems 25(10), 1005–1026 (2010)
13. Chen, Y.N., Tseng, T.L., Chen, C.C., Huang, C.C.: Rule Induction Based on an Incremental Rough Set. Expert Systems with Applications 36(9), 11439–11450 (2009)
14. Guan, Y., Wang, H.: Set-valued information systems. Information Sciences 176(17), 2507–2525 (2006)
15. Qian, Y.H., Dang, C.Y., Liang, J.Y., Tang, D.W.: Set-valued ordered information systems. Information Sciences 179(16), 2809–2832 (2009)
16. Chen, H.M., Li, T.R., Zhang, J.B.: A method for incremental updating approximations based on variable precision set-valued ordered information systems. In: Proc. GrC, pp. 96–101 (2010)

Comparative Analysis on Margin and Fuzzy Rough Sets Based Feature Selection

Hong Shi and Xiaoyun Zhang

Tianjin University, Tianjin 300072, P.R. China
serena@tju.edu.cn

Abstract. Feature selection methods obtain their optimal feature subsets by a strategy of weighting features according to their contribution to classification. Margin and rough sets are widely discussed in feature evaluation these years. However, no work has been contributed to compare their performance. In this paper, we introduce four feature weighting algorithms. WDL-MFD and FD-ranking, are designed based on fuzzy rough sets. They use fuzzy dependency as their feature evaluation criterion. While the other two, Simba and Relief, use margin to measure the significance of the features. In our work, we give a theoretical and empirical analysis and experimentally compare the two kinds of feature evaluation techniques, and demonstrate their detailed difference and connection. The experimental results show there is no significant difference among the performance obtained with different techniques.

Keywords: Feature selection, feature evaluation, fuzzy dependency, margin, feature weights.

1 Introduction

Feature selection plays an increasingly important role in pattern recognition and machine learning. Overfitting may occur if we train a model with a small sample described by lots of features. In fact, usually most of the features provide no useful information for predicting the classes of samples. Moreover, feature selection becomes particularly necessary in high-dimensional data analysis. Recently, Feature selection methods with gene expression data have obtained intensive research [11,12,13].

In essential, Feature selection is a preprocessing issue of building a feature selection criterion and finding a search strategy in pattern recognition and machine learning. Researchers have proposed many effective search algorithms, for example, the forward/backward greedy search algorithm, the branch-and-bound procedure, and the floating search methods, they are feasible with many practical applications. But the optimal feature subset cannot be found by each of them [3], yet it has been proved that exhaustive search to discover the optimal solution is a NP-hard problem. For these reasons, research works tend to feature weighting strategies to rank features by their weights. Feature weight derives directly from a feature evaluation criterion or be learned by optimizing a feature evaluation function, improvement of algorithm performance is expected.

J.T. Yao et al. (Eds.): RSCTC 2012, LNAI 7413, pp. 370–379, 2012.

Anyway, choosing a proper evaluation function for feature selection is the core issue. In our discussion, we concern on the feature selection criterion. Various of feature measures have been used or developed this years, such as distance measures [4], information measures [7], correlation measures, consistency measures, dependency measures, etc.

Margin is a geometric measure with a distance perspective to evaluate the confidence of a classifier to its decision, the simple and successful algorithm Relief weights a specified feature by the margin it induces. Simba is also a typical margin based method, which constructs an optimization objective function to optimize feature weights by maximizing the hypothesis margin.

With the successful application in handling the inconsistent problems based on rough set theory, dependency, as one of the effective measures in feature reduction has aroused widespread attention [8]. Dependency is defined as the ratio of positive region over the universe, where positive region is the sample sets that can be determinate classified. For further study to generalize with heterogenous data, numerical, fuzzy. Fuzzy dependency has been proposed with an extensive notion: the proportion of fuzzy consistent ones over the whole universe. As fuzzy dependency reflects the usefulness of features to decisions, then obviously, more useful of a feature, much bigger weight it possess. We will show detail strategies with FD-ranking and WDL-MFD algorithms in section 2.

In our work, we compare the margin principle-based and fuzzy dependency-based feature selection algorithms with sophisticated classifier CART and RBF-SVM, we analyze the variation of the trained feature weights by the different methods, as we all know, a bigger feature weight indicates a better discriminating power. With the learned weights, combining with filter and wrapper methods [2], we schedule our experiments, we rank the features in a descending order of the weights, based on $10 - fold$ cross-validation technique, we obtain the feature subset where the classification accuracy gets the biggest value. We compare and analyze which algorithm selects a subset that makes the best expression to decision as approximate as the original data does.

In the remaining parts of the paper, we will describe the principles of the four algorithms in detail with Section 2, followed by the experiment results on performance of these algorithms and comparative analysis in Section 3, finally, we give the conclusion in Section 4.

2 Algorithm Review and Discussion

In this section, margin based and fuzzy dependency based methods are separately discussed, first, we present two different ideas of building evaluation functions with margin principle, as called Relief and Simba. Similarly, two feature weighting methods with fuzzy dependency have been created in FD-ranking and WDL-MFD.

2.1 Margin Based Algorithm Relief and Simba

The main idea of Relief is to iteratively learn feature weights by their distinguishability between self class of a randomly picked sample and other classes. The margin expression $|x - NM(x)| - |x - NH(x)|$ reflects the confidence of a classifier to predict sample x. When a sample point is far away from the heterogeneous classes and much near to samples in the same class, the classification certainty gets high value, otherwise, low. Algorithm description of Relief is shown in Table 1.

Table 1. Algorithm Relief

```
Initiate the weight vector: w = 0;
for j = 1 : T                      /*T is the iterative number*/
    randomly choose a sample x;
    find the nearest miss NM(x) and nearest hit NH(x);
    for i = 1 : N                  /*N is the number of the features*/
        w_i = w_i + |x_i − NM(x)_i| − |x_i − NH(x)_i|;
    end
end
```

The novelty of the Simba algorithm [5] is that it extends the hypothesis-margin formula as follows:

$$\theta_S^w = \frac{1}{2}\left(\|x - NM(x)\|_w - \|x - NH(x)\|_w\right) \quad . \tag{1}$$

where S is a sample set, x is a point in it, near miss NM and near hit NH of x are both found in S. w is the weight vector on the feature set, $\|z\|_w = \frac{1}{\sum_i w_i}\sqrt{\sum_i (w_i z_i)^2}$. Note that, w takes a real value. Then the evaluation function is developed as:

$$e(w) = \sum_{x \in U} \theta_{U \backslash x}^w(x) \quad . \tag{2}$$

where U is the training set, other notations say definitions above.

Given a dataset, $e(w)$ can be regarded as a single variable function with weight w, then problem of finding the weight vector is changed into an optimization issue of maximizing $e(w)$, since $e(w)$ is continuously differentiable at almost everywhere, gradient ascent search strategy is used. Calculation formula of the gradient of $e(w)$ is:

$$(\nabla e(w))_i = \frac{1}{2}\sum_{x \in U}\left(\frac{(x_i - NM(x)_i)^2}{\|x - NM(x)\|_w} - \frac{(x_i - NH(x)_i)^2}{\|x - NH(x)\|_w}\right) w_i \quad . \tag{3}$$

In each iteration, every feature weight is updated by adding a variation Δ_i to the original weight with $w_i = w_i + \Delta_i$. Set the step size in the gradient ascent search to 1, then $\Delta_i = (\nabla e(w))_i$.

In algorithmic terms, Simba has the same computational complexity $O(TNM)$ with Relief, where T is the iteration number, N is the feature number, and M is sample number of the dataset. In Simba, evaluation function also updates in every step with the new learned weights, this is a superiority over Relief. Relief is a little blind to learn weights on separated features, ignoring whether or not there will be a performance improvement with the combined feature subset, things may be better with Simba, for the evaluation function with the optimization technique may produce a global optimal weight. Both of this two methods may yield with some redundant features.

2.2 Fuzzy Dependency Based Algorithm FD-Ranking and WDL-MFD

Pawlak rough sets generate fundamental granules with a rigid equivalence relation to handle classification-inconsistency problem this years, and show well performance particularly with nominal data, with high demand of numerical or fuzzy information processing, a set of fuzzy rough models with fuzzy similarity relations have been developed. At present, combining kernel methods with fuzzy rough sets is a hot topic, for more information, interested readers can refer to [10]. The two algorithms we will present are based on a generalized model of fuzzy rough sets [6] proposed by Yeung, Chen, et al.

$$\underline{R_S}X(x) = \inf_{y\in U} S(N(R(x,y)), X(y)); \quad \overline{R_T}X(x) = \sup_{y\in U} T(R(x,y), X(y)) \quad . \tag{4}$$

where R is a fuzzy equivalent relation defined on U. $R(x,y)$ is the membership of y to the fuzzy equivalent class of x, i.e. $[x]_R(y) = R(x,y)$. T is a triangular norm, and S is its dual. The two terms $\underline{R_S}X(x)\backslash\overline{R_T}X(x)$ represent the membership of x to the fuzzy lower approximation of X or to the fuzzy upper approximation of X.

Definition 1. *given a classification learning issue, k is $T-$ equivalence relation on U computed with Gaussian function $k(x,y)$ in feature space* $\mathrm{B} \subseteq \mathrm{A}$. *$U$ is divided into $\{d_1, d_2, \cdots, d_N\}$ with the decision attribute. The fuzzy dependency of* $\mathrm{D}(\mathrm{D} = \bigcup_{i=1}^{N} d_i)$ *on* B *is defined as*

$$r_B(D) = \frac{|\bigcup_{i=1}^{N} \underline{kd_i}|}{|U|} \quad . \tag{5}$$

where $|\bullet|$ is the cardinality of a subset, applying the membership function of fuzzy lower approximation in model (4), we get

$$r_B(D) = \frac{1}{n}\sum_{x\in U} 1 - \exp\left(-\frac{\|x - NM(x)\|^2}{\sigma}\right) \quad . \tag{6}$$

with kernel function $k(x,y)$ being the fuzzy similarity relations, the membership of a sample x to the fuzzy lower approximation of the same class can be interpreted as the distance of x to the nearest sample $NM(x)$ in other classes in kernel space, then, fuzzy dependency is the average distance on total samples.

Fuzzy dependency reflects the ratio of fuzzy classification consistent samples over the universe. It plays an increasingly important role in feature selection. Now we introduce two methods using fuzzy dependency to weight features. First, FD-ranking, we will give a brief illustration of the algorithm through figure 1 and compare it with Relief.

figure 1 describes a two-class problem in a 2-dimensional real space, suppose $'*'$ stands for the first class, and $'\circ'$ represent the second class. we use sample x to show how FD-ranking learns feature weights.

- first, choose one feature, for example, $f1$;
- find the nearest miss $NM(x)$ of sample x in the second class;

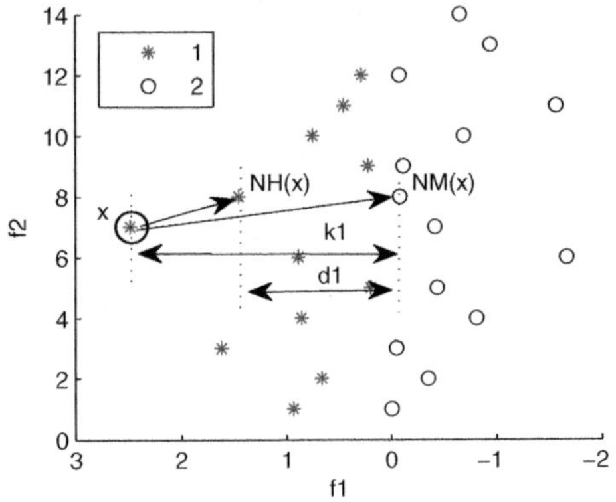

Fig. 1. diagram form for FD-ranking and Relief

- calculate $1 - k(x_1, NM(x)_1)$, this reflects how much does feature1 contribute to the fuzzy dependency, the value is associated to $k1$;
- loop on the whole samples, accumulate the value of the above expression, assign it to the weight of feature1 as w_1.
- the same way for the weight of feature2 w_2.

In weight learning mechanisms, FD-ranking and Relief are the same. But a small difference shown in figure 1: Relief find two samples: $NM(x)$ and $NH(x)$, directly use the horizontal distance $d1$ as the increment of w_1. In fact, the expression $1 - k(x_1, NM(x)_1)$ in FD-ranking can be interpreted as a distance function in kernel space.

Fuzzy Dependency-ranking calculates the fuzzy dependency of decision to a certain feature on the whole sample set, then directly uses this value as the feature's weight, and the next loop for the next feature, while Relief randomly picks one sample in each iteration, with this sample, we get all weights' increment for the next sample. in fact, FD-ranking and Relief have similar learning mechanisms. Also, the computational complexity of FD-ranking algorithm is $O(NM^2)$, equivalent to Relief.

Inspired by Simba, WDL-MFD, a weighted distance learning algorithm via maximizing fuzzy dependency between features and decision, proposed in previous work [9], is a desirable method. Gradient ascent search strategy is introduced into this algorithm for optimizing fuzzy dependency, with the optimized feature weight, an increase of fuzzy dependency has been shown on every experimental dataset.

In each step of the optimization procedure, $(\nabla e(w))_i$ is the ith component in gradient of fuzzy dependency $r_B(D)$ computed on one sample, $(\nabla e(w))_i =$

$$\frac{2}{\sigma}\exp\Big(-\frac{(\|x - NM(x)\|_w)^2}{\sigma \times (\sum_{i=1}^{N} w_i)^2}\Big)\times\left(\frac{(f(x, a_i) - f(NM(x), a_i))^2}{(\sum_{i=1}^{N} w_i)^2} w_i - \frac{(\|x - NM(x)\|_w)^2}{(\sum_{i=1}^{N} w_i)^3}\right) \quad (7)$$

σ is the kernel width, η is the step size in the gradient ascent search, we can easily get their values given a dataset.

Table 2. Algorithm WDL-MFD

```
Initiate the weight vector: w =< 1, 1, ···, 1 >;
compute σ = (1/n) Σ_{x∈U} ||x − NM(x)||;
for j = 1 : M                          /*M is the sample size*/
      for i = 1 : N                    /*N is the number of the features*/
            Δ_i = η (2/σ) (∇e(w))_i;
      end
w = w + Δ;
end
w_i = w_i / Σ w_i;
```

2.3 Comparisons of WDL-MFD and Simba

WDL-MFD, based on fuzzy dependency, has taken fuzzy lower approximation into consideration, without regarding the fuzzy upper approximation, that is to say, algorithms of this kind concern only the samples with fuzzy consistent classification. While Simba, based on margin theory, evaluate the goodness of a feature by comprehensive analysis of both the nearest miss $NM(x)$ and the nearest hit $NH(x)$.

The two methods are similar in terms of algorithmic mechanism, but Simba need to set a proper iterative number T, and randomly pick sample point for each iteration, while in WDL-MFD, iterative number is controlled directly by the number of samples. Every sample is involved in the procedure. There are two parameters in WDL-MFD, σ and η, but not difficult to compute. In computational complexity, WDL-MFD is $O(NM^2)$, equivalent with Simba if we set the iterative number T the same with sample number M.

3 Experimental Analysis

In this section, we will illustrate the behavior of the four algorithms on 11 datasets from UCI machine learning repository [1]. The datasets we prepared, wine and SRBCT are multiclass tasks, others are all binary problems. sick has a large sample size of 2800, while SRBCT is a high-dimensional data with 2308 features based on gene expression, the remaining are normal datasets.

We choose two well-performance classifiers CART and RBF-SVM as training algorithms and use classification accuracy as the metric to evaluate the performance of different algorithms.

First, we focus on two data sets: sonar and SRBCT, we take further observation of the learned feature weights of the four implementations. (We make a preprocess to normalize the max feature weight value to be 1 for facilitating observation.)

In Figure 2, there seems something in common with the four weight value curves on sonar, especially between algorithms WDL-MFD and Relief, which implies that there is no big difference on the feature estimation measures to learn weights. But in reality, the order of features are not the same with different algorithms, this will affects classification behavior on ranking based feature selection algorithms, we will show some details with two tables behind.

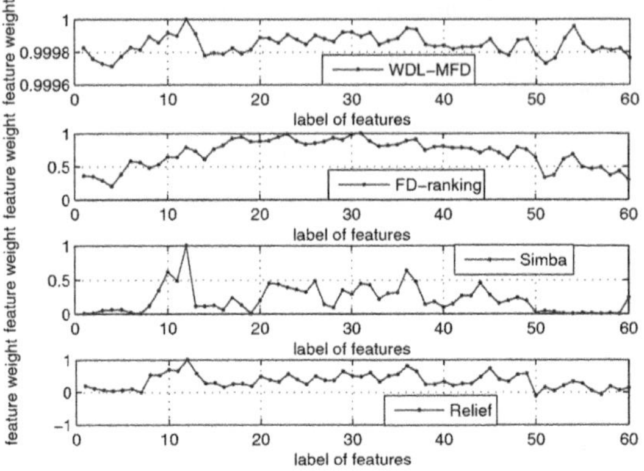

Fig. 2. Feature weights learned with the different four algorithms on sonar

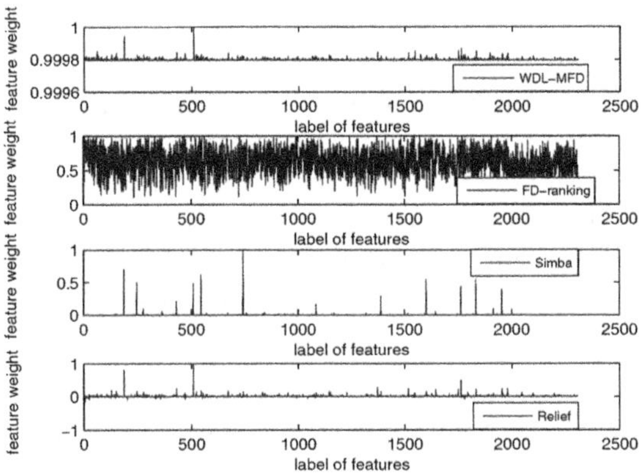

Fig. 3. Feature weights learned with the different four algorithms on SRBCT

Feature weights have a consistent variation with WDL-MFD and Relief in figure 3, only a small number of features gain big weight values, most of the values are sparse. We can see a clearly weight attribution with Simba methods, it performs better than others with SRBCT in our tests. In FD-ranking, feature weights wave a lot, it is odd that it obtains the best performance with CART, while makes little difference with SVM.

As fair as possible, we rank feature weights in a descending order, then add features to a subset one by one, a best feature, two best features, $\cdots$, k best features, at last, we design the classification model through $10 - fold$ cross validation with CART and RBF-SVM, we achieve the optimal number and subset of features where classification accuracy gets the maximum value.

From the two tables below, classification accuracies gain increase after reduction on almost every sample set, except for wine with SVM, classification accuracies have no changes, but features have been reduced more than a half. This indicates effectiveness of all algorithms. Big accuracy improvement on SRBCT has been shown with relatively few features being selected. On CART classifier, FD-ranking method wins Relief by 8 in 11, WDL-MFD equals with Simba and wins Relief, On RBF-SVM, WDL-MFD does better than FD-ranking in 7 datasets and equals in 2 databets, does the same as Simba, and slightly better than Relief. Relief wins FD-ranking this time.

Table 3. Comparison of the four algorithms by classification accuracy and the selected feature number with CART

Data	raw data		WDL-MFD		FD-ranking		Simba		Relief	
	Acc	*f*	*Acc*	*f*	*Acc*	*f*	*Acc*	*f*	*Acc*	*f*
heart	73.7±6.40	13	82.22±7.37	5	82.96±7.65	5	81.85±8.63	6	81.48±6.05	5
hepatitis	91.00±5.45	19	91.00±5.45	18	92.33±4.98	6	91.00±4.46	17	91.00±5.45	15
horse	95.63±3.19	22	96.73±1.73	17	95.90±2.99	20	95.63±3.23	13	96.73±1.73	15
iono	86.45±7.26	3	88.99±7.04	24	89.52±6.90	16	90.35±3.72	20	89.26±5.52	4
sick	98.36±1.24	29	98.43±0.81	21	98.46±1.23	27	98.39±1.22	29	98.43±0.81	25
sonar	71.12±12.82	60	76.93±13.23	26	74.02±9.61	51	77.45±7.67	22	78.86±9.03	26
spam	90.22±3.41	57	90.59±3.81	48	90.57±3.52	39	90.61±3.25	51	90.55±3.81	45
SRBCT	70.17±18.9	2308	94.67±8.64	15	96.00±8.43	301	95.00±11.25	126	94.67±8.64	6
WDBC	90.85±4.59	30	94.74±3.20	5	93.67±3.02	19	93.85±3.98	6	94.56±3.64	5
wine	89.31±6.68	13	95.56±4.38	6	92.15±5.95	4	92.08±4.81	4	92.08±4.81	6
WPBC	69.13±8.56	33	74.16±6.47	4	73.16±7.45	8	77.24±5.10	10	73.16±8.16	8

Table 4. Comparison of the four algorithms by classification accuracy and the selected feature number with RBF-SVM

Data	raw data		WDL-MFD		FD-ranking		Simba		Relief	
	Acc	*f*	*Acc*	*f*	*Acc*	*f*	*Acc*	*f*	*Acc*	*f*
heart	81.11±7.50	13	82.96±5.30	9	81.11±7.50	6	81.48±7.61	4	81.11±7.50	13
hepatitis	83.50±5.35	19	85.50±7.29	5	89.67±7.11	6	85.33±6.13	13	85.83±5.84	8
horse	72.30±3.62	22	88.84±4.19	5	86.71±5.26	3	91.04±5.22	4	88.84±4.19	5
iono	93.79±5.07	34	96.01±3.86	16	94.60±4.77	18	94.89±3.72	14	94.91±4.57	25
sick	93.82±0.24	29	93.93±0.29	22	93.89±0.11	1	93.89±0.11	1	93.89±0.11	1
sonar	85.10±9.48	60	88.45±5.62	50	87.45±5.75	38	88.40±8.93	27	88.95±5.98	57
spam	92.11±2.86	57	92.20±2.82	51	92.13±2.76	49	92.18±2.61	49	92.11±2.86	57
SRBCT	46.00±5.16	2308	81.50±14.83	1	51.00±14.99	2	91.67±14.16	2	81.50±14.83	1
WDBC	98.08±2.25	30	98.08±2.25	20	98.08±2.25	28	98.25±1.84	23	98.08±2.25	22
wine	98.89±2.34	13	98.89±2.34	11	98.89±2.34	9	98.89±2.34	12	98.89±2.34	9
WPBC	80.37±5.33	33	83.42±6.15	18	80.37±5.33	28	80.37±5.33	33	80.89±5.49	15

In order to observe a visualized variation of the classification accuracy performed by the four different algorithms, we illustrate the classification accuracy results on datasets iono and wine, shown in figure 4 and figure 5.

We easily find that classification accuracy doesn't increase at any time, in the beginning, it increases with the feature number and quickly climbing to a max value. With features continuing added, classification accuracy will not grow anymore or even decrease. This phenomenon is so called overfitting, a common problem in feature selection, especialy with high-dimensional data.

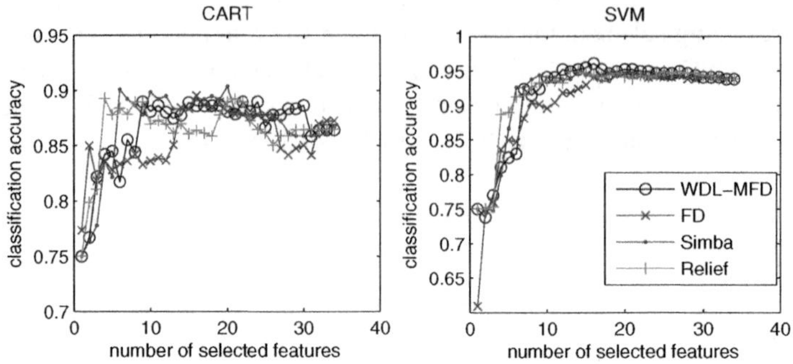

Fig. 4. Variation of classification performances with the number of selected features on iono

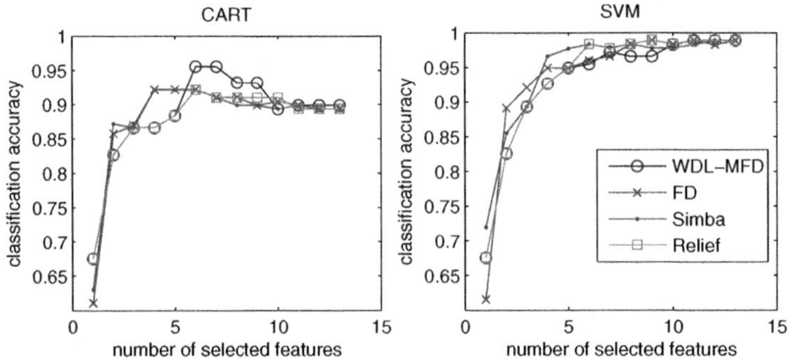

Fig. 5. Variation of classification performances with the number of selected features on wine

4 Conclusion

In this paper, we compare four feature weighting algorithms, two are margin based, Relief and Simba, and the other two are fuzzy dependency based, FD-ranking and WDL-MFD. We give concrete demonstration in organizing a feature evaluation function and feature weights learning strategies of all the algorithms. We arrange experiments on 11 data to test the performance of these algorithms using classification accuracy as the metric. We show the results in visualized figures. We get a conclusion that both measures based on margin and fuzzy dependency are effective and promising.

The essence of our work lies in weighing gains and losses with margin or fuzzy dependency based feature measures, finding what kind of mechanism or strategy should be combined with to achieve the optimal feature subset, and testing which result is the best approximation of the expression ability on the original data. Our further study will concentrates on eliminating redundant features in the selected subset and special learning problems.

References

1. Asuncion, A., Newman, D.: UCI machine learning repository (2007), http://www.ics.uci.edu/mlearn/MLRepository.html
2. Kohavi, R., John, G.H.: Wrappers for Feature Subset Selection. Artifical Intelligence 97(1-2), 273–324 (1997)
3. Guyon, I., Elisseeff, A.: An Introduction to Variable and Feature Selection. J. Machine Learning Research 3, 1157–1182 (2003)
4. Robnik-Šikonja, M., Kononenko, I.: Theoretical and empirical analysis of ReliefF and RReliefF. Machine Learning 53(1), 23–69 (2003)
5. Gilad-Bachrach, R., Navot, A., Tishby, N.: Margin based feature selection-theory and algorithms. In: Proceedings of the Twenty-first International Conference on Machine Learning. ACM, New York (2004)
6. Yeung, D., Chen, D., Tsang, E., Lee, J., Wang, X.Z.: On the generalization of fuzzy rough sets. IEEE Transactions on Fuzzy Systems 13(3), 343–361 (2005)
7. Hu, Q., Yu, D., Xie, Z.: Information-preserving hybrid data reduction based on fuzzy-rough techniques. Pattern Recognition Letters 27(5), 414–423 (2006)
8. Yao, Y., Zhao, Y.: Attribute reduction in decision-theoretic rough set models. Information Sciences 178(17), 3356–3373 (2008)
9. Hu, Q.H., et al.: Feature Selection via Maximizing Fuzzy Dependency. Fundamenta Informaticae 98, 167–181 (2010)
10. Hu, Q., Yu, D., Pedrycz, W., Chen, D.: Kernelized fuzzy rough sets and their applications. IEEE Transactions on Knowledge and Data Engineering 23(11), 1649–1667 (2011)
11. Sharma, A., Imoto, S., Miyano, S., Sharma, V.: Null space based feature selection method for gene expression data. Int. J. Mach. Learn. & Cyber. (2011), doi: 10.1007/s13042-011-0061-9
12. Boehm, O., Hardoon, D.R., Manevitz, L.M.: Classifying cognitive states of brain activity via one-class neural networks with feature selection by genetic algorithms. Int. J. Mach. Learn. & Cyber. 2(3), 125–134 (2011)
13. Tong, D.L., Mintram, R.: Genetic Algorithm-Neural Network (GANN): a study of neural network activation functions and depth of genetic algorithm search applied to feature selection. International Journal of Machine Learning and Cybernetics 1(1-4), 75–87 (2010)

Discovering Novel Knowledge Using Granule Mining

Bin Liu[1,2], Yuefeng Li[1], and Yu-Chu Tian[1]

[1] School of Electrical Engineering and Computer Science
Queensland University of Technology, Brisbane, QLD 4001, Australia
[2] Information Center, Xi'an Shiyou University, Xi'an, China, 710065

Abstract. This paper presents an extended granule mining based methodology, to effectively describe the relationships between granules not only by traditional support and confidence, but by *diversity* and *condition diversity* as well. *Diversity* measures how diverse of a granule associated with the other granules, it provides a kind of novel knowledge in databases. We also provide an algorithm to implement the proposed methodology. The experiments conducted to characterize a real network traffic data collection show that the proposed concepts and algorithm are promising.

Keywords: Granule mining, Rough set, Decision rule, Association Rule.

1 Introduction

Rough set theory describes decision rules by decision table, which compresses databases into granules and reveals the associations between granules [3,4]. The advantage of using decision rules is to reduce the two-steps of association mining into one process. However, it lacks accuracy and flexibility to deal with the associations between data granules in databases [2].

Granule mining [1,2] is a novel theory that interprets decision rules in terms of association rules. It formally describes the process of finding interesting granules, as well as the corresponding associations between granules in a database. Granule mining also proposes building association mappings for efficiently discovering the interesting association rules in different size granules.

In this paper, we further extend granule mining to describe the relationships between granules not only by traditional support and confidence, but by *diversity* and *condition diversity* as well. *Diversity* measures how diverse of a granule associated with the other granules, it provides a kind of novel knowledge in databases. *Condition diversity* is to extract rules to interpret a granules's *diversity*. We also present an algorithm for efficiently implementing the proposed concepts. We use the proposed methodology to describe the characteristics of network traffic, with a promising result.

The remainder of the paper is structured as follows. Section 2 presents some preliminary concepts. In section 3, we propose the new concepts of *diversity* and *condition diversity*. The algorithm is presented in Section 4. The experimental result is presented in Section 5. It is followed by conclusion in the last section.

J.T. Yao et al. (Eds.): RSCTC 2012, LNAI 7413, pp. 380–387, 2012.

2 Preliminaries

In this section, we introduce some preliminary concepts such as granule, decision table, decision rules, association mapping etc. [3,2].

2.1 Granule, Decision Table, Decision Rule

Formally, a transaction database can be described as an information table (T, V^T), where T is the set of transactions in which each transaction is a set of items, and $V^T = \{a_1, a_2, \ldots, a_n\}$ is a set of attributes for all transactions in T.

Let B be a subset of V^T. B determines a binary relation $I(B)$ on T. The family of all equivalence classes of $I(B)$, that is denoted by T/B, are referred to $B - granules$ [3]. The class in T/B induced by t is denoted by $B(t)$.

Definition 1. *Let $g = B(t)$ be a granule induced by t. Its covering set*

$$coverset(g) = \{t' | t' \in T, B(t') = B(t) = g\}$$

The *support* or *frequency* of granule g is the cardinality of $coverset(g)$ such that $sup(g) = |coverset(g)|$.

Definition 2. *Given two granules $g_1 \in T/B_1$, $g_2 \in T/B_2$, if $B_1 \subset B_2$ and $coverset(g_1) \supseteq coverset(g_2)$, we say g_1 is a generalized granule of g_2, and use $g_1 \succ g_2$ to denote the generalized relationship between g_1 and g_2.*

The tuple (T, V^T, C, D) is called a decision table if $C \cap D = \emptyset$ and $C \cup D \subseteq V^T$, where C is a set of condition attributes and D is a set of decision attributes. The granules determined by C,D and $C \cup D$ are called $C - granules$(condition granules), $D - granules$(decision granules), and basic $granules$, respectively.

A condition granule cg and a decision granule dg form a decision rule $cg \rightarrow dg$. Usually, a minimum support threshold min_sup and a minimum confidence threshold min_conf are specified to select the interesting decision rules.

2.2 Association Mapping

The relationships between condition granules and decision granules can be described as basic association mappings($BAMs$).

Definition 3. *Let (T, V^T, C, D) be a decision table, T/C, T/D and $T/(C \cup D)$ are $C - granules$, $D - granules$ and basic granules. For each condition granule $cg \in T/C$, its basic association mapping is*

$$\Gamma_{CD}(cg) = \{(dg, sup(cg \wedge dg)) | (cg \wedge dg) \in T/(C \cup D)\}$$

where $sup(cg \wedge dg)$ is the support of granule $cg \wedge dg$.

$\Gamma_{CD}(cg)$ includes all the decision granules that have relationships with the condition granule cg. Specially, $\Gamma_{CD}(cg) = \emptyset$ if $D = \emptyset$.

The *support* of cg can be calculated from its association mapping

$$sup(cg) = \sum_{(dg_i, sup) \in \Gamma_{CD}(cg)} sup(cg \wedge dg_i)$$

We can set a minimum support min_sup to select the significant condition granules.

For each $(dg, sup(cg \wedge dg)) \in \Gamma_{CD}(cg)$, cg and dg forms a decision rule $cg \rightarrow dg$, $sup(cg \rightarrow dg) = sup(cg \wedge dg)$ and $conf(cg \rightarrow dg) = \frac{sup(cg \wedge dg)}{sup(cg)}$. We can set a minimum confidence threshold min_conf to select the interesting rules.

2.3 Generalized Association Mapping

Let D_h be a subset of D, T/D_h is an set of decision granules $D_h - granules$. The relationships between $C - granules$, $D_h - granules$ can be represented by high level association mappings Γ_{CD_h}. Γ_{CD_h} can be derived from the basic association mapping Γ_{CD}. We call Γ_{CD_h} generalized association mappings of Γ_{CD}.

Lemma 1. *Let $\Gamma_{CD_h}(cg)$ and $\Gamma_{CD}(cg)$ are two association mappings, $cg \in T/C$, $D_h \subset D$. For each element $(d_h g, sup(cg \wedge d_h g)) \in \Gamma_{CD_h}(cg)$,*

$$sup(cg \wedge d_h g) = \sum_{(cg \wedge d_h g) \succ (cg \wedge dg_i)} sup(cg \wedge dg_i) \tag{1}$$

where $(dg_i, sup(cg \wedge dg_i)) \in \Gamma_{CD}(cg)$. From Γ_{CD_h}, we can generate a set of rules, which can reveal the knowledge of cg in a high level.

3 Diversity

In this section, we introduce the concept of *diversity* of a granule, we then present the concept of *diversity* and *condition diversity* of a rule.

3.1 Diversity of Granule

Definition 4. *Given a condition granule cg and its basic association mapping $\Gamma_{CD}(cg)$, where $cg \in T/C, C \cap D = \emptyset$. The diversity of condition granule cg is defined as the cardinality of set $\Gamma_{CD}(cg)$,*

$$divs(cg) = |\Gamma_{CD}(cg)|$$

The *diversity* measures how diverse is a condition granule connecting with its decision granules. The higher the *diversity* value, the more diverse a condition granule is, the more significant a condition granule is. Specially, $divs(cg) = 1$ if $\Gamma_{CD}(cg) = \emptyset$. In this situation, no granule has relationship with cg.

Note that we always calculate diversity value of a condition granule cg according to its basic association mapping $\Gamma_{CD}(cg)$ since the cardinality of $\Gamma_{CD_h}(cg)$ only reflect cg's diversity in the generalized tier. Similarly, considering that granule $cg \wedge d_h g$ is a condition granule, $(cg \wedge d_h g) \in T/(C \cup D_h)$, its diversity is $|\Gamma_{(C \cup D_h)(D-D_h)}(cg \wedge d_h g)|$. Specially, for a basic granule $(cg \wedge dg) \in T/(C \cup D)$, its diversity $divs(cg \wedge dg) = 1$ because $\Gamma_{(C \cup D)(\emptyset)}(cg \wedge dg) = \emptyset$. Generally, the diversity of a basic granule is 1 because no granule has relationship with a basic granule.

The diversity of a granule $(cg \wedge d_h g) \in T/(C \cup D_h)$ can be calculated according to the diversity of all its basic granules $(cg \wedge dg_i) \in T/(C \cup D)$.

Lemma 2. *Let $\Gamma_{CD_h}(cg)$ and $\Gamma_{CD}(cg)$ are two association mappings, $\Gamma_{CD}(cg)$ is a basic association mapping, $cg \in T/C$, $D_h \subset D$. For each element $(d_h g, sup(cg \wedge d_h g)) \in \Gamma_{CD_h}(cg)$,*

$$divs(cg \wedge d_h g) = |\Gamma_{(C \cup D_h)(D-D_h)}(cg \wedge d_h g)|$$
$$= \sum_{(cg \wedge d_h g) \succ (cg \wedge dg_i)} 1 = \sum_{(cg \wedge d_h g) \succ (cg \wedge dg_i)} divs(cg \wedge dg_i) \quad (2)$$

In Section 2.2 we set a minimum threshold *min_sup* to select the interesting granules. We can also set a minimum threshold *min_divs*, to select the significant condition granules. A condition granule is deemed significant if its *support* is larger than *min_sup* or its *diversity* is larger than *min_divs*.

3.2 Condition Diversity of Rule

Definition 5. *A condition granule cg and a decision granule dg form a rule $cg \rightarrow dg$. Its diversity is defined as*

$$divs(cg \rightarrow dg) = divs(cg \wedge dg).$$

Its condition diversity *is defined as*

$$cond\_divs(cg \rightarrow dg) = \frac{divs(cg \wedge dg)}{divs(cg)} \quad (3)$$

The $cond\_divs(cg \rightarrow dg)$ is a ratio of granule $(cg \wedge dg)$'s *diversity* and granule cg's *diversity*. The higher the *cond_divs*, the more likely the decision granule causes the condition granule behaving diverse. In Section 2.2, we set a *min_conf* to select the interesting rules. We also can set a minimum *condition diversity* threshold, *min_conf*, to select the rules that have strong contributions to the condition granule's diversity. In this context, minimum *confidence* and minimum *condition diversity* are set to the same value *min_conf*. A rule is an interesting rule if its *cond_divs* or *confidence* is larger than *min_conf*.

4 Algorithms for Granule Mining

In this section, we first introduce a new style association mapping to effectively describe both *support* and *diversity* relationships between granules. We then briefly introduce the algorithms for granule mining according to the new style association mapping and the theory described in the above sections.

4.1 New Style Association Mapping

The *support* and *diversity* relationships between condition granules and decision granules can be represented by a new style association mapping.

Basic Association Mapping. Let (T, V^T, C, D) be a decision table. For each condition granule $cg \in T/C$, its basic association mapping is

$$\Gamma_{CD}(cg) = \{(dg, sup(cg \wedge dg), divs(cg \wedge dg)) | (cg \wedge dg) \in T/(C \cup D)\}$$
$$= \{(dg, sup(cg \wedge dg), 1) | (cg \wedge dg) \in T/(C \cup D) \quad (4)$$

where $sup(cg \wedge dg)$ is the support of granule $cg \wedge dg$, $divs(cg \wedge dg)$ is the diversity of granule $cg \wedge dg$. The diversity of a basic granule $cg \wedge dg_i$ is 1.

Derived Generalized Association Mapping. From the Lemma 1 and Lemma 2, both the support and diversity of a generalized granule $cg \wedge d_h g$ can be calculated from its basic association association mapping. Hence, let $D_h \subset D$, we have

$$\Gamma_{CD_h}(cg) = \{(d_h g, sup(cg \wedge d_h g), divs(cg \wedge d_h g)) | (cg \wedge d_h g) \in T/(C \cup D_h)\}$$
$$= \{(dg, \sum sup(cg \wedge dg_i), \sum divs(cg \wedge dg_i)) | (cg \wedge dg_i) \in T/(C \cup D)\} \quad (5)$$

4.2 Algorithms for Granule Mining

Algorithm 1 outlines the major steps of our proposed method. The input is a set of *Data*. The output is a set of significant condition granules and their interesting rules $Out\_CgsRules$.

Algorithm 1. Granule mining: $NetGmine(Data)$

```
Input: A set of Data;
Output: A set of interesting condition granules and rules Out_CgsRules;
Define attributes set C, D and decision multi-level attributes tree DMAT;
Set threshold min_sup, min_divs, min_conf;
Out_CgsRules = null;
BAMs = GenBAMs(Data, C, D); /* Generate the basic association mappings;*/
foreach Γ_CD(cg) in BAMs do
    if sup(cg) ≥ min_sup or divs(cg) ≥ min_divs then
        /* Select the interesting rules in Γ_CD(cg) whose conf or cond_divs values*/
        /* are larger than min_conf and add them into CgRules; */
        CgRules = Prune_LessConfRules(Γ_CD(cg), min_conf);
        /* Recursively discover interesting rules according to Γ_CD(cg) and DMAT;*/
        CgRules = CgRules ∪ DecGM(Γ_CD(cg), DMAT.child);
        Out_CgsRules = Out_CgsRules + CgRules;
    end
end
Output Out_CgsRules;
```

In the algorithm, the decision multi-level attributes tree $DMAT$ is designed to recursively deriving generalized association mappings and interesting rules. $DMAT.child$ is a sub-tree of $DMAT$ and $DMAT.sibling$ is a sibling tree of $DMAT$. $DMAT.attributes$ represents a set of attribute names in that node. The attributes set of root node of $DMAT$ is set to D.

In the algorithm 1, function $GenBasicAM(Data, C, D)$ generates basic association mappings; function $Prune\_LessConfRules(\Gamma_{CD}(cg), min\_conf)$ extracts the interesting rules in the association mapping. We do not present the details of the two functions for saving the space. Function $DecGM(\Gamma_{CD}(cg), DMAT)$ recursively derives different level generalized association mappings and rules, the function is presented in algorithm 2.

Algorithm 2. Recursively deriving generalized association mappings and rules: $DecGM(\Gamma_{CD}(cg), DMAT)$

```
Input: Association Mapping Γ_CD(cg), Multi-level Attributes Tree DMAT;
Output: A set of rules CgRules ;
CgRules = null;
while DMAT ≠ null do
    D_h = DMAT.attributes; /*Get a set of high level decision attributes;*/
    Γ_CD_h(cg) = null; /*Initialize this level association mapping;*/
    foreach (dg, sup(cg ∧ dg), divs(cg ∧ dg)) in Γ_CD(cg) do
        d_h g = GenGranule(dg, D_h); /*Generate a high level decision granule;*/
        if (d_h g, sup(cg ∧ d_h g), divs(cg ∧ d_h g)) ∉ Γ_CD_h(cg) then
            Γ_CD_h(cg) = Γ_CD_h(cg) ∪ {(d_h g, sup(cg ∧ dg), divs(cg ∧ dg))};
        end
        else
            (d_h g, sup(cg ∧ d_h g), divs(cg ∧ d_h g)) =
            (d_h g, sup(cg ∧ d_h g) + sup(cg ∧ dg), divs(cg ∧ d_h g) + divs(cg ∧ dg)) ;
        end
    end
    CgRules = CgRules ∪ Prune_LessConfRules(Γ_CD_h(cg), min_conf);
    if DMAT.child ≠ null then
        /* Recursively generate next level association mappings and rules; */
        CgRules = CgRules ∪ DecGM(Γ_CD_h(cg), DMAT.child);
    end
    DMAT = DMAT.sibling; /*Move to the sibling node of the DMAT;*/
end
return CgRules;
```

5 Experiments

There are several purposes of the experiments. The first is to evaluate the effect of the thresholds to select the interesting condition granules and rules. Another is to conduct a result case study to show the effectiveness of the proposed method.

The experimental datasets we used are *MAWI* data traces, which can be downloaded from http://mawi.wide.ad.jp/mawi/samplepoint-B/20060303/. The datasets are the four largest 15-minute data files on 03/03/2006. Table 1 lists the characteristics of the datasets.

In the experiments, we select five attributes to represent the features of packets, which are source IP address(SrcIP), source port(SrcPrt), destination IP address (DestIP), destination port(DestPrt) and protocol(Prot). We designate *SrcIP* as condition attribute and the rest attributes as decision attributes, we try to discover significant hosts from the traffic data and to extract interesting rules for the significant hosts.

5.1 Number of Hosts and Rules

Table 2 and Table 3 list the number of significant hosts and interesting rules discovered by the proposed method. The 'Novel' columns represent the novel hosts or rules discovered after using the *diversity* measure. From the table we can see, a large number of novel hosts and novel rules are discovered. *Diversity* reveals a kind of novel knowledge.

Table 1. Characteristics of the selected datasets

ID	Time Captured	Packets Number	SrcIP Number
A	19:45-20:00	12,938,715	76,734
B	20:00-20:15	10,874,733	79,287
C	22:00-22:15	10,444,069	81,395
D	22:15-22:30	11,552,731	89,495

Table 2. Number of Hosts

	$min\_sup = 7000$	$min\_divs = 400$	Novel
A	152	147	103
B	161	138	94
C	174	150	111
D	168	159	118

Table 3. Number of Rules($min\_conf = 0.3$)

	$min\_sup = 7000$	$min\_divs = 400$	Novel
A	219	213	168
B	233	210	160
C	238	223	181
D	257	239	196

5.2 Case Study

Table 4 lists three results discovered from Dataset A. We briefly explain them to show their effectiveness to understand the behaviors of network traffic.

The first host and its rule discover a DoS attack, the host sent $5,606,840(support)$ packets to host 19.51.190.128's $64,996(diversity)$ ports. The second host is a scan host, it connects $58,816(diversity)$ hosts' port 1433, each packet for one host $(support/diversity)$. The third host and its rules show that rule that has high $confidence$ can has low *condition diversity*, and vice versa. $Diversity$ helps to understand the behaviors of network traffic.

Table 4. Some interesting results

ID	Condition	support	diversity	Decision	$conf$	$cond\_divs$
1	srcIP=215.35.248.109;	5,606,931	64,996	srcPrt=2893;prot=UDP; destIP=19.51.190.128;	1.00	1.00
2	srcIP=207.89.143.152;	58,816	58,816	srcPrt=6000;prot=tcp; destPort=1433;	1.00	1.00
3	srcIP=137.32.36.66;	70,662	12,555	prot=UDP;destPrt=53;	0.12	0.67
				srcPrt=25;prot=tcp;	0.84	0.29

6 Conclusion

This paper extends the theory of granule mining. It proposed to discover the knowledge in databases not only according to support and confidence, but also by diversity and condition diversity. An algorithm was also proposed to efficiently implement the proposed methodology. Experiments performed in real network traffic have shown that granule mining provides a promising methodology for knowledge discovery in databases.

References

1. Li, Y., Zhong, N.: Interpretations of association rules by granular computing. In: IEEE International Conference on Data Mining, Melbourne, Florida, USA, pp. 593–596 (2003)

2. Li, Y., Yang, W., Xu, Y.: Multi-tier granule mining for representations of multidimensional association rules. In: Proceedings of 6th IEEE International Conference on Data Mining, pp. 953–958 (2006)
3. Pawlak, Z.: Rough sets and intelligent data analysis. Information Sciences 147(1-4), 1–12 (2002)
4. Yao, Y.Y., Zhao, Y., Maguire, R.B.: Maguire, R.B.: Explanation Oriented Association Mining Using Rough Set Theory. In: Wang, G., Liu, Q., Yao, Y., Skowron, A. (eds.) RSFDGrC 2003. LNCS (LNAI), vol. 2639, pp. 165–172. Springer, Heidelberg (2003)

Feature Weighting Algorithm Based on Margin and Linear Programming

Wei Pan, Peijun Ma, and Xiaohong Su

School of Computer Science and Technology
Harbin Institute of Technology, Harbin 150001, China
panwei@hit.edu.cn

Abstract. Feature selection is an important task in machine learning. In this work, we design a robust algorithm for optimal feature subset selection. We present a global optimization technique for feature weighting. Margin induced loss functions are introduced to evaluate features, and we employs linear programming to search the optimal solution. The derived weights are combined with the nearest neighbor rule. The proposed technique is tested on UCI data sets. Compared with Simba and LMFW, the proposed technique is effective and efficient.

1 Introduction

Feature selection plays an important role in machine learning and pattern recognition for reducing store space and computational complexity [1]. In recent years, it is widely applied in the domains such as image categorization [2], character recognition [3] and gene classification[4].

Generally speaking ,the feature selection method can be divided into the filter mode [5]and wrapper mode[6] depending on constructed mode. In the wrapper mode, the feature selection method evaluates the candidate features with a classification technique. However, the filter mode compute the quality of features with an independent functions, including distances [1,7] and the mutual information [8,9,10], and so on.

Margin is widely used to evaluate feature quality in the last decade. Margin can be understood as a generalized distance measure between different classes. We can get good classification models through maximizing margin and minimizing margin induced classification loss. As so far, there have been several feature selection methods based on the margin, such as G-filp [1], Simba[1] [11], Relief[12], and E-Relief[13]. G-filp minimizes margin-loss to maximize the margin with greedy search. It can calculate both continuous and discrete loss functions. Relief and its extended algorithms compute the margin in the feature space and use the margin as the weights of features. Simba makes some improvements to Relief. It calculates the weight of each feature with a gradient descent method, and then iteratively updates weights by minimizing the loss. All these methods evaluate features through Hypothesis-margin [1] which is computed with the distance from a random sample point x in sample set to a different class of sample

J.T. Yao et al. (Eds.): RSCTC 2012, LNAI 7413, pp. 388–396, 2012.

point from x nearest to classification surface and the distance from x to the same class of sample point as x nearest to classification surface.

However, there are two problems with the above algorithms. First, they construct classification loss functions only by using the distance among different class of sample points which near the decision boundary. Besides, only the enumeration method can obtain the optimal solution to a nonconvex function. Both the gradient descent algorithm and the iteration process are approximate solutions.

In order to solve these problems, Li proposed a feature selection algorithm based on the nearest neighbor classification loss margin [11]in 2009, which used Euclidean distance to calculate distances between samples. First, the algorithm divides samples into several neighborhood. Then, it computes the classification loss by using the distances between sample points within the same classes and out of classes. This algorithm also calculates feature weights with a gradient descent algorithm. Later, Weinberger proposed a new evaluation strategy, and design the LMNN [14] algorithm. It constructs classification loss by using the same idea as Li's [11], but there are three pieces of differences. First, it uses Mahalanobis distance, instead of Euclidean distance, when calculating the margin. Second, it uses a method named SDP to calculate the covariance matrix in Mahalanobis distance in acquiring the feature weights. Third, it removes the sample that doesn't obtain losses for reducing time complexity. Besides, Chen used an expression by adding squares of sample distances with weight factors , which is solved from the covariance matrix in Mahalanobis distance [15], and also uses it to reconstruct classification loss functions used in [11,14], then calculate feature weighting through linear programming.

Based on the work [14,15], the performance of nearest neighbor classification is improved. But some drawbacks still exist. First they use hinge loss function, which is sensitive to noises. Here we give a more effective expression of nearest neighbor classification loss through geometric analysis to margin loss and use soft margin Strategy, then we use this expression to reconstruction classification loss functions in [15], and transfer the classification loss function to a linear programming issue. Then we calculate the optimal feature weights. Some experiments are presented to compare the proposed technique with Simba and LMFW [15].

2 Margin-Loss for Nearest Neighbor Classification

The initial idea of LMFW algorithm is to minimize the distance between sample points in the same class and maximize the margins between different classes. In order to describe LMFW algorithm, we review the classification loss function.

In the supervised learning, assume there're N samples in training set S. $\{x_i, y_i\}_{i=1}^{N} \in \mathrm{R}^N$; y_i is the class label of x_i. In S, $y_i \in \{1, 2, \ldots, \ell\}$, so we can express the M features of each sample x_i as $x_i=\{x_{i1}, x_{i2}, \ldots, x_{iM}\}$, and use the element $\tau_{ij}=\{0,1\}$ in matrix Γ. If x_i and x_j belong to the same class and τ_{ij}=1; otherwise, τ_{ij}=0.

Definition 1. The objective neighborhood which contains random sample x_i and the samples with the same class label as x_i.

So we can use a matrix $\boldsymbol{C}$ which defined as objective neighborhood matrix to mark which samples are belonging to the same objective neighborhood. $c_{ij} \in \{0,1\}$ $(i,j=1,2,\ldots,N)$are elements in C and if x_i and x_j belong to the same objective neighborhood then $c_{ij}=1$, otherwise $c_{ij}=0$.

Definition 2. Assume S is a training set, and x_i and x_j are samples in S, and w is a weight vector of the feature. the distance from x_i to x_j is computed with

$$\boldsymbol{D}^2 = (x_i - y_j)^T \boldsymbol{Q}(x_i - y_j) = (x_i - y_j)^T \boldsymbol{W}^T \cdot \boldsymbol{W}(x_i - y_j) = \sum_{k=1}^{M} g_k^2 (x_{ik} - x_{jk})^2. \quad (1)$$

In the formula (1), $\boldsymbol{Q}$ is the covariance matrix of x_i and x_j, and $\boldsymbol{Q} = \boldsymbol{W}^T \cdot \boldsymbol{W}$, $\boldsymbol{W}$ is an diagonal matrix, $\boldsymbol{T}$ is transposition, $W_{kk}=g_k$ $(k=1,2,\ldots,M)$. g_k is a weighting factor.

By substituting $g_k^2 = w_k$, we can get the objective function of LMFW.

$$\varepsilon(\mathbf{w}) = \sum_{ij}^{N} c_{ij} \sum_{k=1}^{M} w_k (x_{ik} - x_{jk})^2 + \kappa \sum_{ijp}^{N} c_{ij} (1 - \tau_{ip}) \gamma_{ijp} \quad (2)$$

$$\gamma_{ijp} = \hbar(1 + \sum_{k=1}^{M} w_k [(x_{ik} - x_{jk})^2 - (x_{ik} - x_{pk})^2]) \quad (3)$$

In the objective function, constant $\kappa > 0$ is used to balance weights of different terms. And c_{ij} is an element of $\boldsymbol{C}$, τ_{ij} is defined as an element of Γ, also we define γ_{ijp} as margin loss in formula (2), $\hbar(z) = max(z,0)$ is the standard Hinge loss function in formula (3).

If we minimize $\varepsilon(\boldsymbol{w})$, so that in (2), the first term minimizes difference samples in objective neighborhood using the sum of squares with weighting based on neighborhood, and the second term maximizes losses in the objective neighborhoods.

3 Soft-Margin-Loss Evaluate and Linear Programming

As the hinge loss is sensitive to noises, it cannot give the precise estimation of classification confidence if the raw data are mixed up with noises. So soft-margin-loss is introduced in support vector machine. Here we introduce soft-margin into feature evaluation.

3.1 Evaluating Features with Soft-Margin-Loss

Hypothesis-margin is computed as the distance difference of a sample point x to its x (nearmiss) and its x (nearhit). In Fig 1., we select 3 neighboring points as an

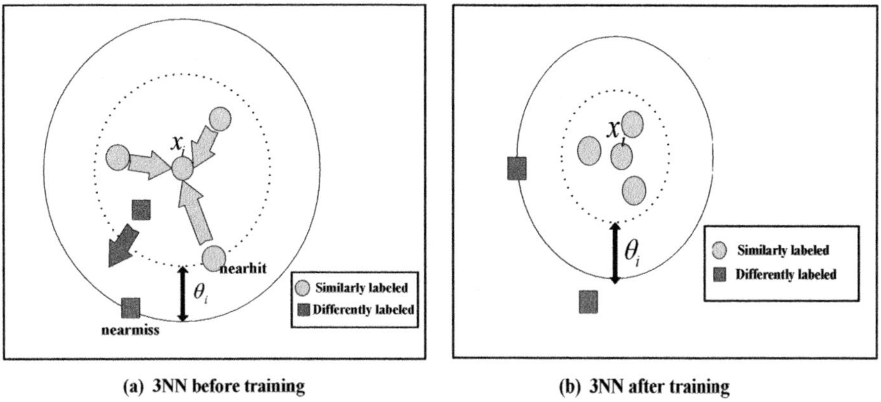

Fig. 1. Comparison of 3NN: illustration of one input's neighborhood before training (left) versus after training (right)

example to analyze the geometric meaning of $\hbar$ which stands for the soft-margin-loss. Before training, if there is no samples with different class label appearing in the objective neighborhood of x_i, $\hbar$ means that the distance from sample with different class label to x_i should not exceed the sum between Hypothesis-margin θ_i and the distance from x_i to points in the objective neighborhood. Unfortunately, it will cause deviation to classification when samples with different class labels appear in the objective neighborhood of x_i. But we can calculate the nearmiss and nearhit of x_i to avoid the deviations. If the distance between x_i and its nearmiss is less than the distance between and its nearhit, there are some samples with different class labels appearing in the objective neighborhood of x_i in this case. Then, we can eliminate these samples and calculate from the rest samples for obtaining a more precise θ_i. After training, as we can see in subgraph (2) of Fig.1, the objective neighborhood of the 3 neighboring points is narrow and samples with different class label are pushed away to the location whose distance to the objective neighborhood is θ_i.

Definition 3. Assume S is a training set and $x_i\{i=1,2,\ldots,N\}$ is a sample in S. So the classification loss function of x_i is

$$\varepsilon(x_i) = \sum_{j}^{N} c_{ij} \sum_{k=1}^{M} w_k (x_{ik} - x_{jk})^2 + \mu \sum_{jp}^{N} c_{ij}(1 - \tau_{ip})\xi_{ijp} \tag{4}$$

$$\xi_{ijp} = \hbar(\theta_i + \sum_{k=1}^{M} w_k[(x_{ik} - x_{jk})^2 - (x_{ik} - x_{pk})^2]) \tag{5}$$

In the above function, constant μ is a balancing weight. c_{ij} is the element of C and τ_{ip} is defined as an element of τ. The difference is that we define ξ as soft-margin-loss and $\hbar(z)=max(z,0)$ as hinge loss based on soft-margin-loss. Furthermore, θ_i

is calculated as follows: θ_i=$|(x_i - nearmiss(x_i))^2$-$(x_i - nearhit(x_i)^2)|$, where $nearhit(x_i)$ is the point with the same class label of x_i and $nearmiss(x_i)$ is the point with the different class label of x_i. both of which are nearest to x_i and can be easily obtained from C and τ.

Definition 4. Assume there are N sample points in the training set $\boldsymbol{S}$, the classification loss function is defined as

$$\varepsilon(\boldsymbol{S}) = \sum_{i=1}^{N} \varepsilon(x_i) \tag{6}$$

$$\varepsilon(\boldsymbol{S}) = \sum_{ij}^{N} c_{ij} \sum_{k=1}^{M} w_k (x_{ik} - x_{jk})^2 + \nu \sum_{ijp}^{N} c_{ij}(1 - \tau_{ip})\xi_{ijp} \tag{7}$$

The classification loss function above contains two terms. The first is used to punish the large distance between samples and its objective neighborhood, which suggests that samples in the same neighborhood should be closer after training. The second is used to punish small distance between samples and samples with different classification labels, which suggests sample points with different classifications appearing in the neighborhood should not exceed the sum θ_i and distances between sample points in the objective neighborhood.

3.2 Linear Programming Based on Soft-Margin-Loss

According to Formula (7), we can solve it by transferring the loss function to a linear optimization problem with constraint conditions. Now we give the linear programming model of the problem as following.

$$min \sum_{ij}^{N} c_{ij} \sum_{k=1}^{M} w_k (x_{ik} - x_{jk})^2 + \nu_1 \sum_{ijp}^{N} c_{ij}(1 - \tau_{ip})\xi_{ijp} + \nu_2 \sum_{k=1}^{M} w_k \tag{8}$$

$$\textbf{e.t.} \sum_{k=1}^{M} w_k\{(x_{ik} - x_{pk})^2 - (x_{ik} - x_{pk})^2\} \geq \theta_i - \xi_{ijp}, \xi_{ijp} \geq 0, w_k \geq 0. \tag{9}$$

In Formula(8), ν_1 and ν_2 are positive constants used to control importance levels of three terms and can be calculated through cross validation. ξ_{ijp} is a loose variable. The first term in the function is the sum of sample distance in the neighborhood; the second is the margin-loss; the last is the nonnegative constraint.

Linear programming can efficiently solve convex optimization problems with a large scale of variables and constraint conditions. And the computational complexity of linear programming models is M+$k{\cdot}N^2$, where M is the feature dimensions, k is the number of neighboring samples with different class labels and N is the number of training samples. In fact, most of loose variables ξ_{ijp} can not get positive values so that the hinge loss is zero. In addition, loose variables ξ_{ijp}

are sparse in linear programming. So we can significantly improve the speed. Finally, it is notable that we only have to compute those sample points which have different classification tags appearing in objective neighborhoods. We can reduce classification errors of these samples after the feature weights are optimized.

4 Experimental Analysis

We give a series of numerical experiments to compare the size of the selected features, error rates before and after feature selection and running time of the algorithms in order to check their efficiency. Eight datasets are downloaded from UCI machine learning database [16] for testing Simba [1], LMFW [15].

The detailed information about data is listed in Table 1. In the following table, there are some missing attribute values in Heart, Autos, Soybean, Spam data sets which are replaced by average attribute values counted by values in the same column as missing attribute values. Besides, we normalize the data before experiments because of different dimensions of features. And in the experiment, we use a linear programming toolbox in MATLAB to realize the MLLP and LMFW algorithms, and our experiment platform includes a Window XP SP3 system, an AMD x4640 3.0G CPU, MATLAB2010, and memory size is 2G.

Table 1. Description of datasets

DataSet	Samples	Attributes	Missing	Classes
Wdbc	569	30	no	2
Wine	178	13	no	3
Soybean	683	34	yes	19
Iono	351	34	no	2
Heart	303	13	yes	5
Autos	204	23	yes	6
Spam	4601	57	yes	2
Mfeat	2000	649	no	10

We divide each data set into training set and test set, and then compute feature weights using Simba, LMFW, MLLP with the training set and also validate the selected features based on KNN classifier and ten-fold cross validation. We get the classification error rate of KNN.

Table 2 shows the size of the selected features and the MLLP feature subset selected in the descending order in terms of weights acquired by MLLP, Simba and LMFW. ν_1 and ν_2 take values in $\{0.0001,\ 0.001,\ 0.01,\ 0.1,\ 0.5,\ 1,\ 5,\ 10,\ 100,\ 1000,\ 10000\}$.

Besides, we give the classification error rates of KNN(k=1) in Table 3. the performance of the raw data sets is listed in the first column of Table 3. Table 4 shows the CPU's running time of the algorithms which are averaged by 100 times' testing for each data set.

Table 2. Comparison of feature reduction rate(%)

DataSet	Attributes	Simba	LMFW	MLLP
Wdbc	30	26.67	10	73.33
Wine	13	53.85	30.77	30.77
Soybean	34	5.88	58.2	55.88
Iono	34	35.29	76.47	88.24
Heart	13	7.69	15.38	23.08
Autos	23	82.61	73.91	91.3
Spam	57	14.04	61.4	61.4
Mfeat	649	78.27	83.98	86.13
Average	–	38.04	51.34	63.77

Now we conclude the superiority compared with the other two algorithms. First in Table 2, MLLP can select more effective feature subsets. The feature dimensions decrease 63.77%, while the other two decrease 38.04% and 51.34%. So the MLLP's reduction rate of feature dimensions increase 12.5% and 25.7%, respectively. Observing the results in Table 3, we can find that the MLLP's classification error rate decrease between 1% and 2.3%, which is not a significant improvement comparing with the other two algorithms. Finally, in Table 4, it is obvious that the running time of MLLP and LMFW is much less than Simba.

Table 3. Comparison of classification error rate(%)

DataSet	Before Feature Selecting	Simba	LMFW	MLLP
Wdbc	4.56	3.69	3.87	3.68
Wine	5.14	1.74	2.78	1.67
Soybean	8.92	8.92	5.6	5.17
Iono	13.6	9.95	7.9	7.38
Heart	24.44	23.33	22.22	19.26
Autos	39.38	35.2	29.28	28.01
Spam	11.43	11.56	12.08	11.39
Mfeat	1.9	2.75	2.3	1.62
Average	13.67	12.14	10.75	9.77

Figure 2 shows the performance curves when the selected features are added one by one. We can see that all the features chosen by the three algorithms can produce good performance. The features producing large weights can also obtain good classification accuracy Also the 3 algorithms are able to catch important features as we can see in Fig 2., which the classification precision enhance first, and then it may decrease or be maintained according to the increasing attribute values. But it is easily to find the superior of the algorithm we promote because of using less number of features to acquire the same high level of classification precision as the other two.

Table 4. Comparison of Running Time(Seconds)

DataSet	Simba	LMFW	MLLP
Wdbc	6.453	0.138	0.14
Wine	0.625	0.094	0.172
Soybean	10.594	0.437	0.297
Iono	2.391	0.063	0.109
Heart	1.344	0.11	0.094
Autos	0.891	0.187	0.531
Spam	440.236	2.031	5.011
Mfeat	325.782	0.475	0.312

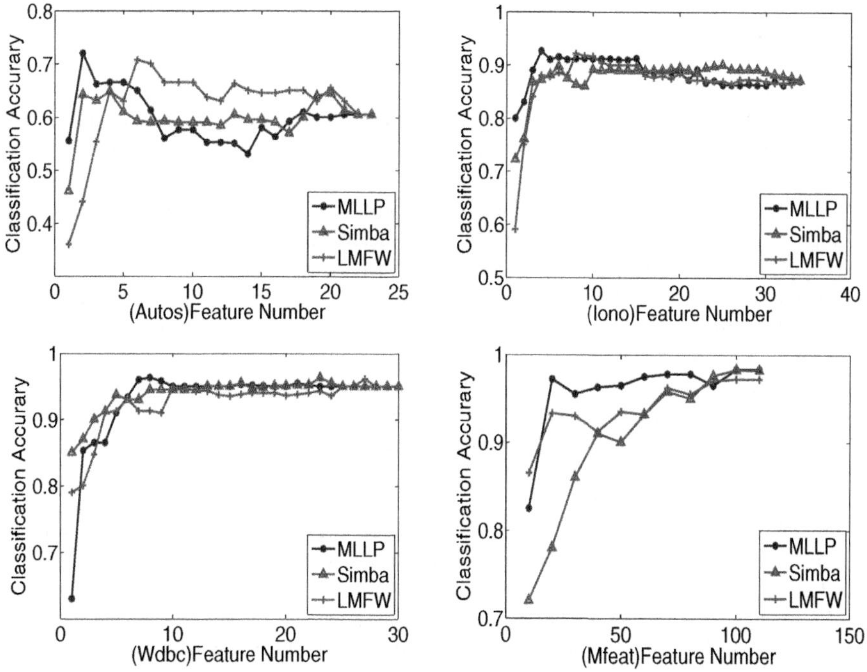

Fig. 2. Comparative of 1NN Classification accuracy

5 Conclusion

We give a method to calculate feature weights based on margin loss and linear programming in this work. Tested with the numerical experiments, the proposed MLLP algorithm exhibit some superiority including high efficiency and classification accuracies and large reduction rate.

Acknowledgments. This work is supported by National Natural Science Foundation of China under Grants 61175027.

References

1. Gilad-Bachrach, R., Navot, A., Tishby, N.: Margin based feature selection–theory and algorithms. In: Proceedings of the 21st International Conference on Machine Learning, p. 40 (2004)
2. Chen, M., Ebert, D., Hagen, H., Laramee, R.S.: Data Information and Knowledge in Visualization. Computer Graphics and Applications, 12–19 (2009)
3. Liu, C., Jaeger, S., Nakagawa, M.: Offline Recognition of Chinese Characters: the State of Art. IEEE Transcation on Pattern Analysis and Machine Intelligence 2, 198–213 (2004)
4. Saeys, Y., Inza, I., Larranaga, P.: A review of feature selection techniques in bioinformatics. Bioinformatics 19, 2507–2517 (2007)
5. Liu, H., Yu, L.: Toward integrating feature selection algorithms for classification and clustering. IEEE Transactions on Knowledge and Data Engineering 17, 494–502 (2005)
6. Kohavi, R., John, G.: Wrapper for feature subset selection. Artifical Intelligence, 234–273 (1997)
7. Pal, M.: Margin-based feature selection for hyperspectral data. International Journal of Applied Earth Observation and Geoinformation 11, 212–220 (2009)
8. Peng, H., Long, F., Ding, C.: Feature Selection Based on Mutual Information: Criteria of Max-Dependency, Max-Relevance, and Min-Redundancy. IEEE Transactions on Pattern Analysis and Machine Intelligence 8, 1226–1236 (2005)
9. Huang, D., Chow, T.W.S.: Effective feature selection scheme using mutual information. Neurocomputing 63, 325–343 (2005)
10. Liu, H., Sun, J., Liu, L., Zhang, H.: Feature selection with dynamic mutual information. Pattern Recognition 42, 1330–1339 (2009)
11. Li, Y., Lu, B.-L.: Feature selection based on loss-margin of nearest neighbor classification. Pattern Recognition 42, 1914–1921 (2009)
12. Kononenko, I.: Estimating Attributes: Analysis and Extensions of RELIEF. In: Bergadano, F., De Raedt, L. (eds.) ECML 1994. LNCS, vol. 784, pp. 171–182. Springer, Heidelberg (1994)
13. Sun, Y.: Iterative RELIEF for Feature Weighting: Algorithms,Theories, and Applications. IEEE Transations on Pattern Analysis and Machine Intelligence 6, 1–17 (2007)
14. Weinberger, K.Q., Blitzer, J., Saul, L.K.: Distance Metric Learning for Large Margin Nearest Neighbor Classification. Journal of Machine Learning Research, 207–244 (2009)
15. Chen, B., Liu, H., Chai, J., Bao, Z.: Large Margin Feature Weighting Method via Linear Programming. IEEE Transactions on Knowledge and Data Engineering 10, 1475–1486 (2009)
16. Merz, C.J., Merphy, P.: UCI repository of machine learning databases [OB/OL] (1996), `http://www.ics.uci.edu/~mlearn/MLRRepository.html`

Fuzzy Rough Decision Trees

Shuang An[1,2] and Qinghua Hu[2]

[1] Northeastern University, Shenyang 110819, P.R. China
anshuang_001@163.com
[2] Tianjin University, Tianjin 300072, P.R. China
huqinghua@hit.edu.cn

Abstract. Fuzzy rough sets are widely studied and applied in the domain of machine learning and data mining these years. In this work, this theory is used to design a fuzzy rough decision tree algorithm which can be used to deal with the cognitive uncertainties such as vagueness and ambiguity associated with human thinking and perception. In our algorithm, both selecting nodes and splitting branches in constructing the tree are based on fuzzy rough set theory. Especially, the current branching point is determined by pureness of the two branches, where the pureness is based on fuzzy lower approximation. The comparison results show that our decision tree algorithm is equivalent to or outperforms some popular decision tree algorithms.

1 Introduction

Since the concept of fuzzy rough sets was originally introduced, fuzzy rough set theory has been applied in many fields for handling fuzziness or uncertainty of the real-valued or fuzzy data sets [7]. This theory is claimed to be an important mathematical tool for granular computing and uncertainty reasoning in the past decade [10,14,22].

Decision trees are one of the most popular studied methods in domains of machine learning, pattern recognition and data mining [16]. Decision tree method is comprehensible and interpretable, and syncretizes feature-selection mechanism [9]. The represented performance by decision trees is close to or even outperforming other state-of-the-art methods [8]. Decision tree algorithms have been applied in classification and regression [2,17]. Classification trees are one of the most wildly used methods, and its goal is to find an accurate mapping from instance space to label space. ID3 is a typical algorithm for generating decision trees for classification [16]. Cognitive uncertainties, such as vagueness and ambiguity, have been incorporated into the knowledge induction process by using fuzzy decision trees [25]. The fuzzy ID3 can generate fuzzy decision trees without much computation [3]. It has the great matching speed and is especially suitable for large-scale learning problems [9,25].

Many methods have been developed for constructing decision trees and these methods are very useful in building knowledge-based expert systems [4,13,19,21]. For crisp classification problem, constructing a decision tree contains selecting

J.T. Yao et al. (Eds.): RSCTC 2012, LNAI 7413, pp. 397–404, 2012.

nodes and pruning strategy, and the number of branches for a selected node is decided by the number of the attribute values. For the data described by real-valued or fuzzy attributes, selecting nodes, splitting branches and pruning technique are three necessary factors in building a decision tree. In this work, a fuzzy rough decision tree algorithm is generated by using fuzzy rough sets for dealing with classification problem on real-valued or fuzzy data sets. Thereinto, both selecting nodes and splitting branches are based on fuzzy lower approximation operator.

The architecture of the paper is shown as follows. Section 2 reviews the basic theory of fuzzy rough sets. In Section 3, we introduce a fuzzy based on fuzzy rough decision tree algorithm whose performance will be tested in Section 4. Finally, Section 5 shows some conclusions.

2 Basic Notations of Fuzzy Rough Sets

Given a nonempty universe U, R is a fuzzy binary relation on U. If R satisfies reflexivity ($R(x,x)=1$), symmetry ($R(x,y)=R(y,x)$) and sup-min transitivity ($R(x,y) \geq \text{supmin}_{z\in U}\{R(x,z), R(z,y)\}$), we say R is a fuzzy equivalence relation which can be used to measure the similarity between any two objects. The fuzzy equivalence class $[x]_R = r_{i1}/x_1 + r_{i2}/x_2 + ... + r_{in}/x_n$ is the fuzzy granule induced by sample x and fuzzy equivalence relation R on U, where $[x]_R(y) = R(x,y)$ for all $y \in U$. Based on fuzzy equivalence relations fuzzy rough sets were first introduced by Dubois and Prade [7].

Definition 1. *Let U be a nonempty universe, R be a fuzzy equivalence relation on U and $F(U)$ be the fuzzy power set of U. Given a fuzzy set $F \in F(U)$, the lower and upper approximations of F are defined as*

$$\begin{cases} \underline{R}F(x) = \inf\limits_{y\in U} \max\{1 - R(x,y), F(y)\}, \\ \overline{R}F(x) = \sup\limits_{y\in U} \min\{R(x,y), F(y)\}. \end{cases} \tag{1}$$

Later, some models of fuzzy rough sets were introduced based on fuzzy logic operators which are summarized as follows [12,23,24,26,18,11].

$$\begin{aligned} &(1)\text{T-upper approximation operator}: \overline{R_T}A(x) = \sup_{u\in U} T(R(x,u), A(u)); \\ &(2)\text{S-lower approximation operator}: \underline{R_S}A(x) = \inf_{u\in U} S(N(R(x,u), A(u)); \\ &(3)\sigma\text{-upper approximation operator}: \overline{R_\sigma}A(x) = \sup_{u\in U} \sigma(N(R(x,u)), A(u)); \\ &(4)\vartheta\text{-lower approximation operator}: \underline{R_\vartheta}A(x) = \inf_{u\in U} \vartheta(R(x,u), A(u)). \end{aligned} \tag{2}$$

Although models of fuzzy rough sets were defined with different operators, the essence of the lower and upper approximations are the same. If A is a crisp set, $\underline{R_S}A(x)$ and $\underline{R_\vartheta}A(x)$ can be used to measure memberships of objects belonging to A definitely, and $\overline{R_T}A(x)$ and $\overline{R_\sigma}A(x)$ can be used to measure memberships of objects belonging to A probably.

With the definition of fuzzy rough sets, the membership of a sample $x \in U$ belonging to the positive region of the decision D in feature subset B is defined as

$$\mathrm{POS}_B(D)(x) = \sup_{X \in U/D} \underline{R}_B(X)(x). \tag{3}$$

And the fuzzy dependency of D on B with fuzzy rough sets, denoted by $\mathrm{FD}_B(D)$, is defined as

$$\mathrm{FD}_B(D) = \frac{\sum_{x \in U} \mathrm{POS}_B^R(D)(x)}{|U|}. \tag{4}$$

In Section 3, we select nodes and branch points of trees with the lower approximation and dependency function of fuzzy rough sets.

3 Constructing Decision Trees Based on Fuzzy Rough Sets

Decision tree is one of the popular methods for data mining due to its comprehension and interpretability. It is an effective and efficient tool for building classifiers, extracting rules and designing regression models from a set of objects [2]. In classification learning, ID3 algorithm proposed by Quinlan [16] was used to forecasting labels of objects described by symbolic data. In order to simulate the fuzzy reasoning, fuzzy decision trees regarded as a generalization of the crisp case are studied in [15]. It first gets fuzzified data from original data set by some methods, such as clustering, and then constructing decision trees on the fuzzy information table. Fuzzy ID3 is the state-of-art method for making fuzzy decision tree [3,25], whose architecture is similar to ID3. The advantage of fuzzy decision trees is that they can naturally handle different types of attributes (e.g., numerical and categorical), due to which many researchers were focusing on constructing fuzzy decision trees.

In this paper, we propose a decision tree based on fuzzy rough set theory, named fuzzy rough decision trees (FRDT). The architecture of our decision tree has two advantages: one is that FRDT can be used to classification problems with both symbolic and numerical data, and the other one is FRDT can make trees on original data directly without the fuzzification process of data.

There are three basic issues in developing a greedy algorithm for learning decision trees on the data set with continuous attributes: selecting nodes, determining branching criterion and pruning strategies, where selecting nodes play an important roles in building effective and efficient decision trees. In prior decision tree models, information gain is a good measure for selecting nodes, and Shannon's information entropy is usually taken as a measure to compute information gain brought by an attribute selected.

Our decision tree adopts dependency of fuzzy rough sets as the criterion of selecting nodes. For data sets with continuous attributes, fuzzy decision tree first fuzzifies data into several semantic values, the number of which is the number of branches. In this work, FRDT takes two-branch tree, and branching values are also selected with fuzzy rough set theory. In the architecture of FRDT, we set the stopping criterion by restricting the number of nodes and rate of the most objects to avoid overfitting and growing a giant tree. The detail process of constructing a fuzzy rough decision tree is described as follows.

Selecting nodes: Each attribute is evaluated by fuzzy dependency, and the attribute with the maximal importance is selected as the current node (containing root node) of tree.

Branching principle: Let it supposed that C_1 is selected as the current node N_1. Given a attribute value $C_1(x_i)$ (i=1,2,...,n, n is the number of samples on the current node), which means the value of sample x_1 on attribute C_1, the sample set on the current node is divided into two branches.

Table 1. Sample distribution on two branches

	$Class_1$	$Class_2$
$Branch_1$	$\|B_1C_1\|$	$\|B_1C_2\|$
$Branch_2$	$\|B_2C_1\|$	$\|B_2C_2\|$

Table 1 illustrates the sample distribution on N_1, where $|B_iC_j|$ stands for the number of samples on $branch_k$ from $class_j$ (k=1,2; j=1,2). For any $C_1(x_i)$, we use the following measure

$$\sum_{j=1}^{2}\left|\sum_{i=1}^{|B_jC_1|+|B_jC_2|}(-1)^{label(x_i)}\cdot L\_APP\_MEM(x_i)\right| \tag{5}$$

to evaluate the quality of branching off at $C_1(x_i)$. $L\_APP\_MEM(x_i)$ is lower approximation membership of sample x_i belonging to its own class, and $label(x_i)$ is the class label of sample x_i.

The above formula (7) can be simplified as

$$\sum_{j=1}^{2}\left|\sum_{p=1}^{|B_jC_2|} L\_APP\_MEM(x_p)-\sum_{q=1}^{|B_jC_1|} L\_APP\_MEM(x_q)\right| \tag{6}$$

This is to compute the absolute error between sum of lower approximation membership of samples coming from one class and other classes. This strategy is to find a value, about which most samples come from one class. And most samples with attribute values smaller than the selected value come from another class. The value satisfies this condition is considered as a good branching point. With a attribute value, we can gain a evaluation result. The branching value is determined via the following formula.

$$\max_{x \in X_1} \left\{ \sum_{j=1}^{2} \left| \sum_{p=1}^{|B_j C_2|} L\_APP\_MEM(x_p) - \sum_{q=1}^{|B_j C_1|} L\_APP\_MEM(x_q) \right| \right\}. \quad (7)$$

Here, X_1 is the sample set on the current node.

Stopping criterion: Obviously, if all the samples in a node belong to the same class, the tree should stop growing up; in order to avoid overfitting training data, we also stop growing the tree if fuzzy dependency of the selected feature is smaller than a threshold ε.

The pseudocode of the FRDT is shown as Algorithm 1.

Table 2. Algorithm Description

```
Tree Growing of Fuzzy Rough Decision Tree
---------------------------------------------------------------------------
Input: training set S described with m features i.e. {f1, f2, ..., fm}
Output: fuzzy rough decision tree T

TreeGrow(S,δ)
Initialize (T,S);
r = min(ClassRate(L),ClassRate(R));
if (r ≤ δ)
      LeafNode(S);                 /*a leaf node is generated*/
                                   /*S is the training set of current node*/
      Return(T);
else
      MarkTreeNode(S);             /*CurrentNode= argmax_f(FD(f_i))*/
                                   /*FD(f_i) is the fuzzy dependency of f_i*/
      FindBranchPoint(S);          /*the branch point is selected with formula (7)*/
      LeftBranch(T) = TreeGrow(S_L,δ);     /*tree growing on the left branch*/
      RightBranch(T) = TreeGrow(S_R,δ);    /*tree growing on the right branch*/
end
```

4 Experimental Analysis

In this section, we perform experiments to test the performance of the fuzzy rough decision tree. We first show the architecture of the FRDT on an artificial data set containing 130 samples described by 13 features with 3 classes. We use 88 samples to construct a two branches fuzzy rough decision tree which is shown as Fig 1, and test accuracy of left samples is 96.8%. In Fig 1, a_{13}, a_2, a_1 and a_4 are features, and numbers on each leaf nodes are the numbers of samples.

Next we conduct experiments on seven data sets from UCI [1] to test the performance of FRDT. The description of data sets is shown as Table 3. The experiment is performed by taking 10-fold cross-validation. In order to evaluate the efficiency of FRDT, we compare the classification accuracies of data sets with

NN [6,5], C4.5 [17], CART [2], fuzzy ID3 [3] and FDTBFRT (fuzzy decision tree based on fuzzy rough sets) [21]. Comparison of classification performance of different decision trees is shown in Table 4.

In Table 4, 'N' is the number of rules and 'Acc' is the average test accuracy. It is shown that FRDT can produce the highest accuracy on WDBC, sonar, diabetes and rice. FDTBFRT can produce the highest classification accuracy on WPBC, ionosphere and lungcancer. As a whole, our new decision tree has the highest average classification accuracy on all the data sets of all the trees.

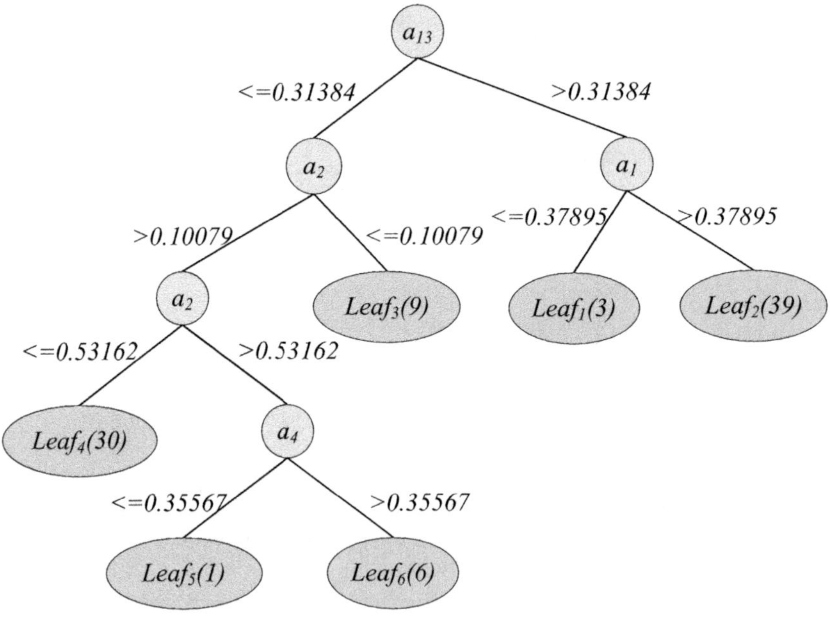

Fig. 1. FRDT

Table 3. Summaries of data sets

Data	Samples	Features	Classes
WDBC	569	30	2
WPBC	198	33	2
sonar	208	60	2
ionosphere	351	34	2
diabetes	768	8	2
lungcancer	137	7	2
rice	105	5	2

Table 4. Classification accuracies (%) of data sets

Data	FRDT		NN	C4.5		CART		FUZZYID3		FDTBFRT	
	Acc	*N*	*Acc*	*Acc*	*N*	*Acc*	*N*	*Acc*	*N*	*Acc*	*N*
WDBC	96.2	7	95.8	93.3	13	93.1	9	93.7	3	95.3	10
WPBC	78.3	9	66.2	73.2	16	70.6	1	76.0	4	83.5	7
sonar	87.1	8	87.1	75.9	18	74.0	10	71.9	6	78.1	10
ionosphere	84.7	16	86.4	91.5	18	87.3	3	66.4	4	88.6	109
diabetes	89.3	10	70.6	75.2	20	74.4	13	78.4	4	83.5	11
lungcancer	87.7	7	59.4	72.3	2	75.2	2	90.0	6	91.4	17
rice	89.6	8	80.0	76.9	3	80.8	2	82.7	3	83.3	6
AVG.	87.6	9	77.9	79.8	13	79.3	6	79.9	5	86.2	24

5 Conclusion

Fuzzy rough sets were introduced as a mathematical tool to deal with uncertainty of continuous or fuzzy data sets. This theory has attracted much attention from the various domains in recent years. In this work, we apply fuzzy rough sets to design a decision tree called fuzzy rough decision tree which is generated to be a classification tree for real-valued or fuzzy data sets.

Our tree is composed of three processes: selecting nodes, splitting branches and pruning technique. The current node is selected to be the feature with maximal fuzzy dependency. The branching point is determined with maximal absolute error between the sum of lower approximation membership of samples coming from one class and the other class. This strategy is to find a value, above which most samples come from one class. And most samples with attribute values smaller than the selected value come from another class. For a leaf node, if the proportion of samples from some class less than a given threshold this leaf will be pruned. Finally, we operate some experiments on seven data sets with two classes to test the performance of the proposed fuzzy rough decision tree, and the comparison analysis results show that our algorithm is outperform CART, C4.5, fuzzy ID3 and FDTBFRT on a certain extent for classification problem.

Acknowledgments

References

1. Blake, C.L., Merz, C.J.: UCI Repository of Machine Learning Databases (1998), `http://www.ics.uci.edu/~mlearn/MLRepository.html`
2. Breiman, L., Friedman, J., Stone, C., Olshen, R.: Classification and Regression Trees. Chapman and Hall/CRC, Boca Raton (1984)
3. Cios, K.J., Sztandera, L.M.: Continuous ID3 Algorithm with Fuzzy Entropy Measures. In: Proceeding on IEEE International Conference on Fuzzy Systems, San Diego, CA, pp. 469–476 (1992)
4. Cornelis, C., Jensen, R., Martin, G.H., Slezak, D.: Attribute Selection with Fuzzy Decision Reducts. Information Sciences 180, 209–224 (2010)

5. Cover, T.M., Hart, P.E.: Nearest Neighbor Pattern Classification. IEEE Transactions on Information Theory 13, 21–27 (1967)
6. Dasarathy, B.V.: Nearest Neighbor NN Norms: NN Pattern Classification Techniques, pp. 1–30. IEEE Computer Society (1990)
7. Dubois, D., Prade, H.: Rough Fuzzy Sets and Fuzzy Rough Sets. General Systems 17, 191–209 (1990)
8. Geurts, P., Ernst, D., Wehenkel, L.: Extremely Randomized Trees. Machine Learning 36, 3–42 (2006)
9. Hullermeier, E., Vanderlooy, S.: Why Fuzzy Decision Trees are Good Rankers. IEEE Transactions on Fuzzy Systems 17, 1233–1244 (2009)
10. Jensen, R., Shen, Q.: Fuzzy-rough Attribute Reduction with Application to Web Categorization. Fuzzy Sets and Systems 141, 469–485 (2004)
11. Mi, J.S., Zhang, W.X.: An axiomatic Characterization of a Fuzzy Generalization of Rough Sets. Information Sciences 160, 235–249 (2004)
12. Morsi, N.N., Yakout, M.M.: Axiomatics for Fuzzy Rough Sets. Fuzzy Sets and Systems 100, 327–342 (1998)
13. Nguyen, H.S.: Approximate Boolean Reasoning: Foundations and Applications in Data Mining. In: Peters, J.F., Skowron, A. (eds.) Transactions on Rough Sets V. LNCS, vol. 4100, pp. 334–506. Springer, Heidelberg (2006)
14. Tsang, E.C.C., Chen, D.G., Yeung, D.S.: Attributes Reduction Using Fuzzy Rough Sets. IEEE Transactions on Fuzzy Systems 16, 1130–1141 (2008)
15. Olaru, C., Wehenkel, L.: A Complete Fuzzy Decision Tree Technique. Fuzzy Sets and Systems 138, 221–254 (2003)
16. Quinlan, J.R.: Induction of Decision Trees. Machine Learning 1, 81–106 (1986)
17. Quinlan, J.R.: C4.5: Programs for Machine Learning. Morgan Kaufmann, San Francisco (1993)
18. Radzikowska, A.M., Kerre, E.E.: A Comparative Study of Fuzzy Rough Sets. Fuzzy Sets and Systems 126, 137–155 (2002)
19. Safavian, S.R., Landgrebe, D.: A Survey of Decision Tree Classifier Methodology. IEEE Transactions on System Man Cybernet 21, 660–674 (1991)
20. Suarez, A., Lutsko, J.F.: Globally Optimal Fuzzy Decision Trees for Classification and Regression. IEEE Transactions on Pattern Analysis and Machine Intelligence 21, 1297–1311 (1999)
21. Wang, X.Z., Zhai, J.H., Lu, S.X.: Induction of Multiple Fuzzy Decision Trees based on Rough Set Technique. Information Sciences 178, 3188–3202 (2008)
22. Wu, H.Y., Wu, Y.Y., Luo, J.P.: An Interval Type2 Fuzzy Rough Set Model for Attribute Reduction. IEEE Transactions on Fuzzy Systems 17, 301–315 (2009)
23. Wu, W.-Z., Mi, J.-S., Zhang, W.-X.: Generalized Fuzzy Rough Sets. Information Sciences 151, 263–282 (2003)
24. Wu, W.-Z., Mi, J.-S., Zhang, W.-X.: Constructive and Axiomatic Approaches of Fuzzy Approximation Operators. Information Sciences 159, 233–254 (2004)
25. Yuan, Y., Shaw, M.J.: Induction of Fuzzy Decision Trees. Fuzzy Sets and Systems 69, 125–139 (1995)
26. Yueng, D.S., Chen, D.G., Tsang, E.C.C., et al.: On the Generalization of Fuzzy Rough Sets. IEEE Transactions on Fuzzy Systems 13, 343–361 (2005)

Generalized Granulation Model for Data with Multi-complex Values

Xu Tan and Baowen Chen

School of Software, Shenzhen Institute of Info. Tech, 518172 Shenzhen, China
{tanx,chenbw}@sziit.edu.cn

Abstract. In order to establish a better application platform for granular computing, a novel generalized granulation model based on characteristic similarity is constructed in this paper. Considering that in the real-world application, a decision table often contains large amount of different types of complex data, we firstly reform these complex data into unified mathematical descriptions under the probabilistic measures. Then, characteristic similarity relation based on calculations of expectation and variance values, is figured to measure the similarity of each pair of objects with multi-complex attribute values. Lastly, we can get granulation results for all objects in the decision table according to the definition of characteristic similarity matrix. It has been proved that the proposed granulation model is a reasonable extension of Pawlaks equivalence partition model. Finally, examples are given to illustrate the proposed granulation model, which is proved to be effective, feasible and simple.

Keywords: Granular computing, Multi-complex data, Characteristic similarity relation, Probabilistic distribution.

1 Introduction

Granular Computing (GrC) is a novel soft computing method for simulating humans decision thought and problem solving approach, which can also be understood as a way of perception, understanding and representation real world problem together with its solution at different levels of granularity. However, granulation of a real universe of discourse is one of the important aspects whose solution has significant influence to granular computing. Each granule is a clump of objects which are drawn together by indistinguishability, similarity or functionality. By virtue of granular thinking, information in the real world with complexity and uncertainty can be well handled, and robustness analysis can be also developed [1].

In granular computing models, granules are often described into two types: crisp granule and fuzzy granule. Crisp granules are got from granulating a finite universe of discourse through a family of pairwise disjoint subsets under crisp equivalence relation. Such as in the classic Pawlak rough set model, suitable discretization algorithm is commonly needed to partition the value domain of

J.T. Yao et al. (Eds.): RSCTC 2012, LNAI 7413, pp. 405–413, 2012.

real-valued variable into several intervals, and the objects in the same interval are assigned into the same crisp granule. However, this kind of sharp partition is proved to bring information loss[2], and we seldom have a clear-cut partitioning of space in the real world, a certain degree of overlapping among the partitions is common in granulation. Currently, fuzzy granules are considered by extending precise binary equivalence relations to common soft binary relations in most of granular computing models. Ma[3] formulated a covering model by granulating a finite set of a universe of discourse into a family of overlapping granules on the basis of a reflexive relation. Bhatt[4] described fuzzy partitions based on a fuzzy-rough hybrid method. A new granular structure was established based on the definition of tolerance relation in tolerance rough set model, and William discussed the covering based granule model in Ref[5].

However, in many real-world application fields, object under features usually comes with complex formats, such as fuzzy type, literal type, interval type, random type, etc. How to shape feasible size of granules for these complex valued data is the key problem that we should to solve. In recent studies, Hedjazi[6] defined a similarity margin for interval represented data, and classified interval features dataset into granules in soft way. In order to describe proper shape of granules for set-valued decision tables, Qian[7] presented a kind of dominance relation. Wu[8] tried to propose an approach to derive interpretable granules from stochastic data based on rough set model and D-S evidence theory. Furthering, when decision table is given with mixed and multi-types of complex data, we should consider defining proper granular structure to address this problem which is rarely discussed in the past. Hu[9] analyzed granules in decision table with hybrid data under definition of neighborhood relationship, which leads to the limitation in handling two types of data: numerical one and categorical one. To better model reality, Tan[10] considered the situation of decision table with multiple complex formats of data, and tried to depict information granules based on similarity computation which has different formulas for the different types of complex data values. On the basis of existing researches, we want to consider a generalized approach to granulate objects with different complex types of data values in a decision table. This paper is organized as follows: uniform mathematical descriptions are given to describe uncertainties of three familiar types of complex data in section 2. In section 3, the new granulation model is discussed under the definition of characteristic similarity relation. Examples for illustration are presented in section 4. Finally, conclusions come in section 5.

2 Unified Uncertainty Descriptions for Multi-complex Valued Data

In real world applications, attribute values in a decision table are inevitably described with multi-kinds of complex data, and the focus of this section is how to make unified descriptions for these complex types of data, includes interval data, stochastic data and literal data.

Definition 1. Quintuple $T = \{U, C, D, V, f\}$ is called a decision table, where $U = \{o_1, o_2, ..., o_n\}$ is a finite nonempty set of objects, set $C = \{c_1, c_2, ..., c_n\}$ contains conditional attributes reflecting characteristics of objects, D denotes the decision attribute,$V = \bigcup_{a \in C \cup D} V_a$ is the domain of attribute$a$, f is a function such that $f(o, a) \in V_a$ for every $a \in C \cup D$ and $o \in U$, called an information function.

Next, uncertainty descriptions under the unified probabilistic distribution framework will be given to different types of complex-data. 1)$\forall c \in C$, $o \in U$, if $f(o, c)$ takes numeric values in $[v_1, v_2](v_2 > v_1)$, then it can be described as random variable V_c having a uniform distribution in interval $[v_1, v_2]$. 2)$\forall c \in C$, $o \in U$, if $f(o, c)$ takes random values, we can assign appropriate probability distribution function to the random variable V_c, which can be distinguished as discrete one or continuous one, based on the given discourse domain according to the object o. 3)$\forall c \in C$, $o \in U$, if $f(o, c)$ takes literal value ℓ_k (let $L = \{\ell_k | k = 1, 2, ..., l\}$ be the set contains linguistic scale sequences, which represent literal values with increasing intensity, and l stands for quantity of the scale sequences), the literal value ℓ_k can be treated as a fuzzy set $\mu_{\ell_k} : Q \to [0, 1]$ with membership function μ_{ℓ_k} in the universe of discourse $Q = \{h(\ell_k) \in Z | h(\ell_k) = k, \ell_k \in L\}$. Thus, the literal value ℓ_k can be furthering understood as discrete random variable $V_c$having a probability distribution table according to membership degree $\mu_{\ell_k}(h(\ell_k))$in the space Q.

Specially, if V_c takes fixed real value v_c, we can regarded it as random variable V_c obeys uniform distribution in interval $[v_c, v_c]$.

3 Granulation of Objects with Multi-types of Complex Data

We present below the similarity comparison between each object, which contains different kinds of complex values under different conditional attributes. Thus, we can get the feasible way to divide objects of the decision table into a series of granules.

Based on analyses in Section 2, comparison between two objects in decision table can be converted to comparison between two m-dimensional random vectors. Due to the dependence among the m random variables(it's means that the m conditional attributes exhibit some degree of correlation with each other), and the obedience of different random variable to different types of probability distribution, the two objects should be compared under each random variable(conditional attribute) one by one. Additionally, inspired by the fact that expectation and variance are two important numerical characteristics for each random variable, we can compare the m pairs of expectation and variance values to assess similarities between two objects.

Given an object takes discrete value under conditional attribute $c \in C$, we can make the assumption that random variable V_c takes value in the discrete space $\{v_c^1, v_c^2, ..., v_c^z, ...\}$, and corresponding probability values are given by $\{p_c^1, p_c^2, ..., p_c^z, ...\}$, then the expectation and variance can be calculated by:

$$E(V_c) = \sum_z v_c^z \cdot p_c^z \qquad D(V_c) = \sum_z [v_c^z - E(V_c)]^2 p_c^z \tag{1}$$

As to an object takes continuous value under conditional attribute $c \in C$, we can make the assumption that continuous random variableV_c takes value in $[v_c^1, v_c^2]$, and corresponding probability density function is $p(v_c)$, then the expectation and variance can be calculated by:

$$E(V_c) = \int_{v_c^1}^{v_c^2} v_c \cdot p(v_c) dv_c \qquad D(V_c) = \int_{v_c^1}^{v_c^2} [v_c - E(V_c)]^2 p(v_c) dv_c \tag{2}$$

Given two objects o_i and o_j with multi-complex values $(i, j = 1, 2, ..., n)$, then the comparison between two objects under each conditional attribute $c_t \in C(t = 1, 2, ..., m)$ can be transferred to the comparison between $(E_i(V_{c_t}), D_i(V_{c_t}))$ and $(E_j(V_{c_t}), D_j(V_{c_t}))$. In the following, we will give a new similarity definition between two objects based on calculations of characteristic values.

Definition 2. Given decision table T, two objects $o_i, o_j \in U$ are said to meet relationship of characteristic similarity under conditional attribute $c_t \in C$(noted as $o_i Sim_{c_t} o_j$), if

$$\frac{1}{2}[(E_i(V_{c_t}) - (E_i(V_{c_t}) + E_j(V_{c_t}))/2)^2 + (E_j(V_{c_t}) - (E_i(V_{c_t}) + E_j(V_{c_t}))/2)^2] \\ \leq max\{D_i(V_{c_t}), D_j(V_{c_t})\} \tag{3}$$

Def.2 shows the new similarity measure between two multi-complex valued objects under uncertainty. The basic starting point lies in keeping two uncertain spaces overlapped according to the center of their expectation values. Also, their expectation values are expected to be closer.

Property 1. $\forall o_i, o_j \in U$, $c_t \in C$, characteristic similarity relation $o_i Sim_{c_t} o_j$ fulfils the properties of reflexivity and symmetry, but may not satisfy transitivity property.

Proof. By the definition 2, we can easily get the conclusions.

Definition 3. Given decision table T, characteristic similarity relation under conditional attribute set $A \subseteq C$ can be defined as $Sim(A) = \{(o_i, o_j) \in U^2 | \forall c_t \in A, o_i Sim_{c_t} o_j\}$. $U/Sim(A) = \{\widetilde{[o_i]}_A | o_i \in U\}$ is a granulated universe of discourse with respect to U of conditional attribute set A, where$\widetilde{[o_i]}_A = \{o_j | (o_i, o_j) \in Sim(A)\}$.

From definition 3, we can write the characteristic similarity matrix $S = [s_{ij}]_{n \times n}$ which describes similarity relationships among n objects based on conditional attribute set $A \subseteq C$. s_{ij} denotes similarity value between object o_i and o_j under conditional attribute set A, where $s_{ij} = 1$, if it meets criteria of the characteristic similarity relation as described in definition 3, otherwise $s_{ij} = 0$.

Theory 1. Pawlaks equality relation is the special case of characteristic similarity relation, and characteristic similarity relation is the extension of equality relation.

Proof. Given decision table T, $\forall o_i, o_j \in U$.
1) if the two objects take fixed real values under $A \subseteq C$, and hold Pawlaks equality relation $Ind(A)$, then $\forall c_t \in A$, we have $f(o_i, c_t) = f(o_j, c_t)$, $E_i(V_{c_t}) = E_j(V_{c_t})$, $D_i(V_{c_t}) = D_j(V_{c_t}) = 0$. Namely, the two objects meet relation $o_i Sim_{c_t} o_j$, $\forall c_t \in A$. Furthering, according to definition 3, the two objects meet the characteristic similarity relation $Sim(A)$.
2) if the two objects take uncertainty values under $A \subseteq C$, Pawlaks equality relation is no more valid. $\forall c_t \in A$, let attribute values of objects o_i and o_j complying with certain probability distributions, and the two expectation values can be infintesimally approached with each. Thus, we can find the value of left side of inequality 3 will infintesimally approach to 0, and the right side value of inequality 3 will be greater than the left side. Moreover, if two random variables take their own expectation values respectively with probabilities approaching to 1.0 (objects o_i and o_j take their fixed values under attribute c_t), then the inequality 3 will degenerate into an identity, and Pawlaks equality relation $Ind(A)$ can be satisfied.

Theory 1 declares that characteristic similarity relation is a more universal relation to gain proper granular description. When data show different forms of uncertainties in reality, binary equivalence relation will not be applicable, and the characteristic similarity relation can build relaxed relationships for objects described in uncertainty values. Thus, we get the new granular scheme to divide objects with multi-complex data values.

Definition 4. Given decision table T, $\forall Y \in U/Ind(D)$, the lower and upper approximation of Y with regard to $A \subseteq C$ in the characteristic similarity relation induced granular space are

$$\overline{A}(Y) = \{o_i \in U | \widetilde{[o_i]}_A \cap Y \neq \phi\} \qquad \underline{A}(Y) = \{o_i \in U | \widetilde{[o_i]}_A \subseteq Y\}$$

respectively.

Theory 2. Given decision table T, let $A_1 \subseteq A_2 \subseteq C$, we have $\bigcup_{Y \in U/IND(D)} \underline{A_1}(Y) \subseteq \bigcup_{Y \in U/IND(D)} \underline{A_2}(Y) \subseteq \bigcup_{Y \in U/IND(D)} \overline{A_2}(Y) \subseteq \bigcup_{Y \in U/IND(D)} \overline{A_1}(Y)$.

Proof. From definition 3, let $A_1 \subseteq A_2 \subseteq C$ and $U/Sim(A_2) = \{X_1, X_2, ..., X_w\}$, we have $U/Sim(A_1) = \{X_1, ..., X_i \cup X_j, ..., X_{j-1}, X_j, ..., X_w\}$, namely $\underline{U/Sim(A_2)} \subseteq U/Sim(A_1)$. From definition 4, $\forall Y \in U/Ind(D)$, we have $\overline{A_2}(Y) \subseteq \overline{A_1}(Y)$ and $\underline{A_1}(Y) \subseteq \underline{A_2}(Y)$. In addition, $\bigcup_{Y \in U/IND(D)} \underline{A_1}(Y) \subseteq$

$\bigcup_{Y \in U/IND(D)} \underline{A_2}(Y)$ and $\bigcup_{Y \in U/IND(D)} \overline{A_2}(Y) \subseteq \bigcup_{Y \in U/IND(D)} \overline{A_1}(Y)$ will be satisfied. Furthering, based on concepts of the upper and lower approximations, we can get the conclusions.

To sum up, granulation based on characteristic similarity relation is a kind of soft partition which formulates a set of overlapping granules. Whats more, these granules exhibit various kinds of sizes and shapes for different combinations of conditional attributes. Thus, a reasonable approximation result can be obtained to a given data set in practice.

4 Example Illustrations

To elucidate the proposed granulation model, example on real-world dataset has been performed, in which data are represented by different types of uncertainty values. Table1 shows the decision table of evaluating qualities of tobacco leaves[11]. According to different producing areas, 5 types of tobacco leaf samples(o_1 to o_5) are showed in object set U. Conditional attributes c_1 to c_4, which respectively represent chroma, aroma quality, puff number and tar content, are some preliminary indexes to evaluate the quality of tobacco leaves. Decision attribute D indicates quality degrees of tobacco leaf samples, are divided into grades of top, middle and low. According to statistics of multi-batch tobacco leaves in different producing areas, objects under attribute chroma are given literal values based on standard grading scalesdense, strong, moderate, weak, light; objects under attribute aroma quality are given literal values based on standard grading scales perfect, good, normal, poor; objects under attribute puff number are given random values in a given interval within a normal distribution, whose distribution parameters are obtained based on statistics of the various tobacco leaves; and objects under attribute aroma quality are given interval values.

Table 1. A multi-complex valued decision table as an example

U	C_1	C_2	C_3	C_4	D
O_1	dense-∼dense	good+∼perfect	[10.1, 11.0]∼N(10.5, 0.09)	[25.8, 31.5]	top
O_2	strong∼dense-	perfect-∼perfect	[9.8, 10.5]∼N(9.9, 0.64)	[29.1, 30.0]	middle
O_3	moderate+∼strong-	good	[10.5, 11.8]∼N(10.7, 0.16)	[28.8, 30.8]	top
O_4	strong	good∼perfect-	[11.2, 12.5]∼N(11.6, 0.36)	[29.8, 30.1]	top
O_5	moderate∼strong-	perfect	[10.2, 12.0]∼N(11.0, 0.25)	[29.5, 31.9]	middle

When data in decision table exhibits various types of complex formats, unified descriptions should be given based on their numerical characteristics.

Take object o_1 in table1 for example, attribute value under chroma is dense-dense. According to the five rank scales of attribute chroma, we can describe this fuzzy literal valued data based on membership distribution $\frac{0.0}{1} + \frac{0.0}{2} + \frac{0.0}{3} + \frac{0.15}{4} + \frac{0.85}{5}$. Furthering, we can regard it as random variable V_{c_1} which takes discrete value in integer range [1,5] under the given probability values. Thus, the corresponding expectation value $E_1(V_{c_1})$=4.85 and variance value $D_1(V_{c_1})$=0.1275. In the same way, as to attribute aroma quality, its can be described as $\frac{0.0}{1}+\frac{0.0}{2}+\frac{0.2}{3}+\frac{0.8}{4}$, and $E_1(V_{c_2})$=3.8, $D_1(V_{c_2})$=0.16. Based on statistic data, tobacco sample o_1 under attribute puff number takes random value in [10.1, 11.0], and obeys normal distribution with N(10.5,0.09). So, we can obtain $E_1(V_{c_3})$=10.5 and $D_1(V_{c_3})$=0.09. Under attribute tar content, interval data [25.8, 31.5] can be treated as random variable V_{c_1} taken discrete value in [25.8, 31.5] with a uniform distribution, then $E_1(V_{c_4})$=28.65 and $D_1(V_{c_4})$=0.475.

Therefore, we can transfer the decision table with multi-types of complex data shown in table 1 to decision table described by data combinations $\{E_i(V_{c_t}), D_i(V_{c_t})\}$, shown in table2. Next, we can start to make comparisons among 5 objects by definition 2. Take the example of comparison between $o_1$and $o_2$based on attribute c_1, we can get $\frac{1}{2}[(E_1(V_{c_1})-(E_1(V_{c_1})+E_2(V_{c_1}))/2)^2+(E_2(V_{c_1})-(E_1(V_{c_1})+E_2(V_{c_1}))/2)^2]=\frac{1}{2}[(4.85-\frac{9.05}{2})^2+(4.2-\frac{9.05}{2})^2]$=0.1056. Meanwhile, $max\{D_1(V_{c_1}),D_2(V_{c_1})\}=max\{0.1275,0.36\}$=0.36. Thus, we obtain the characteristic similarity relation $o_1Sim_{c_1}o_2$. Simultaneously, its easy to get $o_2Sim_{c_1}o_3$. However, theres no characteristic similarity relation between o_1 and o_3 based on attribute c_1, which indicates that transitivity property may not be satisfied in characteristic similarity relation.

Table 2. Unified uncertainty descriptions for table1

U	C_1	C_2	C_3	C_4	D
O_1	{4.85,0.1275}	{3.8, 0.16}	{10.5, 0.09}	{28.65,0.475}	1
O_2	{4.2, 0.36}	{3.9, 0.09}	{9.9, 0.64}	{29.55,0.092}	0
O_3	{3.4, 0.16}	{3.1, 0.15}	{10.7, 0.16}	{29.8,0.167}	1
O_4	{4.0, 0.40}	{3.3, 0.21}	{11.6, 0.36}	{29.95,0.025}	1
O_5	{3.1, 0.29}	{3.85,0.1275}	{11.0, 0.25}	{30.7, 0.20}	0

Let $A=C-\{c_2\}$, through comparative calculations among objects based on each conditional attribute, we can obtain the following characteristic similarity matrices S_1 and S_2 with respect to C and A.

$$S_1=\begin{bmatrix}11010\\11001\\00111\\10111\\01111\end{bmatrix}\qquad S_2=\begin{bmatrix}11010\\11101\\01111\\10111\\01111\end{bmatrix}$$

According to S_1 and S_2, we have granular divisions $U/Sim(C)=\{\{o_1,o_2\},\{o_1,o_4\},\{o_2,o_5\},\{o_3,o_4,o_5\}\}$ and $U/Sim(A)=\{\{o_1,o_2\},\{o_1,o_4\},\{o_2,o_3,o_5\},\{o_3,o_4,o_5\}\}$ respectively. Based on equivalence partitions for the decision attribute $D(U/Ind(D)=\{\{o_1,o_3,o_4\},\{o_2,o_5\}\})$, and the definition 4, the lower and the upper approximations for C on equivalence partitions $\{o_1, o_3, o_4\}$ and $\{o_2, o_5\}$ are calculated as follows:

$$\underline{C}(\{o_1, o_3, o_4\}) = \{o_1, o_4\}, \quad \underline{C}(\{o_2, o_5\}) = \{o_2, o_5\}$$

$$\overline{C}(\{o_1, o_3, o_4\}) = \{o_1, o_2, o_3, o_4, o_5\}, \quad \overline{C}(\{o_2, o_5\}) = \{o_1, o_2, o_3, o_4, o_5\}$$

Also, the lower and the upper approximations for A on equivalence partitions $\{o_1, o_3, o_4\}$ and $\{o_2, o_5\}$ are calculated as follows:

$$\underline{A}(\{o_1, o_3, o_4\}) = \{o_1, o_4\}, \quad \underline{A}(\{o_2, o_5\}) = \{\phi\}$$

$$\overline{A}(\{o_1, o_3, o_4\}) = \{o_1, o_2, o_3, o_4, o_5\}, \quad \overline{A}(\{o_2, o_5\}) = \{o_1, o_2, o_3, o_4, o_5\}$$

Apparently, the conclusion
$\bigcup\limits_{Y \in U/IND(D)} \underline{A}(Y) \subseteq \bigcup\limits_{Y \in U/IND(D)} \underline{C}(Y) \subseteq \bigcup\limits_{Y \in U/IND(D)} \overline{C}(Y) \subseteq \bigcup\limits_{Y \in U/IND(D)} \overline{A}(Y)$
can be validated.

5 Summary and Conclusions

Data in decision table usually comes with different types of complex formats in real-world applications, such as medical, marketing, economical. This means that each object of decision table often takes different kinds of imprecise values corresponding to different conditional attributes, which leads to the difficulties of granulations in a unified form and the difficulties of subsequent computations under the umbrella of granular computing. This paper intends to give a reasonable and generalized granular description model for objects with different types of complex data values. Firstly, unified mathematical descriptions have been constructed for transforming three different types of complex data. Then, by calculating characteristic values of these probabilistic description-oriented data, the characteristic similarity relation among objects is built up. Finally, we got the granulation results for objects under the definition of characteristic similarity matrix. Further more, we discussed the properties and characteristics of this granulation model, and proved that this new granulation model was a reasonable extension of Pawlaks equivalence partition model. Further research will focus on the studies of establishing reasonable granular computing models for practical analyses based on this fundamental work.

Acknowledgments. This work was supported by National Natural Science Foundation of China under grant 71101096, Natural Science Foundation of Guangdong province under grant 10451802904005327 and Shenzhen Basic Research Project for Development of Science and Technology under grant JC201105190819A.

References

1. Pedrycz, W.: Granular computing-the emerging paradigm. Journal of Uncertain Systems 1, 38–61 (2007)
2. Jenson, R., Shen, Q.: Granular computing-the emerging paradigm. Journal of Uncertain Systems. In: Proceedings of IEEE International Conference on Fuzzy Systems, pp. 29–34. IEEE Press, New York (2002)
3. Ma, J.M., Zhang, W.X., Leung, Y.: Granular computing and dual Galois connection. Information Sciences 177, 5365–5377 (2007)
4. Bhatt, R.B., Gopal, M.: On the extension of functional dependency degree from crisp to fuzzy partitions. Pattern Recognition Letters 27, 487–491 (2006)
5. Zhu, W.: Topological approaches to covering rough sets. Information Sciences 177, 1499–1508 (2007)
6. Hedjazi, L., Aguilar-Martin, J., Lann, M.L.: Similarity-margin based feature selection for symbolic interval data. Pattern Recognition Letters 32, 578–585 (2011)
7. Qian, Y., Dang, C., Liang, J.: Set-valued ordered information systems. Information Sciences 179, 2809–2832 (2009)
8. Wu, W.Z., Zhang, M., Li, H.Z.: Knowledge reduction in random information systems via Dempster-Shafer theory of evidence. Information Sciences 174, 143–164 (2005)
9. Hu, Q.H., Zhang, L., Zhang, D.: Measuring relevance between discrete and continuous features based on neighborhood mutual information. Expert Systems with Applications 38, 10737–10750 (2011)
10. Tan, X., Tang, Y.L., Zhang, S.D.: Rough sets based attribute reduction algorithm for hybrid data. Journal of National Univ. of Defense Technology 30, 83–88 (2008) (in Chinese)
11. Tan, X.: Extended rough set models and their applications in quality prediction and evaluation for tobacco leaves. National University of Defense Technology, Changsha (2009)

Granular Approach for Protein Sequence Analysis

Ying Xie[1], Jonathan Fisher[1], Vijay V. Raghavan[2], Tom Johnsten[3], and Can Akkoc[4]

[1] Department of Computer Science, Kennesaw State University, Georgia, USA
[2] Center for Advanced Computer Studies, University of Louisiana at Lafayette, Louisiana, USA
[3] School of Computer and Information Sciences, University of South Alabama, Alabama, USA
[4] Institute of Applied Mathematics, The Middle East Technical University, Ankara, Turkey

Abstract. Granular computing uses granules as basic units to compute with. Granules can be formed by either information abstraction or information decomposition. In this paper, we view information decomposition as a paradigm for processing data with complex structures. More specifically, we apply lossless information decomposition to protein sequence analysis. By decomposing a protein sequence into a set of proper granules and applying dynamic programming to align the position sequences of two corresponding granules, we are able to distribute the calculation of pairwise similarity of protein sequences to multiple parallel processes, each of which is less time consuming than the calculation based on an alignment of original sequences.

1 Introduction

Granular computing can be viewed as a paradigm for information processing, where granules with certain resolutions work as basic units to compute with [1,2,3,4]. Many research works in the area of granular computing focus on information granulation that forms granules from structured data such as information table [5,6,7,8,9]. Different theories and instruments, including fuzzy set [2], rough set [6,7,8,9], quotient space theory [7], and neighborhood system [5], have been applied to model the process of information granulation. Information granulation can be viewed as a bottom-up approach to generate granules via information abstraction or aggregation. In this paper, we focus on the other side of granular computing, which is a top-down process that generates granules via information extraction. More specifically, we are interested in decomposing unstructured data or data with complex structures into granules, such that more structures can be brought to the data or the processing of data can be carried out in a parallel or distributed manner. We further propose a principle of lossless decomposition, which requires that the original data before decomposition can be completely rebuilt from the granules decomposed from the data. In other

J.T. Yao et al. (Eds.): RSCTC 2012, LNAI 7413, pp. 414–421, 2012.

words, granules decomposed from the data should maintain all information of the original data by following the lossless decomposition principle.

In this paper, lossless decomposition is applied to protein sequence analysis. We show that by decomposing a protein sequence into a set of proper granules and applying dynamic programming to align the position sequences of two corresponding granules, we are able to distribute the calculation of pairwise similarity of protein sequences to multiple parallel processes, each of which is less time consuming than the calculation based on an alignments of original sequences.

2 Lossless Decomposition of a Protein Sequence into Granules

The primary structure of a protein is a linear sequence of 20 amino acids, each of which can be represented by a letter such as L or K. An example protein sequence segment is shown as follows: "$\cdots$ILLNQNLVRSIKDSFVVTLISSEVLSF$\cdots$". Since amino acids are basic component of a protein sequence, an intuitive approach of decomposing a sequence is to view each amino acid as a granule. However, this approach loses the order information of the sequence. In other words, the aggregation of all granules decomposed from a protein sequence is unable to reconstruct the sequence. By following the lossless principle of decomposition, we can view granules decomposed from a protein sequence as individual amino acids plus the positions of the corresponding amino acid in the sequence. For instance, the sequence of "ILLNQNLVRSI" can be decomposed into the following 20 granules: <I>:{0, 10}, <L>:{1, 2, 6}, <N>:{3, 5}, <Q>:{4}, <V>:{ 7}, <R>:{8}, <S>:{9}; <A>:{-1}, <D>:{-1}, $\cdots$, where -1 represents the corresponding amino acid does not appear in the protein sequence.

More formally, given a protein sequence S and assuming $\mathbf{A}$ denoting a set of 20 amino acids, we conduct lossless decomposition of S into a set of granules $\mathbf{G} = \{g_{s_x} =< x > |x \in \mathbf{A} : \{p_i\text{: } p_i \text{ is } ith \text{ position of } x \text{ in } S\}\}$, such that $info(S) = info(\mathbf{G})$. This decomposition generates a set of 20 granules. We can further decompose each granule into 20 granules of finer resolution. For instance, given the sequence of "ILLNQNLVRSI", one of the granules <L>:{1, 2, 6} can be further decomposed to the following 20 granules: <LL>:{2}, <LN>:{3}, <LV>:{6}, <LA>:{-1}, <LD>:{-1}, $\cdots$. By this decomposition, the sequence can be represented by a set of 400 granules. More formally, given a protein sequence S and assuming A denoting a set of 20 amino acids, we conduct lossless decomposition of S into a set of granules $\mathbf{G} = \{g_{s_{x_1x_2}} =< x_1x_2 > |x_1 \in \mathbf{A} \,\&\&\, x_2 \in \mathbf{A} : \{p_i\text{: } p_i \text{ is } ith \text{ position of } x_1x_2 \text{ in } S\}\}$, such that $info(S) = info(\mathbf{G})$. If decomposing a protein sequence based on positions of all possible combinations of three consecutive amino acids, we get a set of 8000 granules for each protein sequence; if based on positions of all possible combinations of four consecutive amino acids, we get a set of 16000 granules; more generally, if based on positions of all possible combinations of n consecutive amino acids, we get a set of 20^n granules.

3 Protein Sequence Alignment Based on Granular Representations of Sequences

Evaluating similarity relationship between two given protein sequences is one of fundamental tasks in protein sequence analysis. Similarity in sequences may indicate homology in protein structures or functions. The Needleman-Wunsch algorithm [10] can be used to identify global optimal pairwise sequence alignment by a dynamic programming process. A maximum similarity score can therefore be derived from the optimal pairwise alignment. More specifically, given two protein sequences $R = x_1 x_2 \cdots x_i \cdots x_M$ and $Q = y_1 y_2 \cdots y_j \cdots y_N$, the Needleman-Wunsch algorithm calculate the maximum similarity score that is denoted as $H(M, N)$ by building a scoring matrix H using dynamical programming. The scoring matrix H can be described as follows:

$$\begin{aligned} &H(0,i) = 0 \\ &H(j,0) = 0 \\ &H(i,j) = max \begin{cases} H(i-1,j-1) + w(x_i, y_j) \\ H(i-1,j) + w(x_i, -) \\ H(i,j-1) + w(-, y_j) \end{cases} \end{aligned} \tag{1}$$

where $w(x_i, y_j)$ is the reward(penalty) brought by match(mismatch) between amino acids x_i and y_j; $w(x_i, -)(w(-, y_j))$ is the penalty of deletion(insertion).

As a variation of Needleman-Wunsch algorithm, the Smith-Waterman algorithm [11] uses dynamic programming to find the highest scoring local alignment between two sequences. The scoring matrix of Smith-Waterman algorithm for sequence R and Q can be described as follows:

$$\begin{aligned} &H(0,i) = 0 \\ &H(j,0) = 0 \\ &H(i,j) = max \begin{cases} 0 \\ H(i-1,j-1) + w(x_i, y_j) \\ H(i-1,j) + w(x_i, -) \\ H(i,j-1) + w(-, y_j) \end{cases} \end{aligned} \tag{2}$$

Although these two algorithms are guaranteed to obtain certain optimal alignments between two sequences, the time complexity of both algorithms is $O(MN)$, where M and N are the lengths of the two sequences. If we use either of these two algorithms to search similar sequences for a query sequence in a protein sequence database with K records, the time complexity of the search is $O(MNK)$. In order to speed up the calculation of pairwise similarity/distance between protein sequences, we propose the following method that distributes the process of pairwise sequence alignment to individual granules generated by a lossless decomposition of protein sequences.

Given two protein sequences S_1 and S_2, we conduct lossless decomposition of S_1 and S_2 into sets of granules $\mathbf{G_1}$ and $\mathbf{G_2}$, such that $info(S_1) = info(\mathbf{G_1})$

and $info(S_2) = info(\mathbf{G_2})$. For illustration purpose, we further assume that both decompositions are based on positions of all possible combinations of 2 consecutive amino acids. Therefore, we have

$\mathbf{G_1} = \{gs1_{x_1x_2} =< x_1x_2 > |x_1 \in \mathbf{A} \,\&\&\, x_2 \in \mathbf{A} : \{p_i\text{: } p_i \text{ is } ith \text{ position of } x_1x_2 \text{ in } S_1\}\}$

$\mathbf{G_2} = \{gs2_{x_1x_2} =< x_1x_2 > |x_1 \in \mathbf{A} \,\&\&\, x_2 \in \mathbf{A} : \{q_j\text{: } q_j \text{ is } jth \text{ position of } x_1x_2 \text{ in } S_2\}\}$

Now, for each possible combination of two amino acids x_1 and x_2, get the granule $gs1_{x_1x_2}$ from $\mathbf{G_1}$ and the granule $gs2_{x_1x_2}$ from $\mathbf{G_2}$. The granule $gs1_{x_1x_2}$ can be represented as a position series $p_1, p_2, \cdots, p_i, \cdots, p_m$ where p_i is the *ith* starting position that x_1x_2 appears in S_1; and $gs2_{x_1x_2}$ can be represented as a position series $q_1, q_2, \cdots, q_j, \cdots, q_n$ where q_j is the *jth* starting position that x_1x_2 appears in S_2.

We then define the distance between $gs1_{x_1x_2}$ and $gs2_{x_1x_2}$, which is denoted as $dist(gs1_{x_1x_2}, gs2_{x_1x_2})$, to be a minimal cumulative distance $\curlyvee(p_m, q_n)$ calculated based on an optimal warping path between the position series $p_1, p_2, \cdots, p_i, \cdots, p_m$ and the position series $q_1, q_2, \cdots, q_j, \cdots, q_n$. The optimal warping path can be computed by the dynamic programming process, where the minimal cumulative distance $\curlyvee(p_i, q_j)$ is recursively defined as:

$$\curlyvee(p_i, q_j) = d(p_i, q_j) + min(\curlyvee(p_{i-1}, q_{j-1}), \curlyvee(p_{i-1}, q_j), \curlyvee(p_i, q_{j-1})) \tag{3}$$

For example, assume the position sequence of $gs1_{LN}$ is $\{0, 5, 9, 121, 130\}$, and the position sequence of $gs2_{LN}$ is $\{4, 11, 100\}$. Then, by dynamic programming, the optimal alignment of these two position sequences can be illustrated in the following figure:

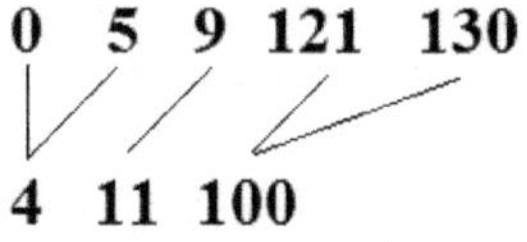

Fig. 1. Alignment between position sequences of two granules

Then the distance between $gs1_{LN}$ and $gs2_{LN}$ can be calculated according to the optimal alignment as $(4-0)+(5-4)+(11-9)+(121-100)+(130-100) = 58$.

Given that the position sequence of a granule is much shorter than the original protein sequence, the alignment between position sequences of two granules by dynamic programming should be much more efficient than the alignment between the original protein sequences. Hence, the calculation of similarity between two protein sequences by pairwise alignment can be distributed to individual granules. For this purpose, we define the distance between the sequence S_1 and the sequence S_2 as the aggregation of the distances between granules, which is shown as follows:

$$dist(S_1, S_2) = \sum_{x_1 \in \mathbf{A}, \mathbf{x_2} \in \mathbf{A}} dist(g_{S1_{x_1x_2}}, g_{S2_{x_1x_2}}) \quad (4)$$

From this definition we can see that the calculation of the distance between two sequences can be distributed to 400 parallel calculations of distances between 400 granule pairs if the lossless decomposition is based on 2 consecutive amino acids. More generally, if the lossless decomposition of a protein sequence is based on n consecutive amino acids, the calculation of the distance between two sequences can be distributed to 20^n calculations of distances between 20^n granule pairs. Assuming the position series for every granule of a protein sequence has the same length, the granular approach for calculating the distance between two protein sequences in a fully distributed computing platform takes approximately only $1/(20^{2n})$ of the time that is required by the global pairwise alignment approach.

Furthermore, we proposal a granular kernel for protein sequence based on the granular approach for pairwise distance calculation as follows:

$$K(S_1, S_2) = e^{-dist(S_1, S_2)} \quad (5)$$

This kernel function enables kernel based learning machines, such as Support Vector Machine (SVM) to be built on the proposed measure of protein sequence distance that can be calculated in a parallel manner.

4 Related Work in Protein Sequence Classification

k-length contiguous subsequence("k-mer") was used by bio-informatics classification methods such as k-spectrum [12]. Classifiers trained on k-mers use these entities and their frequency of occurrence as features. However, a k-mer does not capture the spatial index locations of subsequences.

In the multi-layered Vector Spaces (MLVS) method [13], a feature vector was generated for each ordered pair of amino acids with m steps in a given sequence by segmenting the sequence into multiple equal-length buckets, then counting the number of occurrences of each ordered pair in each bucket of that sequence. Given the fact that protein sequences typically have various lengths, the number of buckets (or the dimension of feature vectors) used in MLVS needs to be unified across sequences. Furthermore, the number of buckets is typically much less the lengths of protein sequences themselves in order to avoid the "curse of dimensionality" in classification. Therefore, the MLVS approach of converting a protein sequence to multiple feature vectors is not lossless. In other words, all those generated feature vectors cannot be re-assembled to the original sequence.

The major contributions of the proposed approach can be summarized as follows: 1) lossless decomposition of a sequence into multiple granules; 2)using dynamic programming to evaluate pairwise distance between two corresponding granules based on their associated position sequences; 3)distributing alignments between two protein sequences to multiple parallel processes, each which aligns much shorter position sequence pairs.

5 Preliminary Experimental Studies

We studied the performance of the proposed approach in protein sequence classification on 53 SCOP protein families. The data set of the 53 SCOP protein families can be downloaded from `http://noble.gs.washington.edu/proj/svmpairwise`. Each of the SCOP families contains a training data set and a testing data set as described in [14]. We simply used 1-nearest neighbor (1NN) approach to predict if a test sequence belongs to the given family or not. More specifically, for each test sequence, we evaluate its similarity with each training sequence, then use the class label of the most similar training sequence as the label for this test sequence. The accuracy rate of the prediction for each family is reported.

We used the following approaches to evaluate similarity between two protein sequences: 1)the Needleman-Wunsch algorithm (NW); 2)the Smith-Waterman

Protein		Accuracy	Rate %		Protein		Accuracy	Rate %	
Family	NW	SW	Single	Pair	Family	NW	SW	Single	Pair
7.3.5.2	99.4872	99.4872	99.3162	76.0114	3.32.1.11	95.7746	98.3568	98.3568	73.4742
2.56.1.2	99.3996	99.072	99.1812	99.5087	3.32.1.13	95.7478	97.5073	97.9472	74.3402
3.1.8.1	98.2691	98.5838	98.8198	99.2919	7.3.6.1	99.6599	99.4331	99.093	80.7256
1.27.1.1	99.5172	99.2414	99.5172	99.5172	7.3.6.2	98.98	98.8623	98.3915	95.8023
1.27.1.2	99.5346	99.4312	99.5863	99.4829	7.3.6.4	98.9796	98.9796	98.5714	80.8163
3.42.1.1	99.0135	98.834	98.2063	97.0403	2.38.4.1	99.1909	99.0291	98.0583	95.6311
1.45.1.2	99.1031	99.1031	99.4021	93.423	2.1.1.1	96.9973	96.2693	96.1783	95.0864
1.4.1.1	98.8597	98.7605	98.2152	93.3069	2.1.1.2	97.0513	97.0513	97.0513	95
2.9.1.2	98.2697	98.4224	98.6768	99.2875	3.32.1.1	97.0052	98.0469	98.0469	73.3073
1.4.1.2	98.8588	98.7161	98.2882	95.8631	2.38.4.3	99.0441	98.6029	98.3088	96.1765
2.9.1.3	99.0014	98.5735	98.1455	99.1441	2.1.1.3	96.8198	96.4664	96.4664	95.7597
1.4.1.3	98.8593	98.4791	99.2395	94.9303	2.1.1.4	96.6667	96.6667	97.3504	96.1539
2.44.1.2	94.4224	92.2905	94.2737	51.3138	2.38.4.5	99.1015	98.832	98.4726	97.7538
2.9.1.4	98.6458	98.4319	99.0021	98.8596	2.1.1.5	96.8652	96.3427	95.7158	95.8203
3.42.1.5	99.0345	98.6207	98.4828	98.2759	7.39.1.2	99.3794	99.2908	98.7589	75.88652
3.2.1.2	98.4768	97.3343	97.639	98.1721	2.52.1.2	99.4531	99.6094	99.4531	99.5313
3.42.1.8	99.1023	98.9228	97.307	97.307	7.39.1.3	99.3351	99.2465	99.0691	85.195
3.2.1.3	97.835	97.1583	98.1055	97.429	1.36.1.2	99.1726	99.1726	98.818	96.9267
3.2.1.4	97.561	96.6899	97.2125	97.7352	3.32.1.8	96.2687	98.1876	98.1876	73.6674
3.2.1.5	98.6063	96.8641	97.7352	98.0836	1.36.1.5	99.1728	98.791	98.1546	97.1365
3.2.1.6	97.0732	96.5854	96.8293	97.0732	7.41.5.1	99.5556	99.5062	99.2099	91.9012
2.28.1.1	97.6683	97.215	97.1179	98.1218	7.41.5.2	99.5556	99.4074	99.0617	97.4321
3.3.1.2	98.5712	98.7619	98.5714	97.7143	1.41.1.2	98.8728	99.1948	98.3897	85.8293
3.2.1.7	97.561	97.8049	97.561	98.2921	2.5.1.1	99.4483	99.1976	99.2477	99.1976
2.28.1.3	98.3373	98.3373	98.5748	93.5867	2.5.1.3	99.3933	99.2278	99.3381	99.283
3.3.1.5	97.8342	97.8342	98.7922	99.0837	1.41.1.5	98.9954	98.609	98.2226	97.9135
7.3.10.1	98.5592	97.5454	95.9712	64.83458	**Average of All**	**98.37638302**	**98.2450566**	**98.18282264**	**92.046**

Fig. 2. Preliminary experimental results[1]

[1] The authors recently corrected a bug in the program used for preliminary experimental studies and generated slightly different results than the values reported in figure 3. The updated results is available at: `https://sites.google.com/site/jrs2012paper/`

algorithm (SW); 3) the proposed granular approach based on single amino acids (Single), and 4) the proposed granular approach based on pairs of amino acids (Pair). For NW and SW, we set the match reward to be 10 and mismatch penalty to be -8. No external scoring matrix is used for this preliminary experimental study. The classification results are summarized in Figure 2.

As can be see in Figure 2, the proposed granular approach based on single amino acids reaches the same level of accuracy rate as the Needleman-Wunsch algorithm and the Smith-Waterman algorithm. In other words, the proposed granular approach is able to distribute the calculation of pairwise similarity to 20 parallel processes without sacrificing accuracy.

The accuracy rate of the proposed granular approach based on pairs of amino acids is around 6% worse than the other three, however, the calculation of similarity of two protein sequences under this setting can be distributed to 400 parallel processes, each of which deals with much short position sequences. Therefore, this approach may be suitable for online analysis of very large scale protein sequence database, where the tradeoff between efficiency and accuracy is necessary.

6 Conclusion and Future Work

In this paper, we studied a lossless decomposition of a protein sequence into a set of granules such that the calculation of pairwise similarity of protein sequences can be distributed to multiple parallel processes, each of which is less time consuming. This study further suggests that lossless or close-to-lossless information decomposition, as a top down paradigm for granular computing, may provide a solution framework for analyzing data set with complex structures. As the next step of our work, we will expand our study in the following ways: 1) compare the proposed method with the blast algorithm, a popular heuristic approach that approximate Smith-Waterman algorithm in sequence search, on both search effectiveness and efficiency; 2) study the way to incorporate external scoring matrix, such as BLOSUM62, in the proposed granular approach; 3) developing adaptive sequence search methods based on the proposed granular approach; 4) apply the lossless decomposition principle to other data sets with complex structures.

References

1. Yao, Y.: The Art of Granular Computing. In: Kryszkiewicz, M., Peters, J.F., Rybiński, H., Skowron, A. (eds.) RSEISP 2007. LNCS (LNAI), vol. 4585, pp. 101–112. Springer, Heidelberg (2007)
2. Zadeh, L.: Some reflections on soft computing, granular computing and their roles in the conception, design and utilization of information/intelligent systems. Soft Computing, 23–25 (1998)
3. Yao, J.T.: Recent Developments in Granular Computing: A Bibliometrics Study. In: Proceedings of IEEE International Conference on Granular Computing, Hangzhou, China, pp. 74–79 (2008)

4. Yao, J.T.: A Ten-Year Review of Granular Computing. In: Proceedings of 2007 IEEE International Conference on Granular Computing, Sillicon Valley, CA, USA, pp. 734–739 (2007)
5. Lin, T.: Granular computing of binary relations I: data mining and neighborhood systems. In: Polkowski, Skowron (eds.) Rough Sets and Knowledge Discovery, pp. 107–121. Physica-Verlag (1998)
6. Lin, T.Y.: A Roadmap from Rough Set Theory to Granular Computing. In: Wang, G.-Y., Peters, J.F., Skowron, A., Yao, Y. (eds.) RSKT 2006. LNCS (LNAI), vol. 4062, pp. 33–41. Springer, Heidelberg (2006)
7. Yao, Y., Liau, C., Zhong, N.: Granular Computing Based on Rough Sets, Quotient Space Theory, and Belief Functions. In: Zhong, N., Raś, Z.W., Tsumoto, S., Suzuki, E. (eds.) ISMIS 2003. LNCS (LNAI), vol. 2871, pp. 152–159. Springer, Heidelberg (2003)
8. Yao, Y.: Information granulation and rough set approximation. International Journal of Intelligent Systems, 87–104 (2001)
9. Pawlak, Z.: Rough Sets: Theoretical Aspects of Reasonging about Data. Kluwer Academic Publishers (1991)
10. Needleman, B., Wunsch, D.: A general method applicable to the search for similarities in the amino acid sequence of two proteins. Journal of Molecular Biology, 443–453 (1970)
11. Smith, F., Waterman, S.: Identification of Common Molecular Subsequences. Journal of Molecular Biology, 195–197 (1981)
12. Leslie, C., Eskin, E., Weston, J., Noble, W.: Mismatch String Kernels for SVM Protein Classification. In: Advances in Neural Information Processing Systems, NIPS 2002, Vancouver, British Columbia, Canada, December 9-14, pp. 1417–1424 (2002)
13. Akkoc, C., Johnsten, T., Benton, R.: Multi-layered Vector Spaces for Classifying and Analyzing Biological Sequences. In: Proceedings of 2011 International Conference on Bioinformatics and Computational Biology, New Orleans, pp. 160–166 (2011)
14. Liao, L., Noble, S.: Combining pairwise sequence similarity and support vector machines for detecting remote protein evolutionary and structural relationships. Journal of Computational Biology, 857–868 (2003)

JRS'2012 Data Mining Competition: Topical Classification of Biomedical Research Papers*

Andrzej Janusz[1], Hung Son Nguyen[1], Dominik Ślęzak[1,2], Sebastian Stawicki[1], and Adam Krasuski[1,3]

[1] Faculty of Mathematics, Informatics and Mechanics, University of Warsaw
Banacha 2, 02-097 Warsaw, Poland
[2] Infobright Inc.
Krzywickiego 34, lok. 219, 02-078 Warsaw, Poland
[3] Chair of Computer Science, Main School of Fire Service
Słowackiego 52/54, 01-629 Warsaw, Poland

Abstract. We summarize the JRS'2012 Data Mining Competition on "Topical Classification of Biomedical Research Papers", held between January 2, 2012 and March 30, 2012 as an interactive on-line contest hosted on the TunedIT platform (`http://tunedit.org`). We present the scope and background of the challenge task, the evaluation procedure, the progress, and the results. We also present a scalable method for the contest data generation from biomedical research papers.

Keywords: Data Mining, Topical Classification, Multi-label Classification, Explicit Semantic Analysis, PubMed, MeSH.

1 Introduction

The JRS'2012 Data Mining Competition was related to multi-label classification problem which is recently one of the hottest topics in machine learning. It is also a topic of interest of an ongoing project, called SONCA (Search based on ONtologies and Compound Analytics), realized at the Faculty of Mathematics, Informatics and Mechanics of the University of Warsaw. It is a part of the project 'Interdisciplinary System for Interactive Scientific and Scientific-Technical Information' (`http://www.synat.pl/`). SONCA is a hybrid database framework application, wherein scientific articles are stored and processed in various forms. SONCA is expected to provide interfaces for intelligent algorithms identifying relations among various types of objects. It extends the typical functionality of scientific search engines by more accurate identification of relevant documents

* This work was partially supported by the grant N N516 077837 from the Ministry of Science and Higher Education of the Republic of Poland, the Polish National Science Centre grant 2011/01/B/ST6/03867 and by the Polish National Centre for Research and Development (NCBiR) under SYNAT – Grant No. SP/I/1/77065/10 in frame of the strategic scientific research and experimental development program: "Interdisciplinary System for Interactive Scientific and Scientific-Technical Information".

J.T. Yao et al. (Eds.): RSCTC 2012, LNAI 7413, pp. 422–431, 2012.

and more advanced synthesis of information. To achieve this, concurrent processing of documents needs to be coupled with an ability to produce collections of new objects using queries specific to analytic database technologies.

Ultimately, SONCA should be capable of answering a user query by listing and presenting resources that correspond to the query *semantically*. In other words, the system should have *understanding* of the intention of the query and of contents of documents stored in the repository, as well as an ability to efficiently retrieve relevant information. The system should be able to use various knowledge sources related to investigated areas of science. It should also allow for independent sources of information about the analyzed objects, such as, e.g., information about scientists who may be identified as articles' authors.

Some developed functionalities of SONCA have been implemented and tested on the biomedical documents, which are freely available from the highly specialized biomedical articles repository called PubMed Central [1]. Rapidly increasing size of the scientific article meta-data and text repositories emphasizes the growing need for accurate and scalable methods for automatic tagging and classification of textual data. For example, medical doctors often search through biomedical documents for information regarding diagnostics, drugs dosage and effects or possible complications resulting from specific treatments. In the queries, they use sophisticated terminology, that can be properly interpreted only with a use of a domain ontology, such as Medical Subject Headings (MeSH) [2]. In order to facilitate the searching process, documents in a database should be indexed with concepts from the ontology. Additionally, the search results could be grouped into clusters of documents, that correspond to meaningful topics matching different information needs. Such clusters should not necessarily be disjoint since one document may contain information related to several topics.

In order to be able to provide semantic relationships between concepts and documents we employ a method called Explicit Semantic Analysis (ESA) [3]. This method associates elementary data entities with concepts coming from a knowledge base. We have field-tested a modified version of the ESA approach on PubMed Central using MeSH (see [4,5]), and found out that while conceptually the method performs really well, it needs introduction of some refinement techniques, that improve semantic tagging. One of the aims of our research and the JRS'2012 Data Mining Competition was to investigate methods and techniques, that allow for accurate predictions of labels given by human experts.

Our competition was hosted on the TunedIT platform[1], which supports automatic evaluation of solutions submitted by multiple participants [6]. It continues the series of contests organized at the RSCTC conferences [7]. The TunedIT platform provides tools such as a leaderboard, which displays preliminary results of the challenge. This allows participants to verify quality of their solution in comparison to others and better adjust their methods. At the same time, the possibility of over-fitting with a solution to the test data by extensive parameter tuning is limited due to separation of preliminary and final test data sets.

---

[1] `http://tunedit.org/challenge/JRS12Contest`

2 Task of the Challenge

Our research group invested a significant amount of time and effort to gather a corpus of documents containing 20,000 journal articles from the PubMed Central repository open-access subset. Each of those documents was labeled by biomedical experts from PubMed with several MeSH *subheadings* which can be viewed as different contexts or general topics discussed in the text. With a use of our automatic tagging algorithm, which is described in the further sections and discussed in more details in [4,8], we associated all the documents with the most related MeSH terms (*headings*). The competition data consisted of information about strengths of those bonds, expressed as numerical values. Intuitively, they can be interpreted as values of a rough membership function that measures a degree in which a term is present in a given text.

The task for the participants of the JRS'2012 Data Mining Competition was to devise algorithms capable of accurately predicting MeSH subheadings (topics) assigned by the experts, based on the association strengths of the automatically generated tags corresponding to MeSH headings. Each document could be labeled with several subheadings and this number was not fixed. In order to ensure that participants who might be not familiar with the biomedicine domain, and with the MeSH ontology in particular, had equal chances as domain experts, the names of the tags and topical classifications were removed from the data. Those names and relations between data columns, as well as a dictionary translating decision class identifiers into MeSH subheadings, could be provided on request for the sake of after-competition research.

The data set was provided in a two-dimensional tabular form as two *tab-separated values* files – a training set and a test set. Each row of those data files represents a single document and, in the consecutive columns, it contains integers ranging from 0 to 1000, expressing association strengths to the corresponding MeSH terms. Additionally, there was available a text file containing labels, whose consecutive rows corresponded to entries in the training data set. Each row of that file was a list of topic identifiers (integers ranging from 1 to 83), separated by commas, which can be regarded as a generalized classification of a journal article. This information was not available for the test set and the task for participants was to predict it using models constructed on the training data.

It is worth noting that, due to a nature of the considered problem, the data sets were highly dimensional - the number of columns roughly corresponded to a size of the MeSH ontology. The data sets are also sparse, since usually only a small fraction of the MeSH terms was assigned to a particular document by our tagging algorithm. Finally, a large number of data columns have little (or even none) non-zero values which is due to a fact that the corresponding concepts were rarely assigned to documents. It was up to participants to decide which of them are still useful for the task.

The quality of label predictions submitted by participants for a single test instance was measured using F_1-*score*, which is defined as a harmonic average of *precision* and *recall*. Let $TrueTopics_i$ denote labels assigned by experts to i-th

Table 1. Seven most represented countries in the JRS'2012 Data Mining Competition

No. participating teams	Country
53	United States
52	People's Republic of China
51	Poland
38	India
30	Russian Federation
13	Spain
13	Afghanistan

test document and let $PredTopics_i$ be a set of predicted labels. Precision of a prediction for the i-th object is defined as:

$$Precision_i = \frac{|TrueTopics_i \cap PredTopics_i|}{|PredTopics_i|}, \quad (1)$$

whereas recall of this prediction is:

$$Recall_i = \frac{|TrueTopics_i \cap PredTopics_i|}{|TrueTopics_i|}. \quad (2)$$

The quality measure used in the competition was an average F_1-*score* over all test documents, which can be defined as:

$$F_1\text{-}score_i = 2 \cdot \frac{Precision_i \cdot Recall_i}{Precision_i + Recall_i}, \quad (3)$$

$$AvgF_1\text{-}score = \frac{\sum_{i=1}^{N} F_1\text{-}score_i}{N} \quad (4)$$

In the above formula N is the total number of test documents.

Participants of the challenge could submit multiple solutions, that were evaluated online on a subset of the test data. The preliminary evaluation scores were constantly made available to submitters and the best result of each participating team was published on the competition's leaderboard. Winners of the competition were decided based on their scores computed on the remaining portion of data, called the final test data set. The division between preliminary and final test sets was not revealed.

3 Summary of the Competition

There was a total of 396 teams with 533 members registered to the challenge. Among them, there ware 126 active teams who submitted at least one solution to the leaderboard. The total number of submissions was 5964. The competition attracted participants from 50 different countries across six continents. Seven most represented countries (with a number of registered teams) are listed in Table 1.

Rank	Team	Time of Submission	Preliminary Result	Final Result
1	+ ULjubljana	Mar 31, 00:00:03	0.535	0.53579
2	+ Kebi	Mar 30, 16:54:39	0.530	0.53343
3	+ D'yakonov Alexander	Mar 31, 00:00:10	0.530	0.53242
4	+ purexa	Mar 30, 23:24:31	0.528	0.53094
5	+ asrk	Mar 15, 08:47:53	0.523	0.53032
6	+ UMoscow	Mar 30, 23:23:50	0.531	0.52939
7	+ Andrew Ostapets	Mar 30, 16:33:56	0.528	0.52885
8	+ Dmitry Kondrashkin	Mar 29, 16:43:55	0.528	0.52871
9	+ MLKD	Mar 30, 18:25:32	0.526	0.52719
10	potapenko	Mar 29, 21:10:25	0.521	0.52174

Fig. 1. A snapshot of competition's leaderboard with the final stands of the top 10 teams, including the winners [9]

The participants were able to significantly improve over the baseline result, which was obtained by assigning five majority classes to all objects in the test set. The improvement in the F_1-*score* exceeded 125% (the best team obtained a result of 0.53579, while the baseline was 0.23721).

From the active teams, 41 agreed to share their approach by sending us a brief report. The winning team solved the challenge by using a meta-learning technique in order to ensemble multiple logistic regression models, a random forest and neural networks [9]. The most interesting of other top-ranked approaches are described in several other papers in the RSCTC 2012 proceedings.

Figure 1 presents the final results of the top 10 teams. Scores of the remaining teams are still available online[2].

To verify potential usefulness of a combination of the top-ranked approaches in the contest we made an experiment in which we merged solutions of five best teams. The new classification of the test documents was constructed by a simple majority voting algorithm. The number of labels to be assigned for each document was decided based on an average length of predictions made by each of the considered teams. Quality of this ensemble was measured using the same evaluation method as in the case of the real contest. As expected, the obtained result, which was 0.536 for the preliminary set and 0.53976 for the final test set,

[2] `http://tunedit.org/challenge/JRS12Contest?m=leaderboard`

word_document (R)		word_stem (S)		mesh_inv_ind (T)		mesh_idf (U)	
◆doc_id	LONG	◆word	VARCHAR	◆stem	VARCHAR	◆stem	VARCHAR
◆word_pos	INTEGER	◆stem	VARCHAR	◆concept	VARCHAR	◆idf	REAL
◆word	VARCHAR			◆tf	REAL		

Fig. 2. The tables used in the case study

was slightly better than the score of the winners of JRS'2012 DMC. It shows that there is definitely a lot of space for further improvements.

4 Semantic Tagging within the SONCA Engine

All the data used in the contest were generated using our current implementation of SONCA. The documents were stored in a relational database in a generic structure using the entity-attribute-value model [10]. For the purpose of analytics, this schema was transformed to a more flattened form. For the sake of clarity of the presentation, let us make some simplifying assumptions.

Let us assume that the information about articles is stored in a table called *document_word*. The table contains columns as follows: *doc_id* – identifier of the document, *word_pos* – an ordinal number of the given word in document, and *word* – a word from the document. Thus, to store a document we need as many rows as there are words in it. Below we will refer to this table as *R*.

The second table which is involved in our calculations is called *word_stem*, and denoted by *S*. The table contains two columns *word* and *stem* – which represent a stem of the given word. A stem is the root of a word. The table stores the information about the stem and the stemming process, which was performed earlier using the standard Porter stemming algorithm[3].

The third table needed to present a semantic tagging process. It is called *stem_inv_ind* and denoted by *T*. The table contains three columns, as follows: *stem, concept* – a name of a concept from MeSH controlled vocabulary and *tf* – a term frequency[4] in the concept description from MeSH.

The last table needed in the calculations is called *mesh_idf*. It is denoted by *U* and contains two columns – *stem* and *idf* – invert document (in this case concept's description) frequency[4]. Figure 2 outlines the introduced tables. Data stored in those tables is used by the ESA method, which aims at determining semantic relationships between documents and (MeSH) concepts.

In our system we want to associate each document with a list of concepts from ontology or knowledge base, such as MeSH [5] and others. This, technically speaking, corresponds to creation of a vector of ontology concepts associated with a document. The vector is constructed in such a way, that each position corresponds to an ontology concept and a numerical value at this position represents a strength

---

[3] http://tartarus.org/~martin/PorterStemmer/
[4] http://en.wikipedia.org/wiki/Tf*idf

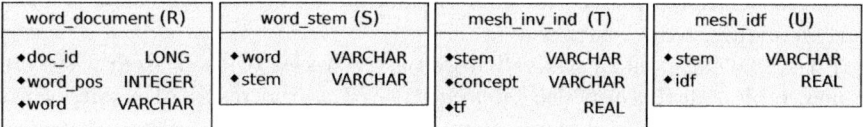

of the association. We describe this calculation with relational calculus formulæ corresponding to SQL queries.

First, the inflected or derived words are reduced to their stems. We create a new table called *word_doc_stemmed* – *R2* as the result of a join of tables *document_word* – *R* and *word_stem* – *S*.

$$R2 \leftarrow \Pi_{(R.doc\_id, S.stem)}(R \underset{word=word}{\bowtie} S) \tag{5}$$

The next task is a calculation of stem frequencies within the documents. We perform this using one table *R2*. The term (stem) frequency is calculated as:

$$R3 \leftarrow \Pi_{(V1.doc\_id, V1.stem, (V1.cnt/V2.cnt\_all) \rightarrow tf)} \Bigg(\rho_{V1}\Big(\gamma_{(R2.doc\_id, R2.stem, R2.count(*) \rightarrow cnt)}(R2)\Big) \underset{doc\_id=doc\_id}{\bowtie} \rho_{V2}\Big(\gamma_{(R2.doc\_id, R2.count(*) \rightarrow cnt\_all)}(R2)\Big)\Bigg) \tag{6}$$

The final step is a calculation of the vectors of concepts associated with the documents and the association strengths.

$$R4 \leftarrow \Pi_{(R3.doc\_id, T.concept, assoc)}\Bigg(\tau_{(assocDESC)}\Big(\gamma_{(R3.doc\_id, T.concept, SUM(\sqrt{R3.tf * T.tf} * U.idf^2) \rightarrow assoc)}\big(R3 \underset{stem=stem}{\bowtie} T \underset{stem=stem}{\bowtie} U\big)\Big)\Bigg) \tag{7}$$

The queries presented above return a complete information, i.e., for each document they give us the levels of association with each and every concept in our ontology (knowledge base). This is both unnecessary and unwanted in practical applications. Empirical experiments show that if we are to present the results to the user, we shall present no more than top-k most associated concepts, with $k \leq 30$. Anything above 30 is likely to produce a perceptual noise. So, as a last step in the calculation we shall prune the result, leaving only the top 30 most associated concepts in each documents' representation.

The calculation of the associations among documents and concepts according to formula (7) is a very time consuming task. From the database viewpoint the problem we want to solve is one of performing a *top-k* analytic, an agglomerative SQL query that involves joins on very large data tables (see [11]). The characteristic property of our task is a possibility to decompose it into smaller subtasks (tasks on sub-tables), given some knowledge about the nature and structure of our data set. The fact that the query-answering can be decomposed into largely independent sub-tasks makes it possible to optimize it by using only top-k instead of all sub-results most of the time. Inasmuch as sub-tasks are

Algorithm 1. Query sharding

```
begin
    N := SELECT DISTINCT doc_id from TABLE
    for doc_id ∈ N do
        run SELECT ... WHERE DOC_ID = doc_id in K threads concurrently
    end
end
```

largely independent from one another, we can also create shards and process them concurrently using, e.g., multiple cores (processors).

By running queries for each of the pieces (documents) separately we achieve an additional profit. We are able to handle queries that require application of the `LIMIT` operator within the `GROUP BY` statement. This functionality was added in the SQL:1999 and SQL:2003 standards by introducing windowing functions and elements of procedural languages. However, these functionalities are not supported by the database solutions utilized in the SONCA framework yet. The ability to limit the processing to only top-k objects (documents) can make a big difference in an execution time [12].

In a case of the query (7), sharding corresponds to creation of separate subqueries for each of the objects, since we know that there is no interference with other objects along the calculation. Objects correspond to documents, and the boundary of an object can be determined by detecting the change of *id* in the column *doc_id*. Thus, each of the considered queries can be decomposed into a series of simpler queries by basing on Algorithm 1.

In the JRS'2012 Data Mining Competition, the data generated by the above-described method was post-processed in order to better fit to the competition's task. After selecting the 30 most associated concepts for each of the processed documents, the associations were extended by a list of neighborhood concepts from the tree structure of MeSH [2]. For a given pair of a document *doc* and a MeSH heading M, the semantic representation of the document was extended by all its relatives with a weight corresponding to the association strength of M to *doc* divided by a total number of relatives of M in MeSH. Since MeSH is organized in a tree-like structure with concepts higher in the hierarchy being more general than those which are lower, children and parents of M were counted separately and their associations were computed with different weights.

Finally, from our corpus of documents collected from PubMed Central open subset [1], we selected 20000 papers for the challenge. We constructed an information system in a form of a flat table having 20000 rows (corresponding to the documents) and 25640 columns. Each attribute represented a MeSH heading which was associated to at least one document in the whole corpus (but not necessarily to one of the documents selected for the contest). We linearly scaled the resulting data so that each association strength was between 0 and 1000, and we rounded the results.

The target labels in the task corresponded to MeSH subheadings pointed out by domain experts. These tags expressed a context for concepts associated with the analyzed texts and were obtained directly from PubMed Central pages.

5 Our Plans for the Future

The traditional search engines are not the best techniques, inasmuch the results related to the keywords of a user query are displayed considering only their literal meaning. Therefore, emergence of new technologies inter alia *Semantic Search* carries a hope for an improvement of search methods. Within semantic search process, the conceptual meaning of the query is considered. However, for a successful implementation of semantic search engines, the semantic text representation (semantic labeling and indexing of documents) are needed.

Labeling documents with concepts that correspond to them semantically has a significant potential. It delivers to a domain expert better methods of navigation across knowledge repositories. However, a manual documents labeling is a very time consuming and expensive method. Especially laborious task is labeling parts of documents like chapters, sections and even sentences. The appearing automatic labeling methods allows to cope with this issue.

The algorithms developed and presented during the contest showed that there is a significant potential in automatic semantic document tagging and indexing. The potential profits from using these methods are as follows: a) improvement of the search engines accuracy by understanding the contextual meaning of texts, b) ability to figure out the most relevant chapter, section or sentence for a user query, c) concept and knowledge matching.

Our current researches are focused on improving the accuracy of our tagging algorithms. In the future research we would also like to investigate the problem of online model updating that utilizes user feedback.

References

1. Roberts, R.J.: PubMed Central: The GenBank of the published literature. Proceedings of the National Academy of Sciences of the United States of America 98(2), 381–382 (2001)
2. United States National Library of Medicine: Introduction to MeSH – 2011 (2011), `http://www.nlm.nih.gov/mesh/introduction.html`
3. Gabrilovich, E., Markovitch, S.: Computing semantic relatedness using wikipedia-based explicit semantic analysis. In: Proceedings of the Twentieth International Joint Conference for Artificial Intelligence, Hyderabad, India, pp. 1606–1611 (2007)
4. Janusz, A., Świeboda, W., Krasuski, A., Nguyen, H.S.: Interactive Document Indexing Method Based on Explicit Semantic Analysis. In: Yao, J.T., et al. (eds.) RSCTC 2012. LNCS (LNAI), vol. 7413, pp. 156–165. Springer, Heidelberg (2012)
5. Ślęzak, D., Janusz, A., Świeboda, W., Nguyen, H.S., Bazan, J.G., Skowron, A.: Semantic Analytics of PubMed Content. In: Holzinger, A., Simonic, K.-M. (eds.) USAB 2011. LNCS, vol. 7058, pp. 63–74. Springer, Heidelberg (2011)

6. Wojnarski, M., Stawicki, S., Wojnarowski, P.: TunedIT.org: System for Automated Evaluation of Algorithms in Repeatable Experiments. In: Szczuka, M., Kryszkiewicz, M., Ramanna, S., Jensen, R., Hu, Q. (eds.) RSCTC 2010. LNCS, vol. 6086, pp. 20–29. Springer, Heidelberg (2010)
7. Wojnarski, M., Janusz, A., Nguyen, H.S., Bazan, J., Luo, C., Chen, Z., Hu, F., Wang, G., Guan, L., Luo, H., Gao, J., Shen, Y., Nikulin, V., Huang, T.-H., McLachlan, G.J., Bošnjak, M., Gamberger, D.: RSCTC'2010 Discovery Challenge: Mining DNA Microarray Data for Medical Diagnosis and Treatment. In: Szczuka, M., Kryszkiewicz, M., Ramanna, S., Jensen, R., Hu, Q. (eds.) RSCTC 2010. LNCS (LNAI), vol. 6086, pp. 4–19. Springer, Heidelberg (2010)
8. Janusz, A., Ślęzak, D., Nguyen, H.S.: Unsupervised similarity learning from textual data. Fundamenta Informaticae (2012)
9. Žbontar, J., Žitnik, M., Zidar, M., Majcen, G., Potočnik, M., Zupan, B.: Team ULjubljana's Solution to the JRS 2012 Data Mining Competition. In: Yao, J.T., et al. (eds.) RSCTC 2012. LNCS (LNAI), vol. 7413, pp. 471–478. Springer, Heidelberg (2012)
10. Kowalski, M., Ślęzak, D., Stencel, K., Pardel, P., Grzegorowski, M., Kijowski, M.: RDBMS Model for Scientific Articles Analytics. In: Bembenik, R., Skonieczny, L., Rybiński, H., Niezgodka, M. (eds.) Intelligent Tools for Building a Scient. Info. Plat. SCI, vol. 390, pp. 49–60. Springer, Heidelberg (2012)
11. Michel, S., Triantafillou, P., Weikum, G.: KLEE: a framework for distributed top-k query algorithms. In: Proceedings of the 31st International Conference on Very Large Data Bases, VLDB 2005, VLDB Endowment, pp. 637–648 (2005)
12. Krasuski, A., Szczuka, M.: Knowledge driven query sharding. In: 16th East-European Conference on Advances in Databases and Information Systems, Poznan, Poland, September 18-21 (2012) (submitted to conference)

A Blending of Simple Algorithms for Topical Classification*

Alexander D'yakonov

Moscow State University, Leninskie Gory, 119991, Moscow, Russia
djakonov@mail.ru

Abstract. Algorithm which has taken the third place in "JRS 2012 Data Mining Competition" among 126 participants is described. The competition was related to the problem of predicting topical classification of scientific publications in a field of biomedicine. The presented algorithm is a combination (blend) of simple classification algorithms: a linear classifier, a k-NN classifier and two SVMs. We build the combination using special estimation matrices. It proves again that combinations have significantly better performance compared to their individual members.

Keywords: topical classification, blending, simple algorithms, text classification, SVD, SVM, k-NN, linear classifier.

1 Introduction

Organizers of "JRS 2012 Data Mining Competition" [1] have gathered a corpus of documents containing 20000 journal articles from the PubMed Central open-access subset [2]. The documents have been labeled by biomedical experts from PubMed with several MeSH subheadings [3]. A label of the document can be viewed as context or topic discussed in the text. One document can have several labels. All the documents are associated with the most related MeSH terms. The competition data consists of information about strengths of those bonds, expressed as numerical values.

The training data consists of a description $m \times n$ matrix X (the feature matrix) and a classification $m \times l$ matrix Y, where $m = 10000$ is a number of articles in the training set, $n = 25640$ is a number of features (MeSH terms), $l = 82$ is a number of classes (topics). The ith row of the matrix X represents the ith journal article from the training set. The jth column of the matrix X contains integers ranging from 0 to 1000, expressing association strength to corresponding MeSH term. The ijth element of the binary matrix Y is equal to 1 iff the ith journal article is related to the jth topic (the article has the jth label).

* This work was supported by the Russian Foundation for Basic Research, project 12-07-00187; by the President of the Russian Federation, project MD-757.2011.9. The author is also grateful to the organizers of "JRS 2012 Data Mining Competition" for running the interesting competition. Finally, we want to thank all the active participants of the challenge for their efforts.

J.T. Yao et al. (Eds.): RSCTC 2012, LNAI 7413, pp. 432–438, 2012.

The test data contains only a $q \times n$ matrix X_2 that describes articles with unknown (for participants) topic classifications, where $q = 10000$ is a number of articles in the test set. The task is to reveal a classification $q \times l$ matrix Y_2 for the test set.

Submissions of participants were evaluated using an average F-score [4] of the predictions:

$$\frac{1}{q}\sum_{i=1}^{q}\frac{2\sum_{j=1}^{l} a_{ij}y_{ij}}{\sum_{j=1}^{l} a_{ij} + \sum_{j=1}^{l} y_{ij}},$$

where $||a_{ij}||_{q\times l}$ was a submitted solution, $||y_{ij}||_{q\times l}$ was the correct classification matrix. The submitted solutions were evaluated on-line on a random subset of the test set ($\sim 10\%$ of all the data), fixed for all participants. The final evaluation was performed after completion of the competition using the remaining part of the test data.

Now we describe the algorithm which has taken the third place in "JRS 2012 Data Mining Competition" with the preliminary result 0.53 and the final result 0.53242 F-score.

2 Combination of Algorithms

The main idea behind our solution is a blending of very simple algorithms, because the usage of different algorithms can sufficiently improve the performance [5], and simple algorithms are more reliable and easy to tune. The same idea is the basis of modern voting classification algorithms, such as Bagging [6] or AdaBoost [7].

Let the $q \times l$ matrix $E = ||e_{ij}||_{q\times l}$ be an output of an algorithm. This matrix may be also a result of an intermediate step of the algorithm. In matrix E the element e_{ij} is equal to estimation of belonging the ith test object (the ith article) to the jth class. The matrix E we will call the estimation matrix. One may easily obtain classification by comparing matrix elements with a threshold: if $e_{ij} > \theta$ then the ith journal article has the jth label. The threshold θ can be calculated by optimizing algorithm performance in local testing (on the training set).

Let $E_1, \ldots, E_k$ be the output estimation matrices from k algorithms. Then, the final classification can be obtained by comparing the matrix

$$c_1E_1 + \cdots + c_kE_k \tag{1}$$

with the threshold θ. The coefficients $c_1, \ldots, c_k$ and the threshold θ can be calculated by gradient descend optimization of F-score on the training set. The same idea (of using the linear combination of estimation matrices) is in the base of an algebraic approach to classification problem solving [8]– [9].

Our experiments show that more complex expressions (than the simple linear combination (1)) can increase performance on 0.5%. In the final solution we use

$$c_1N_{\sqrt{\max}}(E_1) + \cdots + c_kN_{\sqrt{\max}}(E_k) \tag{2}$$

instead of (1), where the operation $N_{\sqrt{\max}}(E)$ divides each element of the matrix E on a root of the maximal element in his row:

$$N_{\sqrt{\max}}(||e_{ij}||) = \left|\left| \frac{e_{ij}}{\sqrt{\max(e_{i1}, \ldots, e_{il})}} \right|\right|.$$

The idea of the deformed linear combination (2) follows from LENKOR technology for data mining problem solving developed by the author [10].

Now we describe machine learning algorithms that we used to calculate estimation matrices from (2).

3 The Linear Classifier over SVDecomposition

It is natural to treat the contest task as a standard linear regression problem [11]. This corresponds to finding a least squares solution C for the equation

$$XC = Y. \tag{3}$$

The matrix

$$X_2 C = X_2 (X^T X)^{-1} X^T Y \tag{4}$$

can be viewed as an estimation matrix, so the last step of this approach is threshold choosing. However the equation (3) is over parameterized (the size of C is $n \times l$) and the matrix $X^T X$ in (4) is singular. The regularization

$$X_2 (X^T X + \varepsilon I)^{-1} X^T Y$$

(it is also known as ridge regression) does not improve performance and takes a lot of time and memory. Therefore it is necessary to "reduce" the matrices X and X_2.

The singular value decomposition (SVD) of the matrix X has the form $X = U \Sigma V^T$ and we can use a low-rank approximation of X:

$$X \simeq U[\,,1{:}r] \cdot \Sigma[1{:}r,1{:}r] \cdot (V[\,,1{:}r])^T,$$

where $V[\,,1{:}r]$ is a submatrix of V formed by the first r columns. We suggest to use the matrices

$$X \cdot V[\,,1{:}r], \quad X_2 \cdot V[\,,1{:}r]$$

as feature matrices for regression problem solving. Then, the estimation matrix is

$$X_2 \cdot V[\,,1{:}r] \cdot \left((X \cdot V[\,,1{:}r])^T \cdot (X \cdot V[\,,1{:}r])\right)^{-1} \cdot (X \cdot V[\,,1{:}r])^T \cdot Y. \tag{5}$$

Our experiments show that the best choice is to take $r = 700$ (see Table 1) and to replace matrices X, X_2 in (5) with matrices $N_{\sqrt{\Sigma}}(X)$, $N_{\sqrt{\Sigma}}(X_2)$, where

$$N_{\sqrt{\Sigma}}(||x_{ij}||) = \left|\left| \frac{x_{ij}}{\sqrt{\sum_{j=1}^{n} x_{ij}}} \right|\right|. \tag{6}$$

Table 1. Performance of SVD+ regression

$r =$	400	500	600	700	800
F-score	47.3	50.0	51.4	**51.6**	51.4

The modified matrices contributed to improvement in result from 0.513 to 0.516. Note that the normalization (6) is not standard (x_{ij} instead of x_{ij}^2 under the sum symbol).

The described algorithm is the best in the blending.

4 k-Nearest Neighbor Algorithm (k-NN)

Even the simplest k-NN method has been shown to perform well in text classification [12]. We use a weighted k-NN algorithm where each point has a weight and neighboring points have a higher vote than the farther points. Firstly, we normalize the feature matrices (X and X_2):

$$||x_{ij}|| \longrightarrow \left|\left|\frac{x_{ij}}{s_j}\right|\right|$$

(each element in the jth column is divided by s_j), where

$$s_j = \frac{1}{l}\sum_{i=1}^{l} \log(s_{ij}+1),$$

$$||s_{ij}||_{q\times l} = Y^T \cdot X.$$

It is tf-idf-like transformation [12] for multilabel task. Some other transformations that we used are listed in Table 2. The second normalization is $N_{\sqrt{\Sigma}}$ (see (6)). Then we use weighted k-NN with $k = 200$, weights

$$w_1 = \frac{k^2}{S_k}, \quad \dots, \quad w_{k-1} = \frac{2^2}{S_k}, \quad w_k = \frac{1^2}{S_k},$$

$$S_k = k^2 + \ldots + 2^2 + 1^2,$$

where w_j is the weight given to the jth neighbor, and similarity function

$$B((x_1,\ldots,x_n),(y_1,\ldots,y_n)) = x_1y_1 + \cdots + x_ny_n.$$

If the second normalization was standard, it would be usual cos-similarity function, which is popular in text classification [12]. Surprisingly, the nonstandard $N_{\sqrt{\Sigma}}$-normalization improves performance by 0.6%.

Table 2. Performance of k-NN

some normalizations	F-score
our normalization	**0.497**
tf-idf	0.470
$s_i = \lvert\{j \in \{1,2,\dots,l\} \mid s_{ij} > 0\}\rvert$	0.489

So, the method outputs a $q \times l$ matrix A_2, where each row is the sum of rows of the matrix Y with the corresponding weights. Similarly on the training set the method outputs a $q \times m$ matrix A (we use leave-one-out [13]). The last step is a linear regression and the final answer (the estimation matrix) of our k-NN algorithm is

$$A_2(A^T A)^{-1} A^T Y.$$

5 Two LIBSVM Algorithms

The last two estimation matrices in (2) are made by LIBSVM package [14]:

```
svmtrain(Y, X, '-t 0 -c 0.01 -h 0')
```

We run two SVM algorithms with linear kernel functions and $c = 0.01$ [15]. The algorithms are different in preliminary normalizations of feature matrices:

$$||x_{ij}|| \longrightarrow \left|\left|\frac{x_{ij}}{\max(x_{i1},\dots,x_{in})}\right|\right|$$

(each element is divided by the maximal element in the row),

$$||x_{ij}|| \longrightarrow \left|\left|\frac{x_{ij}}{\max(x_{1j},\dots,x_{mj})}\right|\right|$$

(each element is divided by the maximal element in the column). These normalizations are chosen to make the algorithms independent even when tuning by the same package.

The reason to use normalizations is very simple: when you tune SVM all features should be in one scale. Unfortunately we did not manage to perform complete research of all the possible normalizations.

After running of LIBSVM package we apply a standard linear regression, as after the k-NN method.

Given that the execution time of LIBSVM was in the neighbourhood of one hour[1], it is reasonable to rely on the faster LIBLINEAR algorithm [17].

[1] On a HP p6050ru computer with Intel Core 2 Quad CPU Q8200 2.33GHz, RAM 3Gb, OS Windows Vista in MATLAB 7.10.0 [16].

6 Conclusions

In this paper we present our solution for the topical classification task of "JRS 2012 Data Mining Competition". We try to use only traditional simple algorithms (a linear regression, a k-NN method, SVMs with linear kernel). The core of our approach is the blending (2) using the estimation matrices. The matrices are constructed by linear regression and the blending is tuned to improve F-score in local tests (on the training set).

We used a similar approach before for winning "ECML/PKDD Discovery Challenge 2011" [10]. In our opinion it is a very good strategy to find "deformed" linear combination of estimation matrices of simple classifiers. The same idea is main in the theory of neural networks [18].

We expect that better results can be obtained by optimizing LIBSVM algorithms and using LIBLINEAR package instead of LIBSVM. Surprisingly, many other simple algorithms do not improve performance of the blending, for example popular in text mining centroid classifier (this algorithm has only 0.42 F-score). The final solution is

$$0.5939N_{\sqrt{\max}}(E_{\text{lin+svd}}) + 0.0011N_{\sqrt{\max}}(E_{\text{knn}}) + \\ 0.2970N_{\sqrt{\max}}(E_{\text{libsvm1}}) + 0.1080N_{\sqrt{\max}}(E_{\text{libsvm1}}) <> \theta = 0.345.$$

In our local testing the maximum is reached at another threshold value (see Table 3). We simply did not manage to submit the solution with $\theta = 0.35$.

Note that the contribution to the solution of the k-NN classifier is very small, however one can be in Top 20 of the competition leaderboard by using the only k-NN algorithm. Without LIBSVM-algorithms the blending has performance only 0.517 F-score.

Table 3. Performance of the blending (in local tests)

$\theta =$	0.335	0.34	0.345	0.350	0.355
F-score	52.72	53.60	53.79	**53.99**	53.79

References

1. http://tunedit.org/challenge/JRS12Contest
2. National Library of Medicine: PubMed Central (PMC): An Archive for Literature from Life Sciences Journals. In: McEntyre, J., Ostell, J. (Eds.): The NCBI Handbook, http://www.ncbi.nlm.nih.gov/books/NBK21087/
3. National Library of Medicine: Introduction to MeSH (2012), http://www.nlm.nih.gov/mesh/introduction.html
4. van Rijsbergen, C.J.: Information Retrieval, 2nd edn. Butterworth (1979)

5. Bauer, E., Kohavi, R.: An Empirical Comparison of Voting Classification Algorithms: Bagging, Boosting, and Variants. Machine Learning 36(1-2), 105–139 (1999)
6. Breiman, L.: Bagging predictors. Machine Learning 24, 123–140 (1996)
7. Freund, Y., Schapire, R.E.: A short introduction to boosting. Journal of Japanese Society for Artificial Intelligence 14(5), 771–780 (1999)
8. Zhuravlev, Y.I.: An Algebraic Approach to Recognition and Classification Problems. In: Problems of Cybernetics, vol. 33, pp. 5–68. Nauka, Moscow (1978); Hafner (1986)
9. Zhuravlev, Yu, I.: Correct Algorithms over Sets of Incorrect (Heuristic) Algorithms: Part II. Kibernetika 6, 21–27 (1977)
10. D'yakonov, A.: Two Recommendation Algorithms Based on Deformed Linear Combinations. In: Proc. of ECML-PKDD 2011 Discovery Challenge Workshop, pp. 21–28 (2011), http://ceur-ws.org/Vol-770/paper5.pdf
11. http://en.wikipedia.org/wiki/Linear_regression
12. Manning, C., Raghavan, P., Schutze, H.: Introduction to Information Retrieval. Cambridge University Press (2008)
13. http://en.wikipedia.org/wiki/Cross-validation_(statistics)
14. Chang, C.-C., Lin, C.-J.: LIBSVM: a library for support vector machines. ACM Transactions on Intelligent Systems and Technology 2(3) (2011), http://www.csie.ntu.edu.tw/~cjlin/libsvm
15. Cortes, C., Vapnik, V.: Support-Vector Networks. Machine Learning, 20 (1995)
16. http://www.mathworks.com
17. Fan, R.-E., Chang, K.-W., Hsieh, C.-J., Wang, X.-R., Lin, C.-J.: LIBLINEAR: A Library for Large Linear Classification. Journal of Machine Learning Research 9, 1871–1874 (2008), http://www.csie.ntu.edu.tw/~cjlin/liblinear
18. Haykin, S.: Neural Networks: A Comprehensive Foundation. Prentice Hall (1999)

F-Measure Maximization in Topical Classification

Weiwei Cheng[1], Krzysztof Dembczyński[2], Eyke Hüllermeier[1], Adrian Jaroszewicz[2], and Willem Waegeman[3]

[1] Department of Mathematics and Computer Science, Marburg University, Germany
{cheng,eyke}@mathematik.uni-marburg.de
[2] Institute of Computing Science, Poznań University of Technology, Poland
{kdembczynski,ajaroszewicz}@cs.put.poznan.pl
[3] Department of Mathematical Modelling, Statistics and Bioinformatics, Ghent University, Belgium
willem.waegeman@ugent.be

Abstract. The F-measure, originally introduced in information retrieval, is nowadays routinely used as a performance metric for problems such as binary classification, multi-label classification, and structured output prediction. In this paper, we describe our methods applied in the JRS 2012 Data Mining Competition for topical classification, where the instance-based F-measure is used as the evaluation metric. Optimizing such a measure is a statistically and computationally challenging problem, since no closed-form maximizer exists. However, it has been shown recently that the F-measure maximizer can be efficiently computed if some properties of the label distribution are known. For independent labels, it is enough to know marginal probabilities. An algorithm based on dynamic programming is then able to compute the F-measure maximizer in cubic time with respect to the number of labels. For dependent labels, one needs a quadratic number (with respect to the number of labels) of parameters for the joint distribution to compute (also in cubic time) the F-measure maximizer. These results suggest a two step procedure. First, an algorithm estimating the required parameters of the distribution has to be run. Then, the inference algorithm computing the F-measure maximizer is used over these estimates. Such a procedure achieved a very satisfactory result in the JRS 2012 Data Mining Competition.

1 Introduction

While being rooted in information retrieval [1], the so-called F-measure is nowadays routinely used as a performance metric for different types of prediction problems, including binary classification, multi-label classification (MLC), and certain applications of structured output prediction, like text chunking and named entity recognition. Compared to measures like the 0-1 loss in binary classification and the Hamming loss in MLC, it enforces a better balance between performance on the minority and the majority class, and it is hence more suitable in the case of imbalanced data, which arises quite frequently in real-world applications.

J.T. Yao et al. (Eds.): RSCTC 2012, LNAI 7413, pp. 439–446, 2012.

The predictive task in the JRS 2012 Data Mining Competition[1] falls into such a category. Generally speaking, this competition concerns the topical classification of biomedical research papers based on the concept information from the MeSH ontology,[2] which are automatically assigned by the tagging system. More precisely, as the training data, there are in total 10000 instances with 25640 features and 83 classes. The values of the features are presented as integers ranging from 0 to 1000, expressing association strengths to corresponding MeSH terms, and the classes correspond to the topic identifiers. There are another 10000 instances as the test data. They share the same format as the training data, except that the class information is not given. Similar to other text classification problems, the data of the JRS competition are very sparse. Consider the training data for example, the most dense feature has 2738 nonzero entries and the most dense class is associated with 2475 instances. The sparseness of the data calls for evaluation metrics like the F-measure. More precisely, the instance-based F-measure is applied in the JRS competition, which we shall discuss later in more details.

The paper is organized as follows. We first introduce the formal setting of multi-label classification and the definition of the instance-based F-measure in Section 2. Inference techniques for F-measure maximization are discussed in Section 3, where we start with the case of independent class labels and then discuss the more general case without the independence assumption. These inference techniques are based on the parameters of the label distribution. We discuss the estimation of such parameters in Section 4. Some empirical evaluations of our approaches are shown in Section 5, prior to the final conclusion in Section 6.

2 Multi-label Learning and Instance-Based F-Measure

The task of the JRS competition is a multi-label learning problem. Let $\mathcal{X}$ denote an instance space, and let $\mathcal{L} = \{\lambda_1, \lambda_2, \ldots, \lambda_m\}$ be a finite set of class labels. An instance $\boldsymbol{x} \in \mathcal{X}$ is (non-deterministically) associated with a subset of labels $L \in 2^{\mathcal{L}}$; this subset is called the set of relevant labels, while the complement $\mathcal{L} \setminus L$ is considered as irrelevant for $\boldsymbol{x}$. It is common to identify L with a binary vector $\boldsymbol{y} = (y_1, y_2, \ldots, y_m)$, where $y_i = 1$ means $\lambda_i \in L$. We denote the set of possible labelings as $\mathcal{Y} = \{0,1\}^m$.

Given a prediction $\boldsymbol{h}(\boldsymbol{x}) = (h_1(\boldsymbol{x}), \ldots, h_m(\boldsymbol{x})) \in \mathcal{Y}$ of an m-dimensional binary label vector $\boldsymbol{y} = (y_1, \ldots, y_m)$, the label vector associated with a single instance, the instance-based F-measure is defined as follows:

$$F(\boldsymbol{y}, \boldsymbol{h}(\boldsymbol{x})) = \frac{2\sum_{i=1}^m y_i h_i(\boldsymbol{x})}{\sum_{i=1}^m y_i + \sum_{i=1}^m h_i(\boldsymbol{x})} \in [0,1] \, , \tag{1}$$

[1] `http://tunedit.org/challenge/JRS12Contest`

[2] `http://www.nlm.nih.gov/mesh/introduction.html`

where $0/0 = 1$ by definition. This measure essentially corresponds to the harmonic mean of precision *prec* and recall *rec*:

$$prec(\boldsymbol{y}, \boldsymbol{h}(\boldsymbol{x})) = \frac{\sum_{i=1}^{m} y_i h_i(\boldsymbol{x})}{\sum_{i=1}^{m} h_i(\boldsymbol{x})}, \quad rec(\boldsymbol{y}, \boldsymbol{h}(\boldsymbol{x})) = \frac{\sum_{i=1}^{m} y_i h_i(\boldsymbol{x})}{\sum_{i=1}^{m} y_i}. \tag{2}$$

One can generalize the F-measure to a weighted harmonic average of these two values, but for the sake of simplicity, we stick to the unweighted mean, which is often referred to as the F1-score or the F1-measure. This variant of the F-measure was also used in the competition.

Modeling the ground-truth as a random variable $\boldsymbol{Y}$, i.e., assuming an underlying probability distribution $p(\boldsymbol{Y})$ on $\{0,1\}^m$, the prediction $\boldsymbol{h}_F^*(\boldsymbol{x})$ that maximizes the expected F-measure is given by

$$\begin{aligned} \boldsymbol{h}_F^*(\boldsymbol{x}) &= \underset{\boldsymbol{h}(\boldsymbol{x}) \in \{0,1\}^m}{\arg\max} \; \mathbb{E}_{\boldsymbol{y} \sim p(\boldsymbol{Y})} \left[F(\boldsymbol{y}, \boldsymbol{h}(\boldsymbol{x})) \right] \\ &= \underset{\boldsymbol{h}(\boldsymbol{x}) \in \{0,1\}^m}{\arg\max} \sum_{\boldsymbol{y} \in \{0,1\}^m} p(\boldsymbol{Y} = \boldsymbol{y}) \, F(\boldsymbol{y}, \boldsymbol{h}(\boldsymbol{x})) \,. \end{aligned} \tag{3}$$

Unfortunately, a closed form of the maximizer $\boldsymbol{h}_F^*(\boldsymbol{x})$ does not exist and a brute-force search is infeasible, as it would require checking all 2^m combinations of prediction vector $\boldsymbol{h}$ and computing a sum over an exponential number of terms for each $\boldsymbol{h}$. However, several algorithms have been introduced recently that compute the F-measure maximizer efficiently.

3 Algorithms for F-Measure Maximization

The problem (3) can be solved via outer and inner maximization [2]. Namely, (3) can be transformed into an inner maximization

$$\boldsymbol{h}^{(k)^*} = \underset{\boldsymbol{h} \in H_k}{\arg\max} \, \mathbb{E}_{\boldsymbol{y} \sim p(\boldsymbol{Y})} \left[F(\boldsymbol{y}, \boldsymbol{h}) \right] , \tag{4}$$

where $H_k = \{\boldsymbol{h} \in \{0,1\}^m \mid \sum_{i=1}^{m} h_i = k\}$, followed by an outer maximization

$$\boldsymbol{h}_F^* = \underset{\boldsymbol{h} \in \{\boldsymbol{h}^{(0)^*}, \ldots, \boldsymbol{h}^{(m)^*}\}}{\arg\max} \mathbb{E}_{\boldsymbol{y} \sim p(\boldsymbol{Y})} \left[F(\boldsymbol{y}, \boldsymbol{h}) \right] . \tag{5}$$

The outer maximization (5) can be done by simply checking all $m+1$ possibilities. The main effort is then required for solving the inner maximization (4).

3.1 Label Independence

By assuming independence of the random variables $Y_1, \ldots, Y_m$, the optimization problem (3) can be substantially simplified. It has been shown independently in [3] and [2] that the optimal solution always contains the labels with the highest marginal probabilities $p_i = P(Y_i = 1)$, or no labels at all. As a consequence, only a few ($m+1$ instead of 2^m) hypotheses $\boldsymbol{h}$ need to be examined.

Furthermore, Lewis [3] has shown that the expected F-measure can be approximated by the following expression under the assumption of independence:[3]

$$\mathbb{E}_{\boldsymbol{y}\sim p(\boldsymbol{Y})}\left[F(\boldsymbol{y},\boldsymbol{h})\right] \simeq \begin{cases} \prod_{i=1}^{m}(1-p_i) & \text{if } \boldsymbol{h}=\boldsymbol{0}\ , \\ \dfrac{2\sum_{i=1}^{m} p_i h_i}{\sum_{i=1}^{m} p_i + \sum_{i=1}^{m} h_i} & \text{if } \boldsymbol{h}\neq\boldsymbol{0}\ . \end{cases} \tag{6}$$

This approximation is exact for $\boldsymbol{h}=\boldsymbol{0}$, and is tractable with $\mathcal{O}(m)$. For $\boldsymbol{h}\neq\boldsymbol{0}$, an upper bound of the error can easily be determined [3]. However, the exact solution can be computed efficiently, as will be explained in more details below.

Jansche [2] and Chai [4] have independently proposed exact procedures for solving the inner maximization (4). The former runs in $\mathcal{O}(m^3)$, while the latter runs in $\mathcal{O}(m^2)$, leading to the overall complexity of $\mathcal{O}(m^4)$ and $\mathcal{O}(m^3)$, respectively. Since both algorithms deliver the same estimate, we focus on Chai's approach here. We refer to it as DP, since it is based on dynamic programming.

Chai [4] has shown that the expected F-measure of $\boldsymbol{h}^{(k)^*}$, the solution of the inner maximization (4) for a given k that assigns ones to k labels with the largest marginal probabilities, can be expressed as follows:

$$\mathbb{E}_{\boldsymbol{y}\sim p(\boldsymbol{Y})}\left[F(\boldsymbol{y},\boldsymbol{h}^{(k)})\right] = 2\prod_{i=1}^{m}(1-p_i)I_1(m)\ ,$$

where $I_1(m)$ is given by the following recurrent equations and boundary conditions:

$$\begin{aligned} I_t(a) &= I_{t+1}(a) + r_t I_{t+1}(a+1) + r_t J_{t+1}(a+1) \\ J_t(a) &= J_{t+1}(a) + r_t J_{t+1}(a+1) \\ I_{k+1}(a) &= 0 \quad J_{m+1}(a) = a^{-1} \end{aligned}$$

with $r_i = p_i/(1-p_i)$. These equations suggest a dynamic programming algorithm of space $\mathcal{O}(m)$ and time $\mathcal{O}(m^2)$ for solving the inner maximization (4) for given k.

3.2 A General Procedure

If the independence assumption is violated, the above methods may produce predictions far away from the optimal one, as shown in [5] by Dembczynski et al. In this paper, the authors have further introduced an exact and efficient algorithm for computing the F-measure maximizer without using any additional assumption on the probability distribution $p(\boldsymbol{Y})$. The algorithm, called general F-measure maximizer (GFM), needs m^2+1 parameters and runs in $\mathcal{O}(m^3)$.

The inner optimization problem (4) can be formulated as follows:

$$\boldsymbol{h}^{(k)^*} = \operatorname*{arg\,max}_{\boldsymbol{h}\in H_k} \mathbb{E}_{\boldsymbol{y}\sim p(\boldsymbol{Y})}\left[F(\boldsymbol{y},\boldsymbol{h})\right] = \operatorname*{arg\,max}_{\boldsymbol{h}\in H_k} \sum_{\boldsymbol{y}\in\{0,1\}^m} p(\boldsymbol{y})\frac{2\sum_{i=1}^{m} y_i h_i}{s_{\boldsymbol{y}}+k}\ ,$$

[3] We henceforth denote **0** and **1** as vectors containing all zeros and ones, respectively.

with $s_{\boldsymbol{y}} = \sum_{i=1}^{m} y_i$. The sums can be swapped, resulting in

$$\boldsymbol{h}^{(k)^*} = \arg\max_{\boldsymbol{h}\in H_k} 2\sum_{i=1}^{m} h_i \sum_{\boldsymbol{y}\in\{0,1\}^m} \frac{p(\boldsymbol{y})y_i}{s_{\boldsymbol{y}}+k} . \tag{7}$$

Furthermore, one can sum up the probabilities $p(\boldsymbol{y})$ for all $\boldsymbol{y}$ with an equal value of $s_{\boldsymbol{y}}$. By using

$$p_{is} = \sum_{\boldsymbol{y}\in\{0,1\}^m : s_{\boldsymbol{y}}=s} y_i p(\boldsymbol{y}) ,$$

one can transform (7) into the following expression:

$$\boldsymbol{h}^{(k)^*} = \arg\max_{\boldsymbol{h}\in H_k} 2\sum_{i=1}^{m} h_i \sum_{s=1}^{m} \frac{p_{is}}{s+k} . \tag{8}$$

As a result, one does not need the whole distribution to solve (4), but only the values of p_{is}, which can be given in the form of an $m \times m$ matrix P with entries p_{is}. For the special case of $k = 0$, we have $\boldsymbol{h}^{(k)^*} = \mathbf{0}$ and $\mathbb{E}_{\boldsymbol{y}\sim p(\boldsymbol{Y})}\left[F(\boldsymbol{y},\mathbf{0})\right] = p(\boldsymbol{Y} = \mathbf{0})$.

If the matrix P and $p(\boldsymbol{Y} = \mathbf{0})$ are given, the solution of (3) is straight-forward. To simplify the notation, let us introduce an $m \times m$ matrix W with elements

$$w_{sk} = \frac{1}{s+k} , \qquad s,k \in \{1,\ldots,m\} . \tag{9}$$

The resulting algorithm needs then to compute the following matrix:

$$\mathrm{F} = \mathrm{PW} ,$$

with entries denoted by f_{ik}. The inner optimization problem (4) can then be reformulated as follows:

$$\boldsymbol{h}^{(k)^*} = \arg\max_{\boldsymbol{h}\in H_k} 2\sum_{i=1}^{m} h_i f_{ik} .$$

The solution for a given $k \in \{1,\ldots,m\}$ is obtained by setting $h_i{=}1$ for the top k largest elements in the k-th column of the matrix F, and $h_i{=}0$ for the rest. The corresponding value of the expected F-measure for $\boldsymbol{h}^{(k)^*}$ has to be stored for being used in the outer maximization. We also need to compute a case in which $k = 0$:

$$\mathbb{E}_{\boldsymbol{y}\sim p(\boldsymbol{Y})}\left[F(\boldsymbol{y},\mathbf{0})\right] = p(\boldsymbol{Y} = \mathbf{0}) .$$

The last step relies on solving the outer maximization (5):

$$\boldsymbol{h}_F^* = \arg\max_{\boldsymbol{h}\in\{\boldsymbol{h}^{(0)^*},\ldots,\boldsymbol{h}^{(m)^*}\}} \mathbb{E}_{\boldsymbol{y}\sim p(\boldsymbol{Y})}\left[F(\boldsymbol{y},\boldsymbol{h})\right] .$$

The complexity of the above algorithm is dominated by the matrix multiplication PW that is solved naively in $\mathcal{O}(m^3)$. The algorithm needs $m^2 + 1$ parameters in total, namely the matrix P and probability $p(\boldsymbol{Y} = \mathbf{0})$.

3.3 Discussion

The DP approach described in Section 3.1 and GFM are characterized by a similar computational complexity, however, the former does not deliver an exact F-measure maximizer if the assumption of independence is violated. On the other hand, the DP approach relies on a smaller number of parameters (m values representing marginal probabilities). GFM needs $m^2 + 1$ parameters, but then computes the maximizer exactly. Since estimating a larger number of parameters is statistically more difficult, it is a priori unclear which method performs better in practice. We are facing here a common trade-off between an approximate method on better estimates (we need to estimate a smaller number of parameters from a given sample) and an exact method on potentially weaker estimates.

4 Learning Parameters of the Distribution

In the above section, we described two inference techniques that compute the F-measure maximizers based on delivered parameters of the label distribution. To estimate these parameters we used two well-known methods for multi-label classification: binary relevance and probabilistic classifier chains.

4.1 Binary Relevance

BR is the simplest approach to multi-label classification. It reduces the problem to binary classification, by training a separate binary classifier $h_i(\cdot)$ for each label. Learning is performed independently for each label, ignoring all other labels. Obviously, BR does not take label dependence into account, but with a proper base classifier it is able to deliver accurate estimates of marginal probabilities. These estimates can be further used as inputs in the DP inference algorithm. BR is, however, not appropriate for GFM.

4.2 PCC

PCC [6] is an approach similar to Conditional Random Fields (CRFs) [7,8], which estimates the joint conditional distribution $p(\boldsymbol{Y} \mid \boldsymbol{x})$. This approach has the additional advantage that one can easily sample from the estimated distribution. The underlying idea is to repeatedly apply the product rule of probability to the joint distribution of the labels $\boldsymbol{Y} = (Y_1, \ldots, Y_m)$:

$$p(\boldsymbol{Y} = \boldsymbol{y} \mid \boldsymbol{x}) = \prod_{i=1}^{m} p(Y_i = y_i \mid \boldsymbol{x}, y_1, \ldots, y_{i-1}) \,. \tag{10}$$

Learning in this framework can be considered as a procedure that relies on constructing probabilistic classifiers for estimating $p(Y_i = y_i | \boldsymbol{x}, y_1, \ldots, y_{i-1})$, independently for each $i = 1, \ldots, m$. By plugging the log-linear model into (10), it can be shown that pairwise dependencies between labels y_i and y_j are modeled.

To sample from the conditional joint distribution $p(\boldsymbol{Y} \mid \boldsymbol{x})$, one follows the chain and picks the value of label y_i by tossing a biased coin with probabilities given by the i-th classifier. From the sample of such observations one can estimate all the parameters required by the GFM algorithm. One can also estimate the marginal probabilities and use the DP algorithm. The result is not necessarily the same as in BR, since we are using a more complex feature space here.

5 Results in the Competition

In this section we report results on the JRS 2012 Data Mining Competition dataset of the methods we discussed in previous sections. Our preprocessing on the competition data is quite straightforward: We simply delete all the empty columns (i.e., zero vectors) in the training data, then the corresponding columns in the test data. The values of features are normalized to $[0, 1]$.

In both BR and PCC we use linear regularized logistic regression from the Mallet package[4] as a base classifier. We tune the regularization parameter for each base classifier independently by minimizing the negative log-likelihood, which should provide better probability estimates. We use 10-fold cross-validation and we choose the regularization parameter from the following set of possible values $\{10^{-5}, 10^{-4}, \ldots, 10^{5}\}$. We use PCC with both inference methods and try different sizes of sample generated from the conditional distribution of a given $\boldsymbol{x}$.

The results of the methods are presented in Table 1. The F-measure is computed over the entire test set delivered by the organizers after the competition. This is a minor difference in comparison to the competition results which are computed over 90% of test examples. The remaining 10% of test examples constitute a validation set that served for computing the scores for the leaderboard during the competition. The last row in the table gives the result of the final method we used in the competition. It relies on averaging over all predictions we computed during the competition. These predictions are the results of the approaches presented in this paper but with different parametrization. In total we gathered 16 predictions and we aggregated them via voting. In this voting procedure we tested different thresholds on the validation set and selected the best one (nine votes from 16).

From the results we can see that there is no big difference among the methods. The voting procedure improves only slightly over BR+DP and PCC+GFM. Interestingly, BR+DP performs here better than PCC+GFM, which suggests independence of the labels. However, one can also observe that PCC+DP loses against other methods. This shows that PCC with the sampling procedure has problems with the accurate estimation of the marginal probabilities. Increasing the sample size improves the results (for both, DP and GFM), but it still seems that BR+DP is the most appropriate method in this case. It is the cheapest one, since it does not require additional sampling in the inference step as PCC does, and gives results only slightly worse than the voting method that averages over many predictions.

---
[4] `http://mallet.cs.umass.edu/`

Table 1. The results of the presented methods obtained on the entire test set. The numbers in parentheses denote the size of the sample in PCC.

Method	F-measure	Method	F-measure
PCC+DP (50)	0.48650	PCC+GFM (50)	0.52286
PCC+DP (200)	0.51979	PCC+GFM (200)	0.53005
PCC+DP (1000)	0.52995	PCC+GFM (1000)	0.53146
BR+DP	0.53279	Voting (final submission)	0.53327

6 Conclusions

The JRS 2012 Data Mining Competition is essentially a multi-label learning problem, where the objective is to optimize the instance-based F-measure. In this paper, we have introduced several theoretically sound methods addressing this optimization problem. We have shown that, although the F-measure maximization becomes significantly simpler under the assumption of independently distributed labels, it can also be accomplished efficiently without this assumption. Our final predictions are produced by a blend of all these methods and have achieved a very satisfactory result, the second place in the competition.

Acknowledgments. Weiwei Cheng and Eyke Hüllermeier are supported by German Research Foundation (DFG). Krzysztof Dembczyński and Adrian Jaroszewicz are supported by the grant 91-515/DS funded by the Polish Ministry of Science and Higher Education. Willem Waegeman is supported as a postdoc by the Research Foundation of Flanders (FWO-Vlaanderen).

References

1. van Rijsbergen, C.J.: Foundation of evaluation. Journal of Documentation 30(4), 365–373 (1974)
2. Jansche, M.: A maximum expected utility framework for binary sequence labeling. In: ACL 2007, pp. 736–743 (2007)
3. Lewis, D.: Evaluating and optimizing autonomous text classification systems. In: SIGIR 1995, pp. 246–254 (1995)
4. Chai, A.: Expectation of F-measures: Tractable exact computation and some empirical observations of its properties. In: SIGIR 2005, pp. 593–594 (2005)
5. Dembczyński, K., Waegeman, W., Cheng, W., Hüllermeier, E.: An exact algorithm for F-measure maximization. In: NIPS 2011, 223–230 (2011)
6. Dembczyński, K., Cheng, W., Hüllermeier, E.: Bayes optimal multilabel classification via probabilistic classifier chains. In: ICML 2010, pp. 279–286 (2010)
7. Lafferty, J., McCallum, A., Pereira, F.: Conditional random fields: Probabilistic models for segmenting and labeling sequence data. In: ICML 2001, pp. 282–289 (2001)
8. Ghamrawi, N., McCallum, A.: Collective multi-label classification. In: CIKM 2005, pp. 195–200 (2005)

IRISA Participation in JRS 2012 Data-Mining Challenge: Lazy-Learning with Vectorization

Vincent Claveau

IRISA – CNRS
Campus de Beaulieu, 35042 Rennes, France
vincent.claveau@irisa.fr

Abstract. In this article, we report on our participation in the JRS Data-Mining Challenge. The approach used by our system is a lazy-learning one, based on a simple k-nearest-neighbors technique. We more specifically addressed this challenge as an opportunity to test Information Retrieval (IR) inspired techniques in such a data-mining framework. In particular, we tested different similarity measures, including one called vectorization that we have proposed and tested in IR and Natural Language Processing frameworks. The resulting system is simple and efficient while offering good performance.

Keywords: Vector space, Vectorization, LSI, k-Nearest Neighbors, Information Retrieval.

1 Introduction

This article describes the IRISA participation in the JRS Data-Mining Challenge. The team was composed of Vincent Claveau, IRISA-CNRS, and was identified as vclaveau. The approach used by our system is a lazy-learning one, relying on a k-nearest-neighbors technique (kNN). In this standard data-mining technique, the object to classify is compared with those from the training set. The closest ones then vote for the classes and the final class(es) are attributed to the new object based on these votes.

Such a technique necessitates to define at least two components: how to compute the similarity between a new object and the training set ones, and how to combine the votes to assign the classes to the new objects. For these two components, we used techniques initially developed in the Information Retrieval (IR) domain. In particular, we show that the similarity measure that we have developed, called vectorization, yields better results than usual similarity measures. It implements a second-order distance based on the use of pivots to build a new vector space representation.

The resulting system does not need any learning step *per se* and is very fast: the full processing (from the processing of the training set to the generation of the results for the test set) takes approximately 5s on a laptop computer. The best score that was obtained during the official evaluation is 0.500. It is ranked

J.T. Yao et al. (Eds.): RSCTC 2012, LNAI 7413, pp. 447–454, 2012.

17th on the leaderboard, with score only 0.02947 points worse than the best score.

The paper is organized as follows: the next section gives some background on how usual techniques from IR can be used to implement kNN. Section 3 presents our vectorization approach to compute second-order similarities. Section 4 details how the classes are predicted from the nearest-neighbors found. The results obtained by our approach and different variants are presented in Section 5, and some conclusive remarks are given in the last section.

2 First Order Similarity for Nearest Neighbors

As we previously said, our whole approach is based on techniques initially developed for Information Retrieval (IR). Thus, we make an analogy between the JRS Challenge data and IR, and more precisely with the vector space model classically used in IR: a vector represents a document (noted d hereafter), and each dimension represents a word. As in IR, instead of computing a distance or similarity directly from the initial vectors, we apply a weighting scheme to the data. This common approach is explained in the two following subsections and is what we call a first order similarity.

2.1 Weighting Schemes

Several weighting schemes have been proposed in IR. Their goal is to give more importance to representative attributes/words of an object/document. The TF-IDF is certainly the most well known of these weighting schemes. It is based on considerations developed in several seminal papers [13,14] and is usually defined as:

$$w_{TF-IDF}(t,d) = TF(t,d) * IDF(t) = tf(t,d) * \log(N/df(t))$$

where $tf(t,d)$ represents the value of the dimension/word t for the vector/document d, N is the total number of vectors and $df(t)$ is the number of vectors having a non-zero dimension t.

Another weighting scheme has been proved much more efficient than the standard TF-IDF for most IR problems. This scheme, called Okapi-BM25, can be seen as a variant of TF-IDF. Its definition is given in Equation 1; it indicates that the weight of word/dimension t in the document/vector d ($k_1 = 2$ and $b = 0.75$ are constants, dl is the length of document, dl_{avg} the average document length).

$$\begin{aligned} w_{BM25}(t,d) &= TF_{BM25}(t,d) * IDF_{BM25}(t) \\ &= \frac{tf(t,d) * (k_1+1)}{tf(t,d) + k_1 * (1 - b + b * dl(d)/dl_{avg})} * \log \frac{N - df(t) + 0.5}{df(t) + 0.5} , \end{aligned} \quad (1)$$

The TF_{BM25} part was initially derived from a probabilistic model of the frequency of terms (dimensions) in the documents (vectors), namely the 2-Poisson model of Harter [15]. This model represents the distribution of terms in the documents as a mixture of two Poisson distributions: one represents the frequency of terms relevant to describe the document, while the other represents the frequency of non relevant ones [9]. In practice in IR, this TF formula is considered as better than the original one because it includes a normalization based on the size of the document. The IDF_{BM25} part is also derived from probabilistic considerations [15] and is quite similar to the empirically set up standard IDF formula.

Note that these weighting schemes do not change the sparsity of the vectors. Every 0-component of the vectors keeps its value to 0. This property is generally exploited to perform very efficient similarity computations (see below).

2.2 Similarity

The data were provided in a vector representation. Each object is thus described as a vector of 25000 dimensions which is very sparse. This sparse vector representation is very often used in IR, and measures like Minkowsky Lp distances are commonly used to compute similarity between two such vectors. For two vectors x and y, Minkowsky distances are defined by equation 2; p is usually chosen as 1 (Manhattan distance), 2 (Euclidean distance) or ∞ (Chebyshev distance), if $p < 1$, Lp is no longer a distance.

$$Lp(x,y) = \sqrt[p]{\sum_i |x_i - y_i|^p} \quad (2)$$

$$cos(x,y) = \frac{\sum_i x_i \cdot y_i}{\|x\| \cdot \|y\|} \quad (3)$$

The cosine similarity (eqn 3) is also very often used in IR and data-mining. Since it is based on the scalar-product of the two vectors, it allows a very efficient computation for sparse vectors since only the components which have non-zero values in both vectors have to be considered. Note that the cosine is equivalent to (i.e. yields the same ordering of neighbours as) the L2 distance if the vectors are normalized: $L2(x,y) = \sqrt{2 - 2 * cos(x,y)}$.

In practice, such distances or similarity measures are computed between the weigthed versions of the vectors (TF-IDF, Okapi or others). More precisely, one vector serves as a query, and its nearest neighbors are the vectors having a minimal distance (or maximal similarity) with it. In IR, it is usual to adopt different weighting schemes for the query vector and the vectors from the collection (training vectors), since the query have some particularities one may want to take into account (for instance, queries in a search engine are often composed of only 2 or 3 words and thus results in a vector much sparser than the text collection ones).

3 Vectorization: Second-Order Similarity

3.1 Principle

However, we have developed a more effective similarity technique, based on a transformation on the initial vector space into another. This transformation, called Vectorization, has been used in various IR and Text Mining tasks [3] where it has shown to provide both a low complexity and accurate results. As any embedding technique, Vectorization aims to project any similarity computation between two objects in a vector space. Its principle is relatively simple. For each document of the considered collection, it consists of computing using an initial similarity measure (e.g., a standard similarity measure like the cosine), some proximity scores to m pivot-objects. These m scores are then gathered into a m-dimensional vector representing the object, as shown in Figure 1.

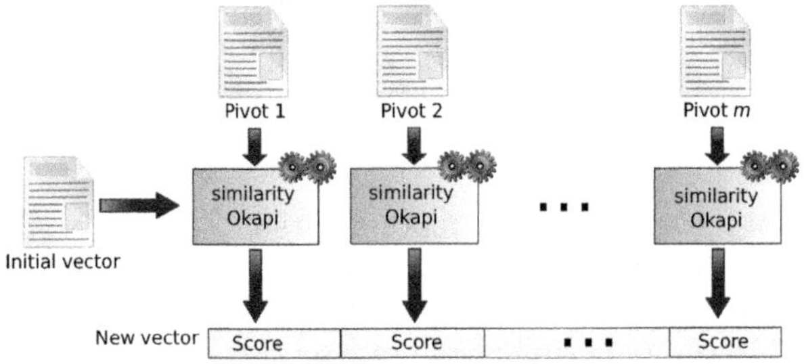

Fig. 1. Vectorization embedding of an example

It is important to notice that the vectorization process changes the representation space. It is not only a space reduction or an approximation of the initial distance as proposed for instance by some authors [2]. It is not either an othogonalization or a linear transform as in Latent semantic Indexing/Analysis (LSI/LSA), probabilistic Latent Semantic analysis (pLSA) [10], Latent Dirichlet Allocation (LDA) [7], principal component analysis (PCA) [1], LDA [6] and even random linear transformations [17]. In these representations, a new vector space is also built, but its dimensions are simply linear combinations of the initial ones.

Changing the representation space, based on the similarity to the pivots, brings up two important properties. Firstly, this embedding helps to reduce the complexity when the initial similarity measure is too expensive to be used in-line [4]. Of course, this property is not really useful in the context of this JRS Challenge but appears as important when vectorization is used for Information Retrieval. Secondly, vectorization will consider two objects as close if they have the same behaviour regarding the pivots, that they are close to same pivots and far for the same pivots. As with LSI or LDA, this indirect comparison makes it possible to pair two objects even if they do not bear common components in the initial vector space.

3.2 Pivots

The pivots can be any objects provided that we are able to compute a similarity between them and the initial vector. They may be artificially created or generated from existing data. In this JRS Challenge context, we have adopted the latter solution: we have built 83 pivots, one per class. Each pivot is simply a sum of a random selection of the vectors belonging to the corresponding class. Each training object is then represented in this 83-dimension space by computing its distance (we used a cosine) to each of the 83 pivots. This embedding provides a more robust representation than the initial 25000-dimension one; indeed, it allows to consider two objects as close, because they are close from the same pivots, even if they were not considered as close in the initial space. This property may be important in the JRS Challenge context in which a topic could be expressed by different MeSH term combinations.

3.3 Similarity

Comparing two 'vectorized' objects can be performed in a very standard way in the new vector space with a L2 distance for example. Yet, it is important to note one property: as we previously underlined it, computing L2 or cosine between two vectors can be very efficiently done when the vectors are sparse since it needs to consider only the dimensions having a non-zero value in each vector. When the vector space is not sparse and has high dimensionality, the cost of computing the distance can be very important.

Yet, many algorithms are available to compute or approximate very efficiently such distances. To save processing time, these techniques either address the completeness of the search, or the accuracy of distance calculation. For instance, the *hashing*-based techniques [16,5] tackle the completeness: the space is divided into portions, and the search is conducted on a subset of these portions. The NV-tree [12] pushes this approach further as it also approximates L2 distances of the portion chosen. Finally, it provides results in $\mathcal{O}(1)$ (ie, a constant time based on a single disk access), whatever the number of vectors in the space. For the experiments presented below and given the small number of pivots and training vectors, such techniques were not necessary; a direct cosine (i.e. equivalent to a direct L2) computation was performed.

4 Vote

Based on the similarity measure described above, the nearest-neighbors of an object can be efficiently retrieved. For each of them, we have a list of its classes and its similarity score. To gain more robustness, we ran different runs. Since the pivots are randomly generated, the results obtained varied from one run to another. Here again, to combine the results, we made an analogy to the IR process. More precisely, we made an analogy to meta-search whose goal is to combine the ranked results of multiple systems. Thus we used one of the most-know combination formula used for meta-search, namely CombMNZ [8,11]. CombMNZ re-orders the classes of the neighbors retrieved by all the systems. Let us note Sys_c the set of systems (or runs in our case) proposing c as a possible class (that is, with a non-zero score) for the considered test object; the CombMNZ score is then defined as:

$$CombMNZ(c) = \sum_{s \in \text{Sys}_c} score(s, i) * |\text{Sys}_c| \qquad (4)$$

It is based on score ($score(s, i)$) associated with each class. In IR, this score is the similarity between the query and the document considered. In our case, it is the sum of similarities with close neighbors belonging to class c as obtained by vectorization. The full process is schematized in Figure 2 The resulting similarity list was cut based on a fixed threshold on the CombMNZ score.

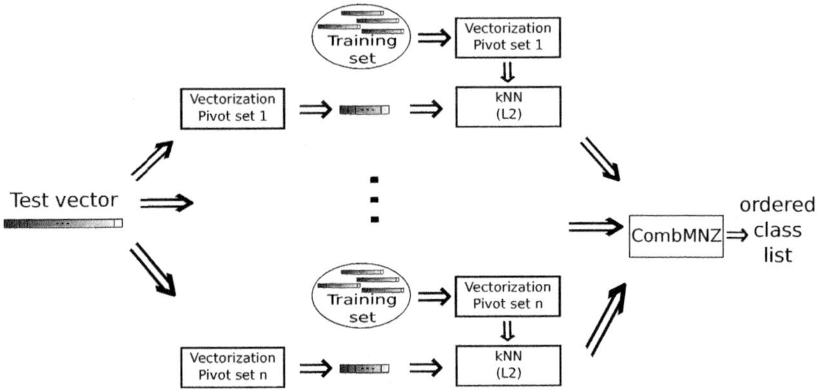

Fig. 2. Complete voting process based on CombMNZ

5 Results

In Table 1, we present different results obtained with the system described in the previous sections. We indicate the results obtained when using different similarity measures instead of vectorization. We thus report the scores obtained by first-order similarities like TF-IDF/L2 and Okapi, as well as the transformation-based similarities LSI and LDA with several sizes of space. These results are expressed in terms of f-measure, as defined by the organisers, and we also indicate the optimal f-measure, that is, the best f-measure that could be obtained if the class list produced by our systems (i.e. by CombMNZ) would have been cut at the best place. In order to have a precise estimate of these measures, we use a 20-fold cross-validation.

Several points are worth noting. Firstly, it seems that the weighting schemes have almost no influence on the results. This result is surprising but difficult to interpret given that no information was given on how the vector values were computed and what they represent.

Another interesting fact is that systems based on a dimensionality reduction, such as LSI, LDA or vectorization, perform better on average than those relying on the initial highly dimensional vector space. Here again, given that no information is given on which dimension represents which MeSH term and how the hierarchical nature of the MeSH was taken into account, these results are difficult to interpret. But, since the 25,000 MeSH index terms used to build the vectors are not independent, it can be supposed that the space reduction helps

Table 1. F-measure and optimal f-measure for different IR-inspired kNN systems

	f-measure	optimal f-measure
no weighting/L2	0.4305	0.5992
TF-IDF/L2	0.4464	0.6023
Okapi	0.4529	0.6142
LSI 83 dims	0.4605	0.6253
LSI 150 dims	0.4795	0.6397
LSI 250 dims	0.4811	0.6435
LSI 350 dims	0.4815	0.6497
LDA 83 dims	0.4087	0.5801
LDA 150 dims	0.4384	0.6057
LDA 250 dims	0.4514	0.6122
LDA 350 dims	0.4482	0.6101
Vectorization 83 pivots	0.5106	0.6915

to match vectors belonging to the same categories even if they are described by different (but dependent) terms.

Finally, among all tested similarity measures, Vectorization performs best. Moreover, the optimal results obtained needs less dimensions than the LSI or LDA techniques, resulting more compact vectors. Several other experiments, not reported here, have been conducted to assess the influence of other parameters. They showed that the number k of neighbors has only a small effect on the results. There is almost no difference for k varying between 3 and 20. Also, different aggregating techniques have been tested beside CombMNZ, such as CombSUM, CombMAX, Condoret vote... All of them yielded lower results, as it has been verified in many IR tasks.

Last, let us note that the official score obtained by our system, computed on the final test data, is 0.50632. It is ranked 17th on the leaderboard, with score only 0.02947 points worse than the best performing system.

6 Conclusions

The approach that we proposed for this JRS Data-Mining Challenge is efficient and yields good results. One of its main characteristics is that it does not rely on a complex Machine Learning approach; it rather uses a lazy learning system inspired by Information Retrieval techniques. In particular, this challenge offered us an opportunity to emphasize the interest of using vectorization to build compact yet precise vector representation of the data in this data-mining framework. Thanks to this representation, the resulting system is very fast while yielding good results.

Many improvements could be done in order to achieve better scores. In particular, one remaining problem was to decide where the list of potential classes provided by CombMNZ should be cut. As we have shown, this choice has a major impact on the results; indeed, if the optimal cut-off value had been chosen for each test object, the results obtained would have reached 0.69. The choice of the evaluation measure (simple f-measure) has made this cut-off step somewhat artificially important regarding the task. Other evaluation measures like the mean average precision (mAP) could have provided a more flexible and informative framework.

References

1. Berry, M., Martin, D.: Principal component analysis for information retrieval. In: Kontoghiorghes, E. (ed.) Handbook of Parallel Computing and Statistics. Statistics: A Series of Textbooks and Monographs (2005)
2. Bourgain, J.: On Lipschitz embedding of finite metric spaces in hilbert space. Israel Journal of Mathematics 52(1) (1985)
3. Claveau, V., Lefvre, S.: Topic segmentation of tv-streams by mathematical morphology and vectorization. In: Procedings of the InterSpeech Conference, Florence, Italy (2011)
4. Claveau, V., Tavenard, R., Amsaleg, L.: Vectorisation des processus d'appariement document-requête. In: 7e Conférence en Recherche d'informations et Applications, CORIA 2010, Sousse, Tunisie, pp. 313–324 (March 2010)
5. Datar, M., Immorlica, N., Indyk, P., Mirrokni, V.: Locality-sensitive hashing scheme based on p-stable distributions. In: Proc. of the 20th ACM Symposium on Computational Geometry, Brooklyn, New York, USA (2004)
6. Blei, D.M., Ng, A.Y., Jordan, M.I.: Latent dirichlet allocation. Journal of Machine Learning Research 3(4-5), 993–1022 (2003)
7. Dumais, S.: Latent semantic analysis. ARIST Review of Information Science and Technology 38(4) (2004)
8. Fox, E., Shaw, J.: Combination of multiple searches. In: Proceedings of the 2nd Text Retrieval Conference (TREC-2), pp. 243–252. NIST Special Publication (1994)
9. Harter, S.: A probabilistic approach to automatic keyword indexing. Journal of the American Society for Information Science 26(6), 197–206 (1975)
10. Hofmann, T.: Probabilistic latent semantic indexing. In: Proc. of SIGIR, Berkeley, USA (1999)
11. Lee, J.: Combining multiple evidence from different properties of weighting schemes. In: Proceedings of the 18th Annual ACM-SIGIR, pp. 180–188 (1995)
12. Lejsek, H., Asmundsson, F., Jónsson, B., Amsaleg, L.: Nv-tree: An efficient disk-based index for approximate search in very large high-dimensional collections. IEEE Trans. on Pattern Analysis and Machine Intelligence 99(1) (2008)
13. Luhn, H.P.: The automatic creation of literature abstracts. IBM Journal on Research and Development 2(2) (1958)
14. Spärck Jones, K.: A statistical interpretation of term specificity and its application in retrieval. Journal of Documentation 28(1) (1972)
15. Spärck Jones, K., Walker, S.G., Robertson, S.E.: Probabilistic model of information retrieval: Development and comparative experiments. Information Processing and Management 36(6) (2000)
16. Stein, B.: Principles of hash-based text retrieval. In: Proc. of SIGIR, Amsterdam, Pays-Bas (2007)
17. Vempala, S.: The Random Projection Method. In: Discrete Mathematics and Theoretical Computer Science, vol. 65. AMS (2004)

On the Homogeneous Ensembling via Balanced Subsets Combined with Wilcoxon-Based Feature Selection

Vladimir Nikulin

Department of Mathematical Methods in Economy,
Vyatka State University, Kirov, Russia
vnikulin.uq@gmail.com

Abstract. Ensembles are often capable of greater prediction accuracy than any of their individual members. As a consequence of the diversity between individual base-learners, an ensemble will not suffer from overfitting. To address a high-dimensionality problem, we developed a special version of Wilcoxon criterion for sparse data, which is very fast. Using this criterion we can compute the matrix of ratings for all features and labels. It is supposed that any particular base-learner will be based on the random subset of features which were selected in accordance with Wilcoxon-based ratings. On the other hand, in many cases we are dealing with an imbalanced data and a classifier which was built using all data has tendency to ignore the minority classes. As a solution to the problem, we propose to consider a large number of relatively small and balanced subsets, where representatives from the both patterns are to be selected randomly.

Keywords: ensembling, boosting, support and relevance vector machines, neural nets, cross validation, classification.

1 Introduction

Development of freely available biomedical databases allows users to search for documents containing highly specialized biomedical knowledge[1]. Rapidly increasing size of scientific literature available through internet, emphasizes the growing need for accurate and scalable methods for automatic tagging and classification of textual data. For example, medical doctors often search through biomedical documents for information regarding diagnostics, drugs dosage and specific treatments. In the queries, they use highly sophisticated terminology, that can be properly interpreted only with a use of a domain ontology, such as Medical Subject Headings (MeSH) [1]. Searching databases for a specific topic can be confusing and ineffective if terms are not used correctly. If authors are reporting studies without using standard controlled vocabulary, indexers might

[1] http://tunedit.org/challenge/JRS12Contest

J.T. Yao et al. (Eds.): RSCTC 2012, LNAI 7413, pp. 455–462, 2012.

not assign the appropriate terminology to represent the studies. In order to facilitate the searching process, documents in a database should be indexed with concepts from the ontology. Additionally, the search results could be grouped into clusters of documents, that correspond to meaningful topics matching different information needs. Such clusters should not necessarily be disjoint since one document may contain information related to several topics.

2 JRS 2012 Data Mining Competition

The data for the Challenge were prepared and presented by the Organisers of the International JRS 2012 Data Mining Competition: "Topical Classification of Biomedical Research Papers", which is a special event within the Joint Rough Sets Symposium in China.

The database for training includes 2 matrices: 1) **L** - binary labels (topic=1, non topic=0) with sizes $n \times k$, and 2) **X** - attributes with sizes $n \times m$, where $n = 10000, m = 25640$ and $k = 83$ - numbers of documents/samples, attributes/features and classes/topics, respectively.

The dataset **T** for testing has the same sizes $n \times m$, where the labels were excluded. The main objective of the Challenge was to predict labels for the test dataset, where any particular document maybe characterized by one or more topics.

As far as we are dealing with high-dimensional data, the most important intermediate task is to compute the matrix of ratings for the attributes and labels **R** with sizes $m \times k$. That means, ratings should be computed specifically for any particular topic.

The data are non-negative and sparse. Therefore, there are much smaller number of positive elements in any column of the matrix **X** compared to n. We shall exploit this property in the following section.

2.1 Wilcoxon-Based Criterion for Sparse Data

Let us drop the label-index in order to simplify notations, because the consideration of any particular label is an identical.

We denote by $\mathcal{N}_a$ a set of all objects/documents from the class a. We used the following separation type criterion [2] (named Wilcoxon) for the selection of the most relevant attributes/terms,

$$WXN(g) = \max(q_{01}(g), q_{10}(g)), \quad (1)$$

where

$$q_{ab}(g) = \sum_{i \in \mathcal{N}_a} \sum_{j \in \mathcal{N}_b} I(x_{ig} < x_{jg}), \quad (2)$$

where I is an indicator function.

Remark 1. The score function WXN can be interpreted as counting for each object having response value a, the number of objects with response b ($a \neq b$) that have smaller expression values, and summing up these quantities.

We have

$$n = n_a + n_b, \tag{3}$$

where $n_a = \#\mathcal{N}_a$ and $n_b = \#\mathcal{N}_b$.

Let us denote by $\mathcal{N}_{a1}$ and $\mathcal{N}_{a0}$ subsets of $\mathcal{N}_a$ with positive and zero elements, $n_{a1} = \#\mathcal{N}_{a1}, n_{a0} = \#\mathcal{N}_{a0}$. By construction, $n_{a1} \ll n_{a0}$.

Definition 1. *Then, we can modify (1) in order to create a special version of the Wilcoxon-based criterion for sparse data*

$$WXN(g) = \max(n_{a0}n_{b1} + q_{ab}(g), n_{b0}n_{a1} + q_{ba}(g)), \tag{4}$$

where

$$q_{ab}(g) = \sum_{i \in \mathcal{N}_{a1}} \sum_{j \in \mathcal{N}_{b1}} I(x_{ig} < x_{jg}). \tag{5}$$

Remark 2. Taking into account that subsets $\mathcal{N}_{a1}$ and $\mathcal{N}_{b1}$ are much smaller compared to $\mathcal{N}_a$ and $\mathcal{N}_b$, the implementation of the Wilcoxon criterion according to (4) and (5) will take significantly less time compared to (2) and (3). The computation of the whole matrix of ratings R (which is quite big) took less than 3min time.

2.2 Generalised Wilcoxon-Based Criterion

One may expect that the quality of the ranking will be improved if we shall take into account in (5) the difference between values of the attributes x_{ig} and x_{jg} :

$$q_{ab}(g) = \sum_{i \in \mathcal{N}_{a1}} \sum_{j \in \mathcal{N}_{b1}} \Phi(x_{jg} - x_{ig}) I(x_{ig} < x_{jg}), \tag{6}$$

where Φ is monotonical and non-negative function, for example, $\Phi(\cdot) = \sqrt{\cdot}$.

2.3 Initial Trimming

Using matrix of ratings $\mathbf{R}$, we conducted some initial trimming of the given matrix $\mathbf{X}$. Firstly, we can compute Q_g maximal rating for any attribute g. The attributes to be selected in the model were identified according to the criterion

$$r_g \geq \frac{Q_g}{\alpha},$$

where α is a regulation parameter. For example, the union of the selected features for all 83 labels includes 14248, 6992 and 4933 attributes in the cases if $\alpha = 25, 11$ and 9.

3 Homogeneous Ensembling with Balanced Random Sets

Ensembles are often capable of greater prediction accuracy than any of their individual members [3]. In many cases, we are dealing with imbalanced data and a classifier which was built using all data has tendency to ignore minority classes. As a solution to the problem, we propose to consider a large number of relatively small and balanced subsets where representatives from the both patterns are to be selected randomly [4].

We conducted experiments with several classification models as base learners: 1) *gbm* [5] and 2) *random Forest* [6] in R. Also, we used 3) *kridge*, 4) *svc* and 5) *Neural Nets* in the Matlab-based package CLOP[2].

3.1 Selection of the Samples

The data are strongly imbalanced with numbers of positive instances ranging from 2 (attribute N77) to 2475 (attribute N40), see Table 1. Suppose, that y_{tg} is a binary label corresponding to the class g, t is an index of the sample. Then, sample x_t will be selected to the training subset according to the condition

$$\xi \leq \xi_1, \; if \; y_{tg} = 1; \quad (7a)$$

$$\xi \leq \xi_0, \; if \; y_{tg} = 0, \quad (7b)$$

where ξ is a standard uniform random variable,

$$\xi_1 = c_1 + (1 - c_1)\frac{n_0}{n_p}, \; \xi_0 = c_0 + (1 - c_0)\frac{n_p}{n},$$

where $c_1 = 0.75, c_0 = 0.025$; $n_0 = 200$ (the selection of the above parameters were based on some qualitative considerations and cannot claim any sort of optimality); and n_p is the number of "positive" samples corresponding to the class g.

3.2 Selection of the Features

To reduce overfitting further, we can add more randomness in the model by selecting the subset of features for any single classifier. The selection algorithm is based on the matrix of ratings **R**, see Section 2.1.

Suppose, we decided to keep in the model top (leading) n_T features, and let us denote by $r_g(n_T)$ the smallest rating qualified for the automatic (guaranteed) acceptance.

In general terms, feature with rating r_g will be accepted to be included in the regression model subject to the following condition

$$\xi \leq \left(\frac{r_g}{r_g(n_T)}\right)^h,$$

where ξ is a standard uniform random variable, h is a regulation parameter. We used the following ranges $20 \leq n_T \leq 60$, $0.35 \leq h \leq 2.0$.

[2] http://clopinet.com/CLOP/

3.3 Selection of the Number of Random Sets

According to Section 3.1, the sample size of any particular set is an increasing function of n_p. Therefore, it appears to be logical to define the number of random sets as a decreasing function of n_p. In most of the cases, we selected the number of random sets as an integer of

$$\sqrt{10\frac{n}{n_p}},$$

with upper limit depending on the speed of the base learner. For example, we can set 150 as an upper limit of the number random sets in the case of *neural nets* or 20 in the case of *gbm*.

4 A Frequentist Approach for the Selection of the Vector of Cut-Off Parameters

As an outcome of the above procedure with homogeneous random sets, we shall compute the matrix of decision functions $\mathbf{S}$, with sizes $n \times k$, which we should transfer to the binary matrix of decision rules $\mathbf{Z}$, with the same sizes.

We shall suppose here that the split of the whole field of 20000 samples between training and test sets was conducted at random. Therefore, it is logical to assume that the distributions (known, also, as prior distributions) of the patterns in the test set are about the same as in training set.

The transformation $\mathbf{S} \Rightarrow \mathbf{Z}$ was conducted according to the following method

$$z_{ij} = \begin{cases} 1 \text{ if } s_{ij} \geq \Delta_j; \\ 0, \text{otherwise,} \end{cases}$$

where cut-off parameter Δ_j was selected as the u_j-th biggest element in the j-th column of the matrix $\mathbf{S}$, where index u_j was calculated according to the matrix of labels $\mathbf{L}$

$$u_j = \sum_{i=1}^{n} l_{ij}.$$

4.1 More Advanced Versions of the Cut-Off Parameters

We had conducted cross-validation experiments specifically for the particular binary labels/topics, and applied more restrictive models (bigger values of the cut-off parameters) against the labels with poorer CV results.

4.2 Dependence between Different Topics

Figure 1 clearly illustrates that there is a strong dependence between different topics, and we are interested to reduce possible contradiction between topics corresponding to the same document (test solution) using the following very simple strategy.

Table 1. List of 32 the most frequent topics, where "I" and "F" indicate an original index of the topic and the number of the corresponding occurrences in the training data

N	I	F	I	F	I	F	I	F
1	47	362	14	583	43	786	39	1133
2	6	380	74	621	75	800	68	1259
3	48	403	58	634	65	832	79	1446
4	66	442	23	651	76	842	41	1474
5	80	448	69	661	50	912	62	1735
6	57	536	20	663	73	933	18	2160
7	21	542	8	688	52	980	44	2435
8	64	548	15	774	46	992	40	2475

Table 2. Top performance in the terms of public LeaderBoard (used only 10% of the test data) corresponding to the single and ensemble models

Model	Score
best	0.501
svc	0.499
neural	0.483
kridge	0.477
blend	0.466
gbm	0.459
RF	0.454

First of all, we shall select the leading predicted topic for any particular document. Next, we shall penalise all the other topics in accordance with the matrix of empirical probabilities, see Figure 1(b), where penalty is a decreasing function of the empirical probability.

5 Ensembling and Blending

We made our first submission to the LeaderBoard on 9th March (that means, more than two months after the competition was started). The best single results was observed with support vector classifier (*svc* function in the Matlab-based package CLOP), where the proper data preprocessing was very essential. We used the following transformation of the features

$$x_{\text{new}} = \left(\frac{x_{\text{given}}}{10}\right)^{0.65}.$$

Based on our initial experience, ensembling produced only some modest progress, and it was much better to concentrate on the improvement of parameter settings for single models.

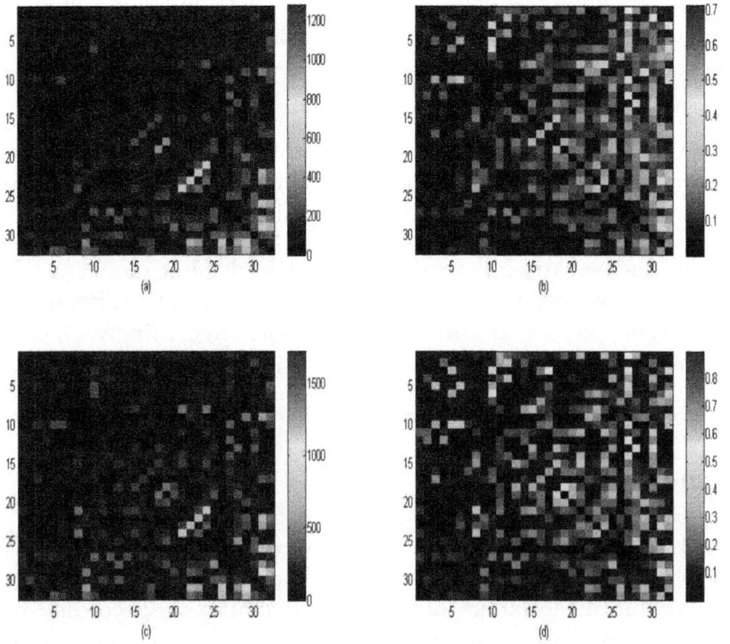

Fig. 1. Hot-maps illustrating frequencies of the different pairs, which were observed together (from 32 the most frequent topics, see Table 1) - left column; empirical conditional probabilities -right column, where top row corresponds to the training data, and second row corresponds to the test data (used solution with the score 0.501 on the public LeaderBoard).

Remark 3. We conducted some CV10 experiments with *gbm* function, and the observed results were quite close to the corresponding result in Table 2.

5.1 On the Blending Strategy

We shall extend here further the ideas of Section 4.2. The matrix $\mathbf{S}$ contains $k = 83$ different solutions, which are dependent and maybe used together in order to produce more advanced solutions. The next fundamental question is how to link (or how to blend) those k different solutions in a most efficient way [7].

We considered here the most straightforward method: split the training data into 2 parts (for example, 70% and 30%), where first part $\mathbf{X}_1$ will be used for training, and the second part $\mathbf{X}_2$ will mimic test dataset $\mathbf{T}$.

The whole blending procedure represents a sequence with two main steps. At the first step, we shall use *svc* (or *kridge*) function applied to $\mathbf{X}_1$ for training, and shall apply the corresponding model to $\mathbf{X}_2$ and $\mathbf{T}$. As a consequence, we shall produce secondary datasets $\mathbf{X}_2^{(2)}$ and $\mathbf{T}^{(2)}$ with k predictors each. At the second step, we shall use *gbm* (or *randomForest*) function applied to $\mathbf{X}_2^{(2)}$ for training, and shall use the corresponding model against $\mathbf{T}^{(2)}$ to produce the final blend solution.

Remark 4. Any predictor in the dataset $\mathbf{X}_2^{(2)}$ corresponds to the particular topic, where the numbers of occurencies in the training set maybe used as the most basic characteristic of the quality of the predictor. We can recommend threshold 200 for the feature selection, which corresponds to 44 predictors.

Remark 5. The best result corresponding to the blending model (see Table 2) is a quite modest. However, the blend solution is a significantly different in the terms of structure compared to the other base solutions. Accordingly, an ensemble with blend solution produced some improvement in the terms of public LeaderBoard score.

6 Concluding Remarks

As it was discussed in Section 4, the whole task for this competition maybe divided into two tasks: 1) computation of the decision functions (matrix **S**), and 2) computation of the binary decision rules (matrix **Z**) according to the given vector of the cut-off parameters, where the first task is a very important and complex.

The second task is rather a trivial one, but require sufficient time for the detailed experiments. The competition criterion maybe very sensitive in relation to the proper selection of the cut-off parameters taking into account the fact that some particular labels are highly imbalanced.

We are proposing to introduce an additional criterion to test separability of the matrix of decision functions **S**. The definition of the criterion is a very simple indeed: an average of the AUCs corresponding to the particular topics or labels.

References

[1] National Library of Medicine: Introduction to MeSH (2012), `http://www.nlm.nih.gov/mesh/introduction.html`
[2] Dettling, M., Buhlmann, P.: Boosting for tumor classification with gene expression data. Bioinformatics 19(9), 63–78 (2003)
[3] Wang, W.: Some fundamental issues in ensemble methods. In: World Congress on Computational Intelligence, pp. 2244–2251. IEEE, Hong Kong (2008)
[4] Nikulin, V.: Classification of imbalanced data with random sets and mean-variance filtering. International Journal of Data Warehousing and Mining 4(2), 63–78 (2008)
[5] Friedman, J.: Greedy function approximation: a gradient boosting machine. Annals of Statistics 29(5), 1189–1232 (2008)
[6] Breiman, L.: Random Forests. Machine Learning 45, 5–32 (2001)
[7] Koren, Y.: The BellKor Solution to the Netflix Grand Prize, Wikipedia, 10 pages (2009)

Representation Learning for Sparse, High Dimensional Multi-label Classification

Ryan Kiros[1], Axel J. Soto[2], Evangelos Milios[2], and Vlado Keselj[2]

[1] Department of Computing Science, University of Alberta
Edmonton, Canada
rkiros@ualberta.ca
[2] Faculty of Computer Science, Dalhousie University
Halifax, Canada
{soto,eem,vlado}@cs.dal.ca

Abstract. In this article we describe the approach we applied for the JRS 2012 Data Mining Competition. The task of the competition was the multi-labelled classification of biomedical documents. Our method is motivated by recent work in the machine learning and computer vision communities that highlights the usefulness of feature learning for classification tasks. Our approach uses orthogonal matching persuit to learn a dictionary from PCA-transformed features. Binary relevance with logistic regression is applied to the encoded representations, leading to a fifth place performance in the competition. In order to show the suitability of our approach outside the competition task we also report a state-of-the-art classification performance on the multi-label ASRS dataset.

Keywords: Multi-label classification, feature learning, text mining.

1 Introduction

Different representations for text corpora have been extensively studied, being TF-IDF and Okapi BM25 two of the most common ways of representing data [9]. The choice of document representation has a major incidence in the performance for tasks such as document classification or retrieval. Particularly, multi-labelled document classification is a research problem that received much less attention in the literature than the single-labelled counterpart. Yet multi-labelled classification is in many cases a more natural approach for document classification tasks [13].

The machine learning community, specially in the area of computer vision, has witnessed the importance of learning feature representations as an alternative to manually configuring the best data representation to feed a prediction method. Learned representations coupled with simple classification methods usually tend to have similar classification accuracy and even overcome other more complex classification methods [3,4,14].

The JRS 2012 Data Mining Competition represented a great opportunity to benchmark and evaluate our hypothesis of whether a learned representation of the data combined with a standard classification approach could have a competitive performance against other approaches for multi-label text classification.

J.T. Yao et al. (Eds.): RSCTC 2012, LNAI 7413, pp. 463–470, 2012.

The learned representation can be thought as a means to both reduce and enhance the original data representation to facilitate learning of the classifier that is used *a posteriori*.

The present paper reports the details of the method that got our highest preliminary score on the competition. Our proposed method is based on a two step feature learning procedure that was motivated by recent work in object recognition [4] and adapted to be used for sparse, high dimensional data. We first embed the inputs using Principal Component Analysis (PCA) and then learn a sparse dictionary that we concatenate with the PCA embeddings for classification. Our final predictions were obtained using binary relevance with a linear logistic regression classifier. In order to further illustrate the benefits of our approach we also show the results on the multi-label SIAM 2007 competition dataset for text mining.

2 Method

Our approach can be divided in three main parts, namely: preprocessing, representation learning and classification.

2.1 Data Preprocessing

Let $X = \{x^{(1)}, \ldots, x^{(m)}\}$ be the set of m training documents where the j-th feature of $x^{(i)}$ is $x_j^{(i)}$, $j = 1 \ldots n$. We first process the data by normalizing each $x^{(i)}$ to $[0, 1]$, where $i = 1 \ldots m$. This is done by dividing each feature by its range (i.e. the difference between the maximum and the minimum value). The data is then rescaled to $[-1, 1]$. Finally, we use a 'regularized' mean centering and variance normalization for each feature j:

$$x_j^{(i)} = \frac{x_j^{(i)} - \mu_j}{\sqrt{\sigma_j^2 + \epsilon}} \tag{1}$$

where the feature means (μ_j) and standard deviations (σ_j) are preserved for use with the test set. For our experiments we use $\epsilon = 0.01$.

2.2 Unsupervised Representation Learning

Given the pre-processed data, we first apply PCA and extract the first k principal components. This calculation is the most expensive procedure for our method, both memory and time wise, due to the size of the covariance matrix. Let $S = \{s^{(1)}, \ldots, s^{(m)}\}$ represent the k-dimensional outputs. We perform one additional processing step of S by centering and normalizing the variances of each individual datapoint $s^{(i)}$. We use the same 'regularized' normalization as in Equation 1 but applied to documents as opposed to features. This step, when applied to images, can be seen as a form of brightness and contrast normalization. Although less motivated for our PCA-transformed data, we found it to be a useful addition to the pipeline.

After this normalization step, the data are now ready to be used for constructing a dictionary (also called prototypes or codebook). Dictionaries have been traditionally used in the area of signal processing for data compression. Common examples to this kind of techniques are vector quantization or k-means [5]. We use orthogonal matching pursuit (OMP) [11], which aims to solve the following optimization problem:

$$\begin{aligned} \underset{D,\hat{s}^{(i)}}{\text{minimize}} \quad & \sum_{i=1}^{m} ||D\hat{s}^{(i)} - s^{(i)}||_2^2 \\ \text{subject to} \quad & ||D^{(j)}||_2^2 = 1 \\ & ||\hat{s}^{(i)}||_0 \leq q \end{aligned} \tag{2}$$

where $D \in \mathbb{R}^{k,d}$ is the dictionary to be learned. The first constraint enforces that the dictionary elements remain normalized (to avoid degeneracy) while the second constraint enforces sparsity, allowing at most q elements of $\hat{s}^{(i)}$ to be non-zero. Our objective is minimized using alternation: first fixing D and minimizing $\hat{s}^{(i)}$, then fixing $\hat{s}^{(i)}$ and minimizing D. We set $q = 1$ for all of our experiments. We chose to use OMP over other methods due to the speed of training, taking only a few minutes to train the dictionary.

As opposed to using the proper representations $\hat{s}^{(i)}$, Coates et al. [4] showed that in the presence of enough labelled data a simple soft activation may perform equally, if not better when applied in an object recognition setting. We follow this approach by encoding that data with a simple rectification unit:

$$\breve{s}^{(i)} = \max\{Ds^{(i)}, 0\}, \tag{3}$$

where the max function is applied componentwise. Finally, we apply another form of 'contrast normalization' over individual datapoints as was previously done with the PCA transformed data. Given the encoded data $\breve{S}$, we concatenate the learned features with the PCA embeddings and we then normalize the data once more using Equation 1. These datapoints, $Z = \{z^{(1)}, \ldots, z^{(m)}\}$, are now ready for training our classifier.

2.3 Classification

To train our model, we used binary relevance with logistic regression incorporating weight decay for regularization. This means that for each single label l we perform the following optimization:

$$\underset{\theta^{(l)}}{\text{minimize}} \quad -\frac{1}{m}[\sum_{i=1}^{m} y^{(i)} \log h_{\theta^{(l)}}(z^{(i)}) + (1 - y^{(i)}) \log(1 - h_{\theta^{(l)}}(z^{(i)}))] + \lambda ||\theta^{(l)}||_F^2 \tag{4}$$

where $h_\theta(z) = 1/(1+\exp(-\theta^T z))$, $y^{(i)}$ is the label of the i-th document (for class l) and $\|\cdot\|_F^2$ stands for the square of the Frobenius norm. Optimization was done using L-BFGS [8] with minFunc[1], often converging in less than 30 iterations per class.

We also experimented using a multi-label SVM with a linear kernel combined with an adaptive thresholding scheme used to maximize F-measure. We found that although the latter was faster to train, it was less stable for parameter selection and often led to an imbalance of either precision or recall.

3 Results

3.1 JRS 2012 Data Set

The competition dataset comprises of 20000 documents out of which half of them were held out for testing (whose labels were not disclosed during the competition time). Each document is represented by a 25640-dimensional vector, where each component represents the association strength to a Medical Subject Heading (MeSH) term.

For our best preliminary result, we used a total of $k = 1600$ principal components and a dictionary size of $d = 3200$. This lead to a final feature vector of size 4800. We used the same parameter λ for each of the classifiers. To obtain λ, we split the training set into 8000 samples for training and 2000 for validation. We then performed a grid search across powers of 2, followed by a more fine-grained search, leading to a chosen parameter of $\lambda = 0.5$. We obtained an F-measure on the validation set of 0.522 and with the same model a final preliminary score of 0.523. We also attempted to find a separate λ_j for each class j, leading to a validation score of 0.527. Unfortunately, this did not generalize to the preliminary result. Our final performance on the test data was **0.53**, resulting in a 5th place finish.

We used the same pipeline as described above throughout the whole competition with all of our improvements coming from modifying the choice of dimensionality reduction, number of bases and normalization steps. Combining the PCA embeddings with the encodings performs much better than either one or the other alone. We found that performance consistently improved by using more bases and principal components.

To further motivate our basis learning approach, we trained the described multi-label SVM on the normalized input data, receiving a validation score of 0.5. When combined with the encodings learned from the PCA embeddings, this increased the score to 0.52. We opted out from further pursuing this approach in the competition due to the high dimensionality of the final feature vectors.

Finally, we experimented with the usefulness of all of our normalization steps and found the most important being the initial regularized normalization as well as the same normalization before training the classifier. In this sense, all of the 'contrast normalization' steps can be safely removed and still result in good performance.

---

[1] `http://www.di.ens.fr/~mschmidt/Software/minFunc.html`

3.2 Additional Results: SIAM 2007 Data Set

In the domain of text mining, a common approach to representing documents is through a bag-of-words (or the more general n-gram) representation. Here, a weight is given for each document-term pair, such as frequency of TF-IDF. Such a representation is often of high dimension with only a few non-zero entries corresponding to the n-grams which are present in the corresponding document. Thus we further evaluate our proposed approach on a text classification task to show that our approach can be utilized in other domains aside from the biomedical classification task from this competition.

We consider the ASRS (Aviation Safety Reporting System) dataset, which was used for the SIAM 2007 competition [10]. ASRS is a collection of roughly 21000 training documents and 7000 testing documents whose labels are 22 dimensional binary vectors indicating the presence of one or more aircraft security issues. Sample classes include weather, fuel emergencies, passenger disruptions, pilot attentiveness and runway obstructions to name a few. The SIAM competition data was sampled from the publicly available ASRS database maintained by NASA[2]. Below is an example of a typical document from the dataset:

"UPON TOUCHDOWN AT NIGHT ON runway _ AT BID THE right land GEAR STRUCK A DEER ON THE runway.I DID NOT SEE THE DEER.THE result DAMAGE WAS THE remove OF THE right GEAR AND THE aircraft settle ON right WING skid OFF THE right SIDE OF runway."

To obtain the initial representation of a document, we first pre-process the data by lowering case and performing stopword removal. We then obtain the 5000 most frequent words for which a document-term matrix of unnormalized TF-IDF values is constructed. The IDF values are obtained from the training set and applied in conjunction with the frequencies calculated on each test point. Model selection was performed in the same way as was done on the JRS competition data using the same number of principal components (1600) and bases (3200).

Table 1. A comparison of (micro-averaged) classification performance on the SIAM ASRS dataset. The first two methods are the results of the competition winner, the first being where the competition score is maximized while the second being where F-measure is maximized.

Method	Precision	Recall	F-Measure	Error
SIAM Winner (Score) [6]	61.53	62.37	61.95	6.80
SIAM Winner (F-Measure) [6]	53.30	**78.55**	63.51	8.01
Normalized Baseline (this paper)	63.15	70.30	66.40	6.31
PCA + OMP (this paper)	**64.25**	71.68	**67.76**	**6.05**

[2] `http://asrs.arc.nasa.gov/`

To test the effectiveness of our approach, we compared our results to the performance of the first place finisher of the SIAM 2007 competition [6]. Table 1 shows our result in comparison to the winner's best approach with respect to the competition evaluation as well as the approach that maximized F-measure. We also included a baseline showing the result when binary relevance is applied directly to the normalized document-term matrix. Our result outperforms the competition winner yielding state-of-the-art results on this dataset[3]. Surprisingly, the baseline without any representation learning is able to also outperform the existing result. We attribute this to the regularized normalization of the features. In particular, result are over 2% worse across all metrics when no regularization ($\epsilon \approx 0$) is used.

3.3 Adapting Precision and Recall

One further observation that was made on both datasets was that the precision and recall can be directly tuned through the logistic regression regularization parameter (λ). More specifically, increasing λ leads to an increase of recall but decrease of precision and equivalently in the opposite when decreasing λ. Figure 1 illustrates this effect on the validation sets of both the JRS competition and ASRS datasets. Moreover, the parameter that led to the best model for the competition was that which had slightly higher recall on the validation set.

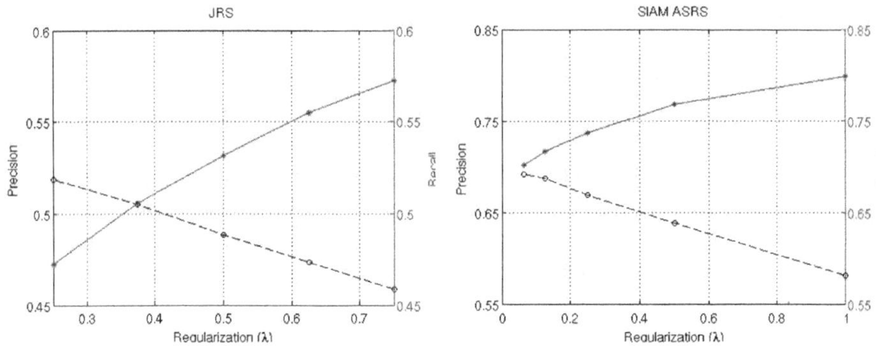

(a) JRS (b) SIAM ASRS

Fig. 1. Graphs illustrating the effect of the logistic regression regularization on precision in black and recall in red (best seen in color)

We note that it is often the case that specific applications may be more interested in maximizing either precision or recall. As a related example to aircraft security, it is much more important for text retrieval systems to have high recall rather than high precision so that all security incidents can be retrieved, even if this may lead to a high false positive rate. Being able to control this effect gives our approach potential use in these types of domains.

[3] We note that other proposed methods have been evaluated from taking random samples from the full ASRS database.

4 Conclusions

In this work we described an unsupervised feature learning methodology in the context of multi-label document classification. Our method obtained the 5th position in the JRS 2012 Data Mining Competition. We also showed the results of applying our method on another challenging and multi-labelled dataset. In this last data set we obtained a state-of-the-art performance, hence indicating the suitability of our approach for multi-label text classification.

There are several possible avenues for future research. One such approach is testing the effectiveness of our method in a semi-supervised setting. More specifically, a dictionary could be trained using both sets of labelled and unlabelled data. Related to this is the use of self-taught learning [12] also known as transfer learning from unlabelled data. One could train a large dictionary on a corpus that have come from a different distribution than the target dataset intended for classification. Such an approach may be effective in tasks where only small amounts of labelled data exist. Our approach could also be used for domain adaptation in situations where two or more datasets have the same label distribution but the target dataset of interest is unlabelled. These situations often occur in sentiment classification tasks when labels may correspond to binary positive and negative opinions or 5-star ratings. Finally, there has been much work on learning deep representations [1,7]. The output encoding features could be used as input to another layer of dictionary training with the first layer bases frozen. Such greedy layer-by-layer training could be used as many times as desired. It is still an open problem what the best approaches to training such architectures are [2].

Throughout our procedure almost all the focus was made on the representation learning and normalization phases, with little effort put towards classification. Binary relevance is flawed in that it does not take into account correlations between labels. It is worth exploring whether improvements can be made by adapting more sophisticated multi-label classification approaches. It also remains as an interesting research question whether the knowledge of a domain ontology, such as the disclosure of the MeSH ontology for the competition dataset, can be used to further improved the classification accuracy.

Acknowledgments. Authors thank The Boeing Company and NSERC for funding of this work.

References

1. Bengio, Y., Lamblin, P., Popovici, D., Larochelle, H.: Greedy layer-wise training of deep networks. Advances in Neural Information Processing Systems 19, 153 (2007)
2. Bengio, Y.: Learning deep architectures for AI. Foundations and Trends in Machine Learning 2(1), 1–127 (2009)
3. Boureau, Y.-L., Bach, F., LeCun, Y., Ponce, J.: Learning mid-level features for recognition. In: IEEE Conference on Computer Vision and Pattern Recognition, pp. 2559–2556 (2010)

4. Coates, A., Ng, A.Y.: The importance of encoding versus training with sparse coding and vector quantization. In: Getoor, L., Scheffer, T. (eds.) International Conference on Machine Learning, pp. 921–928. Omnipress (2011)
5. Gersho, A., Gray, R.M.: Vector quantization and signal compression. Kluwer Academic Publishers, Norwell (1991)
6. Goutte, C.: A probabilistic model for fast and confident categorization of textual documents. In: Berry, M.W., Castellanos, M. (eds.) Survey of Text Mining II, vol. 4, pp. 187–202. Springer (2008)
7. Hinton, G., Osindero, S., Teh, Y.: A fast learning algorithm for deep belief nets. Neural Computation 18(7), 1527–1554 (2006)
8. Liu, D.C., Nocedal, J.: On the limited memory method for large scale optimization. Mathematical Programming 45(3), 503–528 (1989)
9. Manning, C.D., Raghavan, P., Schütze, H.: An Introduction to Information Retrieval. Cambridge University Press (2008)
10. NASA: SIAM 2007 – Aviation Safety Reporting System (ASRS) Challenge Dataset (2007), http://web.eecs.utk.edu/events/tmw07/
11. Pati, Y., Rezaiifar, R., Krishnaprasad, P.: Orthogonal matching pursuit: recursive function approximation with application to wavelet decomposition. In: Asilomar Conference on Signals, Systems and Computers, Pacific Grove, CA, pp. 40–44 (1993)
12. Raina, R., Battle, A., Lee, H., Packer, B., Ng, A.Y.: Self-taught learning, pp. 759–766. ACM Press (2007)
13. Tsoumakas, G., Katakis, I.: Multi-label classification: An overview. International Journal of Data Warehousing & Mining 3(3), 1–13 (2007)
14. Yang, J., Yu, K., Gong, Y., Huang, T.: Linear spatial pyramid matching using sparse coding for image classification. In: IEEE Conference on Computer Vision and Pattern Recognition, pp. 1794–1801 (2009)

Team ULjubljana's Solution to the JRS 2012 Data Mining Competition

Jure Zbontar, Marinka Zitnik, Miha Zidar, Gregor Majcen, Matic Potocnik, and Blaz Zupan

University of Ljubljana, Faculty of Computer and Information Science, Tržažka 25, SI-1000 Ljubljana, Slovenia

Abstract. The task of the JRS 2012 data mining competition was to infer a prediction model capable of associating biomedical journal articles with a subset of topics. Our approach consisted of training a set of base learners, stacking their results, and thresholding the predictions on each label separately. Our method obtained an F-score of 0.53579, which was enough to claim first prize in the competition.

Keywords: multi-label classification, topical classification, sparse datasets, stacking.

1 Introduction

The goal of the JRS 2012 data mining competition was to design an algorithm capable of associating journal articles obtained from PubMed Central [8] with at least one of 83 topics, or labels, as reported in PubMed and defined by manual curation of biomedical experts. Each article was described by 25,640 continuous features with values ranging from 0 to 1,000. The individual feature values measured the degree to which a term was present in a given journal article. In order to ensure that participants who were not familiar with biomedicine had equal chances as domain experts, the names of features and labels were removed. The dataset was split into 10,000 training examples, for which the associated labels were given, and 10,000 test examples, on which our solutions would be evaluated. The dataset was high-dimensional and sparse. There were almost three times as many features as examples and only 0.4% of the feature values contained nonzero entries. The contestants were able to compare their methods on a public leaderboard. The leaderboard results were computed on 10% of the entire test set.

1.1 Evaluation

The submitted solutions were evaluated using the average F-score. First, the F-score of predicting the i-th document was calculated as

J.T. Yao et al. (Eds.): RSCTC 2012, LNAI 7413, pp. 471–478, 2012.

$$\text{precision}_i = \frac{|\{\text{true labels}\}_i \cap \{\text{predicted labels}\}_i|}{|\{\text{predicted labels}\}_i|}$$

$$\text{recall}_i = \frac{|\{\text{true labels}\}_i \cap \{\text{predicted labels}\}_i|}{|\{\text{true labels}\}_i|}$$

$$\text{fscore}_i = 2\frac{\text{precision}_i \cdot \text{recall}_i}{\text{precision}_i + \text{recall}_i} \tag{1}$$

where

- $\{\text{true labels}\}_i$ is the set of labels of document i. These were determined by domain experts.
- $\{\text{predicted labels}\}_i$ is the set of labels of document i predicted by our method.

The average F-score was defined as the average fscore_i over all $m = 10{,}000$ test documents.

$$\text{avg\_fscore} = \frac{\sum_{i=1}^{m} \text{fscore}_i}{m}$$

1.2 Overview

We trained a set of 14 base learners using 5-fold cross validation. The predictions on each fold were stored as they would later be used to build the training set for stacking. The accompanying test set was constructed by retraining each base learner on the entire training set and classifying all the examples in the test set. The 14 individual prediction sets were stacked with a feed forward neural network built on each of the 83 labels independently. The training set for each label consisted of predictions of the 14 base learners for that label. The output of stacking was thresholded by an *ad hoc* procedure which used a greedy algorithm to assign a different threshold to each label prediction. The F-scores obtained by individual components of our solution are summarized in Fig. 1.

2 Base Learners

Different base learners were used to infer classifiers for all 83 labels. The 14 base learners were derived by altering the parameter values of 5 basic models described in this section. As the predictions of these learners were later used in stacking, we aimed at diversifying the classification methods. In order to obtain the F-scores of individual methods we first needed to threshold the predicted probabilities. We chose a threshold value of 0.25. Note that this threshold was used only to determine the F-score of each learner. The input for stacking was constructed from the actual predicted probabilities and not the thresholded predictions. The F-score values reported in Tables 1, 2, 3, 4, and 5 were obtained on the full test set, which was available only after the competition had ended.

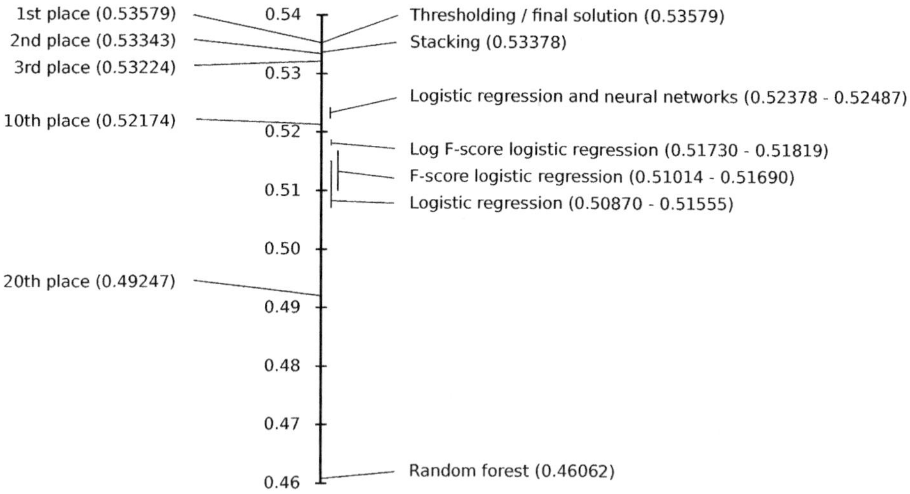

Fig. 1. Leaderboard F-scores of all the base learners, the stacked predictor, and the predictor after optimized thresholding compared to the F-scores obtained by other contestants

2.1 Logistic Regression

We transformed the multi-label classification problem into 83 independent single-label classification problems and predicted the individual labels using L2 regularized logistic regression on the original features. The parameters $\theta \in \mathbb{R}^n$ were obtained by minimizing the cost function $J(\theta)$:

$$J(\theta) = \frac{1}{m} \sum_{i=1}^{m} \left[-y^{(i)} \log h_\theta(x^{(i)}) - (1 - y^{(i)}) \log(1 - h_\theta(x^{(i)})) \right] + \frac{\lambda}{2m} \sum_{j=1}^{n} \theta_j^2 \quad (2)$$

using the L-BFGS optimization algorithm [7], where

- m is the number of training examples,
- $n = 25640$ is the number of features,
- $x^{(i)} \in \{0, 1, \ldots, 1000\}^n$ is the i-th training example,
- $y^{(i)} \in \{0, 1\}$ is 1 if the i-th training example is associated with the label and 0 otherwise,
- λ is the regularization parameter,
- $h_\theta(x)$ is the logistic function, defined as:

$$h_\theta(x) = \frac{1}{1 + e^{-\theta^T x}}$$

The F-scores obtained on the test set are presented in Table 1.

Table 1. Results obtained by logistic regression. The parameter λ is the regularization parameter from (2). The predictions were thresholded at 0.25.

F-score	Parameter
0.50870	$\lambda = 2$
0.51321	$\lambda = 3$
0.51451	$\lambda = 4$
0.51555	$\lambda = 5$

2.2 Logistic Regression and Neural Networks

The approach from Sect. 2.1, in which the multi-label classification problem is split into independent single-label classification problems, is known as the binary relevance method (dubbed PT4 in [6]). Its strong assumption on label independence makes it fast and simple to implement. On the other hand, the predictions made by the binary relevance method are likely to contain label combinations that never co-occur in the training set or, conversely, it might miss obvious combinations. A possible solution would be to run a second learning algorithm on the output of the algorithm from Sect. 2.1 in the hope that it will uncover the dependencies between labels that were missed by the first. Examining the predictions of all 83 labels together, a powerful learning method, like neural networks, should be able to take advantage of the dependencies between labels.

Using 5-fold cross validation we predicted the probabilities of the labels for all training examples using L2 regularized logistic regression. To produce the final predictions, we fed this output into a feed forward neural network with 83 input units, a single hidden layer with 100 units, and 83 output units. The parameter for this method was λ_{nn}, the regularization parameter for the neural network. The regularization parameter for logistic regression, λ_{lr}, was set to 1 and kept constant. The neural network was trained with the backpropagation algorithm, using the L-BFGS algorithm to minimize the cost function. The L-BFGS algorithm was stopped after 400 iterations. Table 2 shows the F-scores obtained by this method. Combining logistic regression and neural networks in this way produced a learning model with the highest F-score of any single base learner.

Table 2. F-score values of logistic regression followed by neural networks. λ_{nn} is the regularization parameter of the neural network. The predictions were thresholded at 0.25.

F-score	Parameter
0.52487	$\lambda_{\text{nn}} = 3$
0.52378	$\lambda_{\text{nn}} = 5$

2.3 F-Score Logistic Regression

Another way around the shortcomings of the binary relevance method is to fit all the parameters $\Theta \in \mathbb{R}^{n\times k}$ in one run of the L-BFGS algorithm. This also allowed us to use a cost function that maximizes the average F-score directly. In order to do so, we first needed to derive a smooth approximation of the F-score that uses probability estimates instead of binary class assignments and is easier to maximize. Noting that

$$\text{fscore}_i = 2\frac{\text{precision}_i \cdot \text{recall}_i}{\text{precision}_i + \text{recall}_i} = 2\frac{|\{\text{true labels}\}_i \cap \{\text{predicted labels}\}_i|}{|\{\text{predicted labels}\}_i| + |\{\text{true labels}\}_i|}$$

we performed the following steps:

1. $|\{\text{true labels}\}_i|$ can be calculated as $\sum_{l=1}^{k} y_l^{(i)}$.
2. Since $h_{\Theta_l}(x^{(i)})$ is the predicted probability of our hypothesis for the l-th label of the i-th document, $|\{\text{predicted labels}\}_i|$ can be approximated by $\sum_{l=1}^{k} h_{\Theta_l}(x^{(i)})$.
3. Similarly, $|\{\text{true labels}\}_i \cap \{\text{predicted labels}\}_i|$ can be approximated by calculating $\sum_{l=1}^{k} h_{\Theta_l}(x^{(i)}) y_l^{(i)}$

giving rise to the equation

$$\text{fscore\_approx}_i = 2\frac{\sum_{l=1}^{k} h_{\Theta_l}(x^{(i)}) y_l^{(i)}}{\sum_{l=1}^{k} [h_{\Theta_l}(x^{(i)}) + y_l^{(i)}]} \tag{3}$$

and the new cost function

$$J(\Theta) = -\frac{1}{m}\sum_{i=1}^{m} \frac{\sum_{l=1}^{k} h_{\Theta_l}(x^{(i)}) y_l^{(i)}}{\sum_{l=1}^{k} [h_{\Theta_l}(x^{(i)}) + y_l^{(i)}]} + \frac{\lambda}{2m}\sum_{j=1}^{n}\sum_{l=1}^{k} \Theta_{jl}^2 \tag{4}$$

where

- $k = 83$ is the number of labels,
- Θ_l is the l-th column of the parameter matrix Θ,
- $y_l^{(i)} \in \{0, 1\}$ is 1 if the i-th training example is associated with label l and 0 otherwise.

The cost function is minimized by computing the partial derivatives with respect to every parameter Θ_{jl} and feeding them into the L-BFGS algorithm. The F-scores obtained by this method are comparable to those of logistic regression from Sect. 2.1 and are presented in Table 3.

2.4 Log F-Score Logistic Regression

The algorithm differs from the one presented in Sect. 2.3 only in the cost function used. The cost function for log F-score logistic regression is obtained by taking

Table 3. The results obtained by F-score logistic regression. The regularization is controlled by adjusting λ from (4). The predictions were thresholded at 0.25.

F-score	Parameter
0.51283	$\lambda = 0.1$
0.51690	$\lambda = 0.2$
0.51524	$\lambda = 0.3$
0.51014	$\lambda = 0.4$

Table 4. The results of log F-score logistic regression. The parameter λ controls the amount of regularization. The predictions were thresholded at 0.25.

F-score	Parameter
0.51819	$\lambda = 0.4$
0.51804	$\lambda = 0.5$
0.51730	$\lambda = 0.6$

the logarithm of (3). Combined with L2 regularization, the new cost function is defined as

$$J(\Theta) = -\frac{1}{m}\sum_{i=1}^{m} \log\left(\frac{\sum_{l=1}^{k} h_{\Theta_l}(x^{(i)}) y_l^{(i)}}{\sum_{l=1}^{k} [h_{\Theta_l}(x^{(i)}) + y_l^{(i)}]}\right) + \frac{\lambda}{2m}\sum_{j=1}^{n}\sum_{l=1}^{k}\Theta_{jl}^2$$

The F-scores shown in Table 4 were higher than the ones obtained by the method in the previous section, despite the fact that the methods are defined in a similar manner.

2.5 Random Forest

Due to computational constraints, we trained a random forest on a subset of features which had at least 50 nonzero entries in the training set. There were slightly over 7,000 such features. An alternative approach included an inference of forests on binarized data (zero vs. nonzero feature value) with information gain-based feature selection. One forest was developed for each of the 83 labels. Forests consisted of 200 trees. The standard tree induction algorithm [4] was used with no pruning other than stopping the inference when a node included fewer than five data instances.

The F-score obtained by random forests (shown in Table 5) was substantially lower than the F-scores acquired by other base learners. We decided to keep random forests in the ensemble anyway, because they seemed to improve the F-score of the stacked solution. A post-competition analysis revealed that this was in fact not the case as the F-scores obtained with and without random forests were practically identical.

Table 5. F-score of random forest. The predictions were thresholded at 0.25.

F-score	Parameter
0.46062	-

3 Stacking

Stacking [5] is a technique for combining predictions of several learners. It was used successfully in past competitions [1,2], most notably the Netflix Prize [3]. As stated in the introduction, the base learners were 5-fold cross-validated on the training set, producing a 10,000 $\times$ 83 table of probability estimates each. As in Sect. 2.1 and 2.5, the multi-label problem was decomposed into 83 independent single-label problems. The training set for the l-th problem was constructed by selecting the predictions for the l-th label from every base learner. A feed forward neural network with 14 input units, 20 units in the hidden layer, and a single output unit was used to obtain the final predicted probabilities. The regularization parameter λ was set to 0.5. Stacking achieved an F-score of 0.53378 on the test set (thresholding the predictions at 0.25), which was a considerable improvement over 0.52487, the best result obtained by a single base learner. Leaving out this crucial step and submitting the base learner with the highest F-score would result in a modest 9-th place in the competition.

4 Thresholding

We adjusted the probability threshold for each label separately, i.e., we learnt the parameters $\theta \in [0,1]^{83}$ and associated an example with label i only if the prediction was greater or equal to θ_i. We decided on a greedy algorithm to fit the parameters θ. Starting with the first label, we chose θ_1, the first threshold, so that it maximized the average F-score. The thresholds not already set by this method were initialized to 0.25. The procedure was repeated on all the labels in sequence. Thresholding resulted in a small improvement of the average F-score which increased from 0.53378 to our final result of 0.53579.

5 Conclusion

We have described the winning solution to the JRS 2012 data mining competition. Our approach is yet another example of how stacking diverse base learners can substantially increase the performance of the overall system. The resulting framework, i.e., the developed method that includes a selection of base learners and continues with stacking and thresholding, can be seen as a general approach to multi-label classification in high-dimensional and sparse datasets. The source code of our solution is available online at `https://bitbucket.org/jzbontar/jrs2012`.

References

1. Yu, H.F., Lo, H.Y., Hsieh, H.P., Lou, J.K., McKenzie, T.G., Chou, J.W., Chung, P.H., Ho, C.H., Chang, C.F., Wei, Y.H., et al.: Feature engineering and classifier ensemble for KDD cup 2010. In: JMLR Workshop and Conference Proceedings (2010)
2. Toscher, A., Jahrer, M.: Collaborative filtering applied to educational data mining. In: KDD Cup (2010)
3. Koren, Y.: The bellkor solution to the netflix grand prize. Netflix prize documentation (2009)
4. Breiman, L.: Random forests. Machine Learning 45(1), 5–32 (2001)
5. Wolpert, D.H.: Stacked generalization. Neural Networks 5(2), 241–259 (1992)
6. Tsoumakas, G., Katakis, I.: Multi label classification: An overview. International Journal of Data Warehousing and Mining 3(3) (2007)
7. Byrd, R.H., Lu, P., Nocedal, J., Zhu, C.: A Limited Memory Algorithm for Bound Constrained Optimization. SIAM Journal on Scientific and Statistical Computing 16(5), 1190–1208 (1995)
8. National Library of Medicine: PubMed Central (PMC): An Archive for Literature from Life Sciences Journals. In: McEntyre J., Ostell J (Eds.): The NCBI Handbook, http://www.ncbi.nlm.nih.gov/books/NBK21087/

Author Index

GPSR Compliance

The European Union's (EU) General Product Safety Regulation (GPSR) is a set of rules that requires consumer products to be safe and our obligations to ensure this.

If you have any concerns about our products, you can contact us on ProductSafety@springernature.com

In case Publisher is established outside the EU, the EU authorized representative is:

Springer Nature Customer Service Center GmbH
Europaplatz 3
69115 Heidelberg, Germany

Batch number: 09490872

Printed by Printforce, the Netherlands